NUMERICAL MATHEMATICS AND SCIENTIFIC COMPUTATION

General Editors

NUMERICAL MATHEMATICS AND SCIENTIFIC COMPUTATION

*P. Dierckx: *Curve and surface fittings with splines*
*H. Wilkinson: *The algebraic eigenvalue problem*
*I. Duff, A. Erisman, and J. Reid: *Direct methods for sparse matrices*
*M.J. Baines: *Moving finite elements*
*J.D. Pryce: *Numerical solution of Sturm–Liouville problems*
K. Burrage: *Parallel and sequential methods for ordinary differential equations*
Y. Censor and S. A. Zenios: *Parallel optimization: theory, algorithms and applications*
M. Ainsworth, J. Levesley, M. Marletta and W. Light: *Wavelets, multilevel methods and elliptic PDEs*
W. Freeden, T. Gervens, and M. Schreiner: *Constructive approximation on the sphere: theory and applications to geomathematics*
Ch. Schwab: *p- and hp- finite element methods: theory and applications to solid and fluid mechanics*
J.W. Jerome: *Modelling and computation for applications in mathematics, science, and engineering*

Monographs marked with an asterisk (*) appeared in the series 'Monographs in Numerical Analysis' which has been folded into, and is continued by, the current series.

p- and *hp*-Finite Element Methods

Theory and Applications in Solid and Fluid Mechanics

CH. SCHWAB

Associate Professor,
Seminar for Applied Mathematics,
ETH Zürich

CLARENDON PRESS • OXFORD

1998

Oxford University Press, Great Clarendon Street, Oxford OX2 6DP
Oxford New York
Athens Auckland Bangkok Bogota Buenos Aires Calcutta
Cape Town Chennai Dar es Salaam Delhi Florence Hong Kong Istanbul
Karachi Kuala Lumpur Madrid Melbourne Mexico City Mumbai
Nairobi Paris São Paolo Singapore Taipei Tokyo Toronto Warsaw
and associated companies in
Berlin Ibadan

Oxford is a registered trade mark of Oxford University Press

Published in the United States by
Oxford University Press Inc., New York

A catalogue record for this book is available from the British Library

Library of Congress Cataloging in Publication Data
(Data available)

ISBN 0 19 850390 3 (Hbk)

Typeset by the author
Printed in Great Britain on acid free paper by
Biddles Ltd, Guildford & King's Lynn

PREFACE

The present lecture notes resulted from a graduate course on finite element analysis which was taught at the Eidgenössische Technische Hochschule Zürich in the winter semester 1995/96. The audience consisted of mathematicians as well as engineers who were interested in learning the mathematical basis for the recent higher-order, *hp* and **spectral element methods**. The course tried to give a reasonably complete overview of the mathematical techniques used to establish *p*- or spectral convergence results for these methods. These techniques are somewhat different from those used for more classical so-called ***h*-version FEM** where convergence is achieved by mesh refinement rather than increasing polynomial degree (*h*-version FEM are very well covered by a number of references, see e.g. [45], [38], [59]; for *p*- or spectral FEM the only texts at present appear to be [30], [35], [78], [144]).

The investigation of the relative merits of higher-order methods over the classical *h*-version approach requires a careful look at the **regularity** of solutions of elliptic boundary value problems. It was known for some time (based on the theory of *n*-widths, [12], [112]) that if the solution of the (second-order, say) boundary value problem belongs to the solution class H^k for some $k > 1$, the *h*-version FEM (with quasiuniform mesh) gives asymptotically optimal approximations, in terms of the number of degrees of freedom (see [12] for more on this). Thus, if the regularity of the solution is measured in terms of the spaces H^k, the *p*- and *hp*-version can never out-perform the classical *h*-version. For elliptic problems with singularities, however, *p*- and *hp*-FEM with properly designed meshes appear to be clearly superior over standard *h*-version FEM (again, with quasiuniform meshes).

The explanation is that solutions to elliptic problems in polygonal or polyhedral domains with analytic data belong to some classes which are substantially smaller than H^k. For example, such solutions are **analytic** in the whole domain except in corners and edges on the boundary. The quantitative description of this regularity requires in fact the control of derivatives of any order of the solution in countably normed weighted Sobolev spaces which take into account the edge- and corner singularities of the domain (the theory itself is due to Babuška and Guo, see [15] and the references there). It is shown that by proper combination of mesh refinement and increasing polynomial degree, the *hp*-FEM can achieve **exponential convergence** for elliptic problems with piecewise analytic solution whereas *h*- or *p*-FEM converge at best algebraically.

A second class of problems where high-order FEM are provably superior to *h*-FEM are **elliptic problems depending on a parameter**. Many such singularly perturbed problems arise in engineering practice – for example plate and shell problems and nearly incompressible materials, to name but a few. Once again standard Sobolev spaces are inadequate to properly characterize the regularity in dependence on the parameter. Here an **asymptotic analysis** of the solution is necessary. For such singularly perturbed problems, it is highly desirable to have **robust** FEM, i.e. methods whose performance does not deteriorate as the problem parameter approaches the critical value (e.g. the plate thickness tends to zero or as the Poisson ratio approaches 0.5). Proving robustness involves some non-standard approximation results, on the one hand for **boundary layers** and on the other hand some **constrained approximations** arising in the FE analysis of the limiting problems at the critical value of the parameter. Such approximations are also closely related to **locking phenomena**.

As with any work touching on current research in an active area, some of the results presented here may be sharpened or generalized in the future; we nevertheless hope to provide a good survey of the theory of *hp*-FEM for various problem classes.

The outline of these notes is as follows. In **Chapter 1**, we introduce some model boundary value problems and discuss different variational formulations for them. A general existence and uniqueness result, for those variational forms, the so-called **inf sup condition**, is proved and applied.

In **Chapter 2**, we introduce the FEM as a general approximation scheme for saddle point problems. We carefully show, following [24], the different roles that test and trial spaces play in ensuring stability and consistency.

Chapter 3 is then devoted to a detailed convergence study of FEM in one dimension, in particular the approximation properties of *p*- and *hp*-FEM. The design of optimal *h*-FEM meshes for certain classes of problems is addressed and the exponential convergence of *hp*-FEM for classes of piecewise analytic solutions is shown. In addition, we show that **boundary layer functions** can be approximated by a proper *hp*-FEM at a robust exponential rate.

Chapter 4 is devoted to the analysis of *h*-FEM with mesh refinement and of *hp*-FEM for the Poisson equation in a polygon. Optimal, algebraic or exponential convergence is proved provided the subspaces are properly designed. Further, we discuss preconditioning of the stiffness matrices arising in the *hp*-FEM.

Chapter 5 addresses the **mixed FEM** for saddle point problems. Here **stability**, in addition to approximability, is the main issue. We review, for the classical **Stokes problem** in a polygon, some techniques for establishing (divergence) stability for *h*-FEM. Then we turn to *p*- and *hp*-FEM. Here the techniques for establishing spectral divergence stability are relatively recent. We show, for the *hp*-FEM subspaces introduced in Chapter 4, that divergence stability holds, but, with a constant which tends to zero algebraically with the number of degrees of freedom; this is reasonable, owing to the exponential consistency of the *hp*-subspaces. In particular, we obtain the exponential convergence of the pressures produced by the mixed *hp*-FE approximation.

Chapter 6 starts with a brief introduction to the theory of elasticity. Numerous different variational formulations for the boundary value problems of linearized elastostatics are given, together with the most frequently used lower-dimensional approximations: membranes, plates and shells. The FE discretization of these problems poses particular challenges due to small thickness, near incompressibility, etc.

Apart from the boundary layers already discussed in Chapter 3, the FE approximation may here be unsatisfactory owing to **constraints** imposed on the FE solution, causing the so-called **locking effect**. This is discussed at some length for shell problems.

The **Appendix A** contains the most important definitions on Sobolev spaces. It is intended as a glossary rather than an introduction to this topic; we present often only the statement of the various trace and embedding theorems without proofs. The reader should be able to understand the main points in Chapter 1 by occasional reference to the Appendix A even without prior knowledge of Sobolev spaces. The language of Sobolev spaces pervades much of applied mathematics and we feel that some familiarity with it is also desirable for engineers, especially since it merely formalizes concepts of finite energy which were intuitively used in the principal virtual work anyhow.

The reader should bear in mind that the material presented here is a rather lopsided presentation of the FEM, it is heavily biased both towards the mathematical aspects and towards *p*- and *hp*-methods; no attempt has been made to duplicate the excellent existing work on *h*-FEM ([45], [37] [38], [12] and the references there). Likewise, no attempt has been made to present in detail data structures and implementational guidelines for the design of *hp*-FE codes with good performance. After pioneering work in the 1970s of B.A. Szabo and his group at Washington University, St. Louis, substantial progress on implementation of adaptive *hp*-FEM was achieved in the late 1980s by J.T. Oden, L. Demkowicz and their coworkers at UT Austin; we mention here only [54], [104] and, for more recent developments, [45]. In the same vein, we have not attempted to present the details of *hp*-FEM convergence theory in three dimensions. Active work is in progress on this; suffice it to say that most results, in particular those in Chapter 4, do extend to three dimensions. We refer to [17] for theory and to [144] for numerical results. Commercial *hp*-FE codes for three dimensional elasticity problems have become available recently, for example PHLEX,[1] PolyFEM,[2] Pro/MECHANICA,[3] STRESSCHECK,[4] STRIPE.[5]

Likewise, no attempt has been made to present the recent results on *hp*-boundary element methods. Here, much progress has been achieved by E.P. Stephan and his group at Hannover and we refer to [91], [76], [140] and the references there for more (see also [57]).

[1]PHLEX is a trademark of COMCO Corp., 7800 School Creek Blvd-Suite 290 E, Austin, TX 78757-7024, USA.

[2]PolyFEM is a trademark of Computer Aided Design Software Inc. Coralville, Iowa, USA.

[3]Pro/MECHANICA is a trademark of Parametric Technology Corp., Waltham, Massachusetts, USA.

[4]STRESSCHECK is a trademark of ESRD Res. Inc., Clayton Plaza, Suite 204, 7750 Clayton Rd., St. Louis, MO 63117, USA.

[5]STRIPE is developed by Flygtekniska Försöksanstalten, Box 11021, S-16111 Bromma, Sweden.

In summary, *hp*-FEM appear to be advantageous only in certain, well-defined situations which nevertheless occur very frequently in engineering: for example problems with piecewise analytic solution and problems in elasticity which exhibit locking (such as plate and shell problems), and problems with boundary layers. Here, the good approximation properties allow one often to achieve accurate solutions with relatively few degrees of freedom, thus, as a rule, direct solvers are used for the solution of the linear systems. In contrast, the very large linear systems arising with h-version FEM require efficient iterative solvers such as the multigrid method; such iterative methods are currently being developed for the p- and *hp*-FEM, applied to $3d$ problems. Despite these omissions, however, we hope that the interested reader will, from the two-dimensional results given here, be able to explore the research literature on the three-dimensional case.

Finally, much of the mathematics presented here is directly or indirectly due to I. Babuška, whom I wish to thank at this point for his guidance over the years, mathematical and otherwise. Thanks are also due to my colleague at the ETH Zürich, Rolf Jeltsch, who made this course and the notes possible. I am also indebted to coworkers at ETH Zürich for numerous suggestions and constructive criticism; in particular, thanks are due to Dr. J. Melenk for his comments on Chapters 1, 2, 3 and 4, and to Mr. D. Schötzau (Dipl. Math.) for reading Chapter 5. Finally, the professional typesetting was done by Mrs. M. Pfister, whom I thank for her patience and efficiency.

Zürich, September 1997 C. Schwab

CONTENTS

1

VARIATIONAL FORMULATION OF BOUNDARY VALUE PROBLEMS

Finite element methods (FEM) are discretizations of boundary value problems in variational form. Therefore, the present chapter is devoted to showing how boundary value problems can be cast into a variational form. This is illustrated for several model problems which we introduce in Section 1. The variational formulations of boundary value problems require the use of **Sobolev spaces** which we do not discuss here (the facts on Sobolev spaces required in these notes are collected in the Appendix A). The reason why we discuss variational formulations of boundary value problems in detail is that a given variational formulation of a boundary value problem is the point of departure (and as such an integral part) of the FEM. In Section 2 we discuss several requirements a variational formulation must satisfy in order to be well-posed and to give rise to a stable FEM. This is exemplified for some model boundary value problems in one and two dimensions. Section 3 then gives a fundamental existence and well-posedness result for variational formulations of boundary value problems, the so-called **inf–sup condition**. This condition will later on in Chapter 2 be seen to ensure the **stability** of general FEM and the invertibility of the stiffness matrix. Several techniques for proving the inf–sup condition are shown. Section 4 is devoted to a characterization of the **trace spaces** for mixed boundary value problems which is necessary for their proper variational formulation. It is also relevant for the variational formulation of contact problems in mechanics. Finally, in Section 5 we present further **mixed variational formulations** of the model problems introduced in Section 1.

1.1 Model problems

We will exemplify the variational or weak formulations of boundary value problems for a few specific problems which, although simple in structure, exhibit many of the phenomena we will encounter later on for more complicated examples from solid and fluid mechanics. Such problems are therefore called **model problems**.

1.1.1 *Axially loaded, elastically supported bar*

Consider the unidirectional deformation of the bar shown in Figure 1.1.

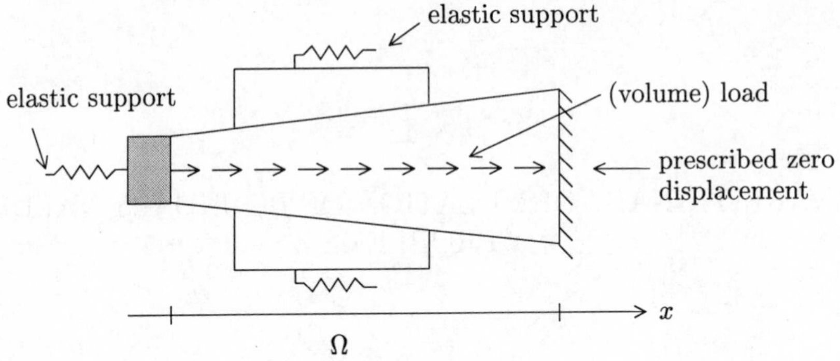

FIG. 1.1. Elastic bar

The **given** (known) **data** are

$f(x)$	applied (volume) load	[force/length]
$E(x)$	modulus of elasticity	[force/length^2]
$A(x)$	cross-sectional area at x	[length^2]
$c(x)$	spring constant of elastic support	[force/length^2]
α_i	spring constants of elastic end support	[force/length]

$\Omega = (0, \ell) = \{x \in \mathbb{R} : 0 < x < \ell\}$ is the domain occupied by the axis of the bar, ℓ is the length of the bar.

The **desired** (unknown) **data** are

$u(x)$	longitudinal displacement
$\sigma(x)$	normal stress in longitudinal direction
$\epsilon(x)$	normal strain in longitudinal direction
$\rho(x)$	support force

A **mathematical model** of the bar is constructed from the following relationships:

a) **Strain displacement relationship:**

$$\epsilon(x) = u'(x) \text{ in } \Omega . \tag{1.1.1}$$

b) **Stress strain relationship:**

$$\sigma(x) = E(x)\, \epsilon(x) = E(x)\, u'(x) \text{ in } \Omega . \tag{1.1.2}$$

c) **Equilibrium:**

$$\frac{d}{dx}\left(A(x)\,\sigma(x)\right) + \rho(x) + f(x) = 0 \text{ in } \Omega\,. \tag{1.1.3}$$

Since the support is elastic, we have $\rho(x) = -c(x)\,u(x)$, i.e., the support force ρ is opposed to the displacement. Therefore (1.1.3) becomes

$$-\frac{d}{dx}\left(A(x)\,\sigma(x)\right) + c(x)\,u(x) = f(x) \text{ in } \Omega\,. \tag{1.1.4}$$

Combining (1.1.2) and (1.1.4), we can eliminate $\sigma(x)$ and obtain one single second-order equation:

$$-\frac{d}{dx}\left(AE(x)\,\frac{du}{dx}\right) + c(x)\,u(x) = f(x) \text{ in } \Omega\,. \tag{1.1.5}$$

Equation (1.1.5) is to be supplemented by boundary conditions on $\partial\Omega$, i.e., at $x = 0$ and at $x = \ell$. The following three types of boundary conditions are typically encountered in engineering practice:

d) **Prescribed displacements:**

$$u(0) = a_1\,, \quad u(\ell) = a_2\,. \tag{1.1.6}$$

e) **Prescribed normal stresses:**

$$\begin{aligned} \sigma_n(0) &= -\sigma(0) = -E(0)\,u'(0) = a_1\,,\\ \sigma_n(\ell) &= \sigma(\ell) = E(\ell)\,u'(\ell) = a_2\,. \end{aligned} \tag{1.1.7}$$

f) **Elastic end supports:**

$$\alpha_1\,u(0) - A(0)\,\sigma(0) = a_1,\ \alpha_2\,u(\ell) + A(\ell)\,\sigma(\ell) = a_2\,. \tag{1.1.8}$$

Here a_1, a_2 and $\alpha_1, \alpha_2 > 0$ are given numbers. In the above, we implicitly assumed that

a) $A, E, \sigma \in C^0(\overline{\Omega}) \cap C^1(\Omega)$,
b) $u \in C^1(\overline{\Omega}) \cap C^2(\Omega)$.

The differentiability in $\Omega = (0, \ell)$ is required for (1.1.1)–(1.1.5) to hold in the pointwise sense, whereas the differentiability in $\overline{\Omega}$, i.e. up to the boundary, is required for (1.1.6)–(1.1.8) to make sense. A solution $u \in C^2(\Omega) \cap C^1(\overline{\Omega})$ is called a **classical solution.**

Remark 1.1 The derivation of the mathematical model (1.1.6)–(1.1.8) of the bar from physical assumptions yields an idealized, one-dimensional model. A physical bar is of course three-dimensional, and the one-dimensional **mathematical model** must be viewed as an **approximation of the underlying**

physical reality. The corresponding **modeling error** obviously cannot be controlled by numerical discretization, e.g., by mesh refinement or other actions. Care must therefore be taken when comparing numerical results obtained from a **discretization of the model** to, say, laboratory measurements. Manipulating a discretization, i.e., a code, so as to reproduce physical experiments does not constitute sound practice. It generally means that at least two distinct errors (possibly more), such as the modeling and the discretization error, cancel each other approximately for the benchmark problems used in the manipulation/parameter tuning.

1.1.2 *Membrane problem*

This problem is the two dimensional analog of (1.1.1)–(1.1.8). We may think now in terms of a thin, three-dimensional sheet which is deforming subject to body and surface forces. No bending stiffness is present. The domain Ω is now a subset of $\mathbb{R}^2$ and coincides with the mid-surface of the membrane (once again, we idealize the membrane as a two-dimensional body). Let us be more specific about the domain Ω. We will admit **curvilinear polygons** which we now define.

Definition 1.2 *A curvilinear polygon $\Omega \in \mathbb{R}^2$ is an open, bounded and connected domain in $\mathbb{R}^2$ such that*

$$\Gamma = \partial\Omega = \bigcup_{j=1}^{M} \overline{\Gamma_j} \tag{1.1.9}$$

where each Γ_j is an open, analytic arc, i.e.,

$$\Gamma_j = \{(\varphi_j(\xi),\, \psi_j(\xi)) \,|\, \xi \in I = (-1,1)\} \tag{1.1.10}$$

where the parametric representations are analytic on $\overline{I}$ and such that

$$|\varphi_j'(\xi)|^2 + |\psi_j'(\xi)|^2 \geq \alpha > 0\,.$$

If φ_j, ψ_j are linear, Γ_j is a straight line segment. If this is the case for $j = 1, \ldots, M$, we refer to Ω simply as a "polygon". The endpoints A_{j-1}, A_j of Γ_j are the **vertices** of Ω, and we identify A_0 with A_N. The internal angle at A_j is ω_j (see Figures 1.2 and 1.3).

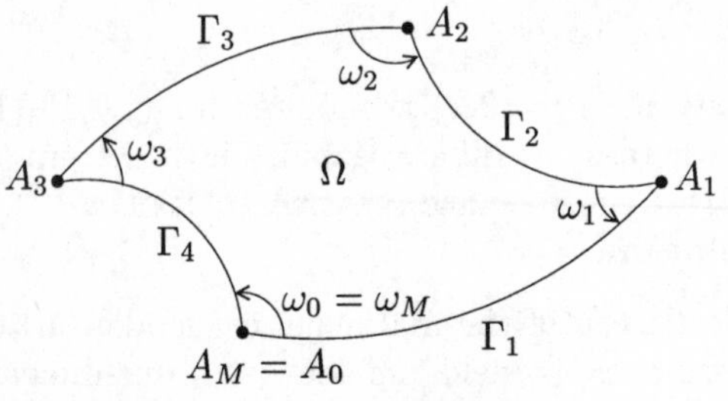

FIG. 1.2. Curvilinear polygon with $M = 4$ sides

The A_j, ω_j and the orientation of Γ_j are shown in Figure 1.2. Note that ω_j is defined as the angle between the tangents to Γ_j and Γ_{j-1} as in Figure 1.3.

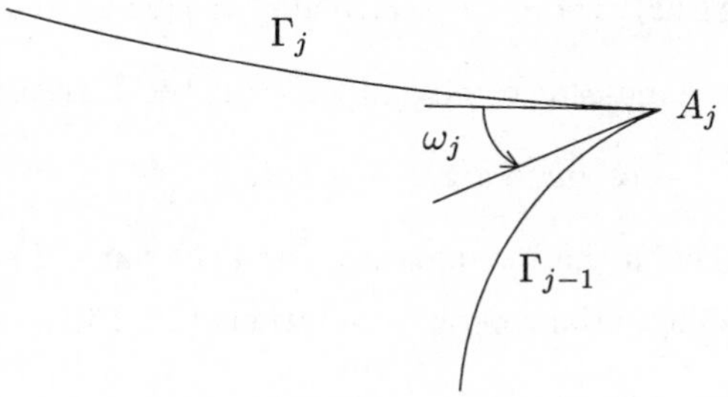

FIG. 1.3. Definition of ω_j

We allow further that

$$0 < \omega_j \leq 2\pi \,, \tag{1.1.11}$$

i.e., Γ_{j-1} and Γ_j and coincide if $\omega_j = 2\pi$ (this is used to model cracks in fracture mechanics).

Exercise 1.3 By referring to the definitions in Appendix A, investigate for which ω_j a curvilinear polygon is a Lipschitz domain.

We assume further that $\Gamma = \partial\Omega = \overline{\Gamma^{[0]}} \cup \overline{\Gamma^{[1]}} \cup \overline{\Gamma^{[2]}}$ where the components are mutually disjoint and are unions of some of the Γ_j. The parts $\Gamma^{[0]}$, $\Gamma^{[1]}$ and $\Gamma^{[2]}$ will later be parts of Γ where, respectively, Dirichlet, Neumann or combined boundary conditions are imposed (cf. (1.1.18)–(1.1.20)).

Most of the domains Ω occurring in practical applications are curvilinear polygons.

The **mathematical model of the membrane** is analogous to that of the elastic bar. We have

a) **Strain displacement relationship**:

$$\epsilon(x) = \mathbf{grad}\, u(x) = (\partial_1 u, \partial_2 u)^\top \text{ in } \Omega \,, \tag{1.1.12}$$

b) **Stress strain relationship**:

$$\sigma(x) = E(x)\,\epsilon(x) = E(x)\,\mathbf{grad}\, u(x) \text{ in } \Omega \,, \tag{1.1.13}$$

c) **Equilibrium**:

$$\operatorname{div}(A(x)\,\sigma(x)) + \rho(x) + f(x) = 0 \text{ in } \Omega \,. \tag{1.1.14}$$

Now $A(x)$, $E(x)$ are 2×2 matrices and $\epsilon(x)$, $\sigma(x)$ are vector functions. We assume elastic support, i.e., $\rho(x) = -c(x)\,u(x)$. With this assumption, (1.1.15) leads to

$$-\operatorname{div}(A(x)\,\sigma(x)) + c(x)\,u(x) = f(x) \text{ in } \Omega \,. \tag{1.1.15}$$

Once again, utilizing (1.1.13) in (1.1.15) we arrive at a differential equation for the displacement only:

$$Lu := -\mathrm{div}\,(AE(x)\,\mathbf{grad}\,u(x)) + c(x)\,u(x) = f(x) \text{ in } \Omega\,. \tag{1.1.16}$$

Notice that (1.1.16) becomes, for $c = 0$ and $AE = 1$, the **Poisson equation**

$$-\mathrm{div}\,(\mathbf{grad}\,u(x)) = -\Delta u(x) = f \tag{1.1.17}$$

which models many phenomena in engineering and the natural sciences.

Remark 1.4 Using differential operator notation, (1.1.16) is sometimes also expressed as

$$-\nabla \cdot (AE(x)\,\nabla u(x)) + c(x)\,u(x) = f(x) \text{ in } \Omega\,.$$

We assume for now $A(x)$, $E(x) \in C^0(\overline{\Omega}) \cap C^1(\Omega)$ componentwise and that $A(x)$, $E(x)$ are symmetric, positive definite.

We supplement (1.1.12)–(1.1.16) with **boundary conditions:**

a) **Prescribed displacement**:

$$u(x) = g^0 \text{ on } \Gamma^{[0]}\,. \tag{1.1.18}$$

b) **Prescribed normal stresses**:

$$\boldsymbol{n}(x) \cdot AE(x)\,\mathbf{grad}\,u(x) = g^1 \text{ on } \Gamma^{[1]}\,. \tag{1.1.19}$$

c) **Elastic support**:

$$\alpha(x)\,u(x) + \boldsymbol{n}(x)\,A(x)\,\sigma(x) = g^2 \text{ on } \Gamma^{[2]}\,. \tag{1.1.20}$$

Here $\boldsymbol{n}(x)$ denotes the exterior unit normal to Ω (see Appendix A) and $\alpha(x) \in C^0(\Gamma)$ is a positive function, $\alpha(x) \geq \alpha_0 > 0$ on Γ.

Several problems arise here. Assume, for example, that $\overline{\Gamma^{[0]}}$ and $\overline{\Gamma^{[1]}}$ abut at some vertex A_j. Then (1.1.18)–(1.1.20) do not specify which boundary condition holds at the vertex. Further, even if $\overline{\Gamma^{[0]}} = \overline{\Gamma^{[2]}} = \varnothing$ and all the data in (1.1.12)–(1.1.15) are in $C^\infty(\overline{\Omega})$ the solution may fail to be in $C^1(\overline{\Omega})$, due to **corner singularities** caused by the vertices A_j (see Chapter 4, Section 4.2 for more details).

In order to resolve these difficulties and to obtain a satisfactory notion of solution we reformulate the model problems, thereby in essence enlarging the class in which solutions are sought. The admissible and proper classes are **Sobolev spaces** of functions with integrable weak (or distributional) derivatives. These classes have, in many cases, physical interpretations as solutions with finite energy. Sobolev spaces and their main properties are described in Appendix A. The solutions thus obtained are called **generalized solutions**.

Many solutions of engineering interest – for example solutions corresponding to concentrated loads, and in particular also finite element solutions – are generalized solutions.

1.2 Generalized solutions. Essential and natural boundary conditions

1.2.1 *Classical solution*

We explain the notion of generalized solution in the context of the one-dimensional model problem (1.1.5) (where a prime now denotes d/dx):

$$-(AE(x)\,u')' + c(x)\,u(x) = f(x) \text{ in } \Omega = (0,1)\,, \tag{1.2.1}$$

with boundary conditions

$$\ell_1(u) = \alpha_1\,u(0) - \beta_1[AEu']\big|_{x=0} = 0\,, \tag{1.2.2}$$

$$\ell_2(u) = \alpha_2\,u(\ell) + \beta_2[AEu']\big|_{x=\ell} = 0\,. \tag{1.2.3}$$

Assume for now

$$A(x),\, E(x) \in C^1(\overline{\Omega}),\; c(x) \in C^0(\overline{\Omega}),\; \alpha_i, \beta_i \geq 0,\; \alpha_i + \beta_i = 1\,. \tag{1.2.4}$$

Given $f \in C^0(\overline{\Omega})$, a **classical solution** of (1.2.1) is a function $u \in C^2(\overline{\Omega})$ satisfying (1.2.1)–(1.2.3) at every $x \in \overline{\Omega}$.

1.2.2 *Generalized formulations*

Generalized or **variational formulations** are obtained by associating with the problem of interest a **bilinear form** $B(u,v)$ and a **linear functional** $F(v)$. The linear functional $F(v)$ typically represents the external forces or loadings applied to the mechanical system and, indeed, will later give rise to the load vector in finite element methods. **Linearity** of $F(\cdot)$ has a natural interpretation:

a) **superposition:** $F(v_1 + v_2) = F(v_1) + F(v_2)$
(i.e., the load applied to the sum of two admissible fields v_1, v_2 equals the sum of the load applied to each of these fields)

and

b) **homogeneity:** $F(\lambda v) = \lambda F(v)$ for every $\lambda \in \mathbb{R}$
(i.e., the load $F(\cdot)$ applied to a multiple λ of an admissible field v equals λ times the load applied to this field).

In mechanics, the fields v are called **virtual displacements**.

The term "bilinear form" means that $B(\cdot,\cdot)$ is linear in each argument, i.e., for fixed u, $B(u,\cdot)$ is a linear functional on v and so is $B(\cdot,v)$ for fixed v.

In a proper variational formulation, both $B(u,v)$ and $F(v)$ must be defined on **properly chosen linear spaces** X and Y, i.e., $B : X \times Y \to \mathbb{R}$, $F : Y \to \mathbb{R}$. Again, in mechanics X and Y represent usually spaces of displacements or virtual displacements of finite energy.

Then a **generalized solution** u_0 of (1.2.1)–(1.2.3) is defined as follows: find $u \in X$ such that

$$B(u,v) = F(v) \text{ for all } v \in Y \, . \tag{1.2.5}$$

Here Y is called **test space** and X **trial space**.

A minimal requirement on $B(u,v)$ and $F(v)$ in order for (1.2.5) to be meaningful is **continuity**: there exist numbers $C_B > 0$ and $C_F > 0$ such that for all $u \in X$, $v \in Y$

$$|B(u,v)| < C_B \|u\|_X \, \|v\|_Y, \quad |F(v)| \leq C_F \, \|v\|_Y \, . \tag{1.2.6}$$

Here $\|u\|_X$, $\|v\|_Y$ denote the norms on X, or Y respectively.

If (1.2.6) holds, we call $F(\cdot)$ a **continuous, linear functional** on Y. The set of all such functionals is denoted by Y' and the smallest constant C_F in (1.2.6) is called the **norm of** $F(\cdot)$; this is written as

$$\|F\|_{Y'} = \sup_{0 \neq v \in Y} \frac{|F(v)|}{\|v\|_Y} \, .$$

Y', equipped with this norm, is itself a complete linear space, the **dual space** of Y.

The generalized formulation (1.2.5) is sometimes also called **variational principle** associated with (1.2.1)–(1.2.3). Many variational principles arise naturally in solid and fluid mechanics as we shall see in the sections below.

A proper generalized formulation (1.2.5) must satisfy the following three **requirements** (besides (1.2.6)):

If u_0 is the classical solution, then $u_0 \in X$ and (1.2.5) holds for $u = u_0$. (1.2.7)

For every continuous, linear functional $F(\cdot)$ on Y, there exists a $u \in X$ satisfying (1.2.5) . (1.2.8)

If u satisfies (1.2.5), there exists a constant C, independent of F, such that the **a priori estimate**

$$\|u\|_X \leq C \, \|F\|_{Y'} = C \sup_{0 \neq v \in Y} \frac{|F(v)|}{\|v\|_Y} \tag{1.2.9}$$

holds.

Remark 1.5 (1.2.9) implies in particular that

$$B(w,v) = 0 \text{ for all } v \in Y \Longrightarrow w = 0 \, . \tag{1.2.10}$$

This, however, means that **a generalized solution** u satisfying (1.2.5) **is unique**. If this were not so, there would be a second solution $\widetilde{u} \in X$, $\widetilde{u} \neq u$ of (1.2.5), and

$$B(u,v) = F(v) \text{ for all } v \in Y \;,$$
$$B(\widetilde{u},v) = F(v) \text{ for all } v \in Y \;.$$

Subtracting, we get

$$0 = B(u,v) - B(\widetilde{u},v) = B(u-\widetilde{u},v) \text{ for all } v \in Y \;.$$

Setting $w = u - \widetilde{u}$, $w \in X$ and, by (1.2.10), $w = 0$, i.e., $u = \widetilde{u}$, a contradiction – hence there cannot be a second generalized solution if (1.2.9) holds.

Remark 1.6 (1.2.7) justifies the notion "generalized solution". Notice that while by (1.2.8) a generalized solution of (1.2.5) always exists, a classical solution u_0 may or may not exist. In particular, not every generalized solution is a classical solution.

Remark 1.7 The a priori estimate (1.2.9) ensures **continuous dependence** on the functional F. Let F_1, F_2 be two functionals and u_1, u_2 the corresponding generalized solutions, i.e.,

$$B(u_1,v) = F_1(v) \text{ for all } v \in Y \;,$$
$$B(u_2,v) = F_2(v) \text{ for all } v \in Y \;.$$

Subtracting these relations from each other, we get

$$B(u_1 - u_2, v) = F_1(v) - F_2(v) =: F(v) \text{ for all } v \in Y$$

and the estimate (1.2.9) implies

$$\|u_1 - u_2\|_X \le C\,\|F\|_{Y'} = C\,\|F_1 - F_2\|_{Y'} \;. \tag{1.2.11}$$

If the change in the loadings is small, so is the change in the solution. We also see that the constant C in the a-priori estimate (1.2.9) measures the sensitivity of the solution with respect to perturbations in the data.

The recipe for obtaining a generalized formulation from (1.2.1)–(1.2.3) is to **multiply** the equation (1.2.1) by a **test function** $v \in Y$ and to **integrate by parts**. We emphasize that **in general there is no unique way to cast a boundary value problem into weak form**; many different, nonequivalent, weak formulations are possible. The following examples illustrate this point. Throughout, the spaces X and Y will be Sobolev spaces and we refer the reader to the Appendix A for their definition and basic properties.

1.2.3 *Generalized formulations of the bar problem*

Let us now return to (1.2.1)–(1.2.3). A first generalized formulation is obtained by multiplying (1.2.1) by $v \in Y := L^2(\Omega)$ and integrating:

$$\int_0^\ell (-(AEu')' + cu)\, v\, dx = \int_0^\ell fv\, dx\,. \tag{1.2.12}$$

Defining

$$B_1(u,v) = \int_0^\ell (-(AEu')' + cu)\, v\, dx\,, \tag{1.2.13}$$

$$F_1(v) = \int_0^\ell fv\, dx\,, \tag{1.2.14}$$

we get the following.

Proposition 1.8 *Assume $AE \in W^{1,\infty}(\Omega)$, $c \in L^\infty(\Omega)$, and $f \in L^2(\Omega)$. Then $B_1(u,v)$ is continuous on $H^2(\Omega) \times L^2(\Omega)$, and $F_1(v)$ is a continuous, linear functional on $L^2(\Omega)$.*

Proof We use the Schwarz inequality and get

$$\begin{aligned}
|B_1(u,v)| &\le \int_0^\ell |(AE)'\, u' + AEu'' + cu|\; |v| dx \\
&\le \left(\|(AE)'\|_{L^\infty} \|u'\|_{L^2} + \|AE\|_{L^\infty} \|u''\|_{L^2} + \|c\|_{L^\infty} \|u\|_{L^2}\right) \|v\|_{L^2} \\
&\le \max\left(\|AE\|_{W^{1,\infty}}, \|c\|_{L^\infty}\right) \left(\|u\|_{L^2} + \|u'\|_{L^2} + \|u''\|_{L^2}\right) \|v\|_{L^2} \\
&\le C\, \|u\|_{H^2}\, \|v\|_{L^2}
\end{aligned}$$

$$\|F\|_{Y'} = \sup_{0 \ne v \in L^2} \frac{|\int_\Omega fv\, dx|}{\|v\|_{L^2}} \le \|f\|_{L^2}$$

(all norms are taken over Ω). □

So far, we have not used the boundary conditions (1.2.2), (1.2.3) in (1.2.13), (1.2.14). The space X_1 for the generalized formulation (1.2.5) is selected as the subspace of $H^2(\Omega)$ of functions satisfying the boundary conditions (1.2.2), (1.2.3):

$$X_1 = \{u \in H^2(\Omega) : \ell_1(u) = \ell_2(u) = 0\}\,. \tag{1.2.15}$$

It is essential here that for $u \in H^2(\Omega)$, the point values of $u'(x)$ at $x = 0$ and $x = \ell$ exist and therefore X_1 is well defined. Thus the first generalized formulation is (1.2.5) with X_1 in (1.2.15), $Y_1 = L^2(\Omega)$ and $B_1(u,v)$, $F_1(u)$.

Assume now that $v \in H^1(\Omega)$ and integrate (1.2.12) by parts, yielding

$$\int_0^\ell (AEu'v' + cuv)dx - (AEu'(x)v(x))\Big|_{x=0}^{x=\ell} = \int_0^\ell fv\,dx\,. \tag{1.2.16}$$

Assume first $\beta_i \neq 0$, $i = 1, 2$ in (1.2.2), (1.2.3). Then

$$(AEu')|_{x=0} = \frac{\alpha_1}{\beta_1}\,u(0), \quad -(AEu')|_{x=\ell} = \frac{\alpha_2}{\beta_2}\,u(\ell)\,.$$

Inserting in (1.2.16) we get

$$B_2(u,v) = F_2(v) \tag{1.2.17}$$

where

$$B_2(u,v) = \int_0^\ell (AEu'v' + cuv)dx + \frac{\alpha_1}{\beta_1}\,u(0)\,v(0) + \frac{\alpha_2}{\beta_2}\,u(\ell)\,v(\ell)\,,$$

$$F_2(v) = \int_0^\ell fv\,dx\,.$$

The form $B_2(u,v)$ is defined on $X_2 \times Y_2 = H^1(\Omega) \times H^1(\Omega)$. Note carefully that $F_2 \neq F_1$ since its domain has been changed. Note also that the boundary conditions can **not** be imposed on X_2 or Y_2 since $u'(0)$, $u'(\ell)$ are not defined for $u \in H^1(\Omega)$.

Exercise 1.9 Assume $AE \in L^\infty(\Omega)$, $c \in L^\infty(\Omega)$, $f \in L^2(\Omega)$. Prove that $|B_2(u,v)| \leq C\|u\|_{H^1}\,\|v\|_{H^1}$, $|F(v)| \leq \|f\|_{L^2}\|v\|_{H^1}$.

In deriving (1.2.17), we assumed $\beta_i \neq 0$, $i = 1, 2$, in (1.2.2), (1.2.3). If $\beta_i = 0$, $i = 1, 2$, $B_2(u,v)$ loses its meaning. Then (1.2.2), (1.2.3) imply that

$$u(0) = u(\ell) = 0\,. \tag{1.2.18}$$

These boundary conditions, also called (homogeneous) Dirichlet conditions, can be enforced by restriction, since $u \in H^1(\Omega)$ has well-defined point values in the sense of trace (cf. Theorem A.21 with $m = n = 1$, $s = 2$, $k = 0$). Therefore we define

$$B_3(u,v) = \int_0^\ell (AEu'v' + cuv)dx, \;\; F_3(v) = \int_0^\ell fv\,dx \tag{1.2.19}$$

on $X_3 \times Y_3$ or Y_3, respectively with $X_3 = Y_3 = H_0^1(\Omega)$. Note again that $F_3 \neq F_2$ since $Y_3 \neq Y_2$.

Finally, assume $v \in H^2$ and integrate (1.2.16) by parts once more to obtain

$$\int_0^\ell u\{-(AEv')' + cv\}dx - (AEu'v)\Big|_{x=0}^{x=\ell} + (AEv'u)\Big|_{x=0}^{x=\ell} = \int_0^\ell fv\,dx \qquad (1.2.20)$$

Assume, for example, that $\beta_1 \neq 0$, $\alpha_2 \neq 0$. Then we may eliminate in (1.2.20) $(AEu')\,|_{x=0} = (\alpha_1/\beta_1)\,u(0)$ and $u(\ell) = -(\beta_2/\alpha_2)\,(AEu')\,|_{x=\ell}$, yielding

$$\begin{aligned}
&\int_0^\ell u\{-(AEv')' + cv\}dx + u(0)\Big\{\frac{\alpha_1}{\beta_1}\,v(0) - A(0)\,E(0)\,v'(0)\Big\} \\
&-A(\ell)\,E(\ell)\,u'(\ell)\Big\{v(\ell) + \frac{\beta_2}{\alpha_2}\,A(\ell)\,E(\ell)\,v'(\ell)\Big\} \\
&= \int_0^\ell u\{-(AEv')' + cv\}dx + \frac{u(0)}{\beta_1}\,\ell_1(v) - \frac{(AEu')(\ell)}{\alpha_2}\,\ell_2(v) \\
&= \int_0^\ell fv\,dx\,.
\end{aligned} \qquad (1.2.21)$$

The natural space for u is now $X_4 = L^2(\Omega)$. Since the functionals $\ell_i(v)$ are well-defined for $v \in H^2(\Omega)$, we select

$$Y_4 = \{v \in H^2(\Omega) : \ell_1(v) = \ell_2(v) = 0\}\,, \qquad (1.2.22)$$

and get the following: find $u \in X_4$ such that

$$B_4(u, v) = F_4(v) \text{ for all } v \in Y_4 \qquad (1.2.23)$$

where

$$B_4(u, v) = \int_0^\ell u\{-(AEv')' + cv\}\,dx, \; F_4(v) = \int_0^\ell fv\,dx\,.$$

In the same way as Proposition 1.8 we verify the following result.

Proposition 1.10 *Assume $AE \in W^{1,\infty}(\Omega)$, $c \in L^\infty(\Omega)$. Then*

$$|B_4(u, v)| \le C\|u\|_{L^2}\,\|v\|_{H^2} \textit{ for all } u \in X_4, v \in Y_4\,.$$

Formulation (1.2.23) allows very general functionals $F_4(v)$. For example, consider the following.

Proposition 1.11 *Let $x_0 \in \overline{\Omega}$. Then $F_4(v) := v'(x_0) \in Y_4'$.*

Proof We must show that

$$|F_4(v)| \le C\,\|v\|_{H^2} \text{ for every } v \in Y_4 .$$

This, however, follows from Theorem A.21 with $n = 1$, $m = s = 2$, $k = 1$. □

Remark 1.12 $F_4(v) = v'(x_0)$ is apparently not in the form given in (1.2.23). However, in the notation of distributions we have

$$F_4(v) = v'(x_0) = \int_0^\ell f v\, dx \text{ with } f(x) = -\delta'(x - x_0)$$

where $\delta(x)$ is the Dirac distribution (point load) at $x = 0$ (and $\int_0^\ell f v\, dx$ is understood as a duality pairing).

1.2.4 *Essential and natural boundary conditions*

Boundary conditions enforced by restriction on the test and/or trial spaces are called **essential boundary conditions.**

All other boundary conditions are called **natural boundary conditions**. They usually enter the variational formulation through the load functional $F(\cdot) \in Y'$.

We emphasize that **the nature of a boundary condition depends on the underlying variational framework** and the restrictions generally differ for test and trial spaces (e.g. X_1, Y_1 and X_4, Y_4).

In the next section we discuss the two-dimensional membrane problem.

1.2.5 *Generalized formulations of the membrane problem*

The problem of obtaining generalized formulation is very similar to the one used for the bar problem. A **first variational formulation** is obtained by multiplying (1.1.16) by $v \in L^2(\Omega) =: Y_5$ and integrating over Ω. This yields

$$B_5(u, v) = \int_\Omega v\, Lu\, dx = \int_\Omega f u\, dx =: F_5(v) \tag{1.2.24}$$

where the differential operator L is as in (1.1.16). Assume for now that

$$g^i = 0 \text{ for } i = 0, 1, 2 \text{ in (1.1.18)–(1.1.20)} .$$

Then, assuming further

$$\left. \begin{array}{l} A, E \in W^{1,\infty}(\Omega) \text{ componentwise, } c \in L^\infty(\Omega) , \\ \alpha \in L^\infty(\Gamma), \partial\Omega \text{ is of class } C^{1,1} \text{ (see Appendix A)} \end{array} \right\} , \tag{1.2.25}$$

we have

$$\begin{aligned} u \in X_5 = \{u \in H^2(\Omega) : u = 0 \text{ on } \Gamma^{[0]},\ \boldsymbol{n} \cdot E\, \textbf{grad}\, u = 0 \\ \text{on } \Gamma^{[1]},\ \alpha\gamma_0 u - \boldsymbol{n} \cdot AE\, \textbf{grad}\, u = 0 \text{ on } \Gamma^{[2]}\} . \end{aligned} \tag{1.2.26}$$

Notice that the boundary conditions make sense since $u \in H^2$ implies that $\gamma_0 u \in H^{3/2}(\Gamma)$ and $\boldsymbol{n} \cdot AE\, \mathbf{grad}\, u \in H^{1/2}(\Gamma)$ by the trace theorem A.17. In particular, all boundary conditions are essential in this formulation.

A **second variational formulation** is obtained by integration by parts. Assume $v \in H^1(\Omega)$. Then

$$B_5(u,v) = \int_\Omega (\mathbf{grad}\, v \cdot AE\, \mathbf{grad}\, u + cuv)\, dx - \int_{\partial\Omega} \gamma_0 v\, \boldsymbol{n} \cdot AE\, \mathbf{grad}\, u\, do\,.$$

The natural space for u and v is now $H^1(\Omega)$. By the trace theorem A.17, $\gamma_0 u$ is still well defined on $H^1(\Omega)$. However, $\mathbf{grad}\, u \in L^2(\Omega)$ now and hence $\boldsymbol{n}(x) \cdot AE(x)$ $\mathbf{grad}\, u$ does **not** have a trace on $\Gamma^{[1]}$. Hence the boundary condition (1.1.19) can**not** be enforced by restriction on the space and must be incorporated into the loading $F(\cdot)$. This yields

$$\int_\Omega fv\, dx = \int_\Omega (\mathbf{grad}\, v \cdot AE\, \mathbf{grad}\, u + cuv)\, dx - \int_{\Gamma^{[1]}} \gamma_0 v\, g^1\, do - \int_{\Gamma^{[2]}} \gamma_0 v (g^2 - \alpha\gamma_0 u)\, do\,. \tag{1.2.27}$$

Sorting terms (i.e., moving terms with known quantities to the right-hand side) we arrive at the following: find

$$\left.\begin{array}{l} u \in X_6 = \{u \in H^1(\Omega) : \gamma_0 u = 0 \text{ on } \Gamma^{[0]}\} =: H^1(\Omega, \Gamma^{[0]}) \\ \text{such that } B_6(u,v) = F_6(v) \text{ for all } v \in Y_6 = X_6 \end{array}\right\}. \tag{1.2.28}$$

Here

$$B_6(u,v) := \int_\Omega (\mathbf{grad}\, v \cdot AE\, \mathbf{grad}\, u + cuv)\, dx + \int_{\Gamma^{[2]}} \alpha(\gamma_0 u)(\gamma_0 v)\, do\,,$$

$$F_6(v) := \int_\Omega fv dx + \int_{\Gamma^{[1]}} (\gamma_0 v)\, g^1\, do + \int_{\Gamma^{[2]}} (\gamma_0 v)\, g^2\, do\,.$$

Now the homogeneous Dirichlet condition (1.1.18) is essential whereas (1.1.19) and (1.1.20) are natural.

Proposition 1.13 *Assume*

$$\left.\begin{array}{l} A, E, c \in L^\infty(\Omega), \alpha \in L^\infty(\Gamma^{[2]}) \textit{ and that} \\ \partial\Omega \textit{ is Lipschitz, } f \in L^2(\Omega),\ g^1 \in L^2(\Gamma^{[1]}),\ g^2 \in L^2(\Gamma^{[2]}) \end{array}\right\}. \tag{1.2.29}$$

Then, for all $u, v \in H^1(\Omega)$

$$|B_6(u,v)| \leq C \, \|u\|_{H^1} \, \|v\|_{H^1}, \; |F_6(v)| \leq C \, \|v\|_{H^1} \, .$$

Proof Follows from the Schwarz inequality and the continuity $\|\gamma_0 u\|_{L^2(\Gamma^{[j]})} \leq C \, \|u\|_{H^1(\Omega)}$ of the trace map γ_0 (cf. Theorem A.11 and Remark A.12). □

Note in particular the weaker assumptions (1.2.29) on the data as compared to (1.2.25). A third variational formulation is given in the following.

Exercise 1.14 Integrating by parts once more in (1.2.27), derive another formulation on $L^2(\Omega) \times$ (a subspace of) $H^2(\Omega)$. Prove an analog of Proposition 1.11, assuming (1.2.25). Which boundary conditions are now natural, which ones are essential?

1.3 Existence of weak solutions. Inf-sup condition

1.3.1 *Theory*

For the model problems in Section 1 several generalized formulations of the form: find $u \in X$ such that

$$B(u,v) = F(v) \text{ for all } v \in Y \tag{1.3.1}$$

were derived for which $B(u,v)$ and $F(v)$ were continuous, i.e., (1.2.6) was satisfied, and such that (1.2.7) holds, i.e., every classical solution (if it exists) is a generalized solution.

Here we address (1.2.8) and (1.2.9), i.e., **existence** of a generalized solution and **a priori estimates**. The following theorem is fundamental for everything that follows in the remainder. For convenience we state the result for spaces of real-valued functions. It is, however, also valid in spaces of complex-valued functions.

Theorem 1.15 *(inf–sup condition). Let* X, Y *be real reflexive Banach spaces with norms* $\| \circ \|_X$ *and* $\| \circ \|_Y$*, respectively. Let further* $B(\cdot,\cdot) : X \times Y \to \mathbb{R}$ *be a bilinear form such that there exist* $C_1, \, C_2 > 0$ *with*

$$|B(u,v)| \leq C_1 \, \|u\|_X \, \|v\|_Y \text{ for every } u \in X, v \in Y \, . \tag{1.3.2}$$

Then the conditions

$$\inf_{0 \neq u \in X} \; \sup_{0 \neq v \in Y} \frac{B(u,v)}{\|u\|_X \, \|v\|_Y} \geq C_2 > 0 \, , \tag{1.3.3}$$

$$\sup_{u \in X} \; B(u,v) > 0 \text{ for every } 0 \neq v \in Y \, , \tag{1.3.4}$$

are necessary and sufficient that for every continuous linear functional F *on* Y *there exists a unique* $u_0 \in X$ *such that*

$$B(u_0, v) = F(v) \text{ for all } v \in Y \, . \tag{1.3.5}$$

If (1.3.3), (1.3.4) *are satisfied, the a priori estimate*

$$\|u_0\|_X \leq \frac{\|F\|_{Y'}}{C_2} = \frac{1}{C_2} \sup_{0 \neq v \in Y} \frac{|F(v)|}{\|v\|_Y} \tag{1.3.6}$$

holds.

Proof We proceed in several steps.

Step 1: Let $u \in X$ be arbitrary, fixed. Then (1.3.2) implies

$$\phi_u(v) := B(u,v) \in Y'$$

since

$$\|\phi_u\|_{Y'} = \sup_{0 \neq v \in Y} \frac{|\phi_u(v)|}{\|v\|_Y} = \sup_{0 \neq v \in Y} \frac{|B(u,v)|}{\|v\|_Y} \leq C_1 \|u\|_X \, .$$

By the definition of Y', there exists a unique $z \in Y'$ such that $\langle z, v \rangle = \phi_u(v) = B(u,v)$ for all $v \in Y$. Denote $z = R(u)$. Then $R : u \to z$ is continuous and linear,

$$R : X \to Y', \ \|R\|_{X \to Y'} \leq C_1 \, .$$

Step 2: We claim that the range of R is closed in Y'. We have

$$\|R(u)\|_{Y'} = \sup_{0 \neq v \in Y} \frac{\langle R(u), v \rangle}{\|v\|_Y} = \sup_{0 \neq v \in Y} \frac{|B(u,v)|}{\|v\|_Y} \geq C_2 \|u\|_X \tag{1.3.7}$$

by (1.3.3). Now let $\{u_n\}_{n=1}^{\infty} \subset X$ be such that $\{R(u_n)\}_{n=1}^{\infty}$ is Cauchy in Y'. Then $\{u_n\}_{n=1}^{\infty}$ is Cauchy in X. To prove it, notice

$$\|R(u_n) - R(u_m)\|_{Y'} = \|R(u_n - u_m)\|_{Y'} \geq C_2 \|u_n - u_m\|_X \, .$$

Hence $R(X)$ is closed in Y'.

Step 3: We claim $R(X) = Y'$.
Assume not. Then $R(X) = \overline{R(X)}^{\|\circ\|_{Y'}} \neq Y'$. By the Hahn–Banach theorem there exists $v_0 \neq 0$ such that $\langle r, v_0 \rangle = 0$ for all $r \in R(X)$. Since Y is reflexive, $v_0 \in Y$ and hence by (1.3.4)

$$0 = \langle r, v_0 \rangle = \langle R(u), v_0 \rangle = B(u, v_u) \text{ for all } u \in X$$

a contradiction to (1.3.4); thus $R(X) = Y'$, i.e., R is onto. This proves (1.3.5).

Step 4: By steps 2 and 3, for every $F \in Y'$, the linear equation $R(u) = F$ has a unique solution and $u = R^{-1}(F)$. By (1.3.7)

$$\|R(u)\|_{Y'} \geq C_2 \|u\|_X \Longrightarrow \|R^{-1}(F)\|_X \leq \frac{1}{C_2} \|F\|_{Y'}$$

i.e., $\|R^{-1}\|_{Y' \to X} \leq 1/C_2$.

This already implies (1.3.6). Alternatively, we may directly refer to (1.3.3) as follows.

Step 4′: Assume $u_0 \neq 0$ and note

$$|B(u_0, v)| = |F(v)| \le \|F\|_{Y'} \, \|v\|_Y \text{ for all } v \in Y$$

whence

$$\begin{aligned} \|F\|_{Y'} &\ge \sup_{0 \neq v \in Y} \frac{|B(u_0,v)|}{\|v\|_Y} = \|u_0\|_X \sup_{0 \neq v \in Y} \frac{|B(u_0,v)|}{\|u_0\|_X \, \|v\|_Y} \\ &\ge \|u_0\|_X \inf_{0 \neq u \in X} \sup_{0 \neq v \in Y} \frac{|B(u,v)|}{\|u\|_X \, \|v\|_Y} \ge C_2 \, \|u_0\|_X \, . \end{aligned}$$

Step 5: The necessity of (1.3.3) follows from (1.3.6). (1.3.5) and (1.3.6) imply (1.3.4). To see this, assume that there is $v_0 \in Y$, $v_0 \neq 0$, with $B(u, v_0) = 0$ for all $u \in X$. By the Hahn–Banach theorem, there is $F \in Y'$ with $F(v_0) \neq 0$. Then

$$0 = B(u, v_0) = F(v_0) \neq 0$$

a contradiction to the solvability (1.3.5). Finally, we have seen in Step 4 that (1.3.5) implies (1.3.6). □

Remark 1.16 To prove the inf–sup condition (1.3.3) the following **general scheme** can be applied:

Given $0 \neq u \in X$, find $v = v_u \in Y$ such that

i) $$\|v_u\|_Y \le \alpha_1 \|u\|_X$$

and such that

ii) $$B(u, v_u) \ge \alpha_2 \|u\|_X^2 \, .$$

Then (1.3.3) holds with $C_2 = \alpha_2/\alpha_1$. Condition (1.3.4) must be verified independently.

Frequently in applications, the space X is a proper subspace of a larger space $\widetilde{X} \supset X$, obtained by enforcement of homogeneous essential boundary conditions, e.g., $X = H_0^1(\Omega) \subset \widetilde{X} = H^1(\Omega)$. Here the following result is useful.

Proposition 1.17 *Let X be a reflexive Banach space and $N \subseteq X$ be a closed subspace. Then N, equipped with the norm of X, is itself a reflexive Banach space. Let in particular $\widetilde{X}$ be a second Banach space and $\ell : X \to \widetilde{X}$ a bounded linear map. Then $N = \{u \in X : \ell(u) = 0\}$ is a closed subspace.*

For a proof, we refer for example to [65], Theorem 1.6.12.

Another important case occurs when the boundary value problem of interest with zero data (i.e., the homogeneous problem) admits nonzero solutions. For example, the homogeneous Neumann problem for the membrane

$$\Delta u = 0 \text{ in } \Omega, \qquad \frac{\partial u}{\partial n} = 0 \text{ on } \partial\Omega$$

admits solutions $u = \text{const.}$ which constitute a closed, linear subspace N of $H^1(\Omega)$. In this case it is useful to consider the **factor space** X/N of X with respect to the space N of nonzero solutions of the homogeneous problem. The set X/N is the set of all **equivalence classes**

$$[u] = \{v \in X : \exists w \in N \ni v = u + w\} .$$

The element u is the **representer** of $[u]$. All vector space operations (addition of elements, scalar multiplication, etc.) are transferred from X to X/N via the representers, i.e.,

$$[u] + [v] := [u + v], \ \alpha[u] := [\alpha u], \ \text{etc.}$$

When working with X/N, we always write u (but mean $[u]$). We have the following result.

Proposition 1.18 *Let X be a reflexive Banach space and N a closed linear subspace. Then X/N is itself a reflexive Banach space with norm*

$$\|u\|_{X/N} := \inf_{v \in N} \|u + v\|_X \, . \tag{1.3.8}$$

For a proof, see e.g. [65]. We will now extend Theorem 1.15 to the case that $B(u, v) = 0$ for some $0 \neq u \in X$ and all $v \in Y$.

Let therefore $B : X \times Y \to \mathbb{R}$ be a continuous bilinear form. There exists a unique linear operator, also denoted by B, from X to Y' such that

$$B(u, v) = \langle Bu, v\rangle_{Y' \times Y} = \langle u, B^* v\rangle_{X \times X'} \text{ for all } u \in X, v \in Y \tag{1.3.9}$$

where $B^* : Y \to X'$ is the adjoint of B.

As intersections of closed spaces,

$$\begin{aligned} \ker B &= \{u \in X : B(u, v) = 0 \text{ for all } v \in Y\} \\ \ker B^* &= \{v \in Y : B(u, v) = 0 \text{ for all } u \in X\} \end{aligned}$$

are closed, linear subspaces of X or Y and, by Proposition 1.18, $X/\ker B$ and $Y/\ker B^*$ are themselves reflexive Banach spaces.

The following result from functional analysis, the closed range theorem, is the basis for the extension of Theorem 1.15.

Proposition 1.19 *Let X, Y be reflexive Banach spaces. Let $B(\cdot,\cdot) : X \times Y \to \mathbb{R}$ be a continuous bilinear form and denote by $B : X \to Y'$ the corresponding linear operator satisfying* (1.3.9). *Then the following are equivalent:*

i) range $(B) \subset Y'$ *is closed*

ii) range $(B^*) \subset X'$ *is closed*

iii) $(\ker B)^\circ := \{F \in X' : F(v) = 0 \text{ for all } v \in \ker B\} = \text{range}\,(B^*)$

iv) $(\ker B^*)^\circ := \{F \in Y' : F(v) = 0 \text{ for all } v \in \ker B^*\} = \text{range}\,(B)$

v) *there exists $\beta > 0$ such that for every $y \in X/\ker B$*

$$\sup_{v \in Y} \frac{B(u,v)}{\|v\|_Y} \geq \beta \|u\|_{X/\ker B} \tag{1.3.10}$$

vi) *for the same $\beta > 0$ as in* v) *and every $v \in Y/\ker B^*$*

$$\sup_{u \in X} \frac{B(u,v)}{\|u\|_X} \geq \beta \|v\|_{Y/\ker B^*} \,. \tag{1.3.11}$$

For a proof, we refer to [152], pp. 205–208. We can now state the announced generalization of Theorem 1.15.

Theorem 1.20 *Let X, Y be reflexive Banach spaces and let $B(\cdot,\cdot) : X \times Y \to \mathbb{R}$ be a continuous bilinear form with associated operator $B : X \to Y'$ as in* (1.3.9) *and $\ker B = \{u \in X,\ B(u,v) = 0 \text{ for all } v \in Y\} \neq \{0\}$. Assume that there exists $C_2 > 0$ such that*

$$\inf_{u \in X/\ker B} \sup_{v \in Y} \frac{B(u,v)}{\|u\|_{X/\ker B}\|v\|_Y} > C_2 > 0 \tag{1.3.12}$$

and

$$\sup_{u \in X} B(u,v) > 0 \text{ for every } 0 \neq v \in Y/\ker B^* \,. \tag{1.3.13}$$

Then for every $F \in Y'$ with $F(v) = 0$ for $v \in \ker B^*$ the problem: find $u \in X$ such that

$$B(u,v) = F(v) \text{ for every } v \in Y \tag{1.3.14}$$

admits a solution $u \in X$ which is unique up to elements $w \in \ker B$ and the **a priori estimate**

$$\|u\|_{X/\ker B} \leq \frac{1}{C_2} \sup_{0 \neq v \in Y} \frac{|F(v)|}{\|v\|_Y} \tag{1.3.15}$$

holds.

Proof We observe that (1.3.12) implies (1.3.10) and hence range $(B) = (\ker(B^*))^\circ = \{F \in Y' : F(v) = 0$ for all $v \in \ker B^*\}$. Therefore the condition $F(v) = 0$ for $v \in \ker B^*$ is necessary for the existence of a solution of (1.3.15).

By Proposition 1.18, $X/\ker B$ and $Y/\ker B^*$ are reflexive Banach spaces and the Theorem is then a consequence of Theorem 1.15. □

Remark 1.21 The condition $F(v) = 0$ for every $v \in \ker B^*$ is in practice a **compatibility condition** for the data of the problem.

The following symmetric variant of the conditions (1.3.3) and (1.3.4) is also frequently found in the literature (see [101]):

$$\sup_{0 \neq v \in Y} \frac{B(u,v)}{\|v\|_Y} \geq C_2 \|u\|_X \text{ for all } u \in X\,, \tag{1.3.16}$$

$$\sup_{0 \neq u \in X} \frac{B(u,v)}{\|u\|_X} \geq C_2 \|v\|_Y \text{ for all } v \in Y\,, \tag{1.3.17}$$

Proposition 1.22 *Conditions* (1.3.3), (1.3.4) *and* (1.3.16), (1.3.17) *are equivalent.*

The proof is left to the reader as an exercise.

Remark 1.23 (on reflexivity). We assumed in Theorem 1.15 that X and Y are reflexive. In practice this is not very restrictive – all Sobolev spaces $W^{k,s}(\Omega)$, $W^{k,s}(\Gamma)$ (cf. Appendix A) are reflexive for $1 < s < \infty$. The spaces $C^{k,\mu}$ as well as $W^{k,s}$ for $s = 1, \infty$ are **not** reflexive, however.

A very important special case of Theorem 1.15 occurs when $X = Y$, as e.g., in (1.2.17), or (1.2.19).

Theorem 1.24 *(Lax–Milgram lemma). Let X be a reflexive Banach space and $B(\cdot,\cdot) : X \times X \to \mathbb{R}$ be a continuous bilinear form which is* **coercive**, *i.e., there exists $C_2 > 0$ such that*

$$B(u,u) \geq C_2 \|u\|_X^2 \textit{ for every } u \in X\,. \tag{1.3.18}$$

Then for every $F \in X'$ there exists a unique solution $u_0 \in X$ of the problem: find $u_0 \in X$ such that

$$B(u_0, v) = F(v) \textit{ for every } v \in X\,. \tag{1.3.19}$$

Moreover, the a priori estimate

$$\|u_0\|_X \leq \|F\|_{X'}/C_2$$

holds.

Proof Observe that (1.3.18) implies (1.3.3) and (1.3.4). □

Corollary 1.25 *Under the assumptions of Theorem 1.24, if in addition the bilinear form is* **symmetric**, *i.e.,*

$$B(u,v) = B(v,u) \ \textit{for every} \ u, v \in X\ , \tag{1.3.20}$$

the solution u_0 of (1.3.19) is the unique minimizer of the quadratic functional (the "potential energy")

$$J(v) = \frac{1}{2}\, B(v,v) - F(v) \tag{1.3.21}$$

over X.

Proof Let u_0 be a critical point of J. Then for every $v \in X$

$$\begin{aligned} 0 &= \frac{d}{d\varepsilon}\, J(u_0 + \varepsilon v)\,|_{\varepsilon=0} \\ &= \frac{d}{d\varepsilon} \left(\frac{1}{2}\, B(u_0,u_0) + \varepsilon B(u_0,v) + \frac{1}{2}\, \varepsilon^2\, B(v,v) - F(u_0) - \varepsilon F(v) \right) \Big|_{\varepsilon=0} \\ &= B(u_0,v) - F(v)\ . \end{aligned}$$

u_0 is a minimizer since J is convex. Uniqueness follows as in Theorem 1.15. □

Remark 1.26 In the symmetric case (1.3.20), the coercivity (1.3.18) implies that $\|u\|_E = \sqrt{B(u,u)}$ is a norm on X, the so-called **energy norm.**

1.3.2 *Examples*

We will now illustrate the abstract theory by casting the model problems from Section 1 into various weak forms. It will become clear that different weak formulations require different regularity of the data and are, in general, not equivalent.

1.3.2.1 *Bar problem*

We discuss generalized formulations of the one-dimensional bar problem. Consider first (1.2.17). We have $X_2 = Y_2 = H^1(\Omega)$, $\Omega = (0,\ell)$ and $B_2(u,v)$, $F_2(v)$ are continuous on $X_2 \times Y_2$ or Y_2, respectively. Also, $H^1(\Omega)$ is a Hilbert space, hence complete and reflexive. We may therefore apply the Lax–Milgram lemma, Theorem 1.24.

Proposition 1.27 *Assume that $AE \in L^\infty(\Omega)$ and that*

$$\begin{aligned} &0 < K_1 \le A(x)\,E(x) \ \ \textit{a.e.}\ x \in \Omega\ , \\ &c(x) \in L^1(\Omega),\ c(x) \ge 0,\ \ \textit{a.e.}\ x \in \Omega,\ \ \alpha_i > 0,\ \beta_i > 0,\ i = 1,2\ . \end{aligned} \tag{1.3.22}$$

Then the generalized formulation (1.2.17) of the bar problem admits a unique solution $u \in H^1(\Omega)$.

Proof We must show coercivity (1.3.18), i.e.,

$$B_2(u,u) \geq C_2 \|u\|^2_{H^1(\Omega)} \quad \text{for all} \quad u \in H^1(\Omega)\,. \tag{1.3.23}$$

Referring to the Lax–Milgram lemma (Theorem 1.24) completes the proof. □

Exercise 1.28 Prove (1.3.12) in the case that $\alpha_1 = \alpha_2 = 0$. Assume

$$\int_0^\ell c(x)dx \geq K_2 > 0\,. \tag{1.3.24}$$

Consider now (1.2.19), i.e., the case when $\beta_i = 0$, $i = 1,2$. Here $X_3 = Y_3 = H_0^1(\Omega) = \{u \in H^1(\Omega) : u(0) = u(\ell) = 0\} \subset H^1(\Omega)$. In order to apply Theorem 1.24, we must show that X_3 is a Banach space. Generally, **all subspaces obtained by enforcement of proper homogeneous essential boundary conditions are themselves reflexive Banach spaces.**

To verify that $X_3 = H_0^1(\Omega)$ is a reflexive Banach space, by Proposition 1.17 we only need to check that $\ell_1(u) = u(0)$, $\ell_2(u) = u(\ell)$ are continuous linear functionals on $H^1(\Omega)$. This, however, is a consequence of the trace theorem A.17.

Proposition 1.29 *Assumptions as in Proposition* 1.27. *Then, for every* $f \in H^{-1}(\Omega)$, *the generalized bar problem* (1.2.19) *admits a unique solution* $u_0 \in H_0^1(\Omega)$.

Proof The continuity of $B_3(u,v)$ is straightforward and left to the reader. Since B_3 is symmetric, we apply Theorem 1.24 with $X = X_3 = H_0^1(\Omega)$. We verify (1.3.18): for $u \in H_0^1(\Omega)$

$$B_3(u,u) = \int_0^\ell (AE(u')^2 + cu^2)dx \geq K_1 \int_0^\ell (u')^2 dx \geq C\|u\|^2_{H^1(\Omega)}$$

by the first Poincaré inequality, Theorem A.25. The assertion now follows from Theorem 1.24. □

Let us now consider the generalized formulation (1.2.23) of the bar problem. Recall

$$X_4 = L^2(\Omega),\ Y_4 = \{v \in H^2(\Omega) : \ell_1(v) = \ell_2(v) = 0\}\,,$$

$$B_4(u,v) = \int_0^\ell u\{-(AEv')' + cv\}dx = \int_0^\ell u\,Lv dx\,.$$

Evidently, we must use Theorem 1.15 here.

Proposition 1.30 *Assume $\beta_i \neq 0$ and that $AE \in W^{1,\infty}(\Omega)$, $c \in L^\infty(\Omega)$ satisfy the assumptions of Proposition* 1.27*. Then for every $F_4 \in (Y_4)'$, there exists a unique $u \in X_4$ such that $B_4(u,v) = F_4(v)$ for all $v \in Y_4$.*

Proof By Proposition 1.8, B_4 is continuous. Using Theorem A.17, we verify $|\ell_i(u)| \leq C_i \|u\|_{H^2}$, $i = 1,2$ where C_i depends on ℓ, $\|AE\|_{L^\infty}$ and on $|\alpha_i|$, $|\beta_i|$; hence by Proposition 1.17, Y_4 is a reflexive Banach space. It remains to establish the inf–sup condition (1.3.3).

Let us apply Remark 1.16. Given $u \in L^2(\Omega) = X_4$, let v_u solve the following problem: find $v_u \in H^1(\Omega) = X_2$ such that

$$B_2(v_u, w) = \int_0^\ell uw\,dx \text{ for all } w \in Y_2 . \tag{1.3.25}$$

By Theorem 1.24, there exists a unique v_u in $H^1(\Omega)$. The next step is crucial. From standard **regularity theory**, $u \in L^2(\Omega)$ and $AE \in W^{1,\infty}(\Omega)$ imply that v_u belongs to $H^2(\Omega)$ and $\ell_i(v_u) = 0$ for $i = 1,2$, i.e., $v_u \in Y_4$. Moreover, the **a priori estimate** $\|v_u\|_{H^2(\Omega)} \leq \alpha_1 \|u\|_{L^2(\Omega)}$ holds and $Lv_u = u$ in $L^2(\Omega)$. Hence $B_4(u, v_u) = \int_0^\ell u\, Lv_u\, dx = \|u\|^2_{L^2(\Omega)}$, so that by Remark 1.16 the inf–sup condition (1.3.3) holds with $C_2 = 1/\alpha_1$. To show (1.3.4), for a given $v \in Y_4$ select the function $u = Lv$.

Proposition 1.27 implies that $Lv \neq 0$ in $L^2(\Omega)$ if $v \neq 0$, hence $B_4(u,v) = \int_0^\ell (Lv)^2 dx > 0$. Now Proposition 1.30 is a consequence of Theorem 1.15. □

Remark 1.31 Generalized solutions $u \in L^2(\Omega)$ are sometimes called "very weak solutions" and the generalized formulation (1.2.23) the "method of transposition". Notice that we need a **regularity result** for the generalized solution of (1.3.25) in order to prove existence of a very weak solution $u \in X_4$.

1.3.2.2 *Membrane problem*

Let us now turn to the **membrane problem** (1.1.16)–(1.1.20). As for the bar problem, we consider first the symmetric weak formulation (1.2.28). We have the following sufficient conditions for existence.

Proposition 1.32 *Assume that $\Omega \subset \mathbb{R}^2$ is a curvilinear polygon,*

$$AE \in [L^\infty(\Omega)]^{2\times 2},\ c \in L^\infty(\Omega),\ \alpha \in L^\infty(\Gamma^{[2]}) . \tag{1.3.26}$$

Assume further **strong ellipticity**, *i.e., there exists $K_1 > 0$ such that*

$$\xi^\top AE(x)\,\xi \geq K_1 |\xi|^2 \text{ for all } \xi \in \mathbb{R}^2,\ \text{a.e. } x \in \Omega \tag{1.3.27}$$

such that

$$c(x) \geq 0 \text{ a.e. } x \in \Omega,\ \alpha(x) \geq 0 \text{ a.e. } x \in \Gamma^{[2]} . \tag{1.3.28}$$

If further $\int_{\Gamma^{[0]}} do > 0$, *then for every* $f \in L^2(\Omega)$, $g^i \in L^2(\Gamma^{[i]})$, $i = 1, 2$, *there exists a unique solution* $u \in H^1(\Omega, \Gamma^{[0]})$ *of* (1.2.28) *and*

$$\|u\|_{H^1(\Omega)} \leq C\big(\|f\|_{L^2(\Omega)} + \|g^1\|_{L^2(\Gamma^{[1]})} + \|g^2\|_{L^2(\Gamma^{[2]})}\big)\,.$$

Proof Let $B_6(\cdot,\cdot)$ and $F_6(\cdot)$ be as in (1.2.28).

Step 1: (continuity of B_6 and F_6)

$$\begin{aligned}
|B_6(u,v)| &\leq \|AE\|_{L^\infty(\Omega)} \int_\Omega |\nabla u|\,|\nabla v| dx + \|c\|_{L^\infty} \int_\Omega |uv| dx \\
&\quad + \|\alpha\|_{L^\infty(\Gamma^{[2]})} \int_{\Gamma^{[2]}} |\gamma_0 u\, \gamma_0 v| do \\
&\leq \|AE\|_{L^\infty(\Omega)} \|\nabla u\|_{L^2(\Omega)} \|\nabla v\|_{L^2(\Omega)} \\
&\quad + \|c\|_{L^\infty(\Omega)} \|u\|_{L^2(\Omega)} \|v\|_{L^2(\Omega)} \\
&\quad + \|\alpha\|_{L^\infty(\Gamma^{[2]})} \|\gamma_0 u\|_{L^2(\Gamma^{[2]})} \|\gamma_0 v\|_{L^2(\Gamma^{[2]})} \\
&\leq C \|u\|_{H^1(\Omega)} \|v\|_{H^1(\Omega)}
\end{aligned}$$

where we used the trace theorem. Likewise

$$\begin{aligned}
|F_6(v)| &\leq \|f\|_{L^2(\Omega)} \|v\|_{L^2(\Omega)} + \|\gamma_0 v\|_{L^2(\Gamma^{[2]})}\, \|g^1\|_{L^2(\Gamma^{[1]})} \\
&\quad + \|\gamma_0 v\|_{L^2(\Gamma^{[2]})} \|g^2\|_{L^2(\Gamma^{[2]})} \\
&\leq C(\|f\|_{L^2(\Omega)} + \|g^1\|_{L^2(\Gamma^{[1]})} + \|g^2\|_{L^2(\Gamma^{[2]})}) \|v\|_{H^1(\Omega)}
\end{aligned}$$

again by the trace theorem A.11 and Remark A.12.

Step 2: (existence). Since $X_6 = Y_6 = H^1(\Omega, \Gamma^{[0]})$, we use the Lax–Milgram lemma, (Theorem 1.24). We check (1.3.18):

$$B_6(u,u) \geq \int_\Omega \nabla u \cdot AE\, \nabla u\, dx$$

by (1.3.28) and by (1.3.27)

$$B_6(u,u) \geq K_1 \int_\Omega |\nabla u|^2 dx = \frac{K_1}{2}\Big(\int_\Omega |\nabla u|^2 dx + C^{-1} \int_\Omega |u|^2 dx\Big)$$

where we used the first Poincaré inequality (Theorem A.25).

Hence, for every $u \in X_6 = Y_6$,

$$B_6(u,u) \geq \frac{K_1}{2} \min(1, C^{-1})\, \|u\|^2_{H^1(\Omega)}\,,$$

i.e., B_6 is coercive on $X_6 \times Y_6$. By the Lax–Milgram lemma, Theorem 1.24, the assertion follows. □

Exercise 1.33 In the variational formulation (1.2.28), which of the boundary conditions are natural and which ones are essential?

We turn now to the nonsymmetric variational formulation of the membrane problem from Exercise 1.14. Since in this formulation all boundary conditions (1.1.18)–(1.1.20) are essential (!), we assume for convenience here

$$g^1 = 0 \text{ on } \Gamma^{[1]}, \quad g^2 = 0 \text{ on } \Gamma^{[2]} . \tag{1.3.29}$$

The next section will deal with inhomogeneous essential boundary conditions. The generalized formulation of the membrane problem is: find $u \in L^2(\Omega) = X_7$ such that

$$B_7(u, v) = F_7(v) \text{ for all } v \in Y_7 . \tag{1.3.30}$$

Here

$$B_7(u, v) = \int_\Omega uL^* v \, dx, \quad F_7(v) = \int_\Omega fv \, dx$$

and

$$Y_7 = \{v \in H^2(\Omega); \gamma_0 v = 0 \text{ on } \Gamma^{[0]}, \boldsymbol{n}(x) \cdot \gamma_0((AE)^\top \nabla v) = 0 \text{ on } \Gamma^{[1]},$$
$$\alpha\gamma_0 v + \boldsymbol{n} \cdot \gamma_0(AE\nabla v) = 0 \text{ on } \Gamma^{[2]}\},$$

where L^*v denotes the **adjoint operator** to L, i.e.,

$$L^*v = -\operatorname{div}((AE)^\top \mathbf{grad}\, v) + cv . \tag{1.3.31}$$

Then analog of Proposition 1.32 is the following

Proposition 1.34 *Assume* (1.3.29) *and further that Proposition* 1.32 *holds for the adjoint problem*

$$\begin{cases} L^*v = f \quad \text{in } \Omega , \\ \gamma_0 v = 0 \text{ on } \Gamma^{[0]}, \quad \boldsymbol{n} \cdot \gamma_0((AE)^\top \nabla v) = 0 \text{ on } \Gamma^{[1]} , \\ \alpha\gamma_0 v + \boldsymbol{n} \cdot \gamma_0\left((AE)^\top \nabla v\right) = 0 \qquad \text{on } \Gamma^{[2]} . \end{cases} \tag{1.3.32}$$

Assume further the following **regularity for the adjoint problem:**

if $f \in L^2$, the generalized solution $v \in H^1(\Omega, \Gamma^{[0]})$ of (1.3.32) *belongs to $H^2(\Omega)$.* (1.3.33)

Then the generalized formulation (1.3.30) *has, for every $F_7 \in (Y_7)'$, a unique very weak solution $u \in L^2(\Omega)$.*

The proof is left as an exercise.

Remark 1.35 (on regularity). The regularity (1.3.33) of the adjoint problem in general does **not** hold when the boundary has corners, or when $\overline{\Gamma^{[0]}} \cap \overline{\Gamma^{[1]}} \neq \varnothing$.

Remark 1.36 Notice that the generalized formulation (1.3.30) will give a priori estimates for $\|u\|_{L^2(\Omega)}$ in terms of the data. A similar argument (the Aubin–Nitsche trick) will be used later to obtain L^2-error estimates for the FE solutions.

1.3.2.3 *Neumann problem*

Consider the membrane problem with $AE = \mathbf{1}$,

$$\Gamma^{[0]} = \Gamma^{[2]} = \varnothing, \quad c = 0 \text{ in } \Omega ,$$

i.e., the pure Neumann problem

$$-\Delta u = f \text{ in } \Omega, \quad \frac{\partial u}{\partial n} = g^1 \text{ on } \Gamma^{[1]} = \partial\Omega . \tag{1.3.34}$$

The symmetric weak formulation reads as follows: find $u \in H^1(\Omega)$ such that

$$B(u,v) = \int_\Omega \nabla u \cdot \nabla v \, dx = F(v) = \int_\Omega fv \, dx + \int_{\partial\Omega} g^1 \gamma_0 v \, do$$

for all $v \in H^1(\Omega)$.

We observe that

$$B(1,v) = B(u,1) = 0$$

for all $u \in X = H^1(\Omega)$ and all $v \in Y = H^1(\Omega)$, i.e., that

$$\ker B = \ker \mathrm{B}^* = \operatorname{span}\{1\} .$$

Since $B(1,1) = 0$, the Lax–Milgram lemma (Theorem 1.24) does not apply, and we must use Theorem 1.20. We obtain for

$$X/\ker B \cong \widetilde{H}^1(\Omega) = \left\{ u \in H^1(\Omega) : \int_\Omega u \, dx = 0 \right\}$$

that

$$B(u,u) = \int_\Omega |\nabla u|^2 dx \geq C_2 \|u\|^2_{H^1}$$

by the second Poincaré inequality (Theorem A.26), from where we get (1.3.12), (1.3.13). By Theorem 1.20, the Neumann problem (1.3.34) has a solution $u \in H^1(\Omega)$, unique up to constants, **provided** the data satisfy the **compatibility condition**

$$0 = F(1) = \int_\Omega f \, dx + \int_{\partial\Omega} g^1 do .$$

1.4 Inhomogeneous essential boundary conditions. The trace spaces $H^{1/2}(\Gamma)$ and $H^{1/2}_{00}(\Gamma^{[1]})$

Essential boundary conditions are those which are imposed on the test/trial spaces. So far, we have considered only examples of **homogeneous** essential boundary conditions – the restriction of the test and trial spaces resulted in closed, linear subspaces where the existence theory applied. Inhomogeneous essential boundary conditions can**not** be treated in this way.

1.4.1 *The trace space* $H^{1/2}(\Gamma)$

Example 1.37 Consider the bar problem (1.2.1) with boundary conditions

$$u(0) = 1,\ u(\ell) = 0 \tag{1.4.1}$$

and assume $AE = 1$, $c = 1$, $\Omega = (0, \ell)$.

We adopt the symmetric variational formulation (1.2.19) with spaces X and Y to be selected, i.e.: find $u \in X$ such that

$$B_3(u, v) = \int_0^\ell (u'v' + uv)dx = F(v) = \int_0^\ell fv\,dx \text{ for all } v \in Y\,. \tag{1.4.2}$$

The choice of X analogous to X_3 would be

$$X = H^1(\Omega) \cap \{u : u(0) = 1,\ u(\ell) = 0\}\,, \tag{1.4.3}$$

whereas Y could again be selected as $H^1_0(\Omega)$. Evidently, however, X **is not a linear subspace of** $H^1(\Omega)$ (since $u_1, u_2 \in X$ does not imply that $u_1 + u_2 \in X$), and the existence theory in Section 3 does not apply. The proper treatment of (1.4.1) is to write

$$u = u_0 + w \tag{1.4.4}$$

where $w \in H^1(\Omega)$ is a (in general not unique) particular solution satisfying the inhomogeneous boundary condition (1.4.1), a so-called **trace lifting**. For example, $w(x) = (1 - x/\ell)$ is suitable.

By superposition then $u_0 \in H^1_0(\Omega)$, and we write (1.4.2) in the following form: find $u_0 \in H^1_0(\Omega)$ such that

$$B_3(u_0, v) = F(v) - B_3(w, v) \text{ for all } v \in H^1_0(\Omega)\,. \tag{1.4.5}$$

Since $w \in H^1(\Omega)$ and B_3 is continuous on $H^1 \times H^1$, problem (1.4.5) admits a unique solution u_0 by the Lax–Milgram lemma, Theorem 1.24 (the reader should verify this in detail). Notice also that even though w is not unique, nevertheless $u_0 + w$ is, since u_0 depends of course on w.

So the admissible Dirichlet boundary conditions are determined by the existence of a suitable trace lifting of the Dirichlet data. The set of all functions in

$L^2(\Gamma)$, $\Gamma = \partial\Omega$, which are boundary values of functions in $H^1(\Omega)$ is denoted by $H^{1/2}(\Gamma)$ and the continuous, linear mapping $\gamma_0 : H^1(\Omega) \to H^{1/2}(\Gamma)$ given by

$$H^1(\Omega) \ni u \longmapsto \gamma_0\, u := u|_\Gamma \in H^{1/2}(\Gamma)$$

is called the **trace operator**. In order to determine a trace lifting, it is of interest to know which functions belong to $H^{1/2}(\Gamma)$.

1.4.2 *The problem of discontinuous Dirichlet data*

Example 1.38 Let $\Omega = (0,1)^2$ and consider the **Dirichlet problem with discontinuous boundary data** on $\Gamma = \partial\Omega$:

$$-\Delta u = f \text{ in } \Omega\,,$$
$$u|_\Gamma = u_0 = \begin{cases} 1 \text{ for } y = 1,\ 0 < x < 1\,, \\ 0 \text{ else}\,. \end{cases} \tag{1.4.6}$$

A symmetric variational formulation analogous to (1.2.28) can be obtained as in Example 1.37 **provided** we can find a lifting (or a particular solution) $w \in H^1(\Omega)$ such that $\gamma_0 w = u_0$. By the trace theorem, Theorem A.17, the trace operator $\gamma_0 : H^1(\Omega) \to H^{\frac{1}{2}}(\Gamma)$ is onto, so that a lifting $w \in H^1(\Omega)$ exists if *and only if* $u_0 \in H^{\frac{1}{2}}(\Gamma)$. This, however, is **not** the case for the discontinuous data u_0 in (1.4.6), according to the next exercises. Therefore, the problem (1.4.6) does not admit a weak solution $u \in H^1(\Omega)$.

Exercise 1.39 Consider $I = (-1,1)$ and $v(x) = \text{sign}(x) = -1$ for $x < 0$ and 1 for $x > 0$. Show that $v \in H^{\frac{1}{2}-\varepsilon}(I)$, $0 < \varepsilon \le \frac{1}{2}$ and that $v \notin H^{\frac{1}{2}}(I)$. Deduce that u_0 in (1.4.6) belongs to $H^{\frac{1}{2}-\varepsilon}(\Gamma)$, and not to $H^{\frac{1}{2}}(\Gamma)$.
Hint: Estimate the Slobodecky norm (A.5.7) of v.

Exercise 1.40 By the Sobolev embedding theorem with $m = n = 1$, $s = 2$, we have for $I = (-1,1)$ that $H^{\frac{1}{2}+\varepsilon}(I) \subset C^0(\overline{I})$ for every $\varepsilon > 0$. Show that $H^{\frac{1}{2}}(I) \not\subset C^0(\overline{I})$, i.e. that $H^{1/2}(I)$ does contain discontinuous functions.
Hint: Consider the function $u(x) = \ln|\ln(e/|x|)|$. Why is $u \in H^{\frac{1}{2}}(I)$?

We deduce that **the problem** (1.4.6) **does not admit a generalized solution in** $H^1(\Omega)$. The remedy is to **weaken the notion of solution** which can be accomplished in different ways. The first variational formulation converts the Dirichlet condition into a natural boundary condition.

Let $v \in H^2(\Omega)$; then multiplication of (1.4.6) by v and twofold integration by parts in (1.4.6) yields

$$\int_\Omega f v\, dx = -\int_\Omega u \Delta v\, dx + \int_\Gamma \left(u\,\frac{\partial v}{\partial n} - v\,\frac{\partial u}{\partial n} \right) do\,.$$

Since no information on $(\partial u/\partial n)$ is given, we select $v|_\Gamma = 0$, i.e., $v \in (H^2 \cap H_0^1)(\Omega)$, and replace u on Γ by the boundary data (1.4.6). Hence we get the so-called **very weak variational formulation** (see also Exercise 1.14 above): find $u \in X_7 = L^2(\Omega)$ such that

$$B_7(u,v) := -\int_\Omega u\Delta v\,dx = F_7(v) := \int_\Omega fv\,dx - \sum_{i=1}^{4}\int_{\Gamma_i} u_0\gamma_1 v\,do \tag{1.4.7}$$

for all $v \in Y_7 := (H^2 \cap H_0^1)(\Omega)$.

Here the Γ_i are the sides of Ω and $\gamma_1 v$ denotes the normal derivative operator which makes sense since we assume now that $v \in H^2(\Omega)$. Notice that now the Dirichlet data u_0 in (1.4.6) has become part of the load functional $F(\cdot)$, i.e., the Dirichlet boundary condition is, in the very weak formulation (1.4.7), a natural boundary condition. Is (1.4.7) an admissible variational formulation for (1.4.6)?

Exercise 1.41 Show that $B_7(\cdot,\cdot)$ in (1.4.7) satisfies the inf–sup condition (1.3.3), (1.3.4) on $X_7 \times Y_7$. Show that F_7 is a continuous, linear functional on $H^2(\Omega)$. (Hint: use Theorem A.17).

Solution: We apply the procedure outlined in Remark 1.16.

Given $0 \neq u \in L^2(\Omega)$, find v_u as solution of the auxiliary problem:

$$-\Delta v_u = u \text{ in } \Omega, \quad v_u = 0 \text{ on } \Gamma\,. \tag{1.4.8}$$

By Proposition 1.32, there exists a unique solution $v_u \in H_0^1(\Omega)$ of this problem. We need $v_u \in H^2(\Omega)$, however, and therefore apply the following **regularity result** (cf. also Proposition 1.34): if the right-hand side u in (1.4.8) belongs to $L^2(\Omega)$ and Ω is a convex polygon, we have

$$v_u \in H^2(\Omega), \quad \|v_u\|_{H^2(\Omega)} \le C(\Omega)\,\|u\|_{L^2(\Omega)}\,.$$

Moreover, by (1.4.8) we also have

$$B(u,v_u) = -\int_\Omega u\,\Delta v_u\,dx = \|u\|^2_{L^2(\Omega)}$$

which implies the inf–sup condition (1.3.3) with constant $1/C(\Omega)$. The proof that $F \in Y'$ requires regularity of f and u_0; for example $f \in L^2(\Omega)$ and $u_0 \in L^2(\Gamma)$ will be sufficient. Note, however, that since $Y \subset H^2(\Omega)$ the regularity for f and u_0 can be considerably weakened.

Remark 1.42 The previous argument shows that the stability of a very weak formulation is intimately connected to the regularity of the weak solution in the domain Ω. For an analysis of (1.4.7) for a general polygon Ω, we refer to [10].

The very weak formulation (1.4.7) has the disadvantage of requiring C^1-conforming test functions for a proper FE discretization. An alternative strategy is to consider u_0 as essential data. By Exercise 1.39, u_0 in (1.4.6) belongs to $H^{1/2-\varepsilon}(\Gamma)$ for $0 < \varepsilon \leq 1/2$. By the trace theorem A.17, the trace operator

$$\gamma_0 : H^{1-\varepsilon}(\Omega) \longmapsto H^{1/2-\varepsilon}(\Gamma) \quad 0 < \varepsilon < 1/2$$

is onto, i.e., the discontinuous data u_0 in (1.4.6) admits a trace lifting $u_0 \in H^{1-\varepsilon}(\Omega)$. Therefore a variational formulation along the lines of Section 1.4.1 requires us to consider $B(u,v) = \int_\Omega \nabla u \cdot \nabla v dx$ on $H_0^{1-\varepsilon}(\Omega) \times H_0^{1+\varepsilon}(\Omega)$ for $0 < \varepsilon < 1/2$. This will lead to a saddle point problem, even though $B(\cdot,\cdot)$ is formally symmetric.

Proposition 1.43 *For $0 < \varepsilon < 1/2$, the bilinear form*

$$B(u,v) = \int_\Omega \nabla u \cdot \nabla v \, dx$$

satisfies the inf–sup condition (1.3.3) *on* $H_0^{1-\varepsilon}(\Omega) \times H_0^{1+\varepsilon}(\Omega)$.

Proof We proceed along the lines of Remark 1.16. Given $u \in H_0^{1-\varepsilon}(\Omega)$, there exists a functional $g_u \in (H_0^{1-\varepsilon}(\Omega))' =: H^{-1+\varepsilon}(\Omega)$ such that

$$\|g_u\|_{H^{-1+\varepsilon}(\Omega)} = \sup_{0 \neq v \in H_0^{1-\varepsilon}(\Omega)} \frac{|g_u(v)|}{\|v\|_{H^{1-\varepsilon}(\Omega)}} = 1$$

and such that $g_u(u) = \|u\|_{H^{1-\varepsilon}(\Omega)}$. Define $G_u := \|u\|_{H^{1-\varepsilon}(\Omega)} g_u$. Then $G_u \in H^{-1+\varepsilon}(\Omega) \subset H^{-1}(\Omega)$ and the Dirichlet problem: find $v_u \in H_0^1(\Omega)$ such that

$$\int_\Omega \nabla v_u \cdot \nabla w dx = G_u(w) \text{ for all } w \in H_0^1(\Omega)$$

has, by the Lax–Milgram lemma, a unique solution $v_u \in H_0^1(\Omega)$.

By elliptic regularity, $G_u \in H^{-1+\varepsilon}(\Omega)$ implies that v_u satisfies

$$v_u \in H_0^{1+\varepsilon}(\Omega), \ \|v_u\|_{H^{1+\varepsilon}(\Omega)} \leq C(\varepsilon,\Omega)\, \|G_u\|_{H^{-1+\varepsilon}(\Omega)} = C(\varepsilon,\Omega)\, \|u\|_{H^{1-\varepsilon}(\Omega)}$$

for $\varepsilon > 0$ sufficiently small and also that

$$B(u, v_u) = \int_\Omega \nabla v_u \cdot \nabla u \, dx = G_u(u) = \|u\|^2_{H^{1-\varepsilon}(\Omega)}$$

by the construction of G_u. This implies (1.3.3) with inf–sup constant $1/C(\varepsilon,\Omega)$. □

We remark that results of this type were first obtained by Nečas in [101] and used in a FE context in [11].

Having obtained a trace lifting $u_0 \in H^{1-\varepsilon}(\Omega)$, we write $u = u_0 + u_1$ with u_1 satisfying

$$u_1 \in H_0^{1-\varepsilon}(\Omega) : \int_\Omega \nabla u_1 \cdot \nabla v \, dx = F(v) = (f, v) - (\nabla u_0, \nabla v) \text{ for all } v \in H_0^{1+\varepsilon}(\Omega)$$

where $(\cdot,\cdot)$ is interpreted as duality bracket. By Proposition 1.43 and Theorem 1.15, there exists a unique u_1, and hence a unique solution $u \in H_0^{1-\varepsilon}(\Omega)$ of (1.4.6).

1.4.3 *The space* $H^{1/2}_{00}(\Gamma^{[1]})$

Consider now the **mixed boundary value problem**

$$\begin{gathered} -\Delta u = f \text{ in } \Omega\,, \\ \gamma_0\, u = u|_{\Gamma^{[0]}} = g^0, \quad \gamma_1 u = \frac{\partial u}{\partial n}\bigg|_{\Gamma^{[1]}} = g^1\,. \end{gathered} \tag{1.4.9}$$

We assume that $\Gamma = \overline{\Gamma^{[0]} \cup \Gamma^{[1]}}$ and that the $\Gamma^{[i]}$ coincide with a union of sides of $\partial\Omega$. We assume further that $g^0 \in H^{\frac{1}{2}}(\Gamma^{[0]})$, $f \in L^2(\Omega)$.

Thus we can find $w \in H^1(\Omega)$ such that $\gamma_0 w = g^0$ on $\Gamma^{[0]}$. We then have the following weak formulation: find $u_0 \in X = \{u \in H^1(\Omega) : \gamma_0 u = 0 \text{ on } \Gamma^{[0]}\}$ such that

$$B(u_0, v) = F(v) = \int_\Omega f v \, dx + \int_{\Gamma^{[1]}} g^1 \gamma_0 v \, do - B(w, v) \tag{1.4.10}$$

for every $v \in Y = X$ with $B(u,v) = \int_\Omega \nabla u \cdot \nabla v \, dx$. By Proposition 1.17 and the continuity of the trace operator, $X \subset H^1(\Omega)$ is a closed linear subspace. The question arises, however, which Neumann data g^1 are admissible in the generalized formulation (1.4.10). Clearly, g^1 must be such that the term

$$\int_{\Gamma^{[1]}} g^1 \gamma_0 v \, do$$

defines a continuous, linear functional on X.

The set of admissible data g^1 is therefore the dual of the set of all traces $\gamma_0 v$ of $v \in X$. Evidently,

$$\gamma_0(X) = \{v \in H^{\frac{1}{2}}(\Gamma) : v \equiv 0 \text{ on } \Gamma^{[0]}\}\,. \tag{1.4.11}$$

It turns out, however, that $\gamma_0(X) \subset H^{\frac{1}{2}}(\Gamma^{[1]})$ with strict inclusion. The range of $\gamma_0(X)$ on $\Gamma^{[1]}$ is denoted by $H^{1/2}_{00}(\Gamma^{[1]})$ and is a proper, closed linear subspace of $H^{1/2}(\Gamma^{[1]})$.

Definition 1.44 *Let $\Omega \subset \mathbb{R}^n$ be a polygon with boundary $\Gamma = \partial\Omega$ and assume $\Gamma = \overline{\Gamma^{[0]} \cup \Gamma^{[1]}}$ where $\int_{\Gamma^{[i]}} do > 0$, $i = 0, 1$. Then*

$$H_{00}^{\frac{1}{2}}(\Gamma^{[1]}) = \{u \in H^{\frac{1}{2}}(\Gamma^{[1]}) : \textit{the zero extension } \widetilde{u} \textit{ of } u \textit{ to all of } \Gamma \textit{ is in } H^{\frac{1}{2}}(\Gamma)\} .$$

The following result characterizes $H_{00}^{\frac{1}{2}}(\Gamma^{[1]})$.

Theorem 1.45 *Let Ω, $\Gamma^{[0]}$, $\Gamma^{[1]}$ be as in Definition 1.44 and let $\rho(x) = \mathrm{dist}(x, \partial\Gamma^{[1]})$. Then*

$$H_{00}^{\frac{1}{2}}(\Gamma^{[1]}) = \{v \in H^{\frac{1}{2}}(\Gamma^{[1]}) \,|\, \rho^{-\frac{1}{2}} v \in L^2(\Gamma^{[1]})\} \tag{1.4.12}$$

and

$$\|v\|^2_{H_{00}^{\frac{1}{2}}(\Gamma^{[1]})} = \|v\|^2_{H^{\frac{1}{2}}(\Gamma^{[1]})} + \int_{\Gamma^{[1]}} \frac{|v|^2}{\rho}\, do \tag{1.4.13}$$

defines a norm in $H_{00}^{\frac{1}{2}}(\Gamma^{[1]})$.

For a proof of this theorem, we refer to [86], Theorem 11.7.

It therefore follows that admissible Neumann data belong to $(H_{00}^{\frac{1}{2}}(\Gamma^{[1]}))'$ and include functions $u \in C^\infty(\Gamma^{[1]})$ which behave like $\rho^{-\frac{1}{2}}$. The spaces $(H_{00}^{\frac{1}{2}}(\Gamma^{[1]}))'$ are used in the variational formulation of fracture mechanics and contact problems.

1.5 Further examples for generalized formulations

All generalized formulations of the bar and the membrane problem considered so far involved an elimination of $\epsilon(x)$ and $\sigma(x)$ from the equations, resulting in a generalized formulation involving $u(x)$ alone. Here we return to the physical laws (1.1.1)–(1.1.3) and (1.1.12)–(1.1.14) and show generalized formulations without elimination of fields, so-called **mixed generalized formulations**. Similar formulations for three-dimensional elasticity will be discussed in Chapter 6 below.

1.5.1 *Mixed formulation of the bar problem I*

Consider the bar problem (1.1.1)–(1.1.3) which is recalled here. In $\Omega = (0, \ell)$ solve

$$\epsilon = u' , \tag{1.5.1}$$

$$\sigma = E\epsilon , \tag{1.5.2}$$

$$-(A\sigma)' = f . \tag{1.5.3}$$

This is augmented by boundary conditions

$$\sigma(0) = 0 \quad u(\ell) = 0 \,. \tag{1.5.4}$$

A generalized formulation is obtained as follows: multiply (1.5.1) by $r(x)$, (1.5.2) by $s(x)$ and (1.5.3) by $v(x)$, integrate over Ω and add. This yields

$$B_8(u, \epsilon, \sigma; r, s, v) = \int_0^\ell ((\epsilon - u')r + (\sigma - E\epsilon)s - A\sigma' v)dx \,.$$

The form B_8 is defined on the pair $X_8 \times Y_8$ of spaces given by

$$X_8 = \{(u, \epsilon, \sigma) : u \in H^1(\Omega),\ u(\ell) = 0;\ \epsilon \in L^2(\Omega);\ \sigma \in H^1(\Omega),\ \sigma(0) = 0\},$$

$$Y_8 = \{(r, s, v) : r \in L^2(\Omega),\ s \in L^2(\Omega),\ v \in L^2(\Omega)\}$$

(in particular all boundary conditions (1.5.4) are essential). The functional F_8 is given by

$$F_8(r, s, v) := \int_\Omega f v \, dx \,,$$

and the generalized formulation becomes: find $(u, \epsilon, \sigma) \in X_8$ such that

$$B_8(u, \epsilon, \sigma; r, s, v) = F_8(r, s, v) \tag{1.5.5}$$

for every $(r, s, v) \in Y_8$.

Proposition 1.46 *Assume in* (1.5.3) *that* $A(x) = A > 0$, $E(x) = E > 0$. *Then for every* $f \in L^2(\Omega)$ *there exists a unique solution of* (1.5.5) *and*

$$\|(u, \epsilon, \sigma)\|_{X_8} := \left(\|u\|^2_{H^1(\Omega)} + \|\epsilon\|^2_{L^2(\Omega)} + \|\sigma\|^2_{H^1(\Omega)}\right)^{\frac{1}{2}} \leq C \, \|f\|_{L^2(\Omega)} \,.$$

Proof We apply Theorem 1.15. It is verified using the trace theorem that $X_8 \subset (H^1 \times L^2 \times H^1)(\Omega)$ is a closed subspace and hence, equipped with the $\| \circ \|_{X_8}$-norm, is a reflexive Banach space by Proposition 1.17. Elementary estimates show B_8 and F_8 are continuous on $X_8 \times Y_8$ and Y_8, respectively.

We verify (1.3.3) and use Remark 1.16. Given $(u, \epsilon, \sigma) \in X_8$, choose

$$r = -au', \quad s = -b\epsilon, \quad v = -c\sigma' \tag{1.5.6}$$

where a, b, c are positive constants at our disposal. Then

$$B_8(u, \epsilon, \sigma; r, s, v) = \int_0^\ell \{-au'\epsilon + a(u')^2 - b\epsilon\sigma + bE\epsilon^2 + cA(\sigma')^2\}dx \,.$$

Since $2|AB| \leq A^2 + B^2$, we get with the Poincaré inequality

$$
\begin{aligned}
2\left|\int_0^\ell u'\epsilon\,dx\right| &\leq \int_0^\ell \left(\alpha(u')^2 + \frac{1}{\alpha}\,\epsilon^2\right) dx, \qquad 0 < \alpha < \infty\,, \\
2\left|\int_0^\ell \epsilon\sigma\,dx\right| &\leq \int_0^\ell \left(\beta\epsilon^2 + \frac{1}{\beta}\,\sigma^2\right) dx \\
&\leq \int_0^\ell \left(\beta\epsilon^2 + \frac{\ell^2}{2\beta}\,(\sigma')^2\right) dx, \qquad 0 < \beta < \infty\,.
\end{aligned}
$$

Hence

$$
\begin{aligned}
B_8(u,\epsilon,\sigma;r,s,v) \geq \int_0^\ell \Big\{ &(u')^2\left(a - \frac{\alpha a}{2}\right) + \epsilon^2\left(bE - \frac{a}{2\alpha} - \frac{b\beta}{2}\right) \\
&+ (\sigma')^2\left(cA - \frac{b\ell^2}{4\beta}\right)\Big\}\,dx\,.
\end{aligned}
$$

We select

$$
\alpha = 1,\quad \beta = E,\quad a = \frac{E}{2},\quad b = 1,\quad c = \frac{1}{2} + \frac{\ell^2}{4\beta A}\,.
$$

Then

$$
\begin{aligned}
B_8(u,\epsilon,\sigma;r,s,v) &\geq \frac{E}{4}\int_0^\ell (u')^2\,dx + \frac{E}{4}\int_0^\ell \epsilon^2\,dx + \frac{A}{2}\int_0^\ell (\sigma')^2\,dx \\
&\geq \alpha_1(A,E)\,\|(u,\epsilon,\sigma)\|^2_{X_8}
\end{aligned}
\tag{1.5.7}
$$

due to the Poincaré inequality (in $\{u \in H^1(\Omega) : u(0) = 0\}$, or $\{u \in H^1(\Omega) : u(\ell) = 0\}$).

It is also verified that $\|(u,\epsilon,\sigma)\|_{X_8} \leq \alpha_2(A,E)\,\|(r,s,v)\|_{Y_8}$, hence we get (1.3.3) with $C_2 = C_2(A,E)$.

Condition (1.3.4) is verified by selecting, for given $0 \neq v \in Y_8$, u_v as in (1.5.6) and by referring to (1.5.7). □

Remark 1.47 We observe that the functional $F_8 = F_8(v)$ in (1.5.5) is quite special. The variational formulation (1.5.5) allows in fact for more general F_8, as for example

$$
F_8(r,s,v) = \int_0^\ell (fv + gr + hs)dx\,.
$$

This corresponds to the problem

$$\epsilon = u' + g, \ \sigma = E\epsilon + h, \ -(A\sigma)' = f$$

where g and h are in $(H^1(\Omega))'$ and can be given interpretations as dislocation and thermostresses, respectively.

1.5.2 *Mixed formulation of the bar problem II*

Integrating by parts in B_8, we get

$$B_9(u,\epsilon,\sigma;r,s,v) = \int_0^\ell (\epsilon r + ur' + s(\sigma - E\epsilon) - A\sigma' v)dx \ , \tag{1.5.8}$$

defined on $X_9 \times Y_9$:

$$\begin{aligned} X_9 &= \{(u,\epsilon,\sigma) : \ u,\epsilon \in L^2(\Omega), \ \sigma \in H^1(\Omega), \ \sigma(0) = 0\} \ , \\ Y_9 &= \{(r,s,v) : \ r \in H^1(\Omega), \ r(0) = 0, \ s,v \in L^2(\Omega)\} \ . \end{aligned}$$

with norms given by

$$\begin{aligned} \|(u,\epsilon,\sigma)\|^2_{X_9} &:= \|u\|^2_{L^2(\Omega)} + \|\epsilon\|^2_{L^2(\Omega)} + \|\sigma\|^2_{H^1(\Omega)} \ , \\ \|(r,s,v)\|^2_{Y_9} &:= \|r\|^2_{H^1(\Omega)} + \|s\|^2_{L^2(\Omega)} + \|v\|^2_{L^2(\Omega)} \ . \end{aligned}$$

As before, we define

$$F_9(v) = \int_0^\ell fv\,dx \ .$$

We have the following result.

Proposition 1.48 *Assume* $A(x) = A > 0$, $E(x) = E > 0$. *Then* $B_9 : X_9 \times Y_9 \to \mathbb{R}$ *satisfies* (1.3.3), (1.3.4) *and the problem: find* $(u,\epsilon,\sigma) \in X_9$ *such that*

$$B_9(u,\epsilon,\sigma;r,s,v) = F_9(v) \ \textit{for every} \ (r,s,v) \in Y_9$$

admits a unique solution.

Proof We follow the pattern of proof for Proposition 1.46. To prove (1.3.3), we proceed as in Remark 1.16. Given $(u,\epsilon,\sigma) \in X_9$, select

$$r = a\int_0^x u\,dx, \ s = -b\epsilon, \ v = c\sigma' \tag{1.5.9}$$

with $a,b,c > 0$ properly chosen. Evidently, $(r,s,v) \in Y_9$. Verification of $B_8(u,\epsilon,\sigma;r,s,v) \geq C_2 \|(u,\epsilon,\sigma)\|^2_{X_9}$ is left as an exercise. □

1.5.3 *An initial value problem*

We show here how to cast an initial value problem into a weak form. This weak formulation can be used as a basis of FE time-stepping schemes.

Let $\Omega = (-1,1)$, $f \in L^2(\Omega)$, and $\lambda > 0$ be given. Consider the initial value problem

$$\lambda^{-1}\dot{u} + \lambda u = f \text{ in } \Omega,\ u(-1) = a\lambda,\ a \in \mathbb{R}\,. \tag{1.5.10}$$

Multiplying by $v \in Y_{10} = \{v \in H^1(\Omega) : v(1) = 0\}$ and integrating by parts, we get the weak formulation

$$B_{10}^{\lambda}(u,v) := \int_{-1}^{1} u(-\lambda^{-1}\,\dot{v} + \lambda v)dt = \int_{-1}^{1} fv\,dt + av(-1) =: F_{10}(v)\,. \tag{1.5.11}$$

Select $X_{10} = L^2(\Omega)$ and define on Y_{10}

$$\|v\|_{Y_{10}^{\lambda}} := \| -\lambda^{-1}\,\dot{v} + \lambda v\|_{L^2(\Omega)}, \quad \lambda > 0\,. \tag{1.5.12}$$

Lemma 1.49 *Define* $\|v\|_{\lambda}$ *by*

$$\|v\|_{\lambda}^2 = \lambda^{-2}\,\|\dot{v}\|_{L^2(\Omega)}^2 + \lambda^2\,\|v\|_{L^2(\Omega)}\,.$$

Then

$$\|v\|_{\lambda} \le \|v\|_{Y_{10}^{\lambda}} \le \sqrt{2}\,\|v\|_{\lambda} \text{ for all } v \in Y_{10}\,. \tag{1.5.13}$$

In particular, $\|\circ\|_{Y_{10}^{\lambda}}$ *is a norm on* Y_{10}.

Proof We compute

$$\begin{aligned}\|v\|_{Y_{10}^{\lambda}}^2 &= \int_{-1}^{1} (-\lambda^{-1}\,\dot{v} + \lambda v)^2\,dt = \int_{-1}^{1} \{\lambda^{-2}(\dot{v})^2 + \lambda^2 v^2\}dt - 2\int_{-1}^{1} \dot{v}\,v\,dt \\ &= \|v\|_{\lambda}^2 - \int_{-1}^{1} (v^2)^{\cdot}\,dt = \|v\|_{\lambda}^2 + (v(-1))^2 \ge \|v\|_{\lambda}^2\,,\end{aligned}$$

and

$$\|v\|_{Y_{10}^{\lambda}}^2 \le 2\{\lambda^{-2}\|\dot{v}\|_{L^2(\Omega)}^2 + \lambda^2\|v\|_{L^2(\Omega)}^2\} = 2\|v\|_{\lambda}^2\,.$$

□

We denote Y_{10} equipped with $\|\cdot\|_{Y_{10}^{\lambda}}$ defined in (1.5.12) by Y_{10}^{λ}.

Proposition 1.50 *For every* $F(v) \in (Y_{10}^{\lambda})'$ *the problem: find* $u_\lambda \in X_{10}$ *such that*

$$B_{10}^{\lambda}(u_\lambda, v) = F(v) \text{ for all } v \in Y_{10}^{\lambda} \tag{1.5.14}$$

has a unique solution and $\|u_\lambda\|_{X_{10}} \le \|F\|_{(Y_{10}^{\lambda})'}$.

Proof We apply Theorem 1.15 and must therefore prove the inf–sup condition (1.3.3). Given $u \in L^2(\Omega)$, there is $v_u \in Y_{10}^\lambda$ such that $u = -\lambda^{-1}\dot{v}_u + \lambda v_u$ and $\|v_u\|_{Y_{10}^\lambda} = \|u\|_{X_{10}}$. Hence

$$\inf_{0\neq u\in X_{10}} \sup_{0\neq v\in Y_{10}^\lambda} \frac{B_{10}^\lambda(u,v)}{\|u\|_{X_{10}}\|v\|_{Y_{10}^\lambda}} \geq 1\,. \tag{1.5.15}$$

Condition (1.3.4) is verified analogously. □

Remark 1.51 Notice that the "initial" condition $u(-1) = a\lambda$ is a **natural** boundary condition in the variational formulation (1.5.14). Notice also that we could have equipped Y_{10} simply with the H^1-norm at the expense of a λ-dependent inf–sup constant in (1.5.15). The norm $\|\cdot\|_{Y_{10}^\lambda}$ corresponds to the "energy" norm in the symmetric case. Moreover, when parabolic equations are discretized in space time dependent problems like (1.5.10) result with arbitrarily large values of $\lambda > 0$. Therefore the λ independence of the inf–sup constant in (1.5.15) is highly desirable.

1.5.4 *The heat equation*

We turn now to the **heat equation**

$$\frac{\partial u}{\partial t} - \Delta u = f \text{ in } D\,, \tag{1.5.16}$$

$$\begin{aligned} u &= 0 \text{ on } I\times\partial\Omega\,,\\ u(-1,\cdot) &= g \text{ in } \Omega\,, \end{aligned} \tag{1.5.17}$$

where $\Omega \subset \mathbb{R}^2$ is a curvilinear polygon, $I = (-1,1)$ and $D = I\times\Omega$. We assume that $f \in L^2(D)$ and $g \in L^2(\Omega)$. By $\dot{v}$ we denote $(\partial v/\partial t)$. Define

$$X_{11} = L^2(I, H_0^1(\Omega)) \tag{1.5.18}$$

and let Y_{11} be the closure of

$$\overset{\circ)}{C} = \{v \in C^\infty(\overline{D}) : \text{ for all } t\in I \quad v(t,x) \in C_0^\infty(\Omega),\ v(1,\cdot) = 0\}$$

in the norm $\|\circ\|_{Y_{11}}$ defined by

$$\|v\|_{Y_{11}}^2 = \int_{-1}^{1} \left(\|\dot{v}\|_{H^{-1}(\Omega)}^2 + \|v\|_{H^1(\Omega)}^2\right) dt\,. \tag{1.5.19}$$

From the definition of $H^{-1}(\Omega)$ we have (cf. Appendix A)

$$\|v\|_{Y_{11}}^2 = \int_{-1}^{1} (\|V\|_{H^1(\Omega)}^2 + \|v\|_{H^1(\Omega)}^2)\, dt \tag{1.5.20}$$

where $V = V(x,t) \in X_{11}$ is the solution of the Dirichlet problem: find $V(\cdot,t) \in H_0^1(\Omega)$ such that

$$\int_\Omega \nabla V \cdot \nabla w \, dx = \int_\Omega \dot{v} w \, dx \quad \text{for all} \quad w \in H_0^1(\Omega) \,. \tag{1.5.21}$$

Now denote by (λ_j, u_j), $u_j \in H_0^1(\Omega)$, an eigenpair of $-\Delta u_j = \lambda_j^2 u_j$ in Ω, $u_j = 0$ on Γ with $(u_j, u_j)_{L^2(\Omega)} = \delta_{ij}$, i.e., a solution of the problem: find $\lambda_j \in \mathbb{C}$ and $0 \neq u_j \in H_0^1(\Omega)$ such that

$$\int_\Omega \nabla u_j \cdot \nabla w \, dx = \lambda_j^2 \int_\Omega u_j w \, dx \quad \text{for all} \quad w \in H_0^1(\Omega) \,. \tag{1.5.22}$$

As it is well-known (see, e.g., [77], [59]), any $u \in H_0^1(\Omega)$ can be expanded in Fourier series, i.e.,

$$u = \sum_{i=1}^{\infty} \alpha_i u_i, \quad \alpha_i = \int_\Omega u u_i \, dx \tag{1.5.23}$$

so that

$$\|u\|_{L^2(\Omega)}^2 = \sum_{i=1}^{\infty} \alpha_i^2, \quad \|\nabla u\|_{L^2(\Omega)}^2 = \sum_{i=1}^{\infty} \lambda_i^2 \alpha_i^2 \,.$$

Thus we may characterize X_{11} as the set of all

$$u(t,x) = \sum_{i=1}^{\infty} \alpha_i(t) \, u_i(x)$$

for which $\| \circ \|_{X_{11}}$ given by

$$\|u\|_{X_{11}}^2 = \int_{-1}^{1} \sum_{i=1}^{\infty} \lambda_i^2 (\alpha_i(t))^2 dt \tag{1.5.24}$$

is finite. Further, for $v \in \overset{\circ)}{C}$ given by

$$v(t,x) = \sum_{i=1}^{\infty} \beta_i(t) \, u_i(x), \quad \beta_i(1) = 0 \tag{1.5.25}$$

a calculation shows that

$$\int_{-1}^{1} \|\dot{v}\|_{H^{-1}(\Omega)}^2 dt = \int_{-1}^{1} \Big(\sum_{i=1}^{\infty} (\dot{\beta}_i(t))^2 \, \lambda_i^{-2} \Big) dt \,.$$

Therefore, for every $v \in \overset{o)}{C}$,

$$\|v\|_{Y_{11}}^2 = \int_{-1}^{1} \Big(\sum_{i=1}^{\infty} (\dot{\beta}_i(t))^2 \, \lambda_i^{-2} + (\beta_i(t))^2 \lambda_i^2 \Big) \, dt < \infty \tag{1.5.26}$$

and Y_{11} is the set of v of the form (1.5.25) such that (1.5.26) holds. By Lemma 1.49, the norm $\| \circ \|_{\widehat{Y}_{11}}$, given by

$$\|v\|_{\widehat{Y}_{11}}^2 = \int_{-1}^{1} \Big(\sum_{i=1}^{\infty} (-\dot{\beta}_i(t)/\lambda_i + \beta_i(t)\lambda_i)^2 \Big) \, dt \tag{1.5.27}$$

is equivalent to $\| \circ \|_{Y_{11}}$ defined in (1.5.26). Define

$$B_{11}(u,v) = \int_{-1}^{1} \int_{\Omega} (\nabla u \cdot \nabla v - u \dot{v}) \, dx \, dt = \int_{-1}^{1} \Big(\sum_{i=1}^{\infty} (\lambda_i^2 \alpha_i \beta_i - \alpha_i \dot{\beta}_i) \Big) \, dt$$

and

$$F_{11}(v) = \int_{-1}^{1} \int_{\Omega} f(t,x) \, v(t,x) dx \, dt + \int_{\Omega} v(-1,x) g(x) \, dx \, .$$

By the Schwarz inequality and the trace inequality, $F_{11} \in (Y_{11})'$. Then the weak formulation of (1.5.16), (1.5.17) is: find $u \in X_{11}$ such that

$$B_{11}(u,v) = F_{11}(v) \text{ for every } v \in Y_{11} \, . \tag{1.5.28}$$

Let us establish the inf–sup condition for $B_{11}(u,v)$.

Proposition 1.52 *There holds for* B_{11}

$$|B_{11}(u,v)| \le \|u\|_{X_{11}} \|v\|_{\widehat{Y}_{11}} \quad \textit{for } u \in X_{11}, \, v \in Y_{11} \tag{1.5.29}$$

and

$$\inf_{0 \neq u \in X_{11}} \; \sup_{0 \neq v \in Y_{11}} \frac{B_{11}(u,v)}{\|u\|_{X_{11}} \|v\|_{\widehat{Y}_{11}}} \ge 1 \, . \tag{1.5.30}$$

Proof

Step 1: We write $u = \sum_i \alpha_i u_i$, $V = \sum_i \beta_i u_i$

$$\begin{aligned} |B_{11}(u,v)| &= \int_{-1}^{1} \Big(\sum_{i=1}^{\infty} \lambda_i \alpha_i (\lambda_i \beta_i - \dot{\beta}_i / \lambda_i) \Big) \, dt \\ &\le \int_{-1}^{1} \Big(\sum_{i=1}^{\infty} (\lambda_i \alpha_i)^2 \Big)^{\frac{1}{2}} \Big(\sum_{i=1}^{\infty} (\lambda_i \beta_i - \dot{\beta}_i / \lambda_i)^2 \Big)^{\frac{1}{2}} \, dt \\ &= \|u\|_{X_{11}} \, \|v\|_{\widehat{Y}_{11}} \, . \end{aligned}$$

Step 2: Given $u \in X_{11}$ in the form $u = \sum_{i=1}^{\infty} \alpha_i(t)\, u_i(x)$, define v_u by

$$v_u := \sum_{i=1}^{\infty} \beta_i(t)\, u_i(x), \; -\lambda_i^{-2}\beta_i + \dot{\beta}_i = \alpha_i \; \text{ and } \; \beta_i(1) = 0\,.$$

Then

$$\begin{aligned} \|v_u\|^2_{\widehat{Y}_{11}} &= \int_{-1}^{1} \Big(\sum_{i=1}^{\infty} (\beta_i \lambda_i - \dot{\beta}_i/\lambda_i)^2 \Big)\, dt \\ &= \int_{-1}^{1} \Big(\sum_{i=1}^{\infty} (\lambda_i \alpha_i)^2 \Big)\, dt = \|u\|^2_{X_{11}} \end{aligned}$$

and

$$\begin{aligned} B_{11}(u, v_u) &= \int_{-1}^{1} \Big(\sum_{i=1}^{\infty} \lambda_i \alpha_i (\lambda_i \beta_i - \dot{\beta}_i/\lambda_i) \Big)\, dt \\ &= \int_{-1}^{1} \Big(\sum_{i=1}^{\infty} (\lambda_i \alpha_i)^2 \Big)\, dt = \|u\|^2_{X_{11}}\,. \end{aligned}$$

□

Corollary 1.53 *By Lemma* 1.49 *we have of course also*

$$|B_{11}(u,v)| \le \sqrt{2}\, \|u\|_{X_{11}} \|v\|_{Y_{11}}\,,$$

$$\inf_{0 \ne u \in X_{11}} \sup_{0 \ne v \in Y_{11}} \frac{B_{11}(u,v)}{\|u\|_{X_{11}} \|v\|_{Y_{11}}} \ge \frac{1}{\sqrt{2}}\,.$$

By Theorem 1.15 we have from (1.5.29), (1.5.30) that for every $F \in (Y_{11})'$ there exists a unique solution $u \in X_{11}$ of (1.5.28) and we have the a priori estimate

$$\|u\|_{X_{11}} \le \|F\|_{\widehat{Y}_{11}} = \sup_{0 \ne v \in Y_{11}} \frac{|F(v)|}{\|v\|_{\widehat{Y}_{11}}} \le \sup_{0 \ne v \in Y_{11}} \frac{|F(v)|}{\|v\|_{Y_{11}}}\,. \tag{1.5.31}$$

1.5.5 *Mixed formulation of the membrane problem*

We return to the membrane problem (1.1.12)–(1.1.14) and give one possible mixed formulation. We assume again homogeneous Dirichlet data, i.e., $g^0 = 0$, and that

$$\Gamma^{[2]} = \varnothing, \qquad \int_{\Gamma^{[0]}} do > 0\,. \tag{1.5.32}$$

Multiplying (1.1.13) and (1.1.14) by $\tau \in L^2(\Omega)$ and $v \in H^1(\Omega, \Gamma^{[0]})$, respectively and integrating by parts yields (denoting by $(u, v)_0$ the integral $\int_\Omega uv\, dx$)

$$(\sigma, \tau)_0 - (E\nabla u, \tau)_0 = 0\,, \tag{1.5.33}$$

$$(A\sigma, \nabla v)_0 + (cu, v)_0 = F(v)\,, \tag{1.5.34}$$

where

$$F(v) = \int\limits_{\Omega} f v dx + \int\limits_{\Gamma^{[1]}} g^1 \gamma_0 v\, do\,.$$

Adding (1.5.33) and (1.5.34), we get the weak formulation: find $(\sigma, u) \in X_{12}$ such that

$$B_{12}(\sigma, u; \tau, v) = F(v) \text{ for all } (\tau, v) \in Y_{12} \tag{1.5.35}$$

where $X_{12} = Y_{12} = L^2(\Omega) \times H^1(\Omega, \Gamma^{[0]})$ is equipped with

$$\|(\sigma, u)\|_{X_{12}} = (\|\sigma\|^2_{L^2(\Omega)} + \|\nabla u\|^2_{L^2(\Omega)})^{1/2}$$

and where

$$B_{12}(\sigma, u; \tau, v) = (\sigma, \tau)_0 - (E\nabla u, \tau)_0 + (A\sigma, \nabla v)_0 + (cu, v)_0\,.$$

By assumption (1.5.32) and the Poincaré inequality, $\| \circ \|_{X_{12}}$ is a norm. To investigate existence and uniqueness for (1.5.35), we need

Proposition 1.54 *Assume that in* (1.5.33), (1.5.34) A, $E \in L^\infty(\Omega)$ *are symmetric and uniformly positive definite, i.e., for a.e.* $x \in \Omega$

$$\xi^\top A(x)\xi \geq \underline{A}|\xi|^2, \quad \xi^\top E(x)\,\xi \geq \underline{E}|\xi|^2, \quad \underline{A},\, \underline{E} > 0, \quad \xi \in \mathbb{R}^2\,.$$

Then $B_{12} : X_{12} \times Y_{12} \to \mathbb{R}$ *is stable, i.e., it satisfies the inf–sup condition* (1.3.3).

Proof We apply Theorem 1.15. Continuity of $B_{12}(\sigma, u; \tau, v)$ on $X_{13} \times Y_{12}$ is easily established and we focus on (1.3.3), (1.3.4).

Given $(\sigma, u) \in X_{12}$, we select $(\tau, v) = (\sigma - M\nabla u, Nu)$ with a symmetric, positive definite matrix M and a positive constant N to be determined. Then

$$\begin{aligned} B_{12}(\sigma, u; \tau, v) &= (\sigma, \sigma)_0 - (\sigma, M\nabla u)_0 - (E\nabla u, \sigma)_0 + (E\nabla u, M\nabla u)_0 \\ &\quad + N(A\sigma, \nabla u)_0 + N(cu, v)_0 \\ &= (\sigma, \sigma)_0 + (\sigma, (NA - M - E)\nabla u)_0 + (\nabla u, ME\nabla u)_0 \\ &\quad + N(cu, u)_0\,. \end{aligned}$$

Selecting $M = NA - E$ and N sufficiently large, $ME = NAE - E^2$ is positive definite and hence

$$\begin{aligned} B_{12}(\sigma, u; \tau v) &= (\sigma, \sigma)_0 + (\nabla u, (NAE - E^2)\,\nabla u)_0 + N(cu, u)_0 \\ &\geq \|\sigma\|^2_{L^2(\Omega)} + K\,\|\nabla u\|^2_{L^2(\Omega)} \geq \min(1, K)\,\|(\sigma, u)\|^2_{X_{12}} \end{aligned}$$

where $K > 0$ is the smallest eigenvalue of $NAE - E^2$. □

1.6 Bibliographical remarks

There are many references on the variational formulation of elliptic boundary value problems and the associated aspects of functional analysis. Regarding functional analysis, we mention here only [6], [39], [152], [120], where also introductions to Sobolev spaces are given. Numerous other references on the modern theory of partial differential equations exist, and we mention here only a few references: [118], [60], [77], [59] which each have only a small part devoted to elliptic problems. Exclusively devoted to elliptic problems in nonsmooth domains are [67] and the classic [100]. The relevance of the inf–sup condition (1.3.2) for the solvability theory of elliptic boundary value problems goes back to [102], [100].

The significance of the inf–sup condition (1.3.3) for the convergence analysis of the FEM was first pointed out by I. Babuška in the fundamental paper [11]. Our proof of Theorem 1.15 is close to the one given there. The extension to factor spaces given in Theorem 1.20 is of course not new in essence, but important for Neumann problems as, e.g., the pure traction problem in linearized elasticity. The space $H^{1/2}_{00}$ and its significance for mixed boundary value problems are known for quite some time, see e.g. [86], Chapter 1. For its use in connection with contact problems see [82]. Finally, the mixed variational formulations of the model problems in Section 5 serve merely as illustrations for this approach – their main impact is in fact in the FEM applied to problems in solid and fluid mechanics, as we will show in Chapters 5 and 6. This is due to the fact that frequently nonlinear constitutive equations are more easily discretized by mixed approaches than by primal (displacement) formulations. The design of stable finite elements in the mixed setting is, as a rule, more delicate than in the primal context (see Chapters 5 and 6 for more).

2

THE FINITE ELEMENT METHOD (FEM): DEFINITION, BASIC PROPERTIES

We introduce the FEM as a general projection method based on a variational formulation of a boundary value problem and establish the terminology of stability, consistency and convergence. We introduce stiffness matrix and load vector and a general approach, based on the discrete inf–sup condition, to assess the stability and consistency of a FEM on a given class of problems quantitatively.

Finally, we present some error estimates for the case that the bilinear form $B(\cdot,\cdot)$ can only be realized approximately by a form $B_N(\cdot,\cdot) \neq B(\cdot,\cdot)$, as is often the case when numerical integration or curvilinear domains are present.

2.1 Approximate solutions

A FEM departs from the variational formulation

$$u_0 \in X \text{ such that } B(u_0, v) = F(v) \text{ for all } v \in Y \tag{2.1.1}$$

of a boundary value problem.

The FEM generates approximate solutions of (2.1.1).

An **approximate solution** u_S is a function

$$u_S(x) = \sum_{i=1}^{N} a_i \, \psi_i(x) \tag{2.1.2}$$

where $\psi_i(x) \in X$, $i = 1, \ldots, N$ are N linearly independent functions in X and a_i are real numbers. The set S of all functions u_S of the form (2.1.2) is a linear space of dimension N contained in X

$$S = \operatorname{span}\{\psi_i(x)\}_{i=1}^{N} \subset X \tag{2.1.3}$$

and the ψ_i are a **basis** of S. For a given space S, there are many possible bases. Denoting by $\{a\}$ the vector $(a_1, a_2, \ldots, a_N)^\top$ and by $\{\psi\} = (\psi_1, \ldots, \psi_N)^\top$, we can write

$$u_S(x) = \{a\}^\top \{\psi\} \,. \tag{2.1.4}$$

A FEM is usually based on special subspaces S which consist of piecewise polynomials.

Example 2.1 (the subspace $S^{\boldsymbol{p},\ell}(\Omega,\mathcal{T})$). We present a subspace which is the basis of an *hp*-FEM in one dimension. Let $\Omega = (a,b)$ and be a nonnegative integer. A **mesh** $\mathcal{T}$ is a partition of Ω into $M(\mathcal{T})$ subintervals $\Omega_j : \mathcal{T} = \{\Omega_j\}_{j=1}^{M(\mathcal{T})}$, $\Omega_j = (x_{j-1}, x_j)$, $a = x_0 < x_1 < \ldots < x_M = b$ where x_i are called **nodal points** and the subdomains $\Omega_j = (x_{j-1}, x_j)$, $j = 1, \ldots, M(\mathcal{T})$ are called **elements**.

By $h_j = (x_j - x_{j-1})$ we denote the **meshwidth** of Ω_j. With each Ω_j we associate a polynomial degree $p_j \geq 1$.

The degrees p_j are combined in the **degree vector** $\boldsymbol{p}$.

Then, for $\ell \geq 1$, we define

$$S^{\boldsymbol{p},\ell}(\Omega,\mathcal{T}) = \{u \in C^{\ell-1}(\Omega) : u\,|_{\Omega_j} \text{ is a polynomial of degree } p_j, \\ \Omega_j \in \mathcal{T},\ j = 1, \ldots, M(\mathcal{T})\} .$$

Evidently,

$$S^{\boldsymbol{p},\ell}(\Omega,\mathcal{T}) \subset H^{\ell}(\Omega) .$$

If $\ell = 1$ we write $S^{\boldsymbol{p}}(\Omega,\mathcal{T})$ for simplicity and if $\boldsymbol{p} = (p, p, \ldots, p)$ (i.e. if the polynomial degree is constant), we write $S^{p}(\Omega,\mathcal{T})$. Of greatest interest are $\ell = 0, 1, 2$ in practice ($\ell = 0$ means that u can be discontinuous between elements).

There are many possible bases for $S^{\boldsymbol{p},\ell}(\Omega,\mathcal{T})$. A particular basis suitable for computations will be presented in Chapter 3.

Exercise 2.2 Calculate $N = \dim(S^{\boldsymbol{p},\ell}(\mathcal{T},\Omega))$ for $\ell = 0, 1$.

Definition 2.3 *Let $S \subset X$ and $V \subset Y$ be subspaces of dimension N. Then the finite element approximation u_S of* (2.1.1) *is defined by*

$$u_S \in S \ \textit{ such that } \ B(u_S, v) = F(v) \ \textit{ for all } \ v \in V . \tag{2.1.5}$$

S is called **trial space** *and V is called* **test space**, *elements of S are* **trial functions**, *elements of V are* **test functions**. (2.1.5) *is called the* **discretization** *of* (2.1.1).

Problem (2.1.1) is equivalent to solving a linear system of N equations in N unknowns. Since $u_S \in S$, we can write

$$u_S = \sum_{i=1}^{N} c_i \psi_i = \{c\}^\top \{\psi\} = \{\psi\}^\top \{c\}$$

and $v \in V$ can be written as

$$v = \sum_{i=1}^{N} b_i \varphi_i = \{b\}^\top \{\varphi\} = \{\varphi\}^\top \{b\} .$$

Hence

$$B(u_S, v) = \{b\}^\top [K] \{c\} , \tag{2.1.6}$$

where the **stiffness matrix** $[K]$ has entries

$$k_{ij} = B(\psi_j, \varphi_i) \, . \tag{2.1.7}$$

Similarly,

$$F(v) = \{b\}^{\top}\{q\}, \quad q_i = F(\varphi_i) \tag{2.1.8}$$

where $\{q\}$ is the **load vector**. Hence (2.1.5) becomes a set of N equations with N unknowns:

$$[K]\{c\} = \{q\} \, . \tag{2.1.9}$$

In order for (2.1.9) to have a unique solution, we require the stiffness matrix $[K]$ to be nonsingular. We will always assume

$$\dim(S) = \dim(V) = N \tag{2.1.10}$$

so that $[K]$ is a square $N \times N$ matrix.

Definition 2.4 $B(\cdot,\cdot)$ *is* **regular on** $S \times V$ *if either*

$$\textit{for every } 0 \neq u \in S \textit{ exists a } v \in V \textit{ such that } B(u,v) \neq 0 \tag{2.1.11}$$

and/or

$$\textit{for every } 0 \neq v \in V \textit{ exists a } u \in S \textit{ such that } B(u,v) \neq 0 \, . \tag{2.1.12}$$

Theorem 2.5

i) $B(\cdot,\cdot)$ *satisfies* (2.1.11) *if and only if it satisfies* (2.1.12).

ii) $B(\cdot,\cdot)$ *satisfies* (2.1.11) *if and only if the stiffness matrix* $[K]$ *is nonsingular.*

Proof We prove ii). Assume (2.1.11) holds and $[K]$ is singular. Then exists $0 \neq \{c\} \in \mathbb{R}^N$ such that $[K]\{c\} = 0$. Let $u = \{c\}^{\top}\{\psi\} \in S$ and $v = \{d\}^{\top}\{\varphi\} \in V$ arbitrary. Then $B(u,v) = \{d\}^{\top}[K]\{c\} = 0$, in contradiction to (2.1.11) and hence $[K]$ is regular.

Conversely, if $[K]$ is regular and (2.1.11) fails, there exists $0 \neq u \in S$ such that $B(u,v) = 0$ for all $v \in V$. Hence, for some $0 \neq \{c\} \in \mathbb{R}^N$

$$0 = \{d\}^{\top}[K]\{c\} \text{ for all } \{d\} \in \mathbb{R}^N, \text{ i.e., } [K]\{c\} = 0 \, .$$

Therefore $[K]$ is singular which is a contradiction and ii) is proved. The same argument shows that (2.1.12) holds if and only if $[K]^{\top}$ is nonsingular. This implies i). □

2.2 Acceptance criteria and refinements

The goal of discretizing (2.1.1) by (2.1.5) is to obtain an approximation u_S of u_0 which satisfies an **acceptance criterion**.

Example 2.6 Given $\Phi \in X'$ with $\Phi(u_0) \neq 0$, and a tolerance $\tau > 0$, an acceptance criterion is given by

$$\frac{|\Phi(u_0) - \Phi(u_S)|}{|\Phi(u_0)|} < \tau . \tag{2.2.1}$$

Φ could be any quantity of interest, as e.g. solution value in a point, moments, lift, stresses, etc.

Example 2.7 Often the **relative error** is of interest:

$$\frac{\|u_0 - u_S\|_X}{\|u_0\|_X} < \tau . \tag{2.2.2}$$

Example 2.8 To evaluate (2.2.1), (2.2.2) the exact solution u_0 must be known. This is unrealistic. In practice one uses therefore computable error estimators

$$Est(u_S, \text{data}) < \tau . \tag{2.2.3}$$

Since Est can only be evaluated **after** u_S has been found, it is also called the **a posteriori** error estimator.

Let $S_i \subset X, V_i \subset Y$ be subspaces of dimension N_i and denote by u_i the corresponding finite element solution. If the acceptance criteria are not met by u_i, we modify S_i and V_i, i.e. we select S_{i+1}, V_{i+1} such that u_{i+1} (hopefully) satisfies the acceptance criteria.

The **refinement operator**

$$(S_{i+1}, V_{i+1}) = \mathcal{R}(S_i, V_i, u_i, A) \tag{2.2.4}$$

is nonlinear, in general. Frequently, but not always, we have $S_i \subset S_{i+1} \subset X$, $V_i \subset V_{i+1} \subset Y$ in which case the refinement $\mathcal{R}$ is **hierarchical**.

Example 2.9 Consider the subspaces $S^{\boldsymbol{p}}(\mathcal{T}, \Omega)$ in Example 2.1. If $\boldsymbol{p} = (1, \ldots, 1)$, we obtain the classical, piecewise linear, Courant finite element space. If S_{i+1} is obtained from S_i by subdivision of (some or every) Ω_i and increasing $M(\mathcal{T})$, we speak of ***h*-refinement**. If S_{i+1} is obtained from S_i by increasing some or all elemental polynomials degrees p_i, we have ***p*-refinement**. A combination of both is called ***hp*-refinement**. All approaches are presently used in commercial FE codes.

A finite element method consists of:

a) variational formulation (2.1.1),
b) discretizations (2.1.5) of (2.1.1) based on pairs of subspaces $S_i \times V_i$,
c) acceptance criteria (2.2.1)–(2.2.3) and
d) refinement operator (2.2.4).

Frequently, especially in classical engineering FE-literature, only a) and b) are considered part of the FEM. This view is narrow and obscures that the FEM is a **method of approximation** of the exact solution u_0 to the adopted variational formulation (2.1.1) of the problem of interest.

2.3 Stability, consistency and convergence

Subtracting (2.1.1) from (2.1.5), we find

$$B(u_0 - u_S, v) = 0 \text{ for all } v \in V . \tag{2.3.1}$$

If $B(\cdot,\cdot)$ is an inner product on a Hilbert space H, (2.3.1) is satisfied by the projection u_S of u_0 onto S. Assuming that $B(\cdot,\cdot)$ is regular on $S \times V$, u_S is uniquely associated with u_0 by (2.3.1) and we write

$$u_S = Pu . \tag{2.3.2}$$

Exercise 2.10 Prove that P is linear and continuous.

Definition 2.11 *Let $\varnothing \neq R \subseteq X$ be a solution set. Then we define the* **stability constant**

$$D(R, S, V, X) := \sup_{u \in R} \frac{||Pu||_X}{||u||_X} . \tag{2.3.3}$$

If $R = X$, we write $D(S, V, X)$ which is nothing but the norm of P.

The stability constant has the following practical significance: Let $R = \{u_0\}$. Then $||u_S||_X = D||u_0||_X$. If D is much larger than 1, small perturbations in the solution u_0 may cause large perturbations in u_S **entirely due to the discretization**. The discretization has poor stability.

Definition 2.12 *Let $u_0 \in X$. Define the* **best approximation error**

$$Z(u_0, S, X) = \inf_{s \in S} ||u_0 - s||_X = ||u_0 - s_0||_X \tag{2.3.4}$$

and

$$C(\{u_0\}, S, V, X) = \frac{||u_0 - u_S||_X}{Z(u_0, S, X)} \geq 1$$

($C = 1$ if $Z = 0$). Let $\varnothing \neq R \subseteq X$ be a solution set. Then we define the **optimality constant**

$$C(R, S, V, X) = \sup_{u_0 \in R} \frac{||u_0 - Pu_0||_X}{Z(u_0, S, X)} . \tag{2.3.5}$$

If $R = X$, we write $C(S, V, X)$.

Evidently, for $u_0 \in R \subseteq X$ we have

$$\begin{aligned}\|u_0 - u_S\|_X &\le C(R,S,V,X)\, Z(u_0,S,X) \\ &\le C(S,V,X)\, Z(u_0,S,X)\,. \end{aligned} \tag{2.3.6}$$

If $C(R,S,V,X) < \infty$, the discretization (2.1.5) is ***C*-quasioptimal** on R; if $R = X$ we simply say it is quasioptimal.

Definition 2.13 *A sequence of discretizations based on pairs* $\{S_i, V_i\}_{i=1}^{\infty}$ *is* **stable** *on (a solution set)* $R \subseteq X$ *if*

$$D(R, S_i, V_i, X) \le K < \infty, \quad i = 1,2,3\dots\,.$$

It is **consistent** *on* R *if for every* $u_0 \in R$

$$Z(u_0, S_i, X) \longrightarrow 0, \quad i = 1,2,3,\dots\,.$$

It is **convergent** *on* R *if for every* $u_0 \in R$

$$C(\{u_0\}, S_i, V_i, X)\, Z(u_0, S_i, X) \to 0, \quad i = 1,2,3\,. \tag{2.3.7}$$

If $R = X$, *we say that* $\{S_i, V_i\}$ *is stable, consistent and convergent, respectively.*

Evidently we have the following result.

Theorem 2.14 *Stability and consistency of* $\{S_i, V_i\}$ *on* R *imply convergence:* $\|u_0 - u_i\|_X \to 0$ *as* $i \to \infty$ *for every* $u_0 \in R$.

Remark 2.15 Stability is sufficient, but not necessary for convergence as (2.3.7) shows. Since $C \ge 1$ always, consistency is necessary for convergence.

The next theorem relates the constants C and D.

Theorem 2.16 *Let* $\varnothing \ne R \subset X$. *Then*

$$D(R,S,V,X) - 1 \le C(R,S,V,X) \le D(R,S,V,X) + 1\,. \tag{2.3.8}$$

Proof Let $0 \ne u \in R$. Then for every $s \in S$

$$\begin{aligned}\|u - Pu\|_X = \|u - s - Pu + s\|_X &\le \|u - s\|_X + \|P(u-s)\|_X \\ &\le \|u - s\|_X + D(\{u\},S,V,X)\|u - s\|_X \\ &\le \|u - s\|_X (1 + D(R,S,V,X))\,.\end{aligned}$$

Since $s \in S$ is arbitrary,

$$\|u - Pu\|_X \le (1 + D(R,S,V,X))\, Z(u,S,X)\,.$$

Hence

$$C(R,S,V,X) = \sup_{u \in R} \frac{\|u - Pu\|_X}{Z(u,S,X)} \le 1 + D(R,S,V,X)\,.$$

The lower bound is obtained as follows:

$$\begin{aligned}\|Pu\|_X &= \|Pu - u + u\|_X \le \|Pu - u\|_X + \|u\|_X \\ &= C(\{u\},S,V,X)\,Z(u,S,X) + \|u\|_X\,.\end{aligned}$$

Since $Z(u,S,X) \le \|u\|_X$,

$$\frac{\|Pu\|_X}{\|u\|_X} \le C(\{u\},S,V,X) + 1\,.$$

Taking the supremum over all $0 \neq u \in R$ completes the proof. □

Corollary 2.17 *Assume that* $\{S_i, V_i\}_{i=1}^{\infty}$ *is a sequence of discretizations of* (2.3.1) *and* $u_i \in S_i$ *the corresponding FE solutions. Then*

$$\|u_0 - u_i\|_X \le (1 + D(\{u_0\}, S_i, V_i, X)) \inf_{s \in S} \|u_0 - s\|_X\,. \tag{2.3.9}$$

Remark 2.18 (on the set R and unstable methods). Constants C and D depend monotonically on R, i.e., if $R_1 \subset R_2 \subset X$, $C(R_1,S,V,X) \le C(R_2,S,V,X)$ etc. A common practice is to test FE methods by **benchmarking**, i.e., to examine their performance for a **finite set** R_{B} of benchmark solutions. For **arbitrary** choice of R_B, such experiments may yield convergent sequences $\{u_i\}$ of discrete solutions, since for every $u_0 \in R_{\mathrm{B}}$ it holds that

$$\|u_0 - u_i\|_X \le (1 + D(R_{\mathrm{B}}, S_i, V_i, X)) \inf_{s \in S} \|u_0 - s\|_X\,,$$

where $D(R_{\mathrm{B}}, S_i, V_i, X)$ may be uniformly bounded, even though $D(S_i, V_i, X) \to \infty$ as $i \to \infty$, i.e., the discretizations based on $\{S_i, V_i\}$ are unstable. Clearly, the quality of FE approximation u_i may then be considerably worse for some $u_0 \notin R_{\mathrm{B}}$. To characterize the set R on which $D(R, S_i, V_i, X) \to \infty$ is difficult in general.

To design a meaningful benchmark for a solution class R of interest one should, for a given sequence $\{S_i, V_i\}$ within the practical range, estimate $D(R, S_i, V_i, X)$ numerically by evaluating D for a **finite** set R_{B} of "worst" elements from R. In general, the finite extremal set R_{B} (if it exists at all) must be found by mathematical analysis and **cannot** be guessed by physical intuition (see also Remarks 2.23 and 2.24 below).

The next two theorems address the practical estimation of $D(S,V,X)$.

Theorem 2.19 *(Lax–Milgram lemma). Let* $X = Y$ *and* $B(u, v) : X \times X \to \mathbb{R}$ *be continuous*

$$|B(u,v)| \le K\|u\|_X \,\|u\|_X \quad \textit{for all } u, v \in X \tag{2.3.10}$$

and coercive

$$B(u,u) \ge C\,\|u\|_X^2 \quad \textit{for all } u \in X\,. \tag{2.3.11}$$

Let $S \subset X$ *be a finite dimensional subspace. Then*

i) $B(u, v)$ *is regular on* $S \times S$

ii)
$$D(S,S,X) \le K/C \tag{2.3.12}$$

iii)
$$\|u_S - u_0\|_X \le K/C \inf_{s\in S} \|u_S - s\|_X\,.$$

Proof

i) For any $0 \neq u \in S$ select $v = u \in S$. Then $B(u, v) \neq 0$ by (2.3.11), and hence $B(u, v)$ is regular.

ii) Let $u \in X$ and $u_S = Pu \in S$ satisfy $B(u - Pu, v) = 0$ for all $v \in V$. Hence $B(Pu, Pu) = B(u, Pu)$. Therefore

$$C\,\|Pu\|_X^2 \le B(Pu,Pu) = B(u,Pu) \le K\,\|u\|_X\,\|Pu\|_X\,.$$

If $u \neq 0$

$$\frac{\|Pu\|_X}{\|u\|_X} \le \frac{K}{C}\,.$$

Taking the supremum over $0 \neq u \in X$ gives (2.3.12).

To prove iii), observe that by (2.3.1)

$$B(u_0 - u_S, u_0) = B(u_0 - u_S, u_0 - s) \quad s \in V\,.$$

Therefore, by (2.3.11)

$$\begin{aligned} C\,\|u_0 - u_S\|_X^2 &\le |B(u_0 - u_S, u_0 - u_S)| = |B(u_0 - u_S, u_0 - s)| \\ &\le K\,\|u_0 - u_S\|_X\,\|u_0 - s\|_X, \quad s \in V\,. \end{aligned}$$

Dividing both sides by $C\,\|u_0 - u_S\|_X$ we get iii). □

Observe that referring to (2.3.9) to prove iii) would have resulted in a weaker bound.

We address now the case when $X \neq Y$.

Theorem 2.20 *(discrete inf–sup condition). Let $B : X \times Y \to \mathbb{R}$ be continuous, i.e.,*

$$|B(u,v)| \leq K \, \|u\|_X \, \|v\|_Y \quad \text{for all} \quad u \in X, v \in Y \, . \tag{2.3.13}$$

Let $S \subset X, V \subset Y$ be subspaces of dimension N such that for every $u \in S$ there exists $v_u \in V$ with

$$|B(u,v_u)| \geq C(S,V) \, \|u\|_X \, \|v\|_Y$$

or, equivalently, that

$$\inf_{0 \neq u \in S} \ \sup_{0 \neq v \in V} \frac{B(u,v)}{\|u\|_X \, \|v\|_Y} \geq C(S,V) > 0 \, . \tag{2.3.14}$$

Then

i) $B(u,v)$ *is regular on* $S \times V$

ii) $D(S,V,X) \leq K/C(S,V)$

iii) $\|u_0 - u_S\|_X \leq \big(1 + K/C(S,V)\big) \inf_{s \in S} \|u_0 - s\|_X .$

Proof

i) We must show that for any $u \in S$ there is a $v \in V$ with $\|v\|_Y = 1$ such that $B(u,v) \neq 0$. By (2.3.14), we have for any $0 \neq u \in S$

$$\sup_{\substack{\|v\|_Y = 1 \\ v \in V}} |B(u,v)| \geq C(S,V) \, \|u\|_X \, .$$

Hence, for every $\varepsilon > 0$, there exists v such that

$$|B(u,v)| \geq (1-\varepsilon) \, C(S,V) \, \|u\|_X > 0 \, .$$

ii) Recall that $u_S = Pu$ satisfies

$$B(u - Pu, v) = 0 \ \text{ for all } \ v \in V \, .$$

By (2.3.14), for any $\varepsilon > 0$ there exists $\bar{v} \in V$ such that

$$|B(Pu,\bar{v})| \geq (1-\varepsilon) \, C(S,V) \, \|Pu\|_X \, \|\bar{v}\|_Y \, .$$

Hence, for any $\bar{v} \in V$

$$(1-\varepsilon) \, C(S,V) \, \|Pu\|_X \, \|\bar{v}\|_Y \leq B(Pu,\bar{v}) = B(u,\bar{v}) \leq K \, \|u\|_X \, \|\bar{v}\|_Y$$

which implies

$$\|Pu\|_X \leq \frac{K \, \|u\|_X}{(1-\varepsilon) \, C(S,V)} \, .$$

Since $\varepsilon > 0$ could be chosen arbitrarily small, ii) follows. Finally, iii) is a direct consequence of (2.3.9). □

We see that the stability of the FEM is guaranteed if the inf–sup condition (2.3.14) holds. Sometimes, in practice, it is however easier to verify the "adjoint" inf–sup condition

$$\inf_{0\neq v\in V} \sup_{0\neq u\in S} \frac{B(u,v)}{\|u\|_X \, \|v\|_Y} \geq \widetilde{C}(S,V) > 0\,. \tag{2.3.15}$$

The next result shows that (2.3.15) implies (2.3.14).

Theorem 2.21 *Assume* (2.3.13), (2.3.15) *and the continuous inf–sup condition*

$$\inf_{0\neq u\in X} \sup_{0\neq v\in Y} \frac{B(u,v)}{\|u\|_X \, \|v\|_Y} \geq C > 0\,, \tag{2.3.16}$$

$$\textit{for all } \; 0 \neq v \in Y : \; \sup_{u\in X} B(u,v) > 0\,. \tag{2.3.17}$$

Then the discrete inf–sup condition (2.3.14) *holds with the stability constant*

$$C(S,V) = \widetilde{C}(S,V)\,.$$

The proof is left to the reader as an exercise.

Remark 2.22 (proving the inf–sup condition). To prove the discrete inf-sup condition, we may proceed as in Remark 1.16; a particular useful technique in the context of mixed FEM is described in Chapter 5, Section 5.5.3.

Remark 2.23 (on computing K and $C(S,V)$). We saw in the preceding theorem that the stability constant $D(S,V,X)$ could be estimated by $K/C(S,V)$. If X, Y are Hilbert spaces with inner products $\langle\cdot,\cdot\rangle_X$ and $\langle\cdot,\cdot\rangle_Y$, respectively, the constants K and $C(S,V)$ are related to singular values of a certain matrix which we derive next.

Let $\psi = \{\psi_i\}$ be a basis of S and $\varphi = \{\varphi_i\}$ a basis of V. Define the (symmetric, positive definite) Gram–matrices

$$[S] = \{\langle\psi_i,\psi_j\rangle_X\}_{i,j=1}^N\,,\; [V] = \{\langle\varphi_i,\varphi_j\rangle_Y\}_{i,j=1}^N\,.$$

Clearly then for $u = \{c\}^\top\{\psi\} \in S$, $\|u\|_X^2 = \{c\}^\top[S]\{c\}$ and for $v = \{d\}^\top\{\varphi\} \in V$, $\|v\|_Y^2 = \{d\}^\top[V]\{d\}$. Then we see that for every $u \in S$, $v \in V$

$$\frac{B(u,v)}{\|u\|_X \, \|v\|_Y} = \frac{\{d\}^\top[K]\{c\}}{(\{c\}^\top[S]\{c\})^{\frac{1}{2}}\,(\{d\}^\top[V]\{d\})^{\frac{1}{2}}}\,.$$

Since $[S]$ and $[V]$ are symmetric and positive definite, they can be diagonalized:

$$[S] = P^\top \Lambda_S P, \quad [V] = Q^\top \Lambda_V Q$$

where $P^\top P = Q^\top Q = [1]$ and Λ_S, Λ_V are positive, diagonal; moreover the matrices

$$[S]^{\pm\frac{1}{2}} = P^\top \Lambda_S^{\pm\frac{1}{2}} P, \quad [V]^{\pm\frac{1}{2}} = Q^\top \Lambda_V^{\pm\frac{1}{2}} Q$$

are symmetric and positive definite. Writing $[S]^{\frac{1}{2}}\{c\} = \{\hat{c}\}$, $[V]^{\frac{1}{2}}\{d\} = \{\hat{d}\}$, we get

$$\frac{B(u,v)}{\|u\|_X \|v\|_Y} = \frac{\{\hat{d}\}^\top [V]^{-\frac{1}{2}} [K] [S]^{-\frac{1}{2}} \{\hat{c}\}}{(\{\hat{d}\}^\top \{\hat{d}\})^{\frac{1}{2}} (\{\hat{c}\}^\top \{\hat{c}\})^{\frac{1}{2}}} \,.$$

It follows that the best values for K and $C(S,V)$ in Theorem 2.20 are given by

$$K = \sup_{0 \neq \hat{c} \in \mathbb{R}^N} \sup_{0 \neq \hat{d} \in \mathbb{R}^N} \frac{\{\hat{d}\}^\top [M] \{\hat{c}\}}{|\{\hat{d}\}| \, |\{\hat{c}\}|} \tag{2.3.18}$$

and

$$C(S,V) = \inf_{0 \neq \hat{c} \in \mathbb{R}^N} \sup_{0 \neq \hat{d} \in \mathbb{R}^N} \frac{\{\hat{d}\}^\top [M] \{\hat{c}\}}{|\{\hat{d}\}| \, |\{\hat{c}\}|} \tag{2.3.19}$$

where $[M] = [V]^{-\frac{1}{2}} [K] [S]^{-\frac{1}{2}}$, and $|\{\hat{c}\}|$ denotes the Euclidean norm of the vector $\{\hat{c}\}$ in $\mathbb{R}^N$. The constants K and C can be characterized in terms of the singular values σ_i of $[M]$. Let

$$[M] = [\widehat{U}]^\top [\Sigma] [\widehat{V}], \quad [\Sigma] = \mathrm{diag}\{\sigma_1, \ldots, \sigma_N\} \,,$$

denote the singular value decomposition of $[M]$. Then $[\widehat{U}]^\top [\widehat{U}] = [\widehat{V}]^\top [\widehat{V}] = [1]$ and hence $|\{\hat{d}\}| = |[\widehat{U}]\{\hat{d}\}|$, $|\{\hat{c}\}| = |[\widehat{V}]\{\hat{c}\}|$. Therefore

$$\frac{\{\hat{d}\}^\top [M] \{\hat{c}\}}{|\{\hat{d}\}| \, |\{\hat{c}\}|} = \frac{\{\hat{d}\}^\top [\Sigma] \{\hat{c}\}}{|\{\hat{d}\}| \, |\{\hat{c}\}|} \,,$$

and from (2.3.18), (2.3.19) we get

$$K = \sup_{|\{c\}|=1} \sup_{|\{d\}|=1} \sum_{i=1}^N \sigma_i \, c_i d_i = \sigma_N \tag{2.3.20}$$

$$C(S,V) = \inf_{|\{c\}|=1} \sup_{|\{d\}|=1} \sum_{i=1}^N \sigma_i \, c_i d_i = \sigma_1 \tag{2.3.21}$$

where we have assumed, as usual, that $0 < \sigma_1 \leq \sigma_2 \leq \sigma_3 \leq \ldots \leq \sigma_N$.

We arrive at

$$D(S,V,X) = \frac{\sigma_N}{\sigma_1} \tag{2.3.22}$$

where σ_1 and σ_N are the smallest and largest singular values of the matrix $[M] = [V]^{-\frac{1}{2}}[K][S]^{-\frac{1}{2}}$, respectively.

Remark 2.24 (On the set R – continued) Let X, Y be Hilbert spaces. Consider sequences $S_i \subset X$ and $V_i \subset Y$ of finite-dimensional subspaces of dimensions $N_i \to \infty$, $i \to \infty$. If $B(u,v)$ is continuous on $X \times Y$, i.e.,

$$|B(u,v)| \le K \, \|u\|_X \, \|v\|_Y \quad u \in X, v \in Y ,$$

it follows from (2.3.20) that the largest singular values σ_{N_i} corresponding to S_i, V_i are uniformly bounded:

$$\sigma_{N_i}^{(i)} \le K < \infty \text{ for all } i .$$

A method of approximation is therefore unstable (in the sense of Definition 2.13) if and only if

$$\sigma_1^{(i)} \to 0 \text{ as } i \to \infty . \tag{2.3.23}$$

In this case the **set R of solutions on which the FEM based on $\{S_i, V_i\}$ is unstable** is associated with the right singular vectors $\{v_j^{(i)}\}$ corresponding to the singular values $\sigma_j^{(i)}$ tending to zero as $i \to \infty$.

Remark 2.25 So far, we have considered in this chapter only the case that the subspaces S_i, V_i are finite dimensional. One can verify, however, (the reader should check this) that **all theorems remain valid also in the more general case when S and V** are closed, possibly infinite dimensional subspaces of X and Y, respectively.

Exercise 2.26 Elaborate Remark 2.23 in the (special) case that $X = Y$, $S = V$ and that $B(u,v) = B(v,u)$, i.e., that $B(\cdot,\cdot)$ is symmetric.

2.4 Variational crimes/nonconforming FEM

Frequently, the FE solution $u_S \in S$ cannot be computed as in (2.1.5), but rather by means of **approximate bilinear forms $B_N(\cdot,\cdot)$, possibly depending on the spaces S_N, V_N**. We discuss error estimates in this case, the so-called **lemmas of Strang**. To this end, we assume here

$$X = Y, \;\; S_N = V_N \subset X . \tag{2.4.1}$$

We assume further that the approximate form $B_N(\cdot,\cdot)$, $F_N(\cdot)$ satisfy

$$\begin{aligned} B_N(\cdot,\cdot) &: S_N \times S_N \to \mathbb{R} \text{ are regular ,} \\ F_N(\cdot) &: \quad S_N \quad\quad\; \to \mathbb{R} \text{ are continuous .} \end{aligned} \tag{2.4.2}$$

Then the FE solutions $u_N \in S_N$ are given by

$$u_N \in S_N : B_N(u_N, v) = F_N(v) \text{ for all } v \in S_N \,. \tag{2.4.3}$$

Theorem 2.27 *(first lemma of Strang). Assume* (2.4.1) *and let* $B_N(\cdot,\cdot)$, $F_N(\cdot)$ *satisfy*

$$B_N(v,v) \geq C_N \, \|v\|_X^2 \qquad \textit{for all } v \in S_N \,, \tag{2.4.4}$$

and

$$|B_N(u,v)| \leq K_N \, \|u\|_X \, \|v\|_X \quad \textit{for all } u, v \in S_N \tag{2.4.5}$$

with $0 < C_N, \, K_N < \infty$. *Then there holds the error estimate*

$$\begin{aligned} \|u_0 - u_N\|_X \leq \inf_{v_N \in S_N} \Big\{ \Big(1 + \frac{K_N}{C_N}\Big) \, \|u_0 - v_N\|_X \\ + \frac{1}{C_N} \sup_{w_N \in S_N} \Big\{ \frac{|B(v_N, w_N) - B_N(v_N, w_N)|}{\|w_N\|_X} \\ + \frac{|F(w_N) - F_N(w_N)|}{\|w_N\|_X} \Big\}\Big\} \,. \end{aligned} \tag{2.4.6}$$

Proof Let $v_N \in S_N$ be arbitrary and set $w_N := u_N - v_N$. Then, from (2.4.4) and (2.4.3) we get

$$\begin{aligned} C_N \, \|u_N - v_N\|_X^2 &\leq B_N(u_N - v_N, u_N - v_N) = B_N(u_N - v_N, w_N) \\ &= B_N(u - v_N, w_N) + [B(v_N, w_N) - B_N(v_N, w_N)] \\ &\quad + [B_N(u_N, w_N) - B(u, w_N)] \\ &= B(u - v_N, w_N) + [B(v_N, w_N) - B_N(v_N, w_N)] \\ &\quad + [F_N(w_N) - F(w_N)] \,, \end{aligned}$$

hence, assuming that $u_N - v_N = w_N \neq 0$,

$$\begin{aligned} \|u_N - v_N\|_X \leq \frac{1}{C_N} \Big\{ K_N \, \|u - v_N\|_X \\ + \frac{|B(v_N, w_N) - B_N(v_N, w_N)|}{\|w_N\|_X} + \frac{|F(w_N) - F_N(w_N)|}{\|w_N\|_X} \Big\} \,. \end{aligned}$$

Taking in the right-hand side first the supremum over all $w_N \in S_N$ and then the infimum over all $v_N \in S_N$, using the triangle inequality we find the assertion

$$\|u_0 - u_N\|_X \leq \|u_0 - v_N\|_X + \|u_N - v_N\|_X \quad \text{for all } v_N \in S_N \,.$$

□

A more general situation occurs if

$$B_N(\cdot,\cdot) : S_N \times S_N \to \mathbb{R}$$

is defined only on $S_N \not\subset X$. In this case we speak of a **nonconforming FEM.** We assume that S_N is equipped with the **mesh-dependent norm** $\|\cdot\|_N$ and satisfies

$$B_N(u,u) \geq C_N \|u\|_N^2 \quad \text{for all} \;\; u \in S_N\,, \tag{2.4.7}$$

$$|B_N(u,v)| \leq K_N \|u\|_N \|v\|_N \text{ for all } \; u \in S_N + X, \text{ for all } \; v \in S_N\,, \tag{2.4.8}$$

$$|F_N(v)| \leq \|F_N\| \, \|v\|_N \qquad \text{for all } v \in S_N \tag{2.4.9}$$

for some $0 < C_N, K_N < \infty$. Then u_N in (2.4.3) is well-defined and we have the following result.

Theorem 2.28 *Assume* (2.4.7), (2.4.8). *Then*

$$\begin{aligned}\|u - u_N\|_N \leq \left(1 + \frac{K_N}{C_N}\right) \inf_{v_N \in S_N} \|u - v_N\|_N \\ + \frac{1}{C_N} \sup_{w_N \in S_N} \frac{|B_N(u,w_N) - F_N(w_N)|}{\|w_N\|_N}\,.\end{aligned} \tag{2.4.10}$$

Proof Let again $v_N \in S_N$. Then (2.4.7) implies

$$\begin{aligned}C_N \|u_N - v_N\|_N^2 &\leq B_N(u_N - v_N, u_N - v_N) \\ &= B_N(u - v_N, u_N - v_N) \\ &\quad + [F_N(u_N - v_N) - B_N(u, u_N - v_N)]\,.\end{aligned}$$

Reasoning as in the proof of Theorem 2.27, we get

$$\|u_N - v_N\|_N \leq \frac{1}{C_N} \left\{ K_N \|u - v_N\|_N + \frac{|B_N(u,w_N) - F_N(w_N)|}{\|w_N\|_N} \right\},$$

and the assertion (2.4.10) follows upon using the triangle inequality as in the proof of Theorem 2.27. □

3

HP-FINITE ELEMENTS IN ONE DIMENSION

The general considerations in Chapter 2 are valid for any family of finite-dimensional subspaces S_i, V_i. The finite element method (FEM) denotes methods of approximation which are based on special families of subspaces $\{S_i, V_i\}$. They are selected based on computational and algorithmic considerations. Here we will discuss the family $S^{\mathbf{p},\ell}(\Omega, \mathcal{T})$ (cf. Chapter 2, Example 2.1) in one dimension. This is mainly to illustrate the principal concepts with a minimum of notational difficulty. All assertions made will have analogs in two and three dimensions – the two-dimensional case will be discussed in the next chapter. In addition, the one-dimensional *hp* convergence analysis presented here is important in its own right, since results in two and three dimensions can be derived from it by tensor product construction. Finally, the results are also relevant for the finite element discretization of boundary integral equation formulations of two-dimensional problems where one-dimensional domains (boundaries) occur naturally.

3.1 The finite element spaces $S^{\mathbf{p},\ell}(\Omega, \mathcal{T})$

3.1.1 *Meshes*

Let $\Omega = (a, b) \subset \mathbb{R}$ be a bounded interval. A **mesh $\mathcal{T}$ on Ω** is a partition of Ω into $M(\mathcal{T})$ open, disjoint subintervals $\Omega_j^{\mathcal{T}}$, $\mathcal{T} = \{\Omega_j^{\mathcal{T}}\}_{j=1}^{M(\mathcal{T})}$, $\Omega_j^{\mathcal{T}} = (x_{j-1}^{\mathcal{T}}, x_j^{\mathcal{T}})$, $a = x_0^{\mathcal{T}} < x_1^{\mathcal{T}} < \ldots < x_M^{\mathcal{T}} = b$. The points $x_j \in \overline{\Omega}$ are **nodal points** (nodes) and the Ω_j are the **elements** of the mesh $\mathcal{T}$. Define

$$h_j^{\mathcal{T}} := x_j^{\mathcal{T}} - x_{j-1}^{\mathcal{T}} \ , \ \ j = 1, \ldots, M \tag{3.1.1}$$

$$h(\mathcal{T}) = \max_{1 \le j \le M(\mathcal{T})} \{h_j^{\mathcal{T}}\} \tag{3.1.2}$$

$h(\mathcal{T})$ is called the **meshwidth** of $\mathcal{T}$. Whenever the underlying mesh $\mathcal{T}$ is clear from the context, we omit the argument $\mathcal{T}$.

3.1.2 *Element mappings*

Each element $\Omega_j \in \mathcal{T}$ can be mapped onto $\widehat{\Omega} = (-1, 1)$, the **reference** or **master element**. We denote the map by Q_j, i.e.,

$$\Omega_j = Q_j(\widehat{\Omega}) \, , \ \ x = Q_j(\xi) \, . \tag{3.1.3}$$

Evidently, a linear mapping Q_j will suffice for (3.1.3)

$$x = Q_j(\xi) = \frac{1}{2}(1-\xi)x_{j-1} + \frac{1}{2}(1+\xi)x_j, \ \xi \in \widehat{\Omega}. \tag{3.1.4}$$

Its inverse is

$$\xi = Q_j^{-1}(x) = \frac{2x - x_j - x_{j-1}}{x_j - x_{j-1}}, \ x \in \Omega_j. \tag{3.1.5}$$

In two dimensions, due to the greater complexity of the geometry, Q_j is frequently nonlinear, giving rise, for example, to so-called **isoparametric elements**. This will be addressed in the next chapter. For now, we will only work with the linear map (3.1.4).

The **Jacobian** of Q_j is constant:

$$\frac{dx}{d\xi} = \frac{1}{2}(x_j - x_{j-1}) = \frac{h_j}{2}. \tag{3.1.6}$$

3.1.3 *Definition of* $S^{\mathbf{p},\ell}(\Omega, \mathcal{T})$

We defined $S^{\mathbf{p},\ell}(\Omega, \mathcal{T})$ in Example 2.1. We will give here an equivalent definition leading directly to the construction of a basis $\{\varphi_j\}$ upon which the FE algorithm can be based.

Definition 3.1 *Let $\Omega = (a,b)$ be an interval, $\mathcal{T}$ be a mesh and $\mathbf{p} = (p_1, \ldots, p_M)$ a vector of polynomial degrees p_j, the degree vector, $\ell \geq 0$ an integer. Let $\mathcal{S}^p$ denote the polynomials of degree p on the master element $\widehat{\Omega}$. Then*

$$S^{\mathbf{p},\ell}(\Omega, \mathcal{T}) = \{u \in H^\ell(\Omega) : u|_{\Omega_j} = s_j(Q_j^{-1}(x)), \ s_j \in \mathcal{S}^{p_j}(\widehat{\Omega})\}.$$

Since Q_j is linear, $u \in S^{\mathbf{p},\ell}(\Omega, \mathcal{T})$ implies that on $\Omega_j \in \mathcal{T}$, u is polynomial of degree p_j. Since $H^\ell(\Omega) \hookrightarrow C^{\ell-1}(\overline{\Omega})$ for $\ell \geq 1$, Definition 3.1 coincides for $\ell \geq 1$ with the one given in Example 2.1, Chapter 2. We define further

$$\begin{aligned} S_0^{\mathbf{p},1}(\Omega, \mathcal{T}) &= S^{\mathbf{p},1}(\Omega, \mathcal{T}) \cap H_0^1(\Omega) \\ &= \{u \in S^{\mathbf{p},1}(\Omega, \mathcal{T}) : u(a) = u(b) = 0\}. \end{aligned} \tag{3.1.7}$$

We saw in Chapter 2 that to compute the stiffness matrix $[K]$, a **basis** $\{\varphi_j\}$ of $S^{\mathbf{p},1}(\Omega, \mathcal{T})$ must be available. We will obtain a basis of $S^{\mathbf{p},1}(\Omega, \mathcal{T})$ by transporting a basis of $\mathcal{S}^p$ on $\widehat{\Omega}$ to Ω_j via Q_j.

3.1.4 *Basis of* $\mathcal{S}^p$

The basis functions of $\mathcal{S}^p$ are denoted by $N_i(\xi)$, $i = 1, \ldots, p+1$ and are called **standard shape functions** or simply **shape functions**. Many different selections of shape functions are possible. We present one, the so-called **hierarchic shape functions** which are particularly suitable for high polynomial degrees.

The selection of shape functions depends also on the degree ℓ of smoothness of $S^{\mathbf{p},\ell}(\Omega,\mathcal{T})$.

If $\ell = 0$, we choose simply the Legendre polynomials

$$N_i(\xi) = L_{i-1}(\xi)\,, \quad i = 1,\dots,p+1\,. \tag{3.1.8}$$

If $\ell = 1$, we select the set

$$\begin{aligned} N_1(\xi) &= \frac{1-\xi}{2}\,, \quad N_2(\xi) = \frac{1+\xi}{2} \\ N_i(\xi) &= \sqrt{\frac{2i-3}{2}} \int_{-1}^{\xi} L_{i-2}(t)\,dt,\ 3 \le i \le p+1\,. \end{aligned} \tag{3.1.9}$$

Note that due to the orthogonality of the Legendre polynomials we have

$$\int_{-1}^{1} \frac{dN_i}{d\xi}\frac{dN_j}{d\xi}\,d\xi = \delta_{ij}\,, \quad i,j \ge 3\,. \tag{3.1.10}$$

It is useful to divide the shape functions into 2 sets, **internal** and **external shape functions**. We say that $N_i(\xi)$ is an **internal shape function of order ℓ** if

$$\frac{d^j N_i}{d\xi^j}(\pm 1) = 0\,, \qquad 0 \le j \le \ell - 1\,. \tag{3.1.11}$$

Shape functions which are not internal are external. For $\ell = 0$, (3.1.11) is void and all shape functions are internal. Note (3.1.11) are 2ℓ conditions on N_i.

An external shape function is **nodal**, if it satisfies exactly $2\ell - 1$ of the conditions (3.1.11), i.e., all but one of the conditions (3.1.11). Nodal shape functions exist only if $p \ge 2\ell - 1$ which we assume in the following example.

Example 3.2 In the set (3.1.9) of shape-functions N_1 and N_2 are nodal, $N_i(\xi)$, $i \ge 3$ are internal.

To make the construction of a basis for $S^{\mathbf{p},\ell}(\Omega,\mathcal{T})$ from the shape functions as simple as possible, the number of internal shape functions should be as large as possible and all external shape functions should be nodal. We denote nodal shape functions by $\overset{0}{N}_i(\xi)$ and internal shape functions by $\overset{1}{N}_i(\xi)$ (the superscript signifying the dimension of the geometry entity the shape function is associated with). Thus, for (3.1.9)

$$\begin{aligned} \overset{0}{N}_1(\xi) &= (1-\xi)/2, \quad \overset{0}{N}_2(\xi) = (1+\xi)/2 \\ \overset{1}{N}_i(\xi) &= \sqrt{\frac{2i+1}{2}} \int_{-1}^{\xi} L_i(t)dt, \quad i = 1,2,\dots,p-1\,. \end{aligned} \tag{3.1.12}$$

Exercise 3.3 Assume $\ell = 2$ and $p \geq 3$. Construct the 4 nodal and $p-3$ internal shape functions.

Solution:

$$\overset{0}{N}_1(\xi) = \frac{1}{4}\,(1-\xi)^2\,(1+\xi),\ \overset{0}{N}_2(\xi) = \frac{1}{4}\,(1-\xi)^2\,(2+\xi)$$

$$\overset{0}{N}_3(\xi) = \frac{1}{4}\,(1+\xi)^2\,(1-\xi),\ \overset{0}{N}_4(\xi) = -\frac{1}{4}\,(1-\xi)\,(2+\xi)^2$$

$$\overset{1}{N}_i(\xi) = \sqrt{\frac{2i+3}{2}}\int_{-1}^{\xi}\int_{-1}^{\eta_1} L_{i+1}(\eta_2)\,d\eta_2\,d\eta_1$$
$$= \sqrt{\frac{2i+3}{2}}\int_{-1}^{\xi}(\xi-\eta)L_{i+1}(\eta)d\eta,\ i = 1,\ldots,p-3\,.$$

3.1.5 *Element shape functions*

The standard shape functions $N_i(\xi)$ induce, via the mappings Q_j, element shape functions on Ω_j.

Definition 3.4 *Let* $\{N_i\}_{j=1}^{p_j+1}$ *be a set of standard shape functions (of order* ℓ *and degree* p*). Then the corresponding element shape functions of order* ℓ *and degree* p_j *are defined by*

$$\nu_i^{[j]}(x) := N_i(Q_j^{-1}(x)),\ x \in \Omega_j,\ i = 1,2,\ldots,p_j+1\,. \tag{3.1.13}$$

We distinguish **external element shape functions**

$$\overset{0}{\nu}{}_i^{[j]}(x) := [Q_j'(-1)]^{i-1}\ \overset{0}{N}_i(Q_j^{-1}(x)),\ i = 1,\ldots\ell\,, \tag{3.1.14}$$

$$\overset{0}{\nu}{}_i^{[j]}(x) := [Q_j'(1)]^{i-\ell-1}\ \overset{0}{N}_i(Q_j^{-1}(x)),\ i = \ell+1,\ldots,2\ell\,, \tag{3.1.15}$$

(assuming that $\overset{0}{N}_i$ *are nodal) and* **internal element shape functions** *of order* ℓ *and degree* p_j

$$\overset{1}{\nu}{}_i^{[j]}(x) := \overset{1}{N}_i(Q_j^{-1}(x))\,, \qquad i = 1,\ldots,p_j+1-2\ell\,. \tag{3.1.16}$$

3.1.6 *Basis functions of* $S^{p,\ell}(\Omega,\mathcal{T})$

We denote basis functions for $S = S^{p,\ell}(\Omega,\mathcal{T})$ by $\varphi_i(x)$. Basis functions associated with node $j = 0,\ldots,M(\mathcal{T})$ are denoted by $\overset{0}{\varphi}{}_i^{[j]}(x)$ and basis functions associated with element Ω_j, $j = 1,\ldots,M(\mathcal{T})$ by $\overset{1}{\varphi}{}_i^{[j]}(x)$. Note that the superscript $[j]$ now indicates either an element or a node number.

If $\ell = 0$, the basis functions φ_i may be discontinuous between elements. Hence there are no external shape functions $\overset{0}{\varphi}{}_i^{[j]}(x)$ and

$$\overset{1}{\varphi}{}_i^{[j]}(x) = \begin{cases} \overset{1}{\nu}{}_i^{[j]}(x) & x \in \Omega_j\,, \\ 0 & \text{else}\,. \end{cases} \tag{3.1.17}$$

If $\ell = 1$, $\varphi_i(x) \in H^1(\Omega)$ and hence φ_i must be continuous (since in one dimension $C^0(\overline{\Omega}) \hookrightarrow H^1(\Omega)$). To achieve it, **corresponding nodal element shape functions of neighboring elements must be joined at node** j, $j = 1, \ldots, M(\mathcal{T}) - 1$. There is only one basis function associated with node j, $\overset{0}{\varphi}{}_1^{[j]}$, given by

$$\overset{0}{\varphi}{}_1^{[j]} = \begin{cases} \overset{0}{\nu}{}_2^{[j]}(x) & x \in \Omega_j \\ \overset{0}{\nu}{}_1^{[j+1]}(x) & x \in \Omega_{j+1} \\ 0 & \text{else}\quad. \end{cases} \tag{3.1.18}$$

This "connecting" of nodal shape functions is exactly corresponding to the element stiffness matrix assembly below.

Every internal element shape function of order 1, $\overset{1}{\nu}{}_i^{[j]}(x)$, $i = 1, 2, \ldots, p-1, j = 1, \ldots, M(\mathcal{T})$ can be extended by zero to Ω to obtain a basis function $\overset{1}{\varphi}{}_i^{[j]}(x)$.

Exercise 3.5 Construct $\overset{0}{\varphi}{}_i^{[j]}$ and $\overset{1}{\varphi}{}_i^{[j]}$ in the case $\ell = 2$. Use Exercise 3.3.

In this way all basis functions φ_i of $S^{\mathbf{p},\ell}(\Omega,\mathcal{T})$ can be obtained and every $u \in S^{\mathbf{p},\ell}(\Omega,\mathcal{T})$ can be written as

$$u = \sum_{i=1}^{N} a_i\,\varphi_i(x)\,. \tag{3.1.19}$$

The coefficients a_i are called **amplitudes** of φ_i and

$$N = \dim\big(S^{\mathbf{p},\ell}(\Omega,\mathcal{T})\big) = \sum_{i=1}^{M(\mathcal{T})} (p_i + 1) - \ell(M(\mathcal{T}) - 1)\,. \tag{3.1.20}$$

Remark 3.6 *The amplitudes a_i associated with* **nodal** *basis functions correspond to function values (or derivatives if $\ell > 1$) of u at the nodes x_i. However, amplitudes a_i associated with* **internal** *basis functions $\overset{1}{\varphi}{}_i^{[j]}(x)$ have no such interpretation. They are, in a sense,* **generalized Fourier coefficients** *of the solution u in Ω_j.*

3.2 Algorithmic pattern of the *hp*-FEM

We will demonstrate the basic steps of the FEM for the model problem of the elastic bar, i.e., for

$$-(AEu')' + cu = f \ \text{ in } \Omega = (a, b) \tag{3.2.1}$$

$$u(a) = u(b) = 0\ . \tag{3.2.2}$$

We assume further

$$AE > 0,\ c \geq 0 \text{ piecewise constant, } a = 0,\ b = 1\ . \tag{3.2.3}$$

We select the variational formulation based on B_3 and $X_3 \times Y_3$: find $u_0 \in H_0^1(\Omega)$ such that

$$\begin{aligned} B_3(u_0, v) &= \int_0^1 (AE(x)\, u_0' v' + c(x)\, u_0 v) dx = F_3(v) \\ &= \int_0^1 f v\, dx\ , \quad v \in H_0^1(\Omega)\ . \end{aligned} \tag{3.2.4}$$

To discretize it, we select $S = V = S_0^{p,1}(\Omega, \mathcal{T}) \subset H_0^1(\Omega)$ where $S_0^{p,1}(\Omega, \mathcal{T}) = S^{p,1}(\Omega, \mathcal{T}) \cap H_0^1(\Omega)$. Thus the FE solution is: find $u_S \in S_0^{p,1}(\Omega, \mathcal{T})$ such that

$$B_3(u_S, v) = F_3(v) \text{ for all } v \in S_0^{p,1}(\Omega, \mathcal{T})\ . \tag{3.2.5}$$

Since B_3 is coercive on $H_0^1(\Omega)$, i.e.,

$$B_3(u, u) \geq \inf_\Omega \{AE(x)\}\, \|u'\|^2_{L^2(\Omega)} \geq C\, \|u\|^2_{H^1(\Omega)}, \quad \text{for all } \ u \in H_0^1(\Omega) \tag{3.2.6}$$

by the first Poincaré inequality, Theorem 2.19 implies that u_S is quasioptimal, i.e.,

$$\|u_0 - u_S\|_{H^1(\Omega)} \leq \frac{\max\{\|AE\|_{L^\infty}, \|c\|_{L^\infty}\}}{\sqrt{C}} \inf_{v \in S_0^{p,1}(\Omega, \mathcal{T})} \|u_0 - v\|_{H^1(\Omega)}\ . \tag{3.2.7}$$

As we saw in Chapter 2, (3.2.5) is equivalent to the linear system

$$[K]\,\{c\} = \{q\}\ . \tag{3.2.8}$$

The efficient evaluation of the stiffness matrix $[K]$, of the load vector $\{q\}$ and the solution of the linear system (3.2.8) is the purpose of a FE code. The **basic algorithmic pattern** of virtually all codes is

i) computation of elemental stiffness matrices and load vectors
ii) static condensation of internal degrees of freedom
iii) assembly of the global stiffness matrix
iv) solution of the linear system for the external amplitudes
v) calculation of the internal amplitudes
vi) calculation of data of interest.

We explain steps i) - iii) and v) in detail, step iv) can be done by Gaussian elimination. In step vi), we shall be interested for example in the **energy norm**

$$\|u_S\|_E = (B_3(u_S, u_S))^{\frac{1}{2}} \tag{3.2.9}$$

of u_S. We drop the subscript "3" at the bilinear form and the load functional from now on.

Step 1: (element stiffness matrices and load vectors). Since $u, v \in S_0^{p,1}(\Omega, \mathcal{T})$, $u, v \in C^0(\overline{\Omega})$. We assume

$$AE(x) = AE_j,\ c(x) = c_j\,, \qquad x \in \Omega_j \tag{3.2.10}$$

i.e., that AE and c are constant on each $\Omega_j \in \mathcal{T}$. The bilinear form can then be written as

$$B(u,v) = \sum_{j=1}^{M(\mathcal{T})} B^{[j]}(u,v) \tag{3.2.11}$$

with the elemental bilinear form given by

$$B^{[j]}(u,v) = \int_{x_{j-1}}^{x_j} (AE_j\, u'v' + c_j\, uv)dx\,. \tag{3.2.12}$$

Let us compute the stiffness matrix of $B^{[j]}(u,v)$, i.e., the **stiffness matrix of** Ω_j. Since on Ω_j

$$u(x)|_{\Omega_j} = s^{[j]}(x) = \sum_{n=1}^{p_j+1} u_n^{[j]}\, \nu_n^{[j]}(x) = \{u^{[j]}\}^\top \{\nu^{[j]}\}\,, \tag{3.2.13}$$

$$v(x)|_{\Omega_j} = t^{[j]}(x) = \sum_{m=1}^{p_j+1} v_m^{[j]}\, \nu_m^{[j]}(x) = \{v^{[j]}\}^\top \{\nu^{[j]}\}\,, \tag{3.2.14}$$

we have

$$B^{[j]}(u,v) = \{v^{[j]}\}^\top [K^{[j]}]\, \{u^{[j]}\}\,, \tag{3.2.15}$$

where the elements $k_{mn}^{[j]}$ of $[K^{[j]}]$ are given by

$$k_{mn}^{[j]} = \int_{x_{j-1}}^{x_j} \left(AE_j \, \frac{d\nu_m^{[j]}}{dx} \, \frac{d\nu_n^{[j]}}{dx} + c_j \, \nu_m^{[j]} \, \nu_n^{[j]} \right) dx \, .$$

Since

$$\nu_m^{[j]}(x) = \nu_m^{[j]}(Q_j(\xi)) = N_m(\xi) \, , \quad \xi \in \widehat{\Omega}$$

we have

$$k_{mn}^{[j]} = \int_{-1}^{1} AE_j \, N'_m \, N'_n (Q'_j)^{-1} \, d\xi + \int_{-1}^{1} c_j \, N_m \, N_n \, Q'_j \, d\xi \, .$$

As Q_j is linear and AE_j, c_j are constant on Ω_j we arrive at

$$k_{mn}^{[j]} = \frac{2}{h_j} \, AE_j \int_{-1}^{1} N'_m \, N'_n \, d\xi + \frac{h_j}{2} \, c_j \int_{-1}^{1} N_m \, N_n \, d\xi \, . \tag{3.2.16}$$

The integrals in (3.2.16) do not depend on j any more and can be evaluated once and for all. Therefore we define the matrix $[\boldsymbol{K}]$ by

$$k_{mn} = \int_{-1}^{1} N'_m \, N'_n \, d\xi \, . \tag{3.2.17}$$

$[\boldsymbol{K}]$ is the **reference element stiffness matrix**. Analogously, we define the **reference element mass matrix** $[\boldsymbol{G}]$ by

$$g_{mn} = \int_{-1}^{1} N_m \, N_n \, d\xi \, . \tag{3.2.18}$$

Notice that $[\boldsymbol{K}]$ and $[\boldsymbol{G}]$ are of size $(p_j+1) \times (p_j+1)$. Since the shape functions are hierarchic, they can be obtained as sections of infinite matrices. Moreover, by (3.2.16) the stiffness matrix of Ω_j can be obtained as a linear combination of $[\boldsymbol{K}]$ and $[\boldsymbol{G}]$:

$$[K^{[j]}] = \frac{2}{h_j} \, AE_j \, [\boldsymbol{K}] + \frac{h_j}{2} \, c_j \, [\boldsymbol{G}] \, . \tag{3.2.19}$$

With the shape functions (3.1.9) we find

$$[\boldsymbol{K}] = \begin{pmatrix} \frac{1}{2} & -\frac{1}{2} & 0 & \dots & 0 \\ & \frac{1}{2} & 0 & \dots & 0 \\ & & 1 & 0 \ldots & 0 \\ & & & \ddots & \vdots \\ \text{sym.} & & & & 0 \\ & & & & 1 \end{pmatrix} . \tag{3.2.20}$$

The mass matrix $[\boldsymbol{G}]$ is, for $p = 4$, given by

$$[\boldsymbol{G}] = \begin{pmatrix} \frac{2}{3} & \frac{1}{3} & -\frac{1}{\sqrt{6}} & -\frac{1}{3\sqrt{10}} & 0 \\ & \frac{2}{3} & -\frac{1}{\sqrt{6}} & -\frac{1}{3\sqrt{10}} & 0 \\ & & \frac{2}{5} & 0 & -\frac{1}{5\sqrt{21}} \\ & & & \frac{2}{21} & 0 \\ \text{sym} & & & & \frac{2}{45} \end{pmatrix} . \tag{3.2.21}$$

In general, $[\boldsymbol{G}]$ is pentadiagonal. Notice that for $m \geq 3$, the internal shape functions satisfy (this follows from orthogonality properties of the Legendre polynomials, see (C.2.5)).

$$N_m(\xi) = \frac{1}{\sqrt{2(2m-3)}} \left(L_{m-1}(\xi) - L_{m-3}(\xi) \right) , \tag{3.2.22}$$

hence

$$g_{mm}(\xi) = \int_{-1}^{1} N_m^2(\xi)\, d\xi = \frac{2}{(2m-1)(2m-5)} , \quad m \geq 3 . \tag{3.2.23}$$

Further, (3.2.22) and the orthogonality of L_m imply that

$$g_{m,m+2} = g_{m+2,m} = \int_{-1}^{1} N_m\, N_{m+2}\, d\xi = \frac{1}{(2m-1)\sqrt{(2m-3)(2m+1)}} \tag{3.2.24}$$

and that $g_{mn} = 0$ otherwise.

To compute the load vector $\{q\}$, observe that with (3.2.14)

$$F^{[j]}(v) = \int\limits_{\Omega_j} fv dx = \{v^{[j]}\}^\top \{q^{[j]}\}$$

and $\{q^{[j]}\} = \{q_m^{[j]} : m = 1, \ldots, p_j + 1\}$ where

$$q_m^{[j]} = \int\limits_{\Omega_j} f \nu_m^{[j]} \, dx \, .$$

Using $x \equiv Q_j(\xi)$ on $\widehat{\Omega}$, we get

$$q_m^{[j]} = \int\limits_{\widehat{\Omega}} f(Q_j(\xi)) \, N_m(\xi) Q_j'(\xi) \, d\xi \tag{3.2.25}$$

which is to be evaluated by a quadrature rule on $\widehat{\Omega}$ (typically a Gauss Legendre formula).

Step 2: (static condensation of internal amplitudes). Let $\ell = 1$. The first step in the condensation procedure is to rewrite (3.2.13), (3.2.14) distinguishing internal and external amplitudes:

$$\begin{aligned} s^{[j]}(x) &= \overset{0}{u}{}_1^{[j]} \, \overset{0}{\nu}{}_1^{[j]} + \overset{0}{u}{}_2^{[j]} \, \overset{0}{\nu}{}_2^{[j]} + \sum_{i=1}^{p_j - 1} \overset{1}{u}{}_i^{[j]} \, \overset{1}{\nu}{}_i^{[j]} \\ &= \{\overset{0}{u}{}^{[j]}\}^\top \{\overset{0}{\nu}{}^{[j]}\} + \{\overset{1}{u}{}^{[j]}\}^\top \{\overset{1}{\nu}{}^{[j]}\} \, , \end{aligned} \tag{3.2.26}$$

$$t^{[j]}(x) \;=\; \{\overset{0}{v}{}^{[j]}\}^\top \{\overset{0}{\nu}{}^{[j]}\} + \{\overset{1}{v}{}^{[j]}\}^\top \{\overset{1}{\nu}{}^{[j]}\} \, . \tag{3.2.27}$$

We partition $[K^{[j]}]$ correspondingly and write

$$B^{[j]}(u_S, v) = \begin{pmatrix} \{\overset{0}{v}{}^{[j]}\} \\ \{\overset{1}{v}{}^{[j]}\} \end{pmatrix}^\top \begin{pmatrix} [K_{00}^{[j]}] \; [K_{10}^{[j]}] \\ [K_{01}^{[j]}] \; [K_{11}^{[j]}] \end{pmatrix} \begin{pmatrix} \{\overset{0}{u}{}^{[j]}\} \\ \{\overset{1}{u}{}^{[j]}\} \end{pmatrix} \tag{3.2.28}$$

$$F^{[j]}(v) = \{\overset{0}{v}{}^{[j]}\}^\top \{\overset{0}{q}{}^{[j]}\} + \{\overset{1}{v}{}^{[j]}\}^\top \{\overset{1}{q}{}^{[j]}\} \, . \tag{3.2.29}$$

Now we observe that all internal amplitudes $\{\overset{1}{u}{}^{[j]}\}$, $\{\overset{1}{v}{}^{[j]}\}$ correspond to shape functions $\{\overset{1}{\nu}{}^{[j]}_m\}$ which are only nonzero within Ω_j. Therefore the definition of the FE solution

$$B(u_S, v) = F(v) \text{ for all } v \in S^{p,1}_0(\Omega, \mathcal{T})$$

implies

$$0 = B(u_S, \overset{1}{\nu}{}^{[j]}_m) - F(\overset{1}{\nu}{}^{[j]}_m) = B^{[j]}(u_S, \overset{1}{\nu}{}^{[j]}_m) - F^{[j]}(\overset{1}{\nu}{}^{[j]}_m), \quad m = 1, \ldots, p_j - 1,$$

or, in matrix form,

external *internal*

$$[K^{[j]}_{01}]\{\overset{0}{u}{}^{[j]}\} + [K^{[j]}_{11}]\{\overset{1}{u}{}^{[j]}\} - \overset{1}{q}{}^{[j]} = 0. \tag{3.2.30}$$

It means that once the external amplitudes $\{\overset{0}{u}{}^{[j]}\}$ are known, the internal ones can be calculated elementwise by

$$\{\overset{1}{u}{}^{[j]}\} = [K^{[j]}_{11}]^{-1} \left(\overset{1}{q}{}^{[j]} - [K^{[j]}_{01}]\{\overset{0}{u}{}^{[j]}\}\right). \tag{3.2.31}$$

We write

$$0 = B(u, v) - F(v) = \sum_{j=1}^{M(\mathcal{T})} B^{[j]}(u, v) - F^{[j]}(v)$$

$$= \sum_{j=1}^{M(\mathcal{T})} \{\overset{0}{v}{}^{[j]}\}^\top [K^{[j]}_{00}]\{\overset{0}{u}{}^{[j]}\} + \{\overset{1}{v}{}^{[j]}\}^\top [K^{[j]}_{11}]\{\overset{1}{u}{}^{[j]}\}$$

$$+ \{\overset{0}{v}{}^{[j]}\}^\top [K^{[j]}_{10}]\{\overset{1}{u}{}^{[j]}\} + \{\overset{1}{v}{}^{[j]}\}^\top [K^{[j]}_{01}]\{\overset{0}{u}{}^{[j]}\}$$

$$- \{\overset{0}{v}{}^{[j]}\}^\top \{\overset{0}{q}{}^{[j]}\} - \{\overset{1}{v}{}^{[j]}\}^\top \{\overset{1}{q}{}^{[j]}\}.$$

Inserting (3.2.31), we get

$$0 = \sum_{j=1}^{M(\mathcal{T})} \{\overset{0}{v}{}^{[j]}\}^\top \left([K^{[j]}_{00}] - [K^{[j]}_{10}][K^{[j]}_{11}]^{-1}[K^{[j]}_{01}]\right)\{\overset{0}{u}{}^{[j]}\}$$

$$- \{\overset{0}{v}{}^{[j]}\}^\top \left(\{\overset{0}{q}{}^{[j]}\} - [K^{[j]}_{10}][K^{[j]}_{11}]^{-1}\{\overset{1}{q}{}^{[j]}\}\right)$$

or

$$0 = \sum_{j=1}^{M(\mathcal{T})} \{\overset{0}{v}{}^{[j]}\}^\top \left([\overset{0}{K}{}^{[j]}]\{\overset{0}{u}{}^{[j]}\} - \{\overset{0}{q}{}^{[j]}\}\right) \tag{3.2.32}$$

where the **condensed element stiffness matrices** and **load vectors** are defined by

$$[\overset{0}{\boldsymbol{K}}{}^{[j]}] = [K_{00}^{[j]}] - [K_{10}^{[j]}]\,[K_{11}^{[j]}]^{-1}\,[K_{01}^{[j]}]\,, \tag{3.2.33}$$

$$\{\overset{0}{\boldsymbol{q}}{}^{[j]}\} = \{\overset{0}{\boldsymbol{q}}{}^{[j]}\} - [K_{10}^{[j]}]\,[K_{11}^{[j]}]^{-1}\,\{\overset{1}{\boldsymbol{q}}{}^{[j]}\}\,, \tag{3.2.34}$$

Note that $[\overset{0}{\boldsymbol{K}}{}^{[j]}]$ is the **Schur complement** of $[K^{[j]}]$ with respect to the internal degrees of freedom.

Remark 3.7 We observe that the computation of $[K^{[i]}]$, $\{\boldsymbol{q}^{[i]}\}$ and the condensation step does not require any information from elements Ω_j, $j \neq i$. It is therefore ideally suited for parallel implementation.

Step 3: (assembly). Observing that

$$\overset{0}{u}{}_1^{[j]} = \overset{0}{u}{}_2^{[j-1]} = u(x_{j-1}),\ \overset{0}{v}{}_1^{[j]} = \overset{0}{u}{}_2^{[j-1]} = v(x_{j-1})\,,$$

$$\overset{0}{u}{}_2^{[j]} = \overset{0}{u}{}_1^{[j+1]} = u(x_j),\quad \overset{0}{v}{}_2^{[j]} = \overset{0}{v}{}_2^{[j+1]} = v(x_j)$$

we may rewrite (3.2.32) in the form

$$\{\overset{0}{\boldsymbol{v}}\}^\top\,([\overset{0}{\boldsymbol{K}}]\,\{\overset{0}{\boldsymbol{u}}\} - \{\overset{0}{\boldsymbol{q}}\}) = 0 \tag{3.2.35}$$

where

$$\{\overset{0}{\boldsymbol{u}}\} = (u(x_1),\, u(x_2), \ldots, u(x_{M-1}))^\top$$

is the vector of nodal variables of the FE solution and $\{\overset{0}{\boldsymbol{v}}\}$ is defined analogously. We see that the **global stiffness matrix** $[\overset{0}{\boldsymbol{K}}]$ is obtained by adding the condensed element stiffness matrices into the global array according to the location of the nodal amplitudes in the global solution vector $\{\overset{0}{\boldsymbol{u}}\}$:

$$[\overset{0}{\boldsymbol{K}}] = \begin{pmatrix} \left[\overset{0}{\boldsymbol{K}}{}^{[1]}\right] & & & 0 \\ & \left[\overset{0}{\boldsymbol{K}}{}^{[2]}\right] & & \\ & & \left[\overset{0}{\boldsymbol{K}}{}^{[3]}\right] & \\ 0 & & & \ddots \end{pmatrix}.$$

Construction of the **global load vector** $\{\overset{0}{q}\}$ proceeds analogously.

Step 4: (solution of the linear system of equations. Work estimate). The assembled global stiffness matrix $[\overset{0}{\boldsymbol{K}}]$ is symmetric and tridiagonal, positive definite and of size $M \times M$. It can therefore be factored by a standard band Cholesky procedure, and then a solution vector can be computed in $O(M)$ operations. Formation of each of the M Schur complements (3.2.33), (3.2.34), takes $O(p_j)$ operations due to the band structure of $[K_{11}^{[j]}]$ so that the total **work for the system solution** is asymptotically

$$C_1\, M + C_2 \sum_{j=1}^{M} p_j =: W(\boldsymbol{p},\mathcal{T})\,. \tag{3.2.36}$$

As we remarked above, the second sum can be reduced if parallel processing is available.

Exercise 3.8 In $\Omega = (-1,1)$, consider the problem

$$\begin{aligned} -u^{(4)} + u^{(2)} + u &= f \ \text{ in } \Omega\,, \\ u(\pm 1) = u'(\pm 1) &= 0\,. \end{aligned}$$

Give a weak formulation and derive stiffness and mass matrices for the p-version FEM based on one element $K = \Omega$, analogous to (3.2.20), (3.2.21), using the shape functions of Exercise 3.3.

3.3 The approximation properties of $S^{\boldsymbol{p},\ell}(\Omega,\mathcal{T})$

We saw in Chapter 2 that the error in the FE approximation $u_S \in S^{\boldsymbol{p},\ell}(\Omega,\mathcal{T})$ can be estimated by

$$\|u_0 - u_S\|_X \le (1 + D(\{u_0\},\ S, V, X))\, Z(\{u_0\}, S, X) \tag{3.3.1}$$

where D is the stability constant and

$$Z(\{u_0\}, S, X) = \inf_{s\in S} \|u_0 - s\|_X \tag{3.3.2}$$

characterizes the approximability of u_0 from the subspace S. Here we characterize Z for $X = L^2(\Omega)$, $H^1(\Omega)$, $S = S^{\boldsymbol{p},\ell}(\Omega,\mathcal{T})$ and various classes of solutions u_0. The analysis follows closely the construction of the space $S^{\boldsymbol{p},\ell}(\Omega,\mathcal{T})$ in the preceding section. First, we investigate the approximation properties of S^p on the reference element $\widehat{\Omega}$, then we transport the estimate on $\widehat{\Omega}$ to each element $\Omega_j \in \mathcal{T}$ via the element mapping Q_j, thereby obtaining an estimate of the global error Z and constructing a global approximation $s_0 \in S$ of $u_0 \in X$ from scaled local error estimates and local approximations, respectively. Our estimates are uniform in the meshwidth h_j as well as in the polynomial degree p_j. This will be used in the next section to derive various classically known convergence estimates for the model problem (3.2.1)–(3.2.4) as special cases.

3.3.1 *Basic approximation properties of $\mathcal{S}^p$ on $\widehat{\Omega} = (-1, 1)$*

Let $u_0 \in L^2(\widehat{\Omega})$. Then u_0 can be expanded into a **Legendre** series

$$u_0(\xi) = \sum_{i=0}^{\infty} a_i\, L_i(\xi)\,, \tag{3.3.3}$$

which converges in $L^2(\widehat{\Omega})$, i.e.,

$$\lim_{p\to\infty} \|u_0 - \sum_{i=0}^{p} a_i\, L_i(\xi)\,\|_{L^2(\widehat{\Omega})} = 0\,. \tag{3.3.4}$$

Multiplying (3.3.3) by $L_j(\xi)$ and integrating over $\widehat{\Omega}$, we get from the orthogonality properties of the $L_i(\xi)$ that

$$a_i = \frac{2i+1}{2} \int_{-1}^{1} u_0(\xi)\, L_i(\xi) d\xi \tag{3.3.5}$$

and

$$\|u_0\|^2_{L^2(\widehat{\Omega})} = \sum_{i=0}^{\infty} \frac{2}{2i+1}\, |a_i|^2\,. \tag{3.3.6}$$

Lemma 3.9 *For every $u_0 \in L^2(\widehat{\Omega})$, we have*

$$Z(u_0,\, \mathcal{S}^p,\, L^2(\widehat{\Omega})) = \Big[\sum_{i=p+1}^{\infty} \frac{2}{2i+1}\, |a_i|^2\Big]^{\frac{1}{2}}\,. \tag{3.3.7}$$

Proof Let $v(\xi) \in \mathcal{S}^p$ be any polynomial of degree p. Then

$$v(\xi) = \sum_{i=0}^{p} c_i\, L_i(\xi)$$

for certain c_i and by (3.3.6)

$$\|u_0 - v\|^2_{L^2} = \sum_{i=0}^{p} \frac{2}{2i+1}\, |a_i - c_i|^2 + \sum_{i=p+1}^{\infty} \frac{2}{2i+1}\, |a_i|^2\,.$$

Therefore $\|u_0 - v\|_{L^2}$ is minimal if and only if $c_i = a_i$, $i = 0, \ldots, p$ and (3.3.7) follows. □

We shall be interested in the behavior of the bound (3.3.7) when u_0 has more regularity than merely belonging to L^2. To make this precise, we need

Lemma 3.10 *Let u_0 be as in (3.3.3). Then, for $k \geq 0$,*

$$\int_{-1}^{1} |u^{(k)}(\xi)|^2 (1-\xi^2)^k \, d\xi = \sum_{i\geq k} |a_i|^2 \, \frac{2}{2i+1} \, \frac{(i+k)!}{(i-k)!} \tag{3.3.8}$$

(which is to be understood to mean that the series converges if and only if the left-hand side is finite).

Proof We first note that

$$\int_{-1}^{1} (1-\xi^2)^k \, L_i^{(k)} \, L_j^{(k)} d\xi = \begin{cases} \dfrac{2}{2i+1} \, \dfrac{(i+k)!}{(i-k)!} & \text{if } i = j\,, \\ 0 & \text{otherwise}\,. \end{cases} \tag{3.3.9}$$

To see this, observe that

$$L_i^{(k)}(\xi) = \frac{(i+k)!}{2^k \, i!} \, P_{i-k}(x;k)$$

where $P_{i-k}(x;k)$ is the Jacobi polynomial of degree $i-k$ with weight $(1-\xi^2)^k$. Then (3.3.9) is a consequence of the orthogonality properties of the Jacobi polynomials (see (C.3.13), (C.3.14) in Appendix C).

Next, let $v \in S^p$, $p \geq k$, be any polynomial. Then $v(\xi) = \sum_{i=0}^{p} c_i \, L_i(\xi)$

$$\begin{aligned} \int_{-1}^{1} (1-\xi^2)^k \, |v^{(k)}(\xi)|^2 \, d\xi &= \int_{-1}^{1} (1-\xi^2)^k \left| \sum_{i=0}^{p} c_i \, L_i^{(k)}(\xi) \right|^2 d\xi \\ &= \sum_{i,j=k}^{p} c_i \, \overline{c_j} \int_{-1}^{1} (1-\xi^2)^k \, L_i^{(k)}(\xi) \, L_j^{(k)}(\xi) \, d\xi \\ &= \sum_{i=k}^{p} \frac{2}{2i+1} \, \frac{(i+k)!}{(i-k)!} \, |c_i|^2 \,. \end{aligned}$$

Finally, we select v to be the partial Legendre series up to order p of u_0. Letting $p \to \infty$, (3.3.8) follows. □

Lemma 3.10 shows that certain weighted sums of Legendre coefficients correspond to weighted Sobolev norms on $\widehat{\Omega} = (-1,1)$. The precise regularity necessary for a certain order of approximation by polynomials is therefore characterized by functions belonging to certain weighted Sobolev spaces which we now define.

Let $0 \le j \le k$ be integers. Then $V_j^k(\widehat{\Omega})$ is the space of all $u_0 \in L^2(\widehat{\Omega})$ for which the expression

$$|u_0|^2_{V_j^k(\widehat{\Omega})} = \sum_{i=j}^{k} \int_{-1}^{1} (1-\xi^2)^i \, |u_0^{(i)}(\xi)|^2 \, d\xi \tag{3.3.10}$$

is finite. For $j > 0$, $|\circ|_{V_j^k}$ is a seminorm which vanishes on $\mathcal{S}^{j-1}$. If $j = 0$, the expression in (3.3.10) is actually a norm and we denote the corresponding space by $V^k(\widehat{\Omega})$.

Theorem 3.11 *Let $u_0 \in V^k(\widehat{\Omega})$, $k \ge 1$. Then u_0 can be written in the form (3.3.3). Let further s_0 be the partial Legendre series*

$$s_0 = \sum_{i=0}^{p} a_i \, L_i(\xi)$$

of u_0. Then we have the approximation error estimate

$$Z(u_0, \mathcal{S}^p, L^2) = \|u_0 - s_0\|_{L^2(\widehat{\Omega})} \le \left[\frac{(p+1-s)!}{(p+1+s)!}\right]^{\frac{1}{2}} |u_0|_{V_s^s(\widehat{\Omega})} \tag{3.3.11}$$

for $0 \le s \le \min(p+1, k)$.

Proof By Lemma 3.9

$$\begin{aligned}(Z(u_0, \mathcal{S}^p, L^2))^2 &= \sum_{i=p+1}^{\infty} \frac{2}{2i+1} \, |a_i|^2 \, \frac{(i+s)!}{(i-s)!} \, \frac{(i-s)!}{(i+s)!} \\ &\le \frac{(p+1-s)!}{(p+1+s)!} \sum_{i=p+1}^{\infty} \frac{2}{2i+1} \, \frac{(i+s)!}{(i-s)!} \, |a_i|^2 \, .\end{aligned}$$

Referring to (3.3.8) completes the proof. □

Theorem 3.11 contains in particular the following asymptotic convergence estimate as $p \to \infty$.

Corollary 3.12 *Let $u_0 \in V^k(\widehat{\Omega})$, $k \ge 1$. Then, as $p \to \infty$, for fixed k,*

$$Z(u_0, \mathcal{S}^p, L^2) = \inf_{s \in \mathcal{S}^p} \|u_0 - s\|_{L^2(\widehat{\Omega})} \le C(k) \, p^{-k} \, |u_0|_{V_k^k(\widehat{\Omega})} \, . \tag{3.3.12}$$

Proof Selecting $s = k$ in (3.3.11) and applying Stirling's formula, we get

$$\frac{(p-k)!}{(p+k)!} \le \left(\frac{\theta}{p}\right)^{2k}, \quad \theta = \left(\frac{e}{2}\right)^{k/p} .$$

This and (3.3.11) imply (3.3.12). □

Remark 3.13 p-asymptotic convergence estimates (i.e., as $p \to \infty$) like (3.3.12) are also called **spectral** convergence estimates. The sharper estimate (3.3.11), however, is required for the convergence analysis of the hp-FEM below.

Theorem 3.11 addressed only the case of L^2-approximation. Ahead, however, we also need approximations in $H^1(\widehat{\Omega})$. In order to join the approximations continuously across element boundaries, we need information on the function values of the approximation s_0 at $\xi = \pm 1$.

Theorem 3.14 *Let $\widehat{\Omega} = (-1, 1)$ and $u_0 \in H^1(\widehat{\Omega})$. Then there exists $s_0 = \pi_p u_0 \in \mathcal{S}^p(\widehat{\Omega})$ such that*

$$s_0(\pm 1) = u_0(\pm 1) , \tag{3.3.13}$$

$$\|u_0' - s_0'\|^2_{L^2(\widehat{\Omega})} = \sum_{i=p}^{\infty} \frac{2}{2i+1} |b_i|^2 , \tag{3.3.14}$$

$$\|u_0 - s_0\|^2_{L^2(\widehat{\Omega})} \le \int_{-1}^{1} \frac{(u_0 - s_0)^2}{1 - \xi^2} \, d\xi = \sum_{i=p}^{\infty} \frac{2}{i(i+1)(2i+1)} |b_i|^2 \tag{3.3.15}$$

where b_i are the Legendre coefficients of u_0', i.e.,

$$b_i = \frac{2i+1}{2} \int_{-1}^{1} u_0'(\xi) \, L_i(\xi) \, d\xi , \quad i = 0, 1, \ldots .$$

Proof We let s_0' be the Legendre series of u_0', truncated after the Legendre polynomial L_{p-1}. Then (3.3.14) follows immediately from Lemma 3.9. We prove (3.3.13). Define

$$s_0(\xi) = \int_{-1}^{\xi} s_0'(\eta) \, d\eta + u_0(-1) .$$

Then $s_0(-1) = u_0(-1)$ and

$$u_0(1) - u_0(-1) = \int_{-1}^{1} u_0'(\xi) \, d\xi = 2b_0 ,$$

$$s_0(1) - s_0(-1) = \int_{-1}^{1} s_0'(\xi) \, d\xi = 2b_0 .$$

Subtracting yields

$$0 = u_0(1) - s_0(1) - (u_0(-1) - s_0(-1)) = u_0(1) - s_0(1) ,$$

which proves (3.3.13).

To prove (3.3.15), observe that

$$u_0(\xi) - s_0(\xi) = \int_{-1}^{\xi} \Big\{ \sum_{i=p}^{\infty} b_i\, L_i(t) \Big\}\, dt = \sum_{i=p}^{\infty} b_i \int_{-1}^{\xi} L_i(t)\, dt$$
$$= \sum_{i=p}^{\infty} b_i\, \psi_i(\xi)$$

where

$$\psi_i = \int_{-1}^{\xi} L_i(t)\, dt\ .$$

The Legendre differential equation $((1-\xi^2)\, L_i')' + i(i+1)\, L_i = 0$ becomes, upon integration,

$$-\psi_i(\xi) = \frac{1}{i(i+1)}\, (1-\xi^2)\, L_i'(\xi)$$

and hence we get

$$\int_{-1}^{1} \frac{1}{1-\xi^2}\, \psi_i(\xi)\, \psi_j(\xi)\, d\xi = \frac{1}{i(i+1)\, j(j+1)} \int_{-1}^{1} (1-\xi^2)\, L_i'(\xi)\, L_j'(\xi)\, d\xi\ .$$

By (3.3.9) therefore

$$\int_{-1}^{1} \frac{1}{1-\xi^2}\, \psi_i(\xi)\, \psi_j(\xi)\, d\xi = \frac{2\delta_{ij}}{i(i+1)(2i+1)}\ ,$$

we get

$$\int_{-1}^{1} |u_0(\xi) - s_0(\xi)|^2\, d\xi \le \int_{-1}^{1} \frac{1}{1-\xi^2}\, |u_0(\xi) - s_0(\xi)|^2\, d\xi$$
$$= \int_{-1}^{1} (1-\xi^2)^{-1} \Big(\sum_{i=p}^{\infty} b_i\, \psi_i(\xi) \Big)^2\, d\xi$$
$$= \sum_{i=p}^{\infty} \frac{2}{i(i+1)(2i+1)}\, |b_i|^2\ .$$

□

Theorem 3.14 allows, together with (3.3.8), to obtain error estimates in terms of the $V^k(\widehat{\Omega})$-norms of u_0.

Corollary 3.15 *Let $u_0 \in H^1(\widehat{\Omega}) \cap V^k(\widehat{\Omega})$ for some $k \geq 1$ and $\widehat{\Omega} = (-1,1)$. Then there exists $s_0 = \pi_p u_0 \in S^p(\widehat{\Omega})$ such that (3.3.13) holds and*

$$\begin{aligned} \|u_0' - s_0'\|_{L^2(\widehat{\Omega})} &\leq \left[\frac{(p-s)!}{(p+s)!}\right]^{\frac{1}{2}} |u_0'|_{V_s^s(\widehat{\Omega})} \\ &\leq C_1(k)\, p^{-k}\, |u_0'|_{V_s^k(\widehat{\Omega})} \end{aligned} \tag{3.3.16}$$

where $0 \leq s \leq \min(p,k)$ is arbitrary and

$$\begin{aligned} \left\|\frac{u_0 - s_0}{\sqrt{1-\xi^2}}\right\|_{L^2(\widehat{\Omega})} &\leq \left[\frac{(p-t)!}{(p+t)!}\,\frac{1}{p(p+1)}\right]^{\frac{1}{2}} |u_0'|_{V_t^t(\widehat{\Omega})} \\ &\leq C_2(k)\, p^{-k}\, |u_0'|_{V_t^k(\widehat{\Omega})} \end{aligned} \tag{3.3.17}$$

for any $0 \leq t \leq \min(p,k)$ and finally, selecting $t = s$,

$$\|u_0 - s_0\|_{H^1(\widehat{\Omega})} \leq \sqrt{2}\left[\frac{(p-s)!}{(p+s)!}\right]^{\frac{1}{2}} |u_0'|_{V_s^s(\widehat{\Omega})} \leq C_3(k)\, p^{-k} |u_0'|_{V_s^k(\widehat{\Omega})}\,. \tag{3.3.18}$$

Proof By (3.3.14), we may estimate as in the proof of Theorem 3.11

$$\|u_0' - s_0'\|^2_{L^2(\widehat{\Omega})} \leq \frac{(p-s)!}{(p+s)!} \sum_{i=p}^{\infty} \frac{2}{2i+1}\,\frac{(i+s)!}{(i-s)!}\,|b_i|^2 = \frac{(p-s)!}{(p+s)!}\,|u_0'|^2_{V_s^s(\widehat{\Omega})}$$

by (3.3.8). This is the first bound in (3.3.16). The second one is obtained from it using $s = \min(p,k)$ and Stirling's formula. The proof of (3.3.17) is analogous and (3.3.18) follows from (3.3.16) and (3.3.17). □

3.3.2 *The rate of convergence of the hp-FEM*

We will now discuss the rate of convergence of the h-version, p-version and the hp-version for the model problem (3.2.1), (3.2.2), with $AE = 1$, $c = 1$ and homogeneous Dirichlet end conditions, i.e., for

$$-u'' + u = f \text{ in } \Omega = (-1,1), \quad u(\pm 1) = 0\,. \tag{3.3.19}$$

We adopt the following symmetric variational formulation: find $u_0 \in X = H_0^1(\Omega)$ such that

$$B(u_0, v) = \int_{-1}^{1} (u_0' v' + u_0 v)dx = \int_{-1}^{1} f v dx =: F(v), \quad v \in H_0^1(\Omega)\,. \tag{3.3.20}$$

For every $f \in L^2(\Omega)$, (3.3.20) admits a unique weak solution u_0 by the Lax–Milgram lemma, Theorem 1.24, and for the FE solutions u_S based on $S = V = S_0^{p,1}(\Omega, \mathcal{T})$ and (3.1.7) we get

$$\|u_0 - u_S\|_{H^1(\Omega)} \leq Z(u_0, S_0^{p,1}(\Omega, \mathcal{T}), H^1) = \inf_{v \in S_0^{p,1}(\Omega,\mathcal{T})} \|u_0 - v\|_{H^1(\Omega)} \, . \quad (3.3.21)$$

Therefore the rate of convergence of the FEM is completely determined by the **approximability** of the exact solution u_0 from the space $S_0^{p,1}(\Omega, \mathcal{T})$ which we study in the present section.

Two cases will be of particular interest: the case of a smooth (analytic) solution u_0 and the **singular solution**

$$u_0(x, x_0) = \begin{cases} 0 & -1 < x < x_0 \\ (x - x_0)^\alpha & x_0 < x < 1 \, . \end{cases} \quad (3.3.22)$$

This behavior is typical for solutions to elliptic problems in curvilinear polygons near corners, as we will see in Chapter 4.

Recall that the ***h*-version FEM** consists in keeping the polynomial degree constant and fixed, i.e.,

$$p_j = p, \quad j = 1, \ldots, M(\mathcal{T}) \quad (3.3.23)$$

and in letting the meshwidth $h(\mathcal{T})$ tend to zero. The ***p*-version FEM** consists in letting the polynomial degrees $p_j \to \infty$ on a fixed mesh $\mathcal{T}$. The ***hp*-version FEM** consists in a combination of both, mesh refinement and increase of p (and hence contains both h-version FEM and p-version FEM as a special case). We will derive now error estimates which are uniform both in h and in p.

We will in particular consider meshes which are **quasiuniform**.

Definition 3.16 *A family of meshes $\{\mathcal{T}_j\}$ in $\Omega = (-1, 1)$ is* **quasiuniform**, *if there exist positive constants c_1, c_2 independent of j such that*

$$0 < c_1 \leq \frac{\max\limits_{1 \leq i \leq M(\mathcal{T}_j)} \{h_i^{\mathcal{T}_j}\}}{\min\limits_{1 \leq i \leq M(\mathcal{T}_j)} \{h_i^{\mathcal{T}_j}\}} = \frac{h(\mathcal{T}_j)}{\underline{h}^{\mathcal{T}_j}} \leq c_2 < \infty \quad (3.3.24)$$

In particular, equidistant nodes $x_j = -1 + 2j/M$ give uniform meshes with $c_1 = c_2 = 1$.

Theorem 3.17 *Let $\Omega = (a, b) \subset \mathbb{R}$ be an interval and let $\mathcal{T}$ be any mesh in Ω. Assume that $u_0 \in H^1(\Omega)$ satisfies*

$$u_0' \in H^{k_i}(\Omega_i) \text{ for } \Omega_i \in \mathcal{T} \text{ and some } k_i \geq 1 \, .$$

Then there exists $s_0 = \pi_{\mathbf{p}} u_0 \in S^{\mathbf{p},1}(\Omega,\mathcal{T})$ *such that*

$$u_0(x_i) = s_0(x_i), \quad i = 0,\dots,M$$

and such that

$$\|u_0' - s_0'\|^2_{L^2(\Omega)} \le \sum_{i=1}^{M} \left(\frac{h_i}{2}\right)^{2s_i} \frac{(p_i - s_i)!}{(p_i + s_i)!} \|u_0'\|^2_{H^{s_i}(\Omega_i)}, \tag{3.3.25}$$

$$\|u_0 - s_0\|^2_{L^2(\Omega)} \le \sum_{i=1}^{M} \left(\frac{h_i}{2}\right)^{2t_i+2} \frac{(p_i - t_i)!}{p_i(p_i+1)(p_i + t_i)!} \|u_0'\|^2_{H^{t_i}(\Omega_i)}, \tag{3.3.26}$$

where $0 \le s_i, t_i \le \min(p_i, k_i)$.

If the polynomial degrees are constant, $p_i = p$, *and the mesh is quasiuniform, we get for* $k = \min_{1\le i\le M}\{k_i\} = s_i = t_i$ *in* (3.3.25), (3.3.26) *that*

$$\|u_0' - s_0'\|_{L^2(\Omega)} \le C\, \frac{h^{\min(p,k)}}{p^k} \|u_0'\|_{H^k(\Omega)}, \tag{3.3.27}$$

$$\|u_0 - s_0\|_{L^2(\Omega)} \le C\, \frac{h^{\min(p,k)+1}}{p^{k+1}} \|u_0'\|_{H^k(\Omega)} \tag{3.3.28}$$

where $h = h(\mathcal{T})$ and C is independent of h and p, but depends on k.

Proof We construct the projection $\pi_{\mathbf{p}} u_0$ elementwise. Let $\Omega_i \in \mathcal{T}$ be any element. A translation and a scaling transforms Ω_i to the reference element $\widehat{\Omega} = (-1,1)$, and the function $u_0 \in H^{k_i+1}(\Omega_i)$ into $\hat{u}_{0i} \in H^{k_i+1}(\widehat{\Omega})$. On $\widehat{\Omega}$, we define the approximation $\hat{s}_{0i} = \pi_{p_i}\hat{u}_0$ as in Theorem 3.14 and obtain from Corollary 3.15 the error estimate

$$\begin{aligned}\|\hat{u}_{0i}' - \hat{s}_{0i}'\|^2_{L^2(\widehat{\Omega})} &\le \frac{(p_i - s_i)!}{(p_i + s_i)!} \int_{-1}^{1} (1-\xi^2)^{s_i}\, |\hat{u}_{0i}^{(s_i+1)}(\xi)|^2\, d\xi \\ &\le \frac{(p_i - s_i)!}{(p_i + s_i)!} \|\hat{u}_{0i}^{(s_i+1)}\|^2_{L^2(-1,1)}\end{aligned}$$

for any $0 \le s_i \le \min(p_i, k_i)$.

Scaling back to Ω_i gives

$$\|u_{0i}' - s_{0i}'\|^2_{L^2(\Omega_i)} \le \left(\frac{h_i}{2}\right)^{2s_i} \frac{(p_i - s_i)!}{(p_i + s_i)!} \|u_{0i}^{(s_i+1)}\|^2_{L^2(\Omega_i)}. \tag{3.3.29}$$

In the same way we get from (3.3.17) for any $0 \le t_i \le \min(p_i, k_i)$

$$\|u_{0i} - s_{0i}\|^2_{L^2(\Omega_i)} \le \left(\frac{h_i}{2}\right)^{2t_i+2} \frac{(p_i - t_i)!}{p_i(p_i+1)(p_i+t_i)!} \|u_{0i}^{(t_i+1)}\|^2_{L^2(\Omega_i)} . \tag{3.3.30}$$

By (3.3.13), we have also

$$s_{0i}(x_{i-1}) = u_{0i}(x_{i-1}), \quad s_{0i}(x_i) = u_{0i}(x_i), \quad 1 \le i \le M .$$

Summing over all elements gives (3.3.25), (3.3.26). The assertions (3.3.27), (3.3.28) follow from this by Stirling's formula. □

Remark 3.18 The error estimates (3.3.25)–(3.3.28) are uniform in h and p. They show that the FEM based on $S^{p,1}(\Omega, \mathcal{T})$ applied to (3.3.20) converges either as the mesh is refined ($h \to 0$) or as the polynomial degree is increased ($p \to \infty$). They also show that for smooth solutions u_0, for which k is large, it is more advantageous to increase p rather than to reduce h at fixed, low p. For, taking $N = \dim(S^{p,1}(\Omega, \mathcal{T}))$ as a work measure, we have from (3.3.28) for h-refinement (at fixed p) for $u_0 \in H^k$

$$\|u_0 - s_0\|_{L^2(\Omega)} \le CN^{-\min(p,k)-1}$$

and for p-refinement (at fixed h)

$$\|u_0 - s_0\|_{L^2(\Omega)} \le CN^{-k-1} .$$

If k is large, it is advantageous to let $N \to \infty$ by raising p.

3.3.3 *Analytic solutions*

Theorem 3.17 shows that for solutions u_0 which are **smooth** in $\overline{\Omega}$, arbitrary high **algebraic** rates of convergence are possible if the polynomial degree p is raised. This is sometimes also referred to as **spectral convergence**. It turns out, however, that the p-version converges in fact **exponentially** if the solution is **analytic** in $\overline{\Omega}$.

Assume that $\Omega = (-1, 1)$ and that u_0 is analytic in $\overline{\Omega} = [-1, 1]$. Then u_0 admits a unique analytic continuation, again denoted by u_0, to some open neighborhood of $\overline{\Omega}$ in the complex plane $\mathbb{C}$. Let $\mathcal{E}_r \supset [-1, 1]$ denote the ellipse with foci ± 1 and semiaxes' sum $r > 1$. We have

Theorem 3.19 *Let u_0' be analytic in the closure of $\mathcal{E}_r$, $r > 1$.*

Then there holds for $s_0 = \pi_p u_0 \in S^p$, $p = 1, 2, 3, \ldots$

$$s_0(\pm 1) = u_0(\pm 1)$$

$$\|u_0' - s_0'\|^2_{L^2(\widehat{\Omega})} \le C(r)\, p\, r^{-2p} \tag{3.3.31}$$

$$\|u_0 - s_0\|^2_{L^2(\widehat{\Omega})} \le C(r)\, p^{-1}\, r^{-2p} . \tag{3.3.32}$$

Proof We have for $p \geq 1$

$$s_0'(\xi) = \sum_{i=0}^{p-1} b_i\, L_i(\xi), \quad b_i = \frac{2i+1}{2} \int_{-1}^{1} u_0'(\xi)\, L_i(\xi)\, d\xi\,.$$

By [52], Theorem 12.4.7, it holds that

$$|b_i| \leq C(r)(2i+1)\, r^{-i} \tag{3.3.33}$$

for some $C(r)$ independent of i. Inserting this bound into (3.3.14), we get

$$\|u_0' - s_0'\|^2_{L^2(\widehat{\Omega})} \leq C^2(r) \sum_{i=p}^{\infty} (2i+1)\, r^{-2i} \leq C^2(r) \int_{x=p}^{\infty} x e^{-\mu x}\, dx$$

where $\mu = 2 \ln r > 0$. Integrating by parts proves (3.3.31).

For (3.3.32) we use (3.3.33) and (3.3.15) and get

$$\|u_0 - s_0\|^2_{L^2(\widehat{\Omega})} \leq C^2(r) \int_{x=p}^{\infty} x^{-1}\, e^{-\mu x}\, dx \leq C^2(r)\, p^{-1} \int_{x=p}^{\infty} e^{-\mu x}\, dx\,.$$

□

This concludes the basic approximation properties of the truncated Legendre series on the reference element $[-1,1]$. At times, however, it is convenient to have an interpolant of u_0 rather than a truncated series expansion. If the interpolation points are judiciously chosen, the interpolants are of similar quality as the best approximations; this is illustrated by the following result which we give without proof. Put

$$\eta = \ln r > 0 \text{ and } M(\eta) = \max_{z \in \mathcal{E}_r} |u_0(z)|\,.$$

Theorem 3.20 *Let u_0 be analytic in $[-1,1]$ and admit an analytic continuation to the ellipse $\mathcal{E}_r \subset \mathbb{C}$ for some $r > 1$. Let $x_j := \cos(j\pi/p)$, $j = 0, \ldots, p$, and let $\widetilde{s}_0$ denote the polynomial of degree p interpolating u_0 in $x_j, j = 0, \ldots, p$. Then there holds*

$$\widetilde{s}_0(\pm 1) = u_0(\pm 1)$$

and for $k = 0, 1, 2, \ldots$

$$\int_{-1}^{1} \frac{|u_0^{(k)}(x) - \widetilde{s}_0^{(k)}(x)|^2}{(1-x^2)^{1/2}}\, dx \leq C(k)\, \frac{M(\eta)}{\sinh(\eta)}\, p^{2k}\, e^{-\eta p}, \quad 0 < \eta < \eta_0 = \ln r$$

where C is independent of η and p.

For a proof, we refer for example to [147].

The preceding theorem implies in particular also estimates for $\|u_0^{(k)} - s_0^{(k)}\|_{L^2(-1,1)}$.

3.3.4 *Singular solutions. h-version FEM*

We address next the rate of convergence of the h-version FEM for the singular solution $u_0(x, x_0)$ in (3.3.22). We observe

$$u_0(x, x_0) \in H^k(\Omega) \text{ if } k < \alpha + \frac{1}{2}\,.$$

This, however, would imply in (3.3.25) no convergence for $\alpha < 3/2$. Since, however, $u_0(x, x_0) \in H^1(\Omega)$ for $\alpha > \frac{1}{2}$, we expect convergence in this range. Thus, we need (3.3.25) for noninteger k. This can be inferred from (3.3.25) by interpolation theory (we refer to Appendix B for background information on interpolation theory).

Lemma 3.21 *Let $k = \kappa + \theta$, $0 < \theta < 1$, $\kappa \geq 1$ an integer. Assume* (3.3.23), (3.3.24), *i.e., that $\mathcal{T}$ is quasiuniform and that the elemental polynomial degrees p_j are equal to p. Then there exists $s_0 \in S^{p,1}(\Omega, \mathcal{T})$ such that*

$$\|u_0 - s_0\|_{H^1(\Omega)} \leq C(k)\, \frac{h^{\min(p,k-1)}}{p^{k-1}}\, \|u_0\|_{[H^\kappa(\Omega), H^{\kappa+1}(\Omega)]_{\theta,\infty}}\,. \tag{3.3.34}$$

Proof All spaces H^k are taken over Ω and we omit this dependence in the proof. Let $0 < \theta < 1$ and let T denote the error operator

$$T : u_0 \longmapsto u_0 - s_0$$

based on the approximation s_0 constructed in the proof of Theorem 3.17. Then (3.3.25) says that T is linear and bounded from $H^s \to H^1$ for $s = \kappa, \kappa + 1$. Therefore, since $k = (1 - \theta)\,\kappa + \theta(\kappa + 1)$, we may use interpolation of linear operators and obtain

$$\|T\|_{[H^\kappa, H^{\kappa+1}]_{\theta,\infty} \to H^1} \leq \|T\|^{1-\theta}_{H^\kappa \to H^1}\; \|T\|^{\theta}_{H^{\kappa+1} \to H^1}\,,$$

and we have for $u_0 \in [H^\kappa, H^{\kappa+1}]_{\theta,\infty}$ that

$$\|u_0 - s_0\|_{H^1} \leq C \left(\frac{h^{\min(p,\kappa)}}{p^\kappa}\right)^{1-\theta} \left(\frac{h^{\min(p,\kappa+1)}}{p^{\kappa+1}}\right)^{\theta} \|u_0\|_{[H^\kappa, H^{\kappa+1}]_{\theta,\infty}}\,.$$

Since $\theta + \kappa = k$, the proof is complete. □

Remark 3.22 The same proof (yielding the same convergence rates) works also for the interpolation space $[H^\kappa, H^{\kappa+1}]_{\theta,q}$ with **any** $q \in [1, \infty]$. However, since $[H^\kappa, H^{\kappa+1}]_{\theta,q} \subseteq [H^\kappa, H^{\kappa+1}]_{\theta,\infty}$, for all q, the selection $q = \infty$ yields the sharpest result. The space $[H^k, H^{k+1}]_{\theta,\infty}$ is a Besov space which characterizes the minimal regularity necessary for a certain rate of approximation by polynomials. For more on Besov spaces and interpolation theory we refer to Appendix B and the references there.

It remains to find the largest k such that $u_0(x,x_0) \in [H^\kappa, H^{\kappa+1}]_{\theta,\infty}$ for $k = \kappa+\theta$, $0 \le \theta \le 1$, $\kappa \in \mathbb{N}_0$.

Lemma 3.23 *Let $\Omega = (a,b)$ be a bounded interval and*

$$u_0(x;\alpha,x_0) = \begin{cases} (x-x_0)^\alpha & x > x_0 \in [a,b) \\ 0 & \textit{else} . \end{cases}$$

Then $u_0 \in [H^\kappa, H^{\kappa+1}]_{\theta,\infty}$ for $\kappa < \kappa+\theta \le \alpha+\frac{1}{2} < \kappa+1$ with $\kappa \in \mathbb{N}_0$.

Proof Since

$$\frac{d^\kappa u_0}{dx^\kappa}(x;\alpha,x_0) = C(\kappa,\alpha)\, u_0(x;\alpha-\kappa,x_0)$$

we need only to check the case $\kappa = 0$. Further, by translation and scaling, it is sufficient to consider

$$x_0 = 0,\ \ \Omega = (0,1),\ \ u_0(x;\beta,0),\ -\frac{1}{2} < \beta < \frac{1}{2}\,.$$

We claim that $u_0(x;\beta,0) \in [H^0,H^1]_{\theta,\infty}$ for $0 \le \theta \le \beta+\frac{1}{2} < 1$. To prove it, note (cf. Appendix B) that

$$\|u_0\|_{[H^0,H^1]_{\theta,\infty}} = \sup_{t>0} t^{-\theta} K(t,u_0),\ \ 0 \le \theta \le 1$$

where the ***K*-functional** is given by

$$K(t,u_0) := \inf_{\substack{u_0=v_0+v_1 \\ v_i \in H^i}} \left(\|v_0\|_{H^0} + t\,\|v_1\|_{H^1}\right),$$

the infimum being taken over all possible splittings of u_0. We obtain an upper bound by constructing a particular splitting. Let $0<\delta<\frac{1}{2}$ be a parameter and let $\psi_\delta(x)$ be the function depicted in Figure 3.1. .

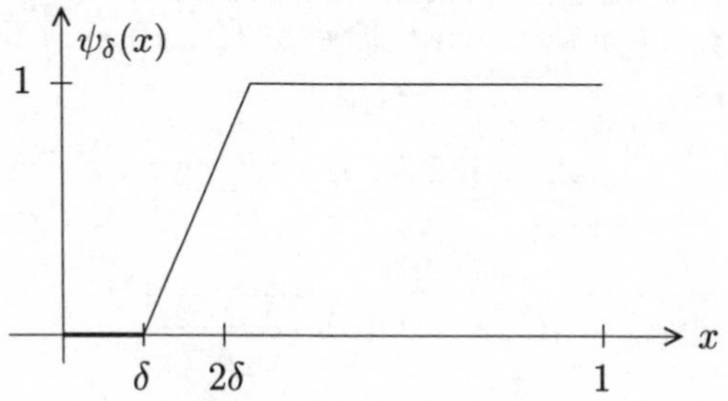

FIG. 3.1. The function $\psi_\delta(x)$

Then we select

$$v_1(x) = \psi_\delta(x)\, u_0(x;\beta,0),\ v_0(x) = u_0 - v_1 .$$

We estimate with $\beta > -\frac{1}{2}$

$$\|v_0\|_{L^2}^2 \le \int_0^{2\delta} |u_0|^2\, dx = \int_0^{2\delta} x^{2\beta}\, dx = \frac{1}{2\beta+1}\, \delta^{2\beta+1}$$

and, using $\beta < \frac{1}{2}$,

$$\|v_1\|_{H^1}^2 \le C\, \delta^{2\beta-1} .$$

Therefore, for $-\frac{1}{2} < \beta < \frac{1}{2}$, we estimate

$$K(t,u_0) \le C_\beta\, (\delta^{\beta+\frac{1}{2}} + t\, \delta^{\beta-\frac{1}{2}}) .$$

For $0 < t < \frac{1}{2}$ we select $\delta = t$, yielding $K(t,u_0) \le C_\beta\, t^{\beta+\frac{1}{2}}$. For $t \ge \frac{1}{2}$ we select $v_1 = 0$ and $v_0 = u_0$. Then

$$\sup_{t>0}\ t^{-\theta}\ K(t,u_0) < \infty$$

if $0 \le \theta \le \beta + \frac{1}{2} \in (0,1)$. □

Combining Lemmas 3.21 and 3.23, we can prove the following result.

Theorem 3.24 *Let* $\Omega = (a,b)$ *and*

$$u_0(x;\alpha,x_0) = \begin{cases} (x-x_0)^\alpha & x > x_0 \in [a,b) \\ 0 & \textit{else} \end{cases} \tag{3.3.35}$$

where $2\alpha \notin \mathbb{N}$. *Let* $\mathcal{T}$ *be quasiuniform with meshwidth* h *and let the degree vector* $\boldsymbol{p}$ *be uniform,* $p_i = p$. *Then there exists* $s_0 \in S^{\boldsymbol{p},1}(\Omega,\mathcal{T})$ *such that*

$$\|u_0 - s_0\|_{H^1(\Omega)} \le C(\alpha)\, \frac{h^{\min(p,\alpha-\frac{1}{2})}}{p^{\alpha-\frac{1}{2}}} , \tag{3.3.36}$$

$$\|u_0 - s_0\|_{L^2(\Omega)}\ \le C(\alpha)\, \frac{h^{\min(p,\alpha-\frac{1}{2})+1}}{p^{\alpha+\frac{1}{2}}} . \tag{3.3.37}$$

If $1 < 2\alpha \in \mathbb{N}$, (3.3.36), (3.3.37) hold with the exponents reduced by an arbitrarily small $\varepsilon > 0$ and $C = C(\alpha,\varepsilon)$.

Proof The case $2\alpha \notin \mathbb{N}$ follows directly from the mentioned lemmas. For $1 < 2\alpha \in \mathbb{N}$, we find with the argument in the proof of Lemma 3.23 that $u_0(x;1/2,0) \in [H^0,H^1]_{\theta,2} = H^\theta(0,1)$ for every $0<\theta<1$ which can be verified by estimation of the Slobodečkij norm of $u_0 = \sqrt{x}$. □

Exercise 3.25 Show that

$$\|\sqrt{x}\|_{H^\theta(0,1)} < \infty \ \text{ for } 0<\theta<1$$

by estimating the Slobodeckij norm of $\sqrt{x}$.

Solution: We have for $\Omega = (0,1)$ and $\beta > -\frac{1}{2}$ that

$$x^\beta \in H^\theta(\Omega) \Longleftrightarrow I(\theta) = \int_0^1\int_0^1 \frac{(x^\beta - y^\beta)^2}{(x-y)^{1+2\theta}}\, dx\, dy < \infty\,.$$

Introducing polar coordinates, we get the bound

$$\begin{aligned} I(\theta) &\le \int_0^{\sqrt{2}}\int_0^{\pi/2} r^{2\beta-1-2\theta}\frac{|\cos^\beta\varphi - \sin^\beta\varphi|^2}{|\cos\varphi - \sin\varphi|^{1+2\theta}}\, r\, dr\, d\varphi \\ &= \int_0^{\sqrt{2}} r^{2(\beta-\theta)} dr \int_0^{\pi/2} \frac{|\cos^\beta\varphi - \sin^\beta\varphi|^2}{|\cos\varphi - \sin\varphi|^{1+2\theta}}\, d\varphi\,. \end{aligned}$$

The first integral exists iff $2(\beta-\theta) > -1$, i.e., for $0<\theta<\beta+1/2$.

We claim that the second integral exists for $\theta < 1$ and all β. To see it, let φ_0 be the only root of the equation $\sin\varphi_0 = \cos\varphi_0$ in the interval $(0,\pi/2)$ and $\delta > 0$ a sufficiently small, fixed parameter. Then, for $\varphi \in (\varphi_0-\delta, \varphi_0+\delta)$, we have

$$|\cos^\beta\varphi - \sin^\beta\varphi| \le c_\beta\,|\varphi-\varphi_0|\,, \quad |\cos\varphi - \sin\varphi| > c|\varphi-\varphi_0|\,.$$

Hence

$$\int_{\varphi_0-\delta}^{\varphi_0+\delta} \frac{|\cos^\beta\varphi - \sin^\beta\varphi|^2}{|\cos\varphi - \sin\varphi|}\, d\varphi \le C\int_0^\delta \xi^{2-1-2\theta}\, d\theta < \infty$$

for $\theta < 1$. Therefore $x^\beta \in H^\theta(0,1)$ if $0<\theta<\beta+1/2$, $\theta<1$. □

3.3.5 *Singular solutions. p-version FEM*

We show now that for the ***p*-version FEM** (i.e., the mesh $\mathcal{T}$ is fixed and $p\to\infty$) and the singular solution u_0 in (3.3.41) one gets **twice** the convergence rate (3.3.36), provided x_0 is a mesh point.

Theorem 3.26 *Let $\Omega = (a, b)$ and $x_0 \in \overline{\Omega}$. Let the mesh $\mathcal{T}$ be fixed and assume that x_0 is a mesh point. Assume further $p_j = p$ for $j = 1, \dots, M(\mathcal{T})$ and that u_0 is a singular solution of the form*

$$u(x) = u_0(x; \alpha, \gamma, x_0) = |\ln|x - x_0||^\gamma |x - x_0|^\alpha \psi(x), \ \alpha > \frac{1}{2}, \ \gamma \geq 0 \quad (3.3.38)$$

where $\psi(x)$ is a C^∞ cut-off function, $\psi \equiv 1$ for $|x - x_0| \in (0, \beta]$, $\psi \equiv 0$ for $|x - x_0| > 2\beta$, $\beta > 0$ sufficiently small. Then there is $u_p \in S^{p,1}(\Omega, \mathcal{T})$ such that

$$||u - u_p||_{H^s(\Omega)} \leq C\, p^{-2(\alpha + \frac{1}{2} - s)} |\ln p|^\gamma, \ 0 \leq s \leq 1 \quad (3.3.39)$$

and

$$u(a) = u_p(a) = 0, \ u(b) = u_p(b) = 0\,, \quad (3.3.40)$$

with C independent of p.

The proof of Theorem 3.20 is somewhat technical and will be given in several lemmas.

It is only necessary to estimate the error stemming from the element containing the singularity; the error from the remaining elements is asymptotically smaller than (3.3.38) since the singular function is analytic there (see Theorem 3.20).

Assume therefore that $\Omega = (-1, 1)$ and that

$$u(x) = (x + 1)^\alpha\, |\ln(1 + x)|^\alpha\, \psi(x), \ \ -1 < x < 0\,. \quad (3.3.41)$$

Divide u by $\xi(x) = x + 1$ and define

$$u_0(x) = u(x)/\xi(x) = |\ln(1 + x)|^\gamma\, (1 + x)^{\alpha - 1}\, \psi(x)\,.$$

Let $\omega \in C^\infty(\mathbb{R})$ be a cut-off function such that

$$\omega(t) = 0 \text{ for } t \leq 1, \ \omega(t) = 1 \text{ for } t \geq 2\,. \quad (3.3.42)$$

Then, for a small parameter $\delta > 0$, we write

$$u_0 = u_0\, \omega^\delta + u_0(1 - \omega^\delta) = v + w, \ \omega^\delta(x) = \omega((1 + x)/\delta)\,. \quad (3.3.43)$$

Our final approximation u_p will be obtained from an approximation to u_0 multiplied by $\vartheta(x)$ to enforce the condition (3.3.40). The approximation to u_0 will be constructed by approximating v in (3.3.43) (which is C^∞ since supp $v \subset (-1 + \delta, 1)$) by a truncated expansion into Jacobi polynomials denoted v_p and by approximating w (which has small support supp $w \subset (-1, -1 + 2\delta)$) by zero.

Remark 3.27 The splitting (3.3.43) of u_0 is similar to what we did in Lemma 3.23 in order to estimate the K-functional of u_0. The remainder of the proof of Theorem 3.26 will, however, never directly appeal to interpolation spaces.

We note the following properties of v and w. For all $k \ge 0$

$$\begin{aligned} \left|\frac{d^k v}{dx^k}\right| &\le C(k)\,|\ln \delta|^{\gamma}\,(1+x)^{\alpha-1-k} && x \in [-1+\delta, 1)\,, \\ &= 0 && x \in (-1, -1+\delta) \end{aligned} \tag{3.3.44}$$

and for $k = 0, 1$

$$\begin{aligned} \left|\frac{d^k(\xi w)}{dx^k}\right| &\le C(k)\,|\ln \delta|\,\ell n(1+x)^{\gamma}\,(1+x)^{\alpha-k} && x \in [-1, -1+2\delta)\,, \\ &= 0 && x \ge -1+2\delta\,. \end{aligned} \tag{3.3.45}$$

Now expand $v(x)$ into Jacobi polynomials, i.e.,

$$v(x) = \sum_{i=0}^{\infty} a_i\,P_i(x)\,, \tag{3.3.46}$$

$P_i(x) = P_i(x;\beta,\beta), \beta > -\frac{1}{2}$ where the index β of the polynomials is to be determined, and where

$$a_i = C_i\,(i+1)\int_{-1}^{1} v(x)\,P_i(x)(1-x^2)^{\beta}\,dx$$

with certain C_i which are uniformly bounded in i. Define

$$v_p(x) = \sum_{i=0}^{p} a_i\,P_i(x)\,, \qquad x \in \Omega\,. \tag{3.3.47}$$

Then the error $v - v_p$ can be bounded similar to that of Legendre series (corresponding to $\beta = 0$). We have the following result.

Lemma 3.28 *Let $V_\beta^\ell(\Omega)$ be the set of all v of the form (3.3.46) for which*

$$|v|_{V_\beta^k(\Omega)} = \Big(\int_\Omega |v^{(k)}(x)|^2\,\omega_{k+\beta}(x)\,dx\Big)^{\frac{1}{2}} < \infty,\ k = 0,1,2,\ldots,\ell\,,$$

with the weight $\omega_\lambda(x) = (1-x^2)^\lambda$. Then, for v_p as in (3.3.47) and $v \in V_\beta^\ell$, we have

$$|v - v_p|_{V_\beta^k(\Omega)} \le C\,p^{-(\ell-k)}|v|_{V_\beta^\ell(\Omega)}, \quad k = 0, 1\,, \tag{3.3.48}$$

$p \ge \ell \ge k$, $C = C(\beta,k,\ell)$ is independent of p and v.

Exercise 3.29 Using the basic properties of the Jacobi polynomials in Appendix C, prove Lemma 3.28. Proceed as in Section 3.1 (i.e., the case $\beta = 0$).

We combine now (3.3.48), (3.3.44) to obtain

Lemma 3.30 *For $k = 0, 1$ and v as in* (3.3.44),

$$|v - v_p|_{V^k_\beta(\Omega)} \le Cp^{-(\ell-k)}\,|\ln\,\delta|^\gamma\,\delta^{\alpha-\frac{1}{2}-\frac{\ell}{2}+\frac{\beta}{2}} \tag{3.3.49}$$

provided ℓ is sufficiently large that $\alpha - \frac{1}{2} - \frac{\ell}{2} + \frac{\beta}{2} < 0$.

Proof By (3.3.48), for $p \ge \ell$,

$$\begin{aligned}|v - v_p|^2_{V^k_\beta(I)} &\le C\,p^{-2(\ell-k)} \int_{-1}^{1} |v^{(\ell)}(x)|^2\,\omega_{\ell+\beta}(x)\,dx \\ &\le C\,p^{-2(\ell-k)}\,|\ln\,\delta|^\gamma \int_{-1+\delta}^{1} (1+x)^{2\alpha-2-2\ell+\ell+\beta}\,dx\end{aligned}$$

where we used (3.3.44). (3.3.49) follows by integration. If $p < \ell$, (3.3.49) still holds possibly with a larger constant. □

For any function $f(x)$ on Ω and $0 < \delta < \frac{1}{4}$, define

$$f_\delta(x) = \begin{cases} f(x - 2\delta) & x > -1 + 2\delta \\ 0 & x < -1 + 2\delta\,. \end{cases}$$

Then for $\xi(x) = 1 + x$ we see that on $\widetilde{\Omega}_j = (-1 + 2\delta, 2\delta)$,

$$|\xi_\delta(x)| \le C(1 - x^2),\ |\xi'_\delta(x)| \le 1\,. \tag{3.3.50}$$

Lemma 3.31 *Let $\delta = p^{-2}$, v satisfy* (3.3.44) *and let v_p be given by* (3.3.47), *with $\beta \ge 2$. Then there exists C independent of p such that for $0 \le s \le 1$*

$$A_s := \|\xi_\delta(v - v_p)\|_{H^s(\widetilde{\Omega}_\delta)} \le C\,|\ln\,p|^\gamma\,p^{-2(\alpha+\frac{1}{2}-s)}\,. \tag{3.3.51}$$

Proof Note that on $\widetilde{\Omega}_j$, $0 < C \le (1 - x^2)/(2\delta)$. Using (3.3.50) we obtain for $\beta \ge 2$

$$\begin{aligned}A_0^2 = \int_{-1+2\delta}^{2\delta} (\xi_\delta(v - v_p))^2\,dx &\le \int_{-1}^{1} \frac{(1-x^2)^\beta}{\delta^{\beta-2}}\,(v - v_p)^2\,dx \\ &\le C\,p^{-2\ell}\,|\ln\,\delta|^{2\gamma}\,\delta^{2\alpha-1-\ell+\beta}/\delta^{\beta-2}\end{aligned}$$

by Lemma 3.30 with $k=0$. Selecting $\delta=p^{-2}$ yields (3.3.51) for $s=0$. In the same way

$$\|\xi_\delta(v-v_p)'\|^2_{L^2(\widetilde{\Omega}_\delta)} = \int_{-1}^{1} \frac{(1-x^2)^{\beta+1}}{\delta^{\beta-1}} (v'-v_p')^2\,dx$$
$$\leq C\,p^{-2(\ell-1)}|\ell n\,\delta|^{2\gamma}\,\delta^{2\alpha-1+\beta-\ell}/\delta^{\beta-1}$$

from Lemma 3.30 with $k=1$. With $\delta=p^{-2}$ we get

$$\|\xi_\delta(v-v_p)'\|_{L^2(\widetilde{\Omega}_\delta)} \leq C\,|\ln\,p|^{2\gamma}\,p^{-4\alpha+2}\,.$$

Finally, using a similar reasoning, we prove

$$\|\xi_\delta'(v-v_p)\|_{L^2(\widetilde{\Omega}_\delta)} \leq C\,|\ln\,p|^{2\gamma}\,p^{-4\alpha+2}\,.$$

This proves (3.3.51) for $s=1$. The general case follows by interpolation. □

Lemma 3.31 allows us to estimate $\|\xi_\delta(v-v_p)\|_{H^s(\widetilde{\Omega}_\delta)}$. Our goal, however, is to estimate $\|\xi(v-v_p)\|_{H^s(\widetilde{\Omega})}$ or, equivalently, $\|\xi_\delta(v_\delta-v_{p\delta})\|_{H^s(\widetilde{\Omega}_\delta)}$. To do this, we need an estimate for $\|\xi_\delta(v_\delta-T_k(v_\delta))\|_{H^s(\widetilde{\Omega}_\delta)}$ where $T_k(v_\delta)$ is a Taylor expansion of v_δ.

Let $v^{[i]}$ be the i th derivative of v, $0\leq i\leq k$, and $k=[\alpha+1/2]$ be the largest integer $\leq \alpha+\frac{1}{2}$. Then $v^{[i]}$ satisfies (3.3.44) with α replaced by $\alpha-i$. Using Lemma 3.31, we get for each $v^{[i]}$ a $v_p^{[i]}$ such that

$$\|\xi_\delta(v^{[i]}-v_p^{[i]})\,\|_{H^s(\widetilde{\Omega}_\delta)} \leq C\,|\ln\,p|^{\gamma}\,p^{-2(\alpha-i+\frac{1}{2}-s)}\,. \tag{3.3.52}$$

Define

$$T_k(v_\delta) = \sum_{i=0}^{k} (-1)^i\,\frac{(2\delta)^i}{i!}\,v^{[i]}\,. \tag{3.3.53}$$

Lemma 3.32 *Let $\delta=p^{-2}$. Then for $k=[\alpha+\frac{1}{2}]$ and $0\leq s\leq 1$,*

$$\|\xi_\delta(v_\delta-T_k(v_\delta))\|_{H^s(\widetilde{\Omega}_\delta)} \leq C\,|\ln\,p|^{\gamma}\,p^{-2(\alpha-s+\frac{1}{2})}\,.$$

Here C is independent of p and δ.

Proof By Taylor's theorem, for any $x\in\widetilde{\Omega}_\delta$, $s=0,1$ we have for some $0\leq\theta_s<2\delta$

$$\left|\frac{d^s}{dx^s}(v_\delta-T_k(v_\delta))(x)\right| \leq C\left|(-2\delta)^{k+1}\,\frac{d^{k+1}}{dx^{k+1}}\,\frac{d^s\,v}{dx^s}\,(x-\theta_s)\right|$$
$$\leq C\,\delta^{k+1}(1+x)^{\alpha-2-s-k}\,|\ln\,\delta|^{\gamma}\,,$$

by (3.3.44). Hence for $\delta = p^{-2}$ we have with (3.3.50)

$$\left\| \frac{d^{1-s}\xi_\delta}{dx^{1-s}} \frac{d^s}{dx^s} (v_\delta - T_k(v_\delta)) \right\|^2_{L^2(\tilde{\Omega}_\delta)}$$

$$\leq C \int_{-1+2\delta}^{2\delta} (1-x^2)^{2s} \, \delta^{2(k+1)} |\ln \delta|^{2\gamma} (1+x)^{2\alpha-4-2s-2k} \, dx$$

$$\leq C \, \delta^{2(k+1)} |\ln \delta|^{2\gamma} \, \delta^{2\alpha-2k-3} ,$$

provided $k > \alpha - 1/2$, which follows from $k = [\alpha + 1/2]$. Put $\delta = p^{-2}$ and use interpolation between $s = 0$ and $s = 1$ to complete the proof. □

Finally, we must estimate $\|\xi_\delta \, w_\delta\|_{H^s(\tilde{\Omega}_\delta)}$.

Lemma 3.33 *Let $\delta = p^{-2}$ and $0 \leq s \leq 1$. Then*

$$\|\xi_\delta w_\delta\|_{H^s(\tilde{\Omega}_\delta)} \leq C \, |\ln p|^\gamma \, p^{-2(\alpha - s + \frac{1}{2})} .$$

Proof Note that

$$\|\xi_\delta w_\delta\|_{H^s(\tilde{\Omega}_\delta)} = \|\xi w\|_{H^s(\tilde{\Omega})} \cdot$$

From (3.3.45) we get

$$\left\| \frac{d^r}{dx^r} (\xi w) \right\|^2_{L^2(\tilde{\Omega})} \leq C \int_{-1}^{-1+2\delta} |\ln(1+x)|^{2\gamma} \, (1+x)^{2(\alpha - r)} \, dx$$

$$\leq C \, |\ln \delta|^{2\gamma} \, \delta^{2(\alpha + \frac{1}{2} - r)}$$

$$= C \, |\ln p|^{2\gamma} \, p^{-4(\alpha + \frac{1}{2} - r)}$$

since $\alpha > 1/2$. □

We can now give the following proof.

Proof of Theorem 3.26 Let $k = [\alpha + 1/2]$ and define

$$Z_{p\delta} := \xi_\delta \Big(\sum_{i=0}^{k} (-1)^i \, \frac{(2\delta)^i}{i!} \, v_p^{[i]} \Big) .$$

Then $Z_{p\delta}$ is a polynomial of degree $p+1$ on $\widetilde{\Omega}_\delta$ and $Z_{p\delta}(-1+2\delta)=0$. Further

$$\begin{aligned}
\|u_\delta - Z_{p\delta}\|_{H^s(\widetilde{\Omega}_\delta)} &= \Big\|\xi_\delta\Big(u_{0\delta} - \sum_{i=0}^{k}(-1)^i\,\frac{(2\delta)^i}{i!}\,v_p^{[i]}\Big)\Big\|_{H^s(\widetilde{\Omega}_\delta)} \\
&\le \|\xi_\delta w_\delta\|_{H^s(\widetilde{\Omega}_\delta)} + \|\xi_\delta\Big(v_\delta - \sum_{i=0}^{k}(-1)^i\,\frac{(2\delta)^i}{i!}\,v^{[i]}\Big)\|_{H^s(\widetilde{\Omega}_\delta)} \\
&\quad + \sum_{i=0}^{k}\frac{(2\delta)^i}{i!}\,\|\xi_\delta\,(v^{[i]}-v_p^{[i]})\|_{H^s(\widetilde{\Omega}_\delta)} \\
&\le C\,|\ln p|^\gamma\,p^{-2(\alpha-s+\frac{1}{2})}\,,
\end{aligned}$$

by referring to Lemmas 3.31, 3.32 and 3.33. □

3.3.6 *Singular solutions. hp-FEM*

In the preceding subsection, we obtained convergence rate estimates $Z(u_0,S,X)$ for the *hp*-version FEM with quasiuniform meshes. For singular solutions u_0 of the type $|x-x_0|^\alpha$, $x_0\in\overline{\Omega}$, we obtained the convergence rate $Z(u_0,S,X)=CN^{-\alpha\pm\frac{1}{2}}$ for $X=L^2(\Omega)$ or $H^1(\Omega)$. If x_0 is a mesh node, we showed also that the *p*-version yields $Z(u_0,S^p(\Omega,\mathcal{T}),H^1)\le CN^{-2\alpha+1}$ where $N=\dim(S)$. Both these rates are **algebraic** in terms of N. We show in this subsection that **exponential convergence** can be achieved if the increase of the polynomial degree is accompanied by a strong, **geometric mesh refinement** towards the singular point x_0. This holds provided the exact solution belongs to the set $\mathcal{B}^\ell_\beta(\Omega)$ which contains solutions of the form $|x-x_0|^\alpha\,\varphi(x)$ where $\varphi(x)$ is analytic at x_0. The reasons to consider such solutions are twofold: on the one hand, they model the singular behavior of solutions to linear, elliptic boundary value problems near corners of polygonal domains, secondly, the solutions of boundary integral equation reformulations of such elliptic problems on polygons are also of this type. We have seen in Theorem 3.17 that the *hp*-FEM converges both as the mesh-width h tends zero as well as when the polynomial degree p tends to infinity. The *rate of convergence* was governed by the *regularity* of the solution, i.e., by how many derivatives of the solution were square integrable. Roughly speaking, the smoother the solution the more rapid the convergence. Nevertheless, the convergence rate that can be achieved this way is always *algebraic* unless the solution is analytic in the whole domain and *p*-refinement is used (cf. Theorem 3.19). Our aim in this section is to show that, for solutions u_0 which are only *piecewise analytic* with singularities at certain points in the domain, *exponential convergence* can be achieved by judicious combination of local mesh refinement and increase of polynomial degree, despite the rather low global regularity of the solution. This is important since in two and three dimensions the solutions of elliptic problems in piecewise analytic domains (such as polygons) are piecewise analytic. In order to prove exponential convergence, however, it does not suffice to control a finite number of derivatives of the solution. We must rather bound the growth of derivatives of arbitrary order. The framework for this will be certain countably normed spaces which we will define next.

Let $\Omega = (0,1)$ and the weight function $\Phi_\beta(x) = x^\beta$ be given. Define the seminorms

$$|u|_{H^{k,\ell}_\beta(\Omega)} = \|\Phi_{\beta+k-\ell}\, D^k u\|_{L^2(\Omega)}\,.$$

Then

$$\|u\|^2_{H^{m,\ell}_\beta(\Omega)} = \begin{cases} \|u\|^2_{H^{\ell-1}(\Omega)} + \sum_{k=\ell}^{m} |u|^2_{H^{k,\ell}_\beta(\Omega)}\,, & \ell > 0 \\ \sum_{k=\ell}^{m} |u|^2_{H^{k,\ell}_\beta(\Omega)}\,, & \ell = 0\,. \end{cases} \tag{3.3.54}$$

Definition 3.34 *$u \in \mathcal{B}^\ell_\beta(\Omega)$ if $u \in H^{m,\ell}_\beta(\Omega)$, $m = \ell, \ell+1, \ldots$ and there exist constants $C > 0$, $d \geq 1$ such that*

$$|u|_{H^{k,\ell}_\beta(\Omega)} \leq C\, d^{k-\ell}\, (k-\ell)! \tag{3.3.55}$$

for $k = \ell, \ell+1, \ldots$.

The set $\mathcal{B}^\ell_\beta(\Omega)$ contains functions that are analytic in $(0,1)$ with possibly an algebraic singularity at $x = 0$. Evidently, $\mathcal{B}^\ell_\beta(\Omega) \subset H^{\ell-1}(\Omega)$ for $\ell \geq 1$.

Exercise 3.35 Consider $u(x) = x^\lambda$. For which λ does u belong to $\mathcal{B}^2_\beta(\Omega)$?

We call a mesh $\mathcal{T}$ **geometrically graded** towards $x_0 = 0$ with **mesh grading factor** $\sigma \in (0,1)$, if

$$\mathcal{T} = \{\Omega_j\}_{j=1}^M,\ \Omega_j = (x_{j-1}, x_j),\ x_j = \sigma^{M-j},\ 1 \leq j \leq M,\ x_0 = 0\,. \tag{3.3.56}$$

Observe that

$$h_j := x_j - x_{j-1} = \lambda\, x_{j-1} \ \text{ for } \ j = 2, \ldots, M-1,\ \lambda = \frac{1-\sigma}{\sigma}\,.$$

The main result of this section is the following

Theorem 3.36 *Let $\widehat{\Omega} = (0,1)$, Ω^M_σ be a mesh that is geometrically graded towards $x_0 = 0$ with grading factor $\sigma \in (0,1)$. Let $u_0 \in \mathcal{B}^\ell_\beta(\widehat{\Omega})$ with $\ell = 1, 2$ and $\Phi_\beta(x) = x^\beta$ for $0 < \beta < 1$. Then there exists $\mu = \mu(\sigma, d)$ (d as in (3.3.55)) such that for $S = S^{\mathbf{p},\ell-1}(\widehat{\Omega}, \Omega^M_\sigma)$ with degree vector $\mathbf{p}$ given by*

$$p_1 = \ell - 1,\ p_j = \max\{\ell, [\mu j]\},\ j = 2, \ldots, M \tag{3.3.57}$$

it holds that

$$Z(u_0, S, H^{\ell-1}) = \inf_{s \in S} \|u_0 - s\|_{H^{\ell-1}(\widehat{\Omega})} \leq C e^{-b\sqrt{N}},\ \ell = 1, 2\,, \tag{3.3.58}$$

where $N = \dim(S^{\mathbf{p},\ell-1}(\widehat{\Omega}, \Omega^M_\sigma))$ and $C > 0$, $b > 0$ are independent of N.

The proof of Theorem 3.36 is given in several steps. Its pattern is very similar to that of the corresponding theorem in two dimensions in the next chapter.

We begin with the following lemma which is a direct consequence of Corollary 3.15.

Lemma 3.37 *Let $\widehat{\Omega} = (-1,1)$ and $u_0 \in H^{p+1}(\widehat{\Omega})$, $p \in \mathbb{N}_0$. Then there exists $s_0 \in \widehat{\Omega}$ and $C > 0$ such that*

$$\|(u_0 - s_0)^{(m)}\|^2_{L^2(\widehat{\Omega})} \leq C\,\frac{(p-s)!}{(p+s+2-2m)!}\,\|u_0^{(s+1)}\|^2_{L^2(\widehat{\Omega})} \tag{3.3.59}$$

for $s \in \mathbb{N}_0$, $s \leq p$, and $m = 0, 1$ if $p \geq 1$ and $s = m = 0$ if $p = 0$. Moreover,

$$u_0(\pm 1) = s_0(\pm 1) \tag{3.3.60}$$

if $p \geq 1$.

By a scaling argument we get the following result.

Lemma 3.38 *Let $J = (a,b) \subset \mathbb{R}$, $h = b - a$ and $u_0 \in H^{p+1}(J)$ for some $p \geq 1$. Then there exists a polynomial s_0 of degree p on J such that for $m = 0, 1$*

$$\|(u_0 - s_0)^{(m)}\|^2_{L^2(J)} \leq \frac{C}{h^{2m}}\left(\frac{h}{2}\right)^{2(s+1)}\frac{(p-s)!}{(p+s+2-2m)!}\,\|u^{(s+1)}\|^2_{L^2(J)} \tag{3.3.61}$$

where $C > 0$ is independent of h, p and u_0 and $s \in \mathbb{N}_0$, $s \leq p$, $m = 0, 1$ if $p \geq 1$, and $s = m = 0$ if $p = 0$. Moreover,

$$u_0(a) = s_0(a), \quad u_0(b) = s_0(b)\ . \tag{3.3.62}$$

if $p \geq 1$.

Proof The estimate (3.3.61) follows from (3.3.59) with the map $Q : \widehat{\Omega} \to J$ given by $x = Q(\xi) = 1/2\,(a + b + \xi h)$. □

Lemma 3.39 *Let $\Omega = (0,1)$, $J = (a,b) \subset \Omega$ and $\lambda > 0$ such that $h = |b - a| = \lambda a$. Then for every $u_0 \in H^{s+1,\ell}_\beta(\Omega)$ exists a polynomial s_0 of degree p on J such that*

$$\|(u_0 - s_0)^{(m)}\|^2_{L^2(J)} \leq C a^{2(\ell-m-\beta)}\frac{\Gamma(p-s+1)}{\Gamma(p+s+3-2m)}\left(\frac{\lambda}{2}\right)^{2s}|u_0|^2_{H^{s+1,\ell}_\beta(\Omega)} \tag{3.3.63}$$

where $1 \leq \ell \leq s+1 \leq p+1$, $s \in \mathbb{R}$, $m = 0, 1$ and (3.3.62) holds.

Proof We have for $\ell \leq s+1$

$$|u|^2_{H^{s+1,\ell}_\beta(\Omega)} \geq \|u^{(s+1)}\Phi_{\beta+s+1-\ell}\|^2_{L^2(J)} \geq a^{2(\beta+s+1-\ell)}\|u^{(s+1)}\|^2_{L^2(J)} ,$$

hence

$$\|u^{(s+1)}\|^2_{L^2(J)} \leq a^{-2(\beta+s+1-\ell)}|u|^2_{H^{s+1,\ell}_\beta(\Omega)} .$$

By Lemma 3.38 we can find s_0 such that

$$\|(u_0-s_0)^{(m)}\|^2_{L^2(J)} \leq Ca^{2(\ell-m-\beta)}\frac{(p-s)!}{(p+s+2-2m)!}\left(\frac{\lambda}{2}\right)^{2s}|u|^2_{H^{s+1,\ell}_\beta(\Omega)} \tag{3.3.64}$$

since $h = \lambda a$ and $s+1 > m$.

Let $\sigma = s+\theta-1$, $0<\theta<1$. Then

$$H^{\sigma+1,\ell}_\beta(\Omega) := (H^{s,\ell}_\beta(\Omega), H^{s+1,\ell}_\beta(\Omega))_{\theta,\infty} .$$

The error operator T, defined by

$$Tu_0 = (u_0 - s_0)^{(m)}$$

is linear and, by (3.3.64), bounded from $H^{s+1,\ell}_\beta(\Omega)$ to $L^2(J)$. By interpolation (cf. Appendix B) it follows that

$$\begin{aligned}
\|T\|^2_{H^{\sigma+1,\ell}_\beta(\Omega)\to L^2(J)} &\leq \|T\|^{2(1-\theta)}_{H^{s,\ell}_\beta\to L^2(J)}\;\|T\|^{2\theta}_{H^{s+1,\ell}_\beta\to L^2(J)} \\
&\leq Ca^{2(\ell-m-\beta)}\left(\frac{\lambda}{2}\right)^{2\sigma}\frac{(p-s)!\,(p-s+1)^{1-\theta}}{(p+s+1-2m)!\,(p+s+2-2m)^{\theta}} \\
&\leq Ca^{2(\ell-m-\beta)}\left(\frac{\lambda}{2}\right)^{2\sigma}\frac{\Gamma(p-s+2-\theta)}{\Gamma(p+s+2-2m+\theta)} \\
&= Ca^{2(\ell-m-\beta)}\left(\frac{\lambda}{2}\right)^{2\sigma}\frac{\Gamma(p+1-\sigma)}{\Gamma(p+1+\sigma+2-2m)}
\end{aligned} \tag{3.3.65}$$

where we used the bounds

$$0 < C_1 \leq \frac{n!\,(n+1)^\theta}{\Gamma(n+1+\theta)} \leq C_2 < \infty, \quad 0<\theta<1 \tag{3.3.66}$$

with C_i independent of n. □

Exercise 3.40 Using Stirling's formula, prove (3.3.55).

We combine the error estimates (3.3.63), applied to each element Ω_j of the geometrically graded mesh Ω^M_σ in (3.3.56).

Lemma 3.41 *Let $\Omega = (0,1)$, $u_0 \in \mathcal{B}^{\ell}_{\beta}(\Omega)$, $\ell = 1,2$. Let the mesh grading factor σ satisfy $0 < \sigma < 1$ and the degree vector be given by $p_1 = \ell - 1$, $p_j \geq \ell$, $j = 2,\ldots,M$. Then there exists $s_0 \in S^{\mathbf{p},\ell-1}(\Omega,\Omega^M_\sigma)$ such that*

$$\|u_0 - s_0\|^2_{H^{\ell-1}(\Omega)} \leq C\left\{h_1^{2(1-\beta)}|u_0|_{H^{\ell,1}_\beta(\Omega_1)} + \sum_{i=2}^{M} x_{i-1}^{2(1-\beta)}\frac{\Gamma(p_i - s_i + 1)}{\Gamma(p_i + s_i + 5 - 2\ell)}\left(\frac{\lambda}{2}\right)^{2s_i}|u_0|^2_{H^{s_i+1,\ell}_\beta(\Omega)}\right\} \tag{3.3.67}$$

for any $\ell - 1 \leq s_i \leq p_i$, $i = 2,\ldots,M$.

Proof We apply Lemma 3.39 on every element $\Omega_i \in \Omega^M_\sigma$ except Ω_1. There exists $s_{0i} \in P_{p_i}(\Omega_i)$ such that

$$\|(u_0 - s_{0i})^{(m)}\|^2_{L^2(\Omega_i)} \leq C\, x_{i-1}^{2(\ell-m-\beta)}\,\frac{\Gamma(p_i - s_i + 1)}{\Gamma(p_i + s_i + 3 - 2m)}\left(\frac{\lambda}{2}\right)^{2s_i}|u_0|^2_{H^{s_i+1,\ell}_\beta(\Omega)}$$

since $u \in \mathcal{B}^{\ell}_{\beta}(\Omega) \Longrightarrow u \in H^{s_i+1,\ell}_\beta(\Omega)$ $(\ell \leq s_i + 1)$.

If $\ell = 2$ and $u \in \mathcal{B}^2_\beta(\Omega)$, it follows that for $i \geq 2$

$$\begin{aligned}\|u_0 - s_{0i}\|^2_{H^1(\Omega_i)} &= \|u - s_{0i}\|^2_{L^2(\Omega_i)} + \|(u - s_{0i})'\|^2_{L^2(\Omega_i)}\\ &\leq C\, x_{i-1}^{2(1-\beta)}\,\frac{\Gamma(p_i - s_i + 1)}{\Gamma(p_i + s_i + 1)}\left(\frac{\lambda}{2}\right)^{2s_i}|u_0|^2_{H^{s_i+1,2}_\beta(\Omega)}\,.\end{aligned}$$

In the case $\ell = 1$ we verify directly that

$$\|u_0 - s_{0i}\|^2_{L^2(\Omega_i)} \leq C x_{i-1}^{2(1-\beta)}\,\frac{\Gamma(p_i - s_i + 1)}{\Gamma(p_i + s_i + 3)}\left(\frac{\lambda}{2}\right)^{2s_i}|u_0|^2_{H^{s_i+1,1}_\beta(\Omega)},\quad i \geq 2\,.$$

It remains to investigate the interval Ω_1. Here we have

$$\|u_0 - s_{01}\|^2_{H^{\ell-1}(\Omega_1)} \leq C h_1^{2(1-\beta)}|u_0|^2_{H^{\ell,1}_\beta(\Omega_1)} \tag{3.3.68}$$

for some $s_{01} \in \mathcal{S}^{\ell-1}(\Omega_1)$, $\ell = 1,2$. Estimate (3.3.68) follows from more general two-dimensional results to be obtained in Chapter 4. Adding the elementwise estimates together gives (3.3.68). Note that in the case $\ell = 2$ the local approximations s_{0i} can be pasted together continuously since $s_{0i}(x_{i-1}) = u_0(x_{i-1})$, $s_{0i}(x_i) = u_0(x_i)$, $i = 1,\ldots,M$. □

We can now turn to the following proof.

Proof of Theorem 3.36 The idea is to analyze the error bound (3.3.67) for particular choices of p_i and s_i.

We start with the estimate (3.3.67) which implies that there exists $s_0 \in S^{p,\ell-1}(\Omega, \Omega_\sigma^M)$ such that

$$\|u_0 - s_0\|^2_{H^{\ell-1}(\Omega)} \leq C\Big\{\sigma^{2(1-\beta)M} + \sum_{i=2}^{M} x_{i-1}^{2(1-\beta)} \frac{\Gamma(p_i - s_i + 1)}{\Gamma(p_i + s_i + 5 - 2\ell)} \times \Big(\frac{\lambda}{2}\Big)^{2s_i} |u_0|^2_{H_\beta^{s_i+1,\ell}(\Omega)}\Big\}. \tag{3.3.69}$$

We claim that for $u_0 \in \mathcal{B}_\beta^\ell(\Omega)$ we have for $\theta \in [0,1]$

$$|u_0|_{H_\beta^{k+\theta,\ell}(\Omega)} \leq C(\ell) d^{k+\theta-\ell}\,\Gamma(k+\theta-\ell+1)\,, \quad k \geq \ell\,. \tag{3.3.70}$$

To see this, observe that from (3.3.55) we have

$$|u_0|_{H_\beta^{k,\ell}(\Omega)} \leq C\, d^{k-\ell}(k-\ell)!, \quad k \in \mathbb{N},\ k \geq \ell\,. \tag{3.3.71}$$

Using interpolation on the identity operator $Tu_0 = u_0$, we get with (3.3.66) the estimate (3.3.71). This then gives

$$|u_0|_{H_\beta^{s+1,\ell}(\Omega)} \leq C(\ell)\, d^s\, \Gamma(s+1), \quad s > 0 \text{ arbitrary}\,.$$

Using this in (3.3.69), we get with

$$x_i - x_{i-1} = \lambda\, x_{i-1} = \frac{1-\sigma}{\sigma}\,\sigma^{M+1-i},\ i = 2, \ldots, M$$

that

$$\|u_0 - s_0\|^2_{H^{\ell-1}(\Omega)} \leq C\,\Big\{\sigma^{2(1-\beta)M} + \sum_{i=2}^{M} \sigma^{2(M-i+1)(1-\beta)} \frac{\Gamma(p_i - s_i + 1)}{\Gamma(p_i + s_i + 5 - 2\ell)}\,\Gamma(s_i+1)^2\Big(\frac{\rho d}{2}\Big)^{2s_i}\Big\} \tag{3.3.72}$$

where $\rho := \max\{1, \lambda\}$.

We select $s_i = \alpha_i p_i$ for $i = 2, \ldots, M$ with $0 < \alpha_i < 1$ to be selected below. Let

$$F(d,\alpha) := \Big(\frac{\alpha d}{2}\Big)^{2\alpha} \frac{(1-\alpha)^{1-\alpha}}{(1+\alpha)^{1+\alpha}}\,, \tag{3.3.73}$$

$F : (1,\infty) \times (0,1) \to \mathbb{R}$. Then there exists a constant $C > 0$ such that

$$\frac{\Gamma(p-s+1)}{\Gamma(p+s+1)}\Big(\frac{\rho d}{2}\Big)^{2s}\,\Gamma(s+1)^2 \leq Cp(F(\rho d, \alpha))^p\,. \tag{3.3.74}$$

This follows from Stirling's formula $\Gamma(n+1) \sim \sqrt{2\pi n}\,(n/e)^n$ and some elementary algebra.

Then (3.3.72), (3.3.73) and (3.3.74) imply

$$\begin{aligned}\|u_0 - s_0\|^2_{H^{\ell-1}(\Omega)} &\le C\Big\{\sigma^{2(1-\beta)M} + \sum_{i=2}^{M} \sigma^{2(M-i+1)(1-\beta)} p_i (F(\rho d, \alpha_i))^{p_i}\Big\} \\ &\le C\sigma^{2(1-\beta)M}\Big\{1 + \sum_{i=2}^{M} \sigma^{2(1-i)(1-\beta)} p_i (F(\rho d, \alpha_i))^{p_i}\Big\}.\end{aligned} \tag{3.3.75}$$

We prove that the curved parentheses are bounded independently of M. To do so, we claim

$$\inf_{0<\alpha<1} F(d,\alpha) = F(d,\alpha_{\min}) < 1, \ \alpha_{\min} = \frac{2}{\sqrt{4+d^2}}\,. \tag{3.3.76}$$

The proof of (3.3.76) is left as an exercise.

Let now $\alpha_i = \max\{1/p_i, \alpha_{\min}\}$, $2 \le i \le M$ with $\alpha_{\min} = 2/\sqrt{4+\rho^2 d^2}$ and define $F_{\min} = F(\rho d, \alpha_{\min}) < 1$.

Select $p_i = \max\{\ell, [\mu i]\}$, $2 \le i \le M$, and

$$\mu > \max\Big\{1, \ \frac{2(1-\beta)\ln\sigma}{\ln F_{\min}}\Big\}. \tag{3.3.77}$$

Define the index i_0 by

$$p_{i0} = 1 + \Big[\frac{1}{\alpha_{\min}}\Big].$$

Then it follows that

$$p_{i0} = [\mu i_0] \le 2 + \frac{1}{\alpha_{\min}}\,. \tag{3.3.78}$$

Inserting (3.3.77) and (3.3.78) into (3.3.75), we find for $\ell = 0, 1$

$$\begin{aligned}\|u_0 - s_0\|^2_{H^{\ell-1}(\Omega)} &\le C\,\sigma^{2(1-\beta)M}\Big\{1 + \sum_{i=2}^{i_0} \sigma^{2(1-i)(1-\beta)} (F_{\min})^{p_i} \\ &\times p_i \max_{1\le i\le i_0} \Big(\frac{F(\rho d, 1/p_i)}{F_{\min}}\Big)^{p_i} + \sum_{i=i_0+1}^{M} \sigma^{2(1-i)(1-\beta)}\, p_i (F_{\min})^{p_i}\Big\}.\end{aligned} \tag{3.3.79}$$

We show that the terms in parentheses are bounded. We have

$$\sigma^{2(1-\beta)(1-i)}\,(F_{\min})^{p_i} = \sigma^{2(1-\beta)}\,\frac{(F_{\min})^{[\mu i]}}{\sigma^{2(1-\beta)i}} \le C\,\sigma^{2(1-\beta)}\,\Big(\frac{(F_{\min})^{\mu}}{\sigma^{2(1-\beta)}}\Big)^{i}.$$

Further, $(F_{\min})^{\mu} < \sigma^{2(1-\beta)}$ due to (3.3.77) and $F_{\min} < 1$. Hence $q := (F_{\min})^{\mu}\,\sigma^{-2(1-\beta)} < 1$ and $\sum_{i>i_0} iq^i < \infty$. Therefore (3.3.79) can be bounded by $C\sigma^{2(1-\beta)M}$.

Since

$$N = \dim S^{p,\ell}(\Omega, \Omega_\sigma^M) \le C\mu M^2, \;\; C \text{ independent of } M, \sigma, \mu\,, \tag{3.3.80}$$

the proof is complete. □

Exercise 3.42 Prove (3.3.76), i.e., that

$$F_{\min} = F(d, \alpha_{\min}), \;\; \alpha_{\min} = \frac{2}{\sqrt{4+d^2}}\,.$$

Exercise 3.43 Prove (3.3.69).

Theorem 3.36 asserted that the *hp*-FEM with *any* geometrically graded mesh Ω_σ^M gives exponential convergence. The question which grading factor $\sigma \in (0,1)$ is best was not answered. For example, meshes based on successive element bisection have $\sigma = 1/2$ and the question arises if the exponential convergence rate (in particular, the constant b in (3.3.58)) can be substantially improved if $\sigma \neq 1/2$. This is indeed the case. By a more careful analysis of the approximation error $Z(u_0, S^{p,1}, H^1)$ for $u_0(x) = x^\alpha(1-x)$, $\alpha \notin \mathbb{N}$, it was shown in [69] that the grading factor

$$\sigma^* = (\sqrt{2}-1)^2 \approx 0.171\ldots$$

is optimal. Notice that σ^* is independent of α. Using σ^* in place of $\sigma = 1/2$ gives errors which are several orders of magnitude smaller, even though the number of degrees of freedom is the same.

3.3.7 *Mesh optimization for h-version FEM*

We saw in the previous subsection that for solutions which are singular at a point $x_0 \in \overline{\Omega}$, strong (geometric) mesh refinement combined with judicious increase of the polynomial degree leads to exponential convergence of the FEM. Here we will be concerned with the following **question**: what are, for a given exact solution u_0, the optimal meshes $\mathcal{T}$ as the polynomial degrees $p_j = p$ are fixed and equal? We will solve this problem exhibiting **asymptotically**, i.e., as $N = \dim(S^{p,1}(\Omega, \mathcal{T})) \to \infty$, optimal mesh sequences.

We begin the discussion with a definition.

Definition 3.44 *Let $\Omega = (a,b) \subset \mathbb{R}$ be a bounded interval. A function $\Gamma : \overline{\Omega} \to [0,1]$ is a* **grading function**, *if*

$$\Gamma \in C^1(\overline{\Omega})\,, \tag{3.3.81}$$

$$1/c \ge \Gamma'(x) \ge c > 0 \;\; in \;\; \Omega\,, \tag{3.3.82}$$

$$\Gamma(a) = 0, \;\; \Gamma(b) = 1\,. \tag{3.3.83}$$

A grading function defines a mesh $\mathcal{T}$ with M elements via

$$\Gamma(x_j) = j/M, \quad j = 0,\ldots,M \,. \tag{3.3.84}$$

A grading function establishes a one-to-one correspondence between the mesh $\mathcal{T}$ and the uniform mesh on $(0,1)$.

Proposition 3.45 *Let $\Gamma : (a,b) \longmapsto (0,1)$ be a grading function and $\{\mathcal{T}_M\}_{M=1}^{\infty}$ the corresponding family of meshes. Then, as $M \to \infty$,*

$$h(\mathcal{T}_M) = \max_{1\le j\le M} \{h_j^{\mathcal{T}_M}\} \to 0 \,. \tag{3.3.85}$$

Proof We have $h_j = x_j - x_{j-1}$. Since $\Gamma' \ge c > 0$, there exists an inverse function $\widehat{\Gamma} : (0,1) \to (a,b)$ and it holds that

$$h_j = \widehat{\Gamma}(j/M) - \widehat{\Gamma}((j-1)/M) = \widehat{\Gamma}'(\xi)/M$$

for some $\xi \in (j-1)/(M, j/M)$. Since the assumption (3.3.82) implies also that $\widehat{\Gamma}'(x) \ge c^{-1} > 0$, the assertion follows. □

To minimize the approximation error over all possible meshes, we re-express it in terms of the grading function $\Gamma(x)$.

Let $\mathcal{T}$ be a mesh in $\Omega = (0,1)$ and $p_j = p$ a constant degree distribution. From (3.3.25) we have

$$\|u_0' - s_0'\|^2_{L^2(\Omega)} \le C^2(p) \sum_{j=1}^{M(\mathcal{T})} h_j^{2\min(p,k)} \int_{\Omega_j} |u_0^{(k+1)}(\xi)|^2 \, d\xi \,.$$

Assuming that u_0 is smooth, k may be selected equal to p and we get

$$\|u_0' - s_0'\|^2_{L^2(\Omega)} \le C^2(p) \sum_{j=1}^{M(\mathcal{T})} h_j^{2p+1} |u_0^{(p+1)}(\xi_j)|^2 \,. \tag{3.3.86}$$

For any grading function Γ we have for some $\widetilde{x}_j \in (x_{j-1}, x_j)$

$$M^{-1} = \Gamma(x_j) - \Gamma(x_{j-1}) = \Gamma'(\widetilde{x}_j)\, h_j \,. \tag{3.3.87}$$

Hence

$$\|u_0' - s_0'\|^2_{L^2(\Omega)} \le C^2(p)\, M^{-2p} \sum_{j=1}^{M} h_j\, |u_0^{(p+1)}(\xi_j)|^2\, |\Gamma'(\widetilde{x}_j)|^{-2p} \,. \tag{3.3.88}$$

Since ξ_j and $\widetilde{x}_j$ are both in Ω_j and Γ' and $u_0^{(p+1)}$ are continuous, we see that the sum in (3.3.88) is in fact a Riemann sum. As $M(\mathcal{T}) \to \infty$, this sum approaches the integral

$$\sum_{j=1}^{M(\mathcal{T})} h_j \, |u_0^{(p+1)}(\xi_j)|^2 \, |\Gamma'(\widetilde{x}_j)|^{-2p} \to \int_a^b (u_0^{(p+1)}(x))^2 \, (\Gamma'(x))^{-2p} dx \tag{3.3.89}$$

and, asymptotically, as $M \to \infty$ we get the bound

$$\|u_0 - s_0\|^2_{H^1(\Omega)} \le C^2(p) M^{-2p} \int_a^b (u_0^{(p+1)}(x))^2 \, (\Gamma'(x))^{-2p} \, dx \, . \tag{3.3.90}$$

Notice that by (3.3.83) we have

$$\int_a^b \Gamma'(x) \, dx = \Gamma(a) - \Gamma(b) = 1 \, . \tag{3.3.91}$$

The problem of mesh optimization is now the following.

Given the exact solution u_0, minimize the bound (3.3.90) **among all possible grading functions Γ satisfying the constraint** (3.3.91).

Let us determine the optimal Γ_0. To derive conditions for Γ_0, let $\Gamma(x) = \Gamma_0(x) + \lambda\eta(x)$ where $\eta(x)$ is a variation and λ a small parameter. Note that (3.3.91) implies

$$\int_a^b \eta'(x) \, dx = 0, \;\; \eta' \in C^0(\overline{\Omega}) \, . \tag{3.3.92}$$

For small λ, we have

$$\begin{aligned} \|u_0' - s_0'\|^2_{L^2(\Omega)} = \frac{C^2(p)}{M^{2p}} \Big\{ & \int_a^b (u_0^{(p+1)}(x))^2 (\Gamma_0'(x))^{-2p} dx \\ & -2\lambda p \int_a^b (u_0^{(p+1)}(x))^2 (\Gamma_0'(x))^{-2p-1} \eta'(x) dx + O(\lambda^2) \Big\} \, . \end{aligned} \tag{3.3.93}$$

To minimize the bound in (3.3.93), its derivative with respect to λ must vanish at $\lambda = 0$. This gives

$$\int_a^b (u_0^{(p+1)}(x))^2 (\Gamma_0'(x))^{-2p-1} \eta'(x) \, dx = 0 \tag{3.3.94}$$

for all $\eta'(x) \in C^0(\overline{\Omega})$ satisfying (3.3.92). Hence (3.3.94) implies that

$$(u_0^{(p+1)}(x))^2 \, (\Gamma_0'(x))^{-2p-1} = C^{-1} = \text{const}$$

or

$$\Gamma_0'(x) = \left[C (u_0^{(p+1)}(x))^2 \right]^{1/(2p+1)} . \tag{3.3.95}$$

The constant C is determined from (3.3.91). This yields

$$\|u_0' - s_0'\|_{L^2(\Omega)} \approx C(p)\, M^{-p} \sqrt{K(u_0)} \tag{3.3.96}$$

where

$$K(u_0) := \int_a^b (u_0^{(p+1)}(x))^2 \, (\Gamma_0'(x))^{-2p} dx .$$

The constant K which depends on Γ_0 and u_0, characterizes the smallest possible error bound (3.3.89) asymptotically achievable, i.e., when M is large.

Remark 3.46 From (3.3.93) we see that a change $\lambda\eta(x)$ of $\Gamma_0(x)$ changes the error by $O(\lambda^2)$. Therefore the error is not very sensitive to the mesh – even roughly optimal meshes (e.g., as produced by adaptive algorithms) will give very close to minimal errors.

Example 3.47 (radical mesh). Let $\Omega = (0, 1)$ and $u_0(x) = x^\alpha - x$ with $\alpha > 1/2$. Then $u_0 \in H_0^1(\Omega)$ and

$$\Gamma_0'(x) = C x^{(2(\alpha - p - 1))/(2p+1)} = C\, x^{C-1}, \ C = \frac{2\alpha - 1}{2p + 1} ,$$

and

$$\Gamma_0(x) = x^{(2\alpha - 1)/(2p+1)} . \tag{3.3.97}$$

The mesh points are explicitly given by

$$x_j = (j/M)^{(p+1/2)/(\alpha - 1/2)}, \ \ j = 0, \ldots, M .$$

We see that as $p \to \infty$, the mesh grading becomes stronger.

Moreover, it can be shown [69] that then

$$\lim_{M \to \infty} M^p |u - u_{\mathrm{FE}}|_{H^1(0,1)} = C(\alpha, p) .$$

Remark 3.48 Strictly speaking, grading functions like (3.3.97) which are singular near the endpoints of an interval are not admissible. The assumption (3.3.81), i.e., that $\Gamma \in C^1(\overline{\Omega})$, was made only for convenience. Weaker assumptions are, in fact, sufficient. Essentially, the grading function Γ must be such that the integrand $u_0^{(p+1)}(x)(\Gamma'(x))^{-p}$ in (3.3.89) is square integrable.

Exercise 3.49 (boundary layer mesh). Let $\Omega = (0,1)$ and consider the boundary layer function $u(x) = \exp(-(1-x)/d)$ where $0 < d \leq 1$ is a parameter. For fixed d and constant polynomial degree p, derive the optimal grading function Γ_0. Give an explicit formula for the mesh points x_i in terms of $M(\mathcal{T})$, p and d. What happens to the optimal mesh as $d \to 0$? As $p \to \infty$?

3.4 Problems with boundary layers. Robustness

3.4.1 *Reaction–diffusion problem*

Our goal in this section is to investigate the *hp*-approximation of **boundary layer functions**.

$$u(x) = \exp(-ax/d), \quad 0 < x < L, \tag{3.4.1}$$

where $d \in (0,1]$ is a small parameter that can approach zero, $a > 0$ is a constant and $L \geq 1$ is a typical length scale of the problem under consideration. We are interested in obtaining convergence estimates that are **robust**, i.e., **uniform** in d, when (3.4.1) is approximated by piecewise polynomials via *p*- and *hp*-FEM.

Boundary layers (3.4.1) arise as solution components in **singularly perturbed elliptic boundary value problems**. We consider the model problem of the elastic bar where now $AE = d^2$ is small. (We indicate the dependence of the solution on the small parameter d by u_d.)

$$L_d u_d = -d^2 u_d''(x) + a^2 u_d(x) = f(x), \quad x \in \Omega = (-1,1), \tag{3.4.2}$$

$$u_d(\pm 1) = \alpha^{\pm}. \tag{3.4.3}$$

The problem (3.4.2), (3.4.3) is also known in the computational fluid dynamics literature as a **reaction–diffusion problem**. A large body of literature has been devoted to the effective resolution of these problems. Most available references analyze the convergence of h-version finite difference or finite element schemes of **fixed** (usually low) polynomial degree in conjunction with various mesh refinements, see, e.g., [32], [65], [122], [150], and the references therein.

If the mesh is quasiuniform (or, more generally, independent of d), either on the whole domain or locally near the boundary, then the optimal (algebraic) decrease in the global error is observed **provided** a condition of the form $h \leq Cd$ is met. Such methods are **nonrobust**, in a sense made precise in Section 3.4.4 ahead. In practical terms, the amount of discretization required with such schemes for satisfactory resolution of the boundary layers may be infeasible when d is very small. On the other hand, strongly graded d-dependent mesh refinement, like the one from [151] presented in Section 3.4.7, does lead to robust convergence, at an optimal rate that is **algebraic** (see Exercise 3.49 and also [32], [132], [150], [151] where this and other graded meshes are discussed).

An alternate approach is to increase the polynomial degree and keep the mesh fixed, i.e., use a *p*-version FEM (or spectral element method). We provide a detailed study of the approximation theory for this method, showing that an **asymptotic superexponential convergence rate** for the error in the energy norm is achieved for $\tilde{p} := p + 1/2 > e/(2d)$. We also provide estimates for this error in the pre-asymptotic phase when d is small, showing that a) for $(3/(4d))^{1/2} \leq \tilde{p} \leq 2d$, the error is bounded by $C \exp(-\tilde{p}^2 d/3)$ and b) for $\tilde{p} \leq K d^{-\frac{1}{2}}$, the error is bounded by Cp^{-1} (numerical experiments in Section 3.4.7 are in agreement with these rates). The results we prove for a single element also hold when a **fixed** mesh with several elements is used. We establish that for the pure p version on fixed meshes, the overall robust rate, uniform in d, is $\mathcal{O}(p^{-1}\sqrt{\ln p})$ and, up to the $\sqrt{\ln p}$ term, this is the best possible. Note that in terms of the number of degrees of freedom, this rate is essentially **twice** the uniform rate of $O(h^{\frac{1}{2}})$ achievable (for the global error) by the h version with quasiuniform meshes (Theorem A.1(ii) of [122]). It is also twice the uniform rate for p-version/spectral element methods that can be established from the results in [42].

Our main result shows that robust exponential rates for the approximation of functions (3.4.1) can be achieved by using, instead of the pure p version, a **variable** mesh with only one more element. More precisely, a **robust exponential rate** can be obtained by using ***hp*-FEM on two elements, where the first one is of size** $O(pd)$. (For problems like (3.4.2)–(3.4.3), three elements are needed, due to boundary layers at either end - see Section 3.4.7.) We call this an hp version since the size (though not the number) of elements changes, as does p. Note that a robust exponential rate is not possible with either the h version or the pure p/spectral version – the estimates obtained in the papers above are all **algebraic**. Finite element computations for (3.4.2)–(3.4.3) presented in Section 3.4.7 confirm the theoretical convergence estimates obtained here and clearly show the dramatic superiority of this robust hp-FEM over low order methods, especially for small values of d.

The outline of this section is as follows. In Section 3.4.2 we present an asymptotic expansion for the solution of the model problem (3.4.2)–(3.4.3) which includes the boundary layers. The proof uses standard techniques and is provided for completeness. In Section 3.4.3, we describe the finite element methods and error measures to be analyzed. We also define the concept of robustness, using a definition from [23]. Section 3.4.4 is devoted to the convergence analysis of the p-version FEM. In Section 3.4.5, we consider an hp-version for which we prove a robust exponential convergence rate in various norms. Finally, in Section 3.4.7 we present numerical experiments comparing, in particular, the p- and hp-version FEMs analyzed here with an h version from [151], based on the asymptotically optimal meshes in Exercise 3.49. We show that the hp version consistently outperforms the other versions and that high accuracy can be achieved with few degrees of freedom for arbitrarily small d (we take values of d as small as 10^{-8}).

Throughout we denote by (u, v) the L_2 inner product. Also, $H_0^1(I) = \{u \in H^1(I) : u(\pm 1) = 0\}$, $H_D^1(I) = \{u \in H^1(I) : u(\pm 1) = \alpha^{\pm}\}$ and $H^{-1}(I) := (H_0^1)^*$, the dual space.

3.4.2 *Regularity of the solution*

The variational formulation of the model problem (3.4.2)–(3.4.3) reads: find $u_d \in H^1_D(I)$ such that

$$B_d(u_d, v) = (f, v) \quad \text{for all } v \in H^1_0(I) . \tag{3.4.4}$$

Here $f \in H^{-1}(I)$ and

$$B_d(u, v) := \int_I \{d^2 u' v' + a^2 uv\} \, dx . \tag{3.4.5}$$

For every $f \in H^{-1}(I)$ the problem (3.4.4) admits a unique solution $u_d \in H^1_D(I)$ and if $f \in H^k(I)$, then $u_d \in H^{k+2}(I) \cap H^1_D(I)$. This regularity, however, is non-uniform in d since in the a priori "shift" estimate

$$\|u_d\|_{k+2} \le C(k, d)\|f\|_k, \quad k = 0, 1, 2, \ldots \tag{3.4.6}$$

the constant C strongly depends on d. The following theorem presents a decomposition of u_d into a smooth part $u_d^M(x)$ and **boundary layers**

$$u_{a,d}(x) = \exp(-a(1 + x)/d), \quad \bar{u}_{a,d}(x) = \exp(-a(1 - x)/d) . \tag{3.4.7}$$

Theorem 3.50 *Let $f \in H^{4m+2}(I)$ for some $m \in \mathbb{N}$. Then*

$$u_d(x) = u_d^m(x) + A_d^m u_{a,d}(x) + B_d^m \bar{u}_{a,d}(x) \tag{3.4.8}$$

where $u_d^m(x)$ satisfies the following regularity estimate uniformly in d for $\ell = 0, 1, \cdots 2m$:

$$|u_d^m|_\ell \le a^{-1}(d/a)^{2m+2-\ell}|f|_{(2m+2)} + 2a^{-2} \sum_{k=0}^{m+1} \left(\frac{d}{a}\right)^{2k} |f|_{(2k+\ell)} . \tag{3.4.9}$$

Further,

$$|A_d^m| + |B_d^m| \le C(a)\Big\{|\alpha^+| + |\alpha^-| + \sum_{k=0}^{m} \left(\frac{d}{a}\right)^{2k} (|f^{(2k)}(-1)| + |f^{(2k)}(+1)|)\Big\} , \tag{3.4.10}$$

where $C(a)$ is independent of m and d.

Proof For $m \in \mathbb{N}$, we define

$$\omega_d^m(x) = \sum_{k=0}^{m} d^{2k} a^{-2k-2} f^{(2k)}(x) . \tag{3.4.11}$$

Then we see, using (3.4.2), that $R_d^m = u_d - \omega_d^m$ satisfies

$$L_d R_d^m(x) = f(x) - L_d\omega_d^m(x) = (d/a)^{2m+2} f^{(2m+2)}(x) =: g(x) . \tag{3.4.12}$$

For m large, we see that ω_d^m will satisfy (3.4.2) up to the correction $g(x)$. However, in general, the boundary conditions (3.4.3) will not be satisfied. We therefore introduce appropriate boundary layer terms to enforce (3.4.3). For this purpose, we define $u_k^{\mathrm{BL}}(x)$ to be the unique solution of

$$L_d u_k^{\mathrm{BL}}(x) = 0 \quad \text{on } I \tag{3.4.13}$$

$$u_k^{BL}(\pm 1) = C_k^{\pm} := \delta_{0,k}\alpha^{\pm} - a^{-2k-2} f^{(2k)}(\pm 1) . \tag{3.4.14}$$

Then with $u_{a,d}$ and $\bar{u}_{a,d}$ as defined in (3.4.7), we may verify that

$$u_k^{\mathrm{BL}}(x) = A_k u_{a,d}(x) + B_k \bar{u}_{a,d}(x) \tag{3.4.15}$$

where

$$A_k = \frac{C_k^- - C_k^+ e^{-2ad}}{1 - e^{-4ad}}, \quad B_k = \frac{C_k^+ - C_k^- e^{-2ad}}{1 - e^{-4ad}} . \tag{3.4.16}$$

Then we write

$$U_m^{\mathrm{BL}}(x) = \sum_{k=0}^{m} d^{2k} u_k^{\mathrm{BL}}(x) = A_d^m u_{a,d}(x) + B_d^m \bar{u}_{a,d}(x) ,$$

where $A_d^m = \sum_{k=0}^m d^{2k} A_k$ and $B_d^m = \sum_{k=0}^m d^{2k} B_k$. We see that

$$|A_d^m| \le \sum_{k=0}^{m} d^{2k}|A_k| \le (1 - e^{-4a})^{-1} \sum_{k=0}^{m} d^{2k}(|C_k^-| + |C_k^+|) \tag{3.4.17}$$

with a similar bound holding for $|B_d^m|$. Equation (3.4.10) follows from (3.4.17) and (3.4.14).

We now define

$$r_d^m = u_d - \omega_d^m - A_d^M u_{a,d} - B_d^m \bar{u}_{a,d} = R_d^m - U_m^{\mathrm{BL}} .$$

Then $r_d^m \in H_0^1(I)$ and r_d^m satisfies (3.4.12). Hence,

$$\|r_d^m\|_E^2 := B_d(r_d^m, r_d^m) = (g, r_d^m) \le \|g\|_0 \, \|r_d^m\|_E .$$

From this we deduce (using (3.4.12)) that

$$\begin{aligned} \|r_d^m\|_0 &\le a^{-1}(d/a)^{2m+2}\|f^{(2m+2)}\|_0 , \\ |r_d^m|_1 &\le a^{-1}(d/a)^{2m+1}\|f^{(2m+2)}\|_0 . \end{aligned} \tag{3.4.18}$$

Since r_d^m satisfies (3.4.12), we may differentiate (3.4.12) successively to obtain, using (3.4.18),

$$\begin{aligned}|r_d^m|_\ell \le a^{-1}(d/a)^{2m+2-\ell}\|f^{(2m+2)}\|_0 \\ + a^{-2}\sum_{k=0}^{\ell-2}(d/a)^{2m+2-\ell+k}\|f^{(2m+2+k)}\|_0\end{aligned} \tag{3.4.19}$$

where $\ell = 0, 1, \ldots 2m$. Moreover, from (3.4.11), we see that for $\ell = 0, 1, \ldots 2m$

$$|\omega_d^m|_\ell \le a^{-2}\sum_{k=0}^{m}(d/a)^{2k}\|f^{(2k+\ell)}\|_0 \, . \tag{3.4.20}$$

Define $u_d^m = \omega_d^m + r_d^m$. Then (3.4.8) holds. Also, using (3.4.19)–(3.4.20), we may establish (3.4.9). □

Remark 3.51 *Suppose $f \in \Pi_{2m+1}(I)$. Then in* (3.4.85), *we have $g(x) \equiv 0$, so that, since $r_d^m \in H_0^1(I)$ satisfies* (3.4.12), *we must have $r_d^m = 0$. Hence, $u_d^m = \omega_d^m$ and it is seen by* (3.4.12) *that $u_d^m \in \Pi_{2m+1}(I)$.*

For any interval $\tilde{I}$ let $\Pi_n(\tilde{I})$ denote the set of polynomials on $\tilde{I}$ of degree $\le n$. The following result follows by Remark 3.51.

Corollary 3.52 *Let $f \in \Pi_{2m+1}(I)$, then $u_d^m \in \Pi_{2m+1}(I)$ in (3.4.8).*

Remark 3.53 Analogous results hold when the Dirichlet conditions (3.4.3) are replaced by Neumann or mixed boundary conditions.

For f smooth enough (i.e., m large enough), we see from Theorem 3.50 that the regularity of u_d (in terms of d) will be determined by the boundary layer terms. We have, in fact, by (3.4.8)–(3.4.10),

$$|u_d|_\ell \le |u_d^m|_\ell + |A_d^m||u_{a,d}|_\ell + |B_d^m||\bar{u}_{a,d}|_\ell \le C(1 + |u_{a,d}|_\ell + |\bar{u}_{a,d}|_\ell) \tag{3.4.21}$$

where the constant C depends upon a, f and $\alpha^\pm$ but is independent of d. For the function $u_{a,d}$, we have for $\ell = 0, 1, 2, \ldots$,

$$|u_{a,d}|_\ell = \left(\frac{d}{a}\right)^{\frac{1}{2}-\ell}\left[\frac{1-e^{-4ad}}{2}\right]^{\frac{1}{2}} \approx Cd^{\frac{1}{2}-\ell} \tag{3.4.22}$$

so that (3.4.22) and its analog for $|\bar{u}_{a,d}|_\ell$, substituted in (3.4.21), gives an upper bound for $|u_d|_\ell$. Since, except for special cases, the coefficients A_d^m, B_d^m are non-zero, we see that the following **equivalence** generally holds

$$|u_d|_\ell \approx C(1 + d^{\frac{1}{2}-\ell}), \quad \ell = 0, 1, \ldots, 2m \, . \tag{3.4.23}$$

To conclude this section, we define the following **solution spaces** which will be used later:

$$H^B_{d,m} = \{u_d | u_d \text{ is a solution of (3.4.2)-(3.4.3) with } f \in H^{4m+2}(I)\,,\ \|f\|_{4m+2} \le B, |\alpha^\pm| \le B\}$$

$$H^B_{d,\Pi_n} = \{u_d | u_d \text{ is a solution of (3.4.2)-(3.4.3) with } f \in \Pi_n(I) \text{ such that all coefficients in } f \text{ are absolutely bounded by } B, |\alpha^\pm| \le B\}\,.$$

3.4.3 *The finite element method*

For any finite-dimensional subspace S of $H^1(I)$, denote $S_D = S \cap H^1_D(I)$, $S_0 = S \cap H^1_0(I)$. A finite element approximation u^S_d of u_d is obtained as usual: find $u^S_d \in S_D$ such that

$$B_d(u^S_d, v) = (f, v) \quad \text{for all } v \in S_0. \tag{3.4.24}$$

For every $d \in (0,1]$ there exists a unique solution $u^S_d \in S_D$ of (3.4.24).

We will be interested in spaces S of piecewise polynomials on I characterized by the mesh-degree combination $(\mathcal{T}, \boldsymbol{p})$, defined as follows. Let the $M+1 \ge 2$ nodal points

$$-1 =: x_0 < x_1 < x_2 < \ldots < x_{M-1} < x_M := 1 \tag{3.4.25}$$

be given, then the mesh $\mathcal{T}$ is defined by

$$\mathcal{T} = \{\Omega_i\}_{i=1}^M, \quad \Omega_i = (x_{i-1}, x_i)\,, \quad h_i = |\Omega_i| = x_i - x_{i-1} \tag{3.4.26}$$

(we will also abuse notation and write $\mathcal{T} = \{x_0, x_1, \ldots, x_M\}$ where convenient).

The degree vector $\boldsymbol{p}$ is defined as before

$$\boldsymbol{p} = (p_1, p_2, \ldots, p_M)\,. \tag{3.4.27}$$

Then

$$S^{\boldsymbol{p},1}(\Omega, \mathcal{T}) = \{u \in H^1(-1,1) : u|_{I_i} \in \Pi_{p_i}(\Omega_i), \quad \Omega_i \in \mathcal{T}\}\,. \tag{3.4.28}$$

Obviously,

$$\begin{aligned} \dim\big(S^{\boldsymbol{p},1}(\Omega,\mathcal{T})\big) &= 1 + \sum_{i=1}^M p_i, \\ N = \dim\big(S^{\boldsymbol{p},1}_D(\Omega,\mathcal{T})\big) &= \dim\big(S^{\boldsymbol{p},1}(\Omega,\mathcal{T})\big) - 2\,. \end{aligned} \tag{3.4.29}$$

By (3.4.4), (3.4.24),

$$B_d(u_d - u^S_d, v) = 0 \quad \text{for all } v \in S^{\boldsymbol{p},1}_0(\Omega, \mathcal{T})\,, \tag{3.4.30}$$

so that $e^S_d = u_d - u^S_d$ satisfies

$$\|e^S_d\|_d = \inf_{\chi \in S^{\boldsymbol{p},1}_D(\Omega,\mathcal{T})} \|u_d - \chi\|_d\,. \tag{3.4.31}$$

Here the **energy norm** $\|\cdot\|_d$, $0 < d \le 1$, is defined by

$$\|v\|_d = (B_d(v,v))^{1/2} = (d^2|v|_1^2 + \|v\|_0^2)^{1/2} \approx d|v|_1 + \|v\|_0 \,. \tag{3.4.32}$$

3.4.4 *Robustness*

The question we wish to explore here is the design of the spaces $S^{\mathbf{p},1}(\Omega, \mathcal{T})$ such that $\|e_d^S\|_d$ has a good convergence rate $g(N) \to 0$ as $N \to \infty$ **independent of** *d*. To do this, we recall the definition of **robustness** from [23].

Definition 3.54 *The FEM for problem* (3.4.24) *using spaces* $S_D^{\mathbf{p},1}(\Omega, \mathcal{T})$ **is robust with uniform order** $g(N)$ *for* $0 < d \le 1$ *with respect to solution sets* $\mathcal{H}_d = H_{d,m}^B$ *(or* H_{d,Π_n}^B*) and error measures* $E_d = \|\cdot\|_d$ *if and only if*

$$\lim_{N\to\infty} \Big(\big(\sup_{d\in(0,1]} \sup_{u_d \in \mathcal{H}_d} E_d(u_d - u_d^N) \big) \frac{1}{g(N)} \Big) = C < \infty \,.$$

Although we concentrate here primarily on the energy norm, other error measures could be considered as well: L_2 norm error estimates obviously follow as a corollary, while the maximum norm error is considered in Corollary 3.77. Note that by (3.4.23), the unscaled H^1 norm of u_d is not bounded uniformly for $d \in (0,1]$, so that we cannot expect robustness with uniform order in this norm (see e.g. estimates (3.4.40) in [122]).

Let $u_d \in H_{d,m}^B$. Using Theorem 3.50 and (3.4.31), we see immediately that for the energy norm,

$$\begin{aligned} E_d(u_d - u_d^N) = \|e_d^S\|_d &\le \inf_{\chi \in S_D^{\mathbf{p},1}} \|(u_d^m + A_d^m u_{a,d} + B_d^m \bar{u}_{a,d}) - \chi\|_d \\ &\le \inf_{\substack{\chi_1 \in S^{\mathbf{p},1} \\ \chi_1(\pm 1) = u_d^m(\pm 1)}} \|u_d^m - \chi_1\|_d \\ &\quad + |A_d^m| \inf_{\substack{\chi_2 \in S^{\mathbf{p},1} \\ \chi_2(\pm 1) = u_{a,d}(\pm 1)}} \|u_{a,d} - \chi_2\|_d \\ &\quad + |B_d^m| \inf_{\substack{\chi_3 \in S^{\mathbf{p},1} \\ \chi_3(\pm 1) = \bar{u}_{a,d}}} \|\bar{u}_{a,d} - \chi_3\|_d \,. \end{aligned} \tag{3.4.33}$$

Assume the space $S^{\mathbf{p},1}(\Omega, \mathcal{T})$ has the following approximation property:

$$\inf_{\substack{\chi \in S^{\mathbf{p},1} \\ \chi(\pm 1) = u(\pm 1)}} \|u - \chi\|_1 \le F(N,k)\|u\|_k, \quad k = 1, 2, \ldots \tag{3.4.34}$$

where $F(N,k)$ is some (optimal) approximation order (i.e., $F(N,k) \to 0$ as $N \to \infty$). Then for $u_d \in H_{d,m}^B$, by Theorem 3.50, the first infimum in (3.4.33) will tend to zero at the rate $KF(N, 2m)$ as $N \to \infty$ where K is a constant independent of d (K only depends upon B and m). Also, we may assume by

Theorem 3.50 that $|A_d^m|, |B_d^m| \leq K$, so that the second infimum in (3.4.33) will decrease at the rate $K\sqrt{a}\, Z(\frac{d}{a}, S^{p,1})$ where

$$\begin{aligned} Z(d, S^{p,1}) &= \inf_{\substack{\chi \in S^{p,1} \\ \chi(\pm 1) = u_{1,d}(\pm 1)}} \|u_{1,d} - \chi\|_d \\ &= \inf_{\substack{\chi \in S^{p,1} \\ \chi(\pm 1) = u_{1,d}(\pm 1)}} \{d^2 |u_{1,d} - \chi|_1^2 + \|u_{1,d} - \chi\|_0^2\}^{\frac{1}{2}} . \end{aligned} \tag{3.4.35}$$

By symmetry about $x = 0$, the last term in (3.4.33) will also have the same bound. Then our FEM will be robust in the sense of Definition 3.54 if and only if $Z(d, S)$ in (3.4.35) can be bounded independently of d, i.e.,

$$\sup_{d \in (0,1]} Z(d, S) \leq G(N) . \tag{3.4.36}$$

In that case, using (3.4.33) - (3.4.36) and Definition 3.54, the *hp*-FEM will be robust with uniform order

$$g(N) = C \max\{F(N, 2m), G(N)\} . \tag{3.4.37}$$

We will use the following related definition.

Definition 3.55 *The spaces $S^{p,1}(\Omega, \mathcal{T})$ will be said to approximate boundary layers $u_{1,d}$* **robustly at the rate** $G(N)$ **in the energy norm** *iff* (3.4.36) *holds.*

Remark 3.56 Our main concern in (3.4.37) is the rate $G(N)$, i.e., finding spaces $S^{p,1}(\Omega, \mathcal{T})$ such that (3.4.36) holds with $G(N) \to 0$ uniformly at a sufficiently fast rate. This is because in general, $G(N)$ will be the dominant term in (3.4.37), the idea being that N is large enough so that $F(N, 2m)$ is sufficiently small. For the *hp* spaces in Section 3.4.5, however, $G(N) \to 0$ exponentially, so that the algebraic rate $F(N, 2m)$ achieved by assuming regularity in terms of finite N will dominate as N becomes sufficiently large. This technical problem could be overcome by restricting the set of solutions $\mathcal{H}_d$ in Definition 3.54 to those for which the first infimum in (3.4.33) decays exponentially (or sufficiently fast). In particular, choosing $\mathcal{H}_d = H^B_{d,\Pi_n}$ will make this infimum vanish for suitable $S^{p,1}(\Omega, \mathcal{T})$ (see Theorem 3.79 ahead). The problem of exponential convergence when $m \to \infty$ was addressed in [93].

Remark 3.57 The FE spaces satisfying (3.4.36) constructed here and the estimates $G(N)$ established for them are also applicable to various other problems where the solution can be decomposed into boundary layers and smooth terms. We refer to [93], [95] and the references there.

3.4.5 *p approximation results*

In this section, we will prove asymptotic error estimates for $Z(d,S)$ given by (3.4.35) as $p \to \infty$, in the case that a single element $I = (-1,1)$ is used, i.e., $S^{p,1}(\Omega, \mathcal{T}) = \mathcal{S}^p(I)$. Our first estimate (3.4.15) will be valid uniformly in d for the range $\tilde{p} > e/2d$. (For any integer k, we write $\tilde{k} = k + 1/2$.) We will also provide separate estimates (again uniform in d) for the pre-asymptotic ranges $\sqrt{3/4d} \le \tilde{p} \le 2d$ and $1 < \tilde{p} < Kd^{-\frac{1}{2}}$. Our final theorem will establish a uniform robustness rate of $Cp^{-1}\sqrt{\ln p}$ for the p version over a fixed mesh, which will be shown to be **optimal** (up to the factor $\sqrt{\ln p}$).

In order to estimate $Z(d,S)$, we will use Theorem 3.14 which we restate here for convenience.

Lemma 3.58 *Let $u, u' \in L_2(I)$ and denote by*

$$b_n = \tilde{n} \int_{-1}^{1} u'(x) P_n(x) dx \tag{3.4.38}$$

the Legendre coefficients of u'. Then there exists $\chi \in \mathcal{S}^p(I)$ such that

$$\chi(\pm 1) = u(\pm 1)\,, \tag{3.4.39}$$

$$\|u' - \chi'\|_{0,I}^2 = \sum_{n=p}^{\infty} \frac{|b_n|^2}{\tilde{n}}\,, \tag{3.4.40}$$

$$\|u - \chi\|_{0,I}^2 \le \sum_{n=p}^{\infty} \frac{|b_n|^2}{n(n+1)\tilde{n}}\,, \tag{3.4.41}$$

$$\|u' - \chi'\|_{0,I} \le \|u' - \xi'\|_{0,I} \tag{3.4.42}$$

for any $\xi \in \mathcal{S}^p(I)$ satisfying $\xi(\pm 1) = u(\pm 1)$.

Remark 3.59 The polynomial χ above is obtained as an antiderivative of the truncated Legendre expansion of u', of degree $p - 1$. While by (3.4.42), this is optimal in the $H^1(I)$ seminorm, it is nonoptimal in the $\|\cdot\|_d$ norm. Nevertheless, Lemma 3.58 is sufficient for our purposes here.

The estimates in (3.4.40), (3.4.41) obviously depend on the size of the Legendre coefficients b_n in dependence on d and n. The following lemma gives precise bounds for these coefficients for our function $u \equiv u_{1,d}$.

Lemma 3.60 *Let $u \equiv u_{1,d}$ and b_n be defined by (3.4.38). Then with $\tilde{n} = n+1/2$,*

$$\left(1-\frac{2\nu_0}{\tilde{n}}\right) \le \frac{b_n(d)}{\phi(n,d)} \le \left(1-\frac{2\nu_0}{\tilde{n}}\right)^{-1}, \quad \text{for } n = 1,2,\ldots \tag{3.4.43}$$

where

$$\phi(n,d) = (-1)^{n+1} \frac{d^{-\frac{1}{2}}\tilde{n}^{\frac{1}{2}}}{(1+z^2)^{\frac{1}{4}}}\, e^{-\tilde{n}(z-\xi(z))}, \quad z = (\tilde{n}d)^{-1}, \tag{3.4.44}$$

$$\xi(z) = (1+z^2)^{\frac{1}{2}} + \ln\left(\frac{z}{1+(1+z^2)^{1/2}}\right) \tag{3.4.45}$$

and

$$\nu_0 = \frac{1}{6\sqrt{5}} + \frac{1}{12} \approx 0.158\,.$$

Proof Using (3.4.38) and the fact that $u'_{1,d} = -d^{-1}u_{1,d}$, we have

$$\begin{aligned} b_n &= -\tilde{n}\, d^{-1}e^{-1d} \int_{-1}^{1} e^{-xd}\, \frac{1}{2^n n!}\, \frac{d^n}{dx^n}((x^2-1)^n)\, dx \\ &= (-1)^{n+1} d^{-n-1} e^{-1d}\, \frac{\tilde{n}}{2^n n!} \int_{-1}^{1} (1-x^2)^n e^{-xd}\, dx\,. \end{aligned}$$

Hence, by formula 3.387 of [66],

$$\begin{aligned} b_n(d) &= (-1)^{n+1} d^{-n-1} e^{-1d}\, \tilde{n} 2^{-n} \sqrt{\pi}\, (2d)^{\tilde{n}}\, I_{\tilde{n}}(d^{-1}) \\ &= (-1)^{n+1} d^{-1/2} \sqrt{2\pi}\, \tilde{n}\;\; e^{-1d}\, I_{\tilde{n}}(d^{-1}) \end{aligned} \tag{3.4.46}$$

where $I_{\tilde{n}}(d^{-1})$ is the modified Bessel function ([66], 8.406). Thus, to obtain the asymptotic behavior of $b_n(d)$, we must investigate $I_{\tilde{n}}(d^{-1})$. To this end, we use asymptotic expansions of $I_\nu(\nu z)$ that are **uniform** for $z > 0$. Such uniformly valid asymptotic expansions have been obtained by F.W.J. Olver in [108].

Let $\nu = \tilde{n} = n + 1/2$ and $z = (\nu d)^{-1}$, then

$$e^{-1d} I_{\tilde{n}}(d^{-1}) = e^{-\nu z}\, I_\nu(\nu z)\,. \tag{3.4.47}$$

It is shown in [108] that

$$e^{-\nu z} I_\nu(\nu z) = \left(\frac{t}{2\pi\nu}\right)^{\frac{1}{2}} e^{-\nu(z-\xi(z))}\, \frac{\sum_{s=0}^{m} \frac{\mathcal{U}_s(t)}{\nu^s} + \epsilon_m(\nu,t)}{1+\epsilon_m(\nu,0)} \tag{3.4.48}$$

where $t = (1+z^2)^{-1/2}$, $m \ge 0$ is an integer, and the $\mathcal{U}_s(t)$ are certain polynomials of degree $3s$ in t (see [108]), the first two of which are given by

$$\mathcal{U}_0(t) = 1, \quad \mathcal{U}_1(t) = (3t - 5t^3)/24\,. \tag{3.4.49}$$

The ϵ_m in (3.4.47) are estimated by ([108])

$$|\epsilon_m(\nu,t)| \le \frac{\nu}{(\nu-\nu_0)} \frac{\mathcal{V}_t^1(\mathcal{U}_{m+1})}{\nu^{m+1}} \tag{3.4.50}$$

where

$$\mathcal{V}_a^b(\mathcal{U}) = \int_a^b |\mathcal{U}'(t)|\, dt \text{ and } \nu_0 = \mathcal{V}_0^1(\mathcal{U}_1) = \frac{1}{6\sqrt{5}} + \frac{1}{12}\,.$$

Simplifying (3.4.46) and using (3.4.47)–(3.4.50) with $m=0$ yields, with $\phi(n,d)$ as in (3.4.44), that

$$b_n = \phi(n,d)\, \frac{1+\epsilon_0(\tilde{n},t)}{1+\epsilon_0(\tilde{n},0)}\,.$$

The assertion then follows since

$$0 < \frac{1+\epsilon_0(\tilde{n},t)}{1+\epsilon_0(\tilde{n},0)} \le \left(1-\frac{2\nu_0}{\tilde{n}}\right)^{-1}\,.$$

□

Remark 3.61 The bound (3.4.43) is sharp in the sense that [108]

$$0.7895 \le 1-\frac{2\nu_0}{\tilde{n}},\ \left(1-\frac{2\nu_0}{\tilde{n}}\right)^{-1} \le 1.2667,\ n\ge 1\,.$$

Lemma 3.60 reduces the description of the asymptotic behavior of $b_n(d)$ to a discussion of the function $\phi(n,d)$. We then obtain the following bounds on the approximation errors (3.4.40), (3.4.41).

Lemma 3.62

$$\|u'-\chi'\|_0^2 \le \sum_{n=p}^{\infty} \theta^+(n,d)\, e^{-2\tilde{n}(z-\xi(z))}\,, \tag{3.4.51}$$

$$\|u-\chi\|_0^2 \le \sum_{n=p}^{\infty} \frac{1}{n(n+1)} \theta^+(n,d)\, e^{-2\tilde{n}(z-\xi(z))}\,, \tag{3.4.52}$$

$$\|u'-\chi'\|_0^2 \ge \sum_{n=p}^{\infty} \theta^-(n,d)\, e^{-2\tilde{n}(z-\xi(z))}\,. \tag{3.4.53}$$

where $z=(\tilde{n}d)^{-1}, \xi(z)$ *is as in* (3.4.45) *and*

$$\theta^{\pm}(n,d) := \left(1-\frac{2\nu_0}{\tilde{n}}\right)^{\mp 2} (d^2+\tilde{n}^{-2})^{-\frac{1}{2}}\,. \tag{3.4.54}$$

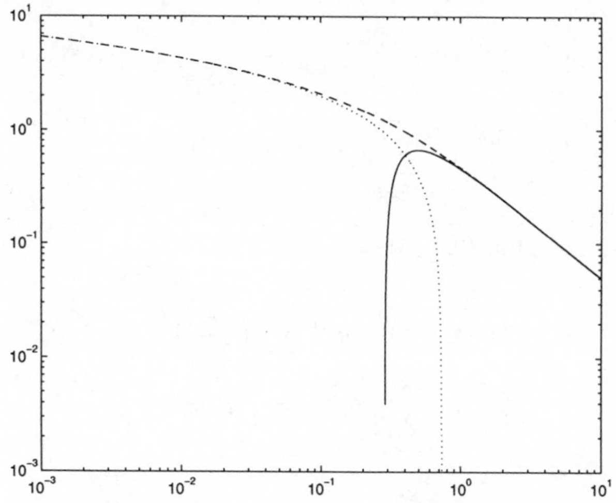

FIG. 3.2. The function $z - \xi(z)$ (- -) and its asymptotes $1/(2z) - 1/(24z^3) - (1 + \ln(z/2))$ ($\cdots$)

As is readily apparent from the expression for $\phi(n, d)$ in (3.4.44), we can expect exponential decay of b_n as $n \to \infty$ **provided** the function $z - \xi(z)$ is positive and of reasonable size. The following lemma provides bounds for $z - \xi(z)$ in terms of the asymptotes shown in Figure 3.2. The proof follows by elementary arguments.

Lemma 3.63 *For any $z > 0$, we have $z - \xi(z) \geq 0$. Moreover, the following bounds hold.*

$$-(1 + \ln(z/2)) \leq z - \xi(z) \leq z - (1 + \ln(z/2)) \tag{3.4.55}$$

$$\frac{1}{2z} - \frac{1}{24z^3} \leq z - \xi(z) \leq \frac{1}{2z} \,. \tag{3.4.56}$$

We now prove an error bound for sufficiently large p ($\tilde{p} > e/2d$).

Theorem 3.64 *Let $r := e/(2\tilde{p}d) < 1$. Then for $u \equiv u_{1,d}$ there exists a polynomial $\chi \in \Pi_p(I)$ such that $\chi(\pm 1) = u(\pm 1)$ and*

$$\|u' - \chi'\|_0 \leq C_1 d^{-1/2} r^{\tilde{p}} (1 - r^2)^{-\frac{1}{2}} \,, \tag{3.4.57}$$

$$\|u - \chi\|_0 \leq C_0 d^{1/2} r^{\tilde{p}} (1 - r^2)^{-\frac{1}{2}} \,. \tag{3.4.58}$$

Here C_i are independent of p and d (numerical values can be read off the proof).

Proof By (3.4.51), (3.4.54), we must estimate the sum

$$S^+ = \sum_{n=p}^{\infty} \left(1 - \frac{2\nu_0}{\tilde{n}}\right)^{-2} (d^2 + \tilde{n}^{-2})^{-\frac{1}{2}} \, e^{-2\tilde{n}(z-\xi(z))} \le \left(1 - \frac{4\nu_0}{3}\right)^{-2} \sum_{n=p}^{\infty} d^{-1} \, e^{-2\tilde{n}(z-\xi(z))} \, . \tag{3.4.59}$$

Using the lower bound in (3.4.55), we get

$$S^+ \le (1 - 4\nu_0/3)^{-2} d^{-1} \sum_{n=p}^{\infty} \left(\frac{e}{2\tilde{n}d}\right)^{2\tilde{n}} = C_1^2 \, d^{-1} \sum_{n=p}^{\infty} r^{2\tilde{n}} = C_1^2 \, d^{-1} r^{2\tilde{p}} (1 - r^2)^{-1} \, . \tag{3.4.60}$$

This is (3.4.57). To prove (3.4.58), we observe that $r < 1$ implies that

$$\frac{1}{p(p+1)} \le \frac{9}{8\tilde{p}^2} \le \frac{9}{8} \left(\frac{4d^2}{e^2}\right) = \frac{9d^2}{2e^2} \, .$$

Hence

$$\frac{1}{p(p+1)} \, S^+ \le C_0^2 \, d \, r^{2\tilde{p}} (1 - r^2)^{-1}, \; C_0^2 = \frac{9}{2e^2} \, C_1^2$$

and (3.4.58) follows. □

Corollary 3.65 *Let* $\Phi(d, S)$ *be as in* (3.4.35). *Then for* $r = e/(2\tilde{p}d) < 1$,

$$\Phi(d, S) \le C_2 d^{1/2} \, r^{\tilde{p}} (1 - r^2)^{-1/2} \, , \quad C_2 = (C_0^2 + C_1^2)^{1/2} \, . \tag{3.4.61}$$

Remark 3.66 The asymptotic rate of convergence with respect to p in (3.4.56) and hence (3.4.61) is optimal up to a constant depending on d, since by (3.4.53),

$$\|u' - \chi'\|_0^2 \ge \sum_{n=p}^{\infty} \theta^-(n, d) \, e^{-2\tilde{n}(z-\xi(z))} =: S^- .$$

Using the upper bound in (3.4.55) yields with $z := (\tilde{n}d)^{-1}$

$$\begin{aligned} S^- &\ge \left(1 - \frac{2\nu_0}{\tilde{p}}\right)^2 \sum_{n=p}^{\infty} (d^2 + \tilde{n}^{-2})^{-\frac{1}{2}} \, e^{-2d} \, r^{2\tilde{n}} \\ &\ge \left(1 - \frac{4\nu_0}{3}\right)^2 e^{-2d} d^{-1} \sum_{n=p}^{\infty} (1 + z^2)^{-1/2} \, r^{2\tilde{n}} \\ &\ge \left(1 - \frac{4\nu_0}{3}\right)^2 (1 + 4/e^2)^{-\frac{1}{2}} \, e^{-2d} d^{-1} \, r^{2\tilde{p}} \, (1 - r^2)^{-1} \, , \end{aligned}$$

since $z \leq (\tilde{p}d)^{-1} < 2/e$. Hence,

$$\|u' - \chi'\|_0 \geq C_3\, e^{-1d}\, d^{-\frac{1}{2}}\, r^{\tilde{p}}(1 - r^2)^{-\frac{1}{2}} .$$

Now χ is the same polynomial as in Lemma 3.58, so that (3.4.42) holds. Comparing the above estimate with (3.4.57), we see that (3.4.57), (3.4.61) are optimal in p for any fixed $d > 0$ as $p \to \infty$.

The estimates in Theorem 3.64 are useful for the case that $\tilde{p}$ is large compared to $e/(2d)$. Such a situation arises in the next section, where this theorem will be applied. In actual practice, if d is small, then it can be difficult to ever be in this asymptotic range of $\tilde{p}$. The computational results in Section 3.4.6 show that convergence is observed in the pre-asymptotic range $\tilde{p} \leq e/(2d)$ as well. Therefore, we now obtain estimates for the rate of decrease of the error in the range $\sqrt{3/(4d)} \leq \tilde{p} \leq e/(2d)$.

In Theorem 3.64, we used the bounds (3.4.55) as well as (3.4.56), the latter being sharper for the range $\tilde{p} \leq e/(2d)$, i.e. $z \geq 2/e$ (see Figure 3.2). It is seen that the two lower bounds for $z - \xi(z)$ in Lemma 3.63 intersect at the root of

$$\frac{1}{2z} - \frac{1}{24z^3} + 1 + \ln\left(\frac{z}{2}\right) = 0 ,$$

i.e., at $z^* = 0.51388...$, which is close to 0.5. Our estimate will therefore be valid for $z \geq 0.5$, i.e., in the extended range $\sqrt{3/(4d)} \leq \tilde{p} \leq 2/d$.

Theorem 3.67 *Assume that $\sqrt{3/(4d)} \leq \tilde{p} \leq 2/d$. Then, for $u \equiv u_{1,d}$, there exists a polynomial $\chi \in \Pi_p(I)$ such that $\chi(\pm 1) = u(\pm 1)$ and*

$$\|u' - \chi'\|_0^2 \leq C_1^2 \left(\tilde{p} + \frac{3}{4d}\right) \exp\left(-\frac{2\tilde{p}^2 d}{3}\right) + C_3\, d^{-1} \left(\frac{e}{4}\right)^{4d} \tag{3.4.62}$$

$$\|u - \chi\|_0^2 \leq \frac{9}{8}\, C_1^2 \tilde{p}^{-2} \left(\tilde{p} + \frac{3}{4d}\right) \exp\left(-\frac{2\tilde{p}^2 d}{3}\right) + \frac{4}{15}\, C_3\, d\left(\frac{e}{4}\right)^{4d} . \tag{3.4.63}$$

Here the constants C_i are independent of p and d (and are given in the proof below).

Proof Let us define the index sets

$$I_1(d) = \{n \in \mathbb{N} : \tilde{p} \leq \tilde{n} < 2d\}, \quad I_2(d) = \{n \in \mathbb{N} : \tilde{n} \geq 2d\} .$$

Then, taking χ to be the polynomial in Lemma 3.58, we have by (3.4.51)

$$\|u' - \chi'\|_0^2 \leq S_1 + S_2, \quad S_i = \sum_{n \in I_i} \theta^+(n, d)\, e^{-2\tilde{n}(z - \xi(z))} .$$

S_1 and S_2 will be estimated using the lower bounds in (3.4.56) and (3.4.55) respectively. First, by (3.4.56),

$$S_1 \leq \sum_{n \in I_1(d)} \theta^+(n,d) \exp\Big(-2\tilde{n}\Big(\frac{1}{2z} - \frac{1}{24z^3}\Big)\Big) \tag{3.4.64}$$

where, since $n \in I_1(d)$, we have $0 < z^{-1} = \tilde{n}d < 2$, so that $0 < z^{-3} < 4z^{-1}$. Hence,

$$\frac{1}{2z} - \frac{1}{24z^3} \geq \frac{1}{2z} - \frac{1}{6z} = \frac{1}{3z}$$

and (3.4.64), (3.4.54) give (with $C_1 = (1 - 4\nu_0/3)^{-1}$ as in (3.4.60))

$$S_1 \leq C_1 \sum_{n \in I_1(d)} \tilde{n}(1 + \tilde{n}^2 d^2)^{-1/2}\, e^{-\frac{2}{3}\tilde{n}^2 d}\,.$$

Now the function $xe^{-\frac{2}{3}x^2 d}$ attains its global maximum at $x = \sqrt{3/(4d)}$ and is decreasing for $x > \sqrt{3/(4d)}$. Hence,

$$\begin{aligned} S_1 &\leq C_1 \sum_{n \in I_1(d)} \tilde{n}\, e^{-\frac{2}{3}\tilde{n}^2 d} \leq C_1 \Big(\tilde{p}e^{-\frac{2}{3}\tilde{p}^2 d} + \int_{\tilde{p}}^{2d} xe^{-\frac{2}{3}x^2 d} dx\Big) \\ &\leq C_1 \Big(\tilde{p} + \frac{3}{4d}\Big) e^{-\frac{2}{3}\tilde{p}^2 d}\,. \end{aligned}$$

For the term S_2, we use (3.4.55). Noting that $\tilde{n}^{-1} \leq d/2$, we have

$$S_2 \leq \sum_{\tilde{n} \geq 2d} \theta^+(n,d) \Big(\frac{e}{2\tilde{n}d}\Big)^{2\tilde{n}} \leq \tilde{C}_3 d^{-1} \sum_{\tilde{n} \geq 2d} \Big(\frac{e}{4}\Big)^{2\tilde{n}}$$

where $\tilde{C}_3 = (1 - \nu_0)^{-2} \left(\frac{4}{5}\right)^{1/2}$. Summing the geometric series leads to the last term in (3.4.62), where $C_3 = \tilde{C}_3(1 - e^2/16)^{-1}$.

For the L_2 estimates, (3.4.52) gives

$$\|u - \chi\|_0^2 \leq \tilde{S}_1 + \tilde{S}_2, \quad \tilde{S}_i = \sum_{n \in I_i(d)} \frac{1}{n(n+1)}\, \theta^+(n,d)\, e^{-2\tilde{n}(z - \xi(z))}, \quad i = 1, 2\,.$$

It is easy to see that

$$\tilde{S}_1 \leq \frac{1}{p(p+1)}\, S_1 \leq \frac{9}{8\tilde{p}^2} S_1\,.$$

Also, for $n \in I_2(d)$, $n(n+1) \geq (16-d^2)/4d^2 \geq 15/4d^2$ for $d \leq 1$, so that

$$\tilde{S}_2 \leq \frac{4d^2}{15} S_2 .$$

The theorem follows. □

Theorem 3.67 leads to the following result.

Corollary 3.68 *For* $\sqrt{3/(4d)} \leq \tilde{p} \leq 2/d$,

$$\Phi(d,S) \leq C_4\, e^{-(\tilde{p}^2 d)/3} + C_5 \left(\frac{e}{4}\right)^{2/d} \tag{3.4.65}$$

where the constants C_i *are independent of* p, d.

Proof Using the definition of $\Phi(d,S)$, we obtain from Theorem 3.67,

$$(\Phi(d,S))^2 \leq C_1^2 \left(\tilde{p} + \frac{3}{4d}\right)\left(d^2 + \frac{9}{8}\,\tilde{p}^{-2}\right)e^{-2/3\,\tilde{p}^2 d} + \frac{19}{15}\, C_3 \left(\frac{e}{4}\right)^{4/d} .$$

Hence, (3.4.65) holds with $C_5^2 = \frac{19}{15} C_3$ and

$$C_1^2\left(\tilde{p}d^2 + \frac{3}{4}\,d + \frac{9}{8}\,\tilde{p}^{-1} + \frac{27}{32}\,\frac{\tilde{p}^{-2}}{d}\right) \leq C_1^2\left(4\tilde{p}^{-1} + \frac{3}{2}\,\tilde{p}^{-1} + \frac{9}{8}\,\tilde{p}^{-1} + \frac{9}{8}\right) \leq C_4^2 .$$

□

We see from Corollary 3.68 that for small d, since the term $C_5(e/4)^{2d}$ is negligible, the first term $C_4 \exp^{-(\tilde{p}^2 d)/3}$ in (3.4.65) will dominate. Hence, the error will decrease at an exponential rate in this range when $\tilde{p}^2 d/3$ is large enough. For $\tilde{p} > e/(2d)$, a better estimate may be provided by Theorem 3.64. When $\tilde{p}^2 d/3$ is small (i.e., $\tilde{p} = Kd^{-\frac{1}{2}}$), the estimate (3.4.65) deteriorates. We will therefore establish another bound, which is valid in the range $1 < \tilde{p} \leq Kd^{-\frac{1}{2}}$. First, we prove the following lemma.

Lemma 3.69 *There exists a unique polynomial* $\chi \in \Pi_p(I)$ *that minimizes* $\|\chi\|_0$ *subject to the constraints* $\chi(\pm 1) = \alpha^{\pm}$. *This polynomial* χ *satisfies*

$$\frac{1}{C}\frac{\max(|\alpha^+|,|\alpha^-|)}{p} \leq \|\chi\|_0 \leq C\frac{\max(|\alpha^+|,|\alpha^-|)}{p} \tag{3.4.66}$$

$$\|\chi\|_1 \leq C \max(|\alpha^+|,|\alpha^-|)p \tag{3.4.67}$$

with $C > 1$ *a constant independent of* $\alpha^{\pm}$, p.

Proof We may write χ in the Legendre series expansion satisfying the end constraints,

$$\chi(x) = \sum_{k=0}^{p} a_k L_k(x), \quad \sum_{k=0}^{p} a_k (\pm 1)^k = \alpha^{\pm} .$$

Introducing Lagrange multipliers for the constraints, we get the minimization problem

$$\min_{\vec{a},\lambda_+,\lambda_-} F(\vec{a}, \lambda_+, \lambda_-) = \sum_{k=0}^{p} \omega_k a_k^2 + \lambda_+ \Big(\sum_{k=0}^{p} a_k - \alpha^+ \Big) + \lambda_- \Big(\sum_{k=0}^{p} a_k (-1)^k - \alpha^- \Big)$$

where $\omega_k = 2/(2k+1)$. Let $A = p(p+1)^2(p+2)/4$. Then it may be shown that the unique minimizer for the above is given by

$$\lambda_{\pm} = \frac{2}{A} \Big(\alpha^{\mp} (-1)^p \, \frac{(p+1)}{2} - \alpha^{\pm} \, \frac{(p+1)^2}{2} \Big)$$
$$a_k = (\alpha^+ + (-1)^k \alpha^-) \, \frac{((p+1)^2 - (-1)^{p+k}(p+1))}{2A\omega_k}$$

from which the bounds for $\|\chi\|_0$ in (3.4.66) follow easily. The bound for $\|\chi\|_1$ follows by the inverse inequality for polynomials, (see Chapter 4, Section 4.6)

$$|\chi|_1 \le C p^{-2} \|\chi\|_0 .$$

□

Theorem 3.70 *Assume that* $1 < \tilde{p} \le K d^{-\frac{1}{2}}$ *for some* K *(which may depend upon* $\tilde{p}$, d*). Then for such* p, d,

$$\Phi(d, S) \le C K p^{-1} \tag{3.4.68}$$

where the constant C *is independent of* K, p *and* d.

Proof We note that for any $\chi \in \Pi_p(I)$ satisfying $\chi(\pm 1) = u_{1,d}(\pm 1)$,

$$\Phi(d, S) \le C(\|u_{1,d}\|_d + d|\chi|_1 + \|\chi\|_0) .$$

We use (3.4.22) to bound $\|u_{1,d}\|_d$ and choose χ as in Lemma 3.69, with $\alpha^{\pm} = u_{1,d}(\pm 1)$. Then we obtain by (3.4.66), (3.4.67),

$$\Phi(d, S) \le C(d^{\frac{1}{2}} + dp + p^{-1}) . \tag{3.4.69}$$

Now since $\tilde{p} \le K d^{-\frac{1}{2}}$, we have $d^{\frac{1}{2}} \le K \tilde{p}^{-1}$. Substituting this in (3.4.69) gives (3.4.68). □

Let us now put together the results of Theorems 3.64–3.70. The following theorem shows that the spaces $S^{p,1}(\Omega, \mathcal{T}) = \Pi_p(I)$ approximate boundary layers

$u_{1,d}$ **robustly** at the rate $G(p) = Cp^{-1}\sqrt{\ln p}$ in the energy norm (in the sense of Definition 3.54). Moreover, the best robust rate possible is Cp^{-1}, so that the result established is optimal up to a factor $\sqrt{\ln p}$.

Theorem 3.71 *Let $S^{p,1}(\Omega, \mathcal{T}) = \Pi_p(I)$. Then*

$$Cp^{-1} \le G(p) = \sup_{d\in(0,1]} \Phi(d,S) \le Cp^{-1}\sqrt{\ln p}$$

where C is a constant independent of p.

Proof Let $d \in (0,1]$ be arbitrary. Suppose first that

$$1 \le \frac{\tilde{p}}{2\sqrt{\ln p}} \le \sqrt{\frac{3}{4d}}, \quad \text{i.e.,} \quad 1 \le \tilde{p} \le \sqrt{\frac{3\ln p}{d}}\,.$$

Then, by Theorem 3.70, with $K = \sqrt{3\ln p}$, we have in this range

$$\Phi(d,S) \le CKp^{-1} \le Cp^{-1}\sqrt{\ln p}\,.$$

Next, for $\sqrt{3/(4d)} \le \tilde{p}/(2\sqrt{\ln p}) \le 2/(2d\sqrt{\ln p})$ we may use Corollary 3.65, by which

$$\Phi(d,S) \le C_4 e^{-\tilde{p}^2 d/3} + C_5\left(\frac{e}{4}\right)^{2/d}.$$

Since $\tilde{p}/(2\sqrt{\ln p}) \ge \sqrt{3/(4d)}$, we have $\tilde{p}^2 d/3 \ge \ln p$, so that $\exp(-\tilde{p}^2 d/3) \le 1/p$. Also, since $2d \ge \tilde{p}$, it follows that $(e/4)^{2d} \le (e/4)^{\tilde{p}} \le p^{-1}$. Hence,

$$\Phi(d,S) \le Cp^{-1} \tag{3.4.70}$$

in this range. Finally, it is easy to see that the estimate for $\Phi(d,S)$ for the range $\tilde{p} > 2d$, given by Theorem 3.64, also satisfies (3.4.70).

To establish the lower bound, we note that by the triangle inequality, for any $\chi \in \Phi_p(I)$ with $\chi(\pm 1) = u(\pm 1)$,

$$\begin{aligned}\Phi(d,S) &\ge \|\chi\|_d - \|u\|_d \\ &\ge \|\chi\|_0 - \|u\|_d.\end{aligned}$$

If $d \to 0$, then $\|u\|_d \to 0$. But, by Lemma 3.69, $\|\chi\|_0 \ge Cp^{-1}$, giving the result. □

Suppose the p version is used with a **fixed** mesh for problems (3.4.24). Then the rate $F(N,k)$ in (3.4.34) satisfies

$$F(N,k) \le Cp^{-(k-1)}$$

so that the first infimum on the right side of (3.4.33) will certainly be less than Cp^{-1}, uniformly in d whenever $k \geq 2$. Using Theorem 3.71 on the whole interval $[-1, 1]$, the infimum involving boundary layers may be uniformly bounded by $Cp^{-1}\sqrt{\ln p}$ (the fact that we have more than one interval can only enhance this rate). Hence, the p version is robust with uniform order $Cp^{-1}\sqrt{\ln p}$. Moreover, this robust rate is optimal up to $\sqrt{\ln p}$, since for $d \to 0$, the approximation of the boundary layer terms in the end intervals cannot be better than Cp^{-1} by Theorem 3.71. We therefore have the following result.

Theorem 3.72 *The p-version with fixed mesh for problems* (3.4.24), $0 < d \leq 1$, *is robust with uniform order* $g(p)$ *satisfying*

$$Cp^{-1} \leq g(p) \leq Cp^{-1}\sqrt{\ln p}$$

with respect to solution sets $H^B_{d,M}$ *(or* H^B_{d,Π_n}*) and error measure the energy norm.*

Remark 3.73 In terms of the number of degrees of freedom N, we see that $g(N) \approx N^{-1}\sqrt{\ln N}$. This is essentially **twice** the best uniform rate of $N^{-1/2}$ that can be attained using the h version with a quasiuniform mesh [122]. Hence, the "doubling" phenomenon for the rate of convergence for the p version which is well known for the case that $(x+1)^\alpha$ type singularities are present at $x = -1$ (see e.g. Theorem 3.26) also occurs when the solution contains boundary layer components of the type $\exp(-(x+1)d)$ at $x = -1$.

3.4.6 *hp approximation results*

In the previous section, we showed that the p-version over a single element yields a superexponential rate of convergence for $\tilde{p} > e/2d$. Also, the error decreases at the (exponential) rate $\exp(-\tilde{p}^2 d/3)$ in the pre-asymptotic range $\sqrt{3/(4d)} \leq \tilde{p} \leq 2/d$ for small d. Unfortunately, in practice both these ranges may be difficult to achieve if d is small and p is restricted ($p \leq 8$ is typical in programs such as MSC/PROBE and STRESSCHECK), so that all that may be observed is the uniform rate of $O(p^{-1}\sqrt{\ln p})$ predicted by Theorems 3.70, 3.71. In this section, we show that if only one extra element of size $O(pd)$ is inserted in the boundary layer, then **robust exponential convergence** is achieved uniformly for $0 < d \leq 1$ as p increases. Since the mesh is changed at each step when p is increased, we call this an *hp*-version FEM (more appropriately, it is an *rp*-version FEM). Naturally, if the polynomial degree p is sufficiently large, we have to allow a transition to the single element mesh analyzed in Theorem 3.64.

A more general question that could be considered is, given N degrees of freedom (N as in (3.4.29)), for what mesh–degree combination Σ (i.e., choice of S_D) is the error minimized? We do not consider this theoretical question here, since the simple two-element mesh below already gives exponential convergence, uniformly in d. This mesh is easier to implement than a general *hp*-version, and moreover, in computational experiments performed using meshes with more elements, we do not achieve better convergence rates (see Section 4.6 and [151]). Note that the mesh–degree combination we propose is similar to the **optimal** mesh–degree combination obtained for a related problem by Scherer in [123] (see Remark 3.76).

The following theorem is our **main result on *hp*-boundary layer approximation** .

Theorem 3.74 *Let* $I = (-1,1)$ *and* $u(x) = u_{1,d} = \exp(-(x+1)d)$. *Let the mesh–degree combination* $\Sigma = (\mathcal{T}, \mathbf{p})$ *be such that for* $p \geq 1$

$$\begin{aligned} &\mathbf{p} = \{p, 1\}, \mathcal{T} = \{-1, -1+\kappa\tilde{p}d, 1\} && \text{if } \kappa\tilde{p}d < 2, \\ &\mathbf{p} = \{p\}, \quad \mathcal{T} = \{-1, 1\} && \text{otherwise} \end{aligned}$$

where $0 < \kappa_0 \leq \kappa < 4/e$ *is a constant independent of* p *and* d. *Then there exists* $u_p \in S^{\mathbf{p},1}(\Omega, \mathcal{T})$ *such that* $u_p(\pm 1) = u(\pm 1)$ *and*

$$\begin{aligned} &\|u - u_p\|_d \leq d^{1/2} C_6\, \alpha^{\tilde{p}}, \quad \|u - u_p\|_0 \leq d^{1/2} C_7\, \alpha^{\tilde{p}}, \\ &\|u' - u_p'\|_0 \leq d^{-1/2} C_8\, \alpha^{\tilde{p}}\,. \end{aligned} \tag{3.4.71}$$

Here the constants are independent of p *and* d *but depend on* κ_0 *and*

$$\alpha := \left\{ \begin{array}{cc} e/(2\tilde{p}d) & \text{if } \kappa\tilde{p}d \geq 2\,, \\ \max\left\{\kappa e/4, e^{-(\kappa-\varepsilon)}\right\} & \text{if } \kappa\tilde{p}d < 2 \end{array} \right\} < 1\,, \tag{3.4.72}$$

with $\varepsilon > \ln p/(2p)$ *arbitrary.*

Proof If $\kappa\tilde{p}d \geq 2$, we have that $r = e/(2\tilde{p}d) < 1$ due to our assumption that $\kappa < 4/e$. Therefore Theorem 3.64 is applicable, and a p-increase in the single element mesh $\mathcal{T} = \{-1, 1\}$ yields exponential convergence with the rate r decreasing with p.

Consider now the case $\kappa\tilde{p}d < 2$, i.e. the two-element mesh $\mathcal{T} = \{-1, -1 + \kappa\tilde{p}d, 1\}$. We assume first that $\tilde{p} \geq 2/\kappa_0$ and construct the function $u_p \in S^{\mathbf{p},1}(\Omega, \mathcal{T})$ elementwise. Denote $I_1 = (-1, a)$ where $a = -1 + \kappa\tilde{p}d$, $\kappa_0 \leq \kappa < 4/e$, and let $s_1 \in \Pi_p(I_1)$. Transforming I_1 to $I = (-1,1)$ we see that for $t = 0, 1$,

$$\int_{-1}^{a} \left(\frac{d^t}{dx^t}(u - s_1)\right)^2 dx = \left(\frac{2}{\kappa\tilde{p}d}\right)^{2t-1} \int_{-1}^{1} \left(\frac{d^t}{dy^t}(\tilde{u} - \tilde{s}_1)\right)^2 dy\,.$$

Here $\tilde{f}(y)$ denotes the image on I of any function $f(x)$ defined on I_1. Consequently, we obtain that $\tilde{u}(y) = \exp(-(y+1)\kappa\tilde{p}/2) = u_{1,\tilde{d}}(y)$ where $\tilde{d} = 2/\kappa\tilde{p}$.

Since $\kappa < 4/e$, we have $r := e/(2\tilde{p}\tilde{d}) = \kappa e/4 < 1$. Now Theorem 3.64 and Corollary 3.65 apply uniformly to functions $u \equiv u_{1,d}$ for all $d \in (0,1]$. Since $\tilde{p} \geq 2/\kappa_0 \geq 2/\kappa$ we have that $\tilde{d} < 1$ and hence Theorem 3.64 and Corollary 3.65

will apply when d is chosen to be $\tilde{d}$. Then, since $r < 1$, we obtain a polynomial $s_p \in \Pi_p(I_1)$ satisfying

$$s_p(-1) = u(-1), \quad s_p(a) = u(a) , \tag{3.4.73}$$

and

$$\left\| \frac{d^t}{dx^t}(u - s_p) \right\|_{0,I_1}^2 \leq C_t^2 \left(\frac{2}{\kappa \tilde{p} d} \right)^{2t-1} \tilde{d}^{1-2t} \frac{r^{2\tilde{p}}}{(1 - r^2)}, \quad t = 0, 1, \tag{3.4.74}$$

$$(d^2 \|u' - s_p'\|_{0,I_1}^2 + \|u - s_p\|_{0,I_1}^2)^{1/2} \leq C_2 \, d^{1/2} \frac{r^{\tilde{p}}}{(1 - r^2)^{1/2}} \, . \tag{3.4.75}$$

This gives the asserted bound on I_1 in the case $\tilde{p} > 2/\kappa_0$. Since this excludes only finitely many values of p, these estimates hold for all p after possibly adjusting the constants C_t, $t = 0, 1, 2$ in (3.4.74), (3.4.75).

As noted in Remark 3.59, the approximation s_p constructed via Lemma 3.58 is optimal in the $|\cdot|_1$ semi-norm but not in the $\|\cdot\|_d$ norm. For fixed $d > 0$, s_p yields the optimal order error as $p \to \infty$, but is suboptimal as $d \to 0$ due to the enforcement of the interpolation condition (3.4.73). Therefore we modify s_p as follows: let $u_p = s_p - s_1 + \tilde{s}_1$ where s_1 is the linear interpolant of $u_{1,d}(x)$ at $x = -1$ and $x = a$ and $\tilde{s}_1$ is a linear function such that $\tilde{s}_1(-1) = u(-1)$ and $\tilde{s}_1(a) = \max\{d^{1/2}u(a), u(1)\}$. Then

$$\begin{aligned} \|u - u_p\|_{d,I_1} &= \|u - (s_p - s_1 + \tilde{s}_1)\|_{d,I_1} \\ &\leq \|u - s_p\|_{d,I_1} + \|s_1 - \tilde{s}_1\|_{d,I_1} \, . \end{aligned} \tag{3.4.76}$$

The first term was estimated in (3.4.75), so we estimate the second term. We have

$$\begin{aligned} \int_{-1}^{a} (s_1 - \tilde{s}_1)^2 dx &\leq \max\nolimits_{-1 \leq x \leq a} |(s_1 - \tilde{s}_1)(x)|^2 \, (1 + a) \\ &\leq |(1 - \sqrt{d})u(a)|^2 \, (1 + a). \end{aligned}$$

Since $1 + a = \kappa \tilde{p} d$ and $u(a) = \exp(-(a + 1)d) = \exp(-\kappa \tilde{p})$, we get

$$\|s_1 - \tilde{s}_1\|_{0,I_1}^2 \leq e^{-2\kappa \tilde{p}} \kappa \tilde{p} d \, .$$

Also,

$$\int_{-1}^{a} (s_1' - \tilde{s}_1')^2 dx \leq (1 - \sqrt{d})^2 \, |u(a)|^2 \, (1 + a)^{-1} \leq \frac{e^{-2\kappa \tilde{p}}}{\kappa \tilde{p} d} \, .$$

Hence

$$d \|s_1' - \tilde{s}_1'\|_{0,I_1} \leq \frac{d^{1/2}}{\sqrt{\kappa \tilde{p}}} e^{-\kappa \tilde{p}}$$

and altogether

$$\begin{aligned}\|s_1 - \tilde{s}_1\|_d &\le d^{\frac{1}{2}}(\kappa\tilde{p} + 1/\kappa\tilde{p})^{\frac{1}{2}} e^{-\kappa\tilde{p}} \\ &\le d^{\frac{1}{2}}(\kappa + 1/\kappa)^{\frac{1}{2}} e^{-(\kappa-\varepsilon)\tilde{p}}\end{aligned} \tag{3.4.77}$$

for any $\varepsilon > (\ln\tilde{p})/(2\tilde{p})$ ($\epsilon = 1/2e$ works for all p). Then, from (3.4.75) - (3.4.77),

$$\|u - u_p\|^2_{d,I_1} \le \tilde{C}_0^2\, d\Big\{\Big(\frac{\kappa e}{4}\Big)^{2\tilde{p}} + e^{-2(\kappa-\varepsilon)\tilde{p}}\Big\}. \tag{3.4.78}$$

Next we consider I_2. Here we select $u_p \in \Pi_1(I_2)$ to be the linear interpolant between $\max\{d^{1/2}u(a), u(1)\}$ at $x = a$ and $u(1)$ at $x = 1$, i.e.,

$$\begin{aligned}u_p(x) = (u(1) - \max\{u(1), \sqrt{d}\,u(a)\})\ &\frac{(x-a)}{1-a} \\ + \max\{u(1), \sqrt{d}\,u(a)\}\ .&\end{aligned} \tag{3.4.79}$$

Now let

$$\frac{2}{\kappa d} \ge \tilde{p} \ge \frac{2}{\kappa d} - \frac{|\ln d|}{2\kappa}\ .$$

Then since $(|\ln d|)/(2\kappa) \le (e^{-1})/(2\kappa d)$, we have $\tilde{p} \ge ((4 - e^{-1}))/(2\kappa d)$ in this range. Also,

$$u(1) \ge \sqrt{d}\,u(a) \ \text{ and } \ 1 - a \le \frac{|\ln d| d}{2} \le \ln\Big(\frac{2\tilde{p}\kappa}{(4-e^{-1})}\Big)\,\frac{d}{2}\ .$$

Hence,

$$\begin{aligned}\int_a^1 u_p^2\,dx \le u^2(1)\frac{|\ln d| d}{2} &\le \frac{d}{2}\ \ln\Big(\frac{2\tilde{p}\kappa}{(4-e^{-1})}\Big)e^{-2\kappa\tilde{p}} \\ &\le \tilde{C}_1\ de^{-2(\kappa-\epsilon)\tilde{p}}\,, \quad \int_a^1 (u_p')^2 = 0\ .\end{aligned} \tag{3.4.80}$$

Next, for $\tilde{p} < 2/(\kappa d) - (|\ell n d|)/(2\kappa)$,

$$u(1) < \sqrt{d}\,u(a) \quad \text{and} \quad \frac{1}{1-a} \le \frac{2}{d|\ln d|}\ .$$

Hence

$$\int_a^1 u_p^2\,dx \le (\sqrt{d}\,u(a))^2\,(1-a) \le 2de^{-2\kappa\tilde{p}}\,, \tag{3.4.81}$$

and

$$\int_a^1 (u_p')^2 dx \le \frac{2d\,u^2(a)}{|\ln d|\ d} \le \frac{2e^{-2\kappa\tilde{p}}}{|\ln d|}\ .$$

For $d \leq e^{-1}$, this gives

$$d^2 \int_a^1 (u_p')^2 dx \leq 2d^2 e^{-2\kappa\tilde{p}}. \tag{3.4.82}$$

For $e^{-1} < d \leq 1$, we have by (3.4.79) that

$$\int_a^1 (u_p')^2 \, dx \leq 2\Big(\frac{(u(1)-u(a))^2}{1-a} + \frac{(1-\sqrt{d})^2 \, u^2(a)}{1-a}\Big) \,.$$

By the mean value theorem, there exists $\xi \in [a,1]$ such that $u(1) - u(a) = u'(\xi)(1-a)$, so that

$$d^2 \int_a^1 (u_p')^2 dx \leq 2e^{-2((\xi+1)/d)}(1-a) + 4d \, \frac{(1-\sqrt{d})^2 \, u^2(a)}{|\ln d|} \tag{3.4.83}$$

$$\leq \tilde{C}_2 d e^{-2\kappa\tilde{p}} \tag{3.4.84}$$

uniformly as $d \to 1$, where $\tilde{C}_2$ may be explicitly evaluated. Hence, we conclude by (3.4.80) - (3.4.84) that

$$\|u_p\|^2_{d,I_2} \leq \tilde{C}_3 \, d \, e^{-2(\kappa-\varepsilon)\tilde{p}} \,. \tag{3.4.85}$$

Also, it is easy to verify that

$$\|u\|^2_{0,I_2} \leq \frac{d}{2} \, e^{-2\kappa\tilde{p}}, \quad \|u'\|^2_{0,I_2} \leq \frac{2}{d} \, e^{-2\kappa\tilde{p}} \tag{3.4.86}$$

so that by (3.4.85), (3.4.86), and the triangle inequality,

$$\|u - u_p\|^2_{d,I_2} \leq \tilde{C}_4 d \, e^{-2(\kappa-\varepsilon)\tilde{p}} \,. \tag{3.4.87}$$

Then the first inequality in (3.4.71) follows from Theorem 3.50, (3.4.78), (3.4.87). The other two inequalities also follow from the estimates above. □

Remark 3.75 The constant κ in Theorem 3.74 could be selected such that $\kappa^* e = 4e^{-\kappa^*}$ which yields $\kappa^* \approx 0.71$. This gives $\alpha \approx e^{-\kappa^*}$ in (3.4.72) when two elements are being used. This value for κ^* is, however, not optimal since it is obtained by minimizing some upper bounds. The optimal choice of κ is numerically addressed in Section 3.4.7 ahead. Using the above value of κ^*, however, the bounds above simplify.

Remark 3.76 The choice of the mesh–degree combination $\Sigma = (\mathcal{T}, \boldsymbol{p})$ used in Theorem 3.74 is similar to that obtained by Scherer in [123]. He considered the best mesh–degree combination (for a fixed number of degrees of freedom N) that would minimize the L^∞ error of best approximation (by discontinuous piecewise polynomials) of the function e^{-x} on the interval $[0,\infty)$. He was able to solve this problem explicitly – the asymptotically optimal Σ was given by $\mathcal{T} = \{0, q_0(p+1), \infty\}$, $\boldsymbol{p} = \{p, 1\}$ where $p = N - 2$ and $q_0 = 0.89548641\ldots$ For this Σ, Scherer showed that the asymptotic L^∞ convergence rate was $e^{-q_0 N} = e^{-q_0(p+2)}$ which (up to an algebraic factor in N) was, asymptotically, the best possible among any mesh-degree combination.

We can also deduce from Theorem 3.74 **pointwise error bounds**

Corollary 3.77 *Under the assumptions of Theorem* 3.74 *we have*

$$\|u - u_p\|_{L^\infty(I)} \le C_9 \alpha^{\tilde{p}} \tag{3.4.88}$$

with α *as in Theorem* 3.74.

Proof This follows from (3.4.71) and the interpolation inequality

$$\|v\|_{L^\infty(I)} \le \sqrt{2}\|v\|^{1/2}_{L_2(I)}\ \|v'\|^{1/2}_{L_2(I)} \quad \text{for all} \quad v \in H^1_0(I)\ .$$

□

Remark 3.78 The estimates in Theorem 3.74, Corollary 3.77 are obtained using polynomials of degree 1 in I_2. They evidently remain valid if I_2 is subdivided and/or the degree p is greater than 1 in I_2.

Theorem 3.74 says that it is sufficient to use two intervals of the type described to resolve boundary layers with a robust exponential convergence rate. As discussed in Section 3.4.2, the solution will typically have other (smoother) components as well. For the approximation of these components, the mesh-degree combination of Theorem 3.74 will typically not be sufficient and will have to be enhanced (e.g., by subdivision or p-increase in element 2). This enhancement will ensure that the rate $F(N,k)$ in (3.4.34), which measures the approximation of these smoother components, is sufficiently rapid. For solutions in $H^B_{d,M}$, the robust rate of convergence $g(N)$ of the hp-version will then be given by (3.4.37), where $G(N)$ represents the exponential rate (3.4.71). As noted in Remark 3.56, the overall rate will be exponential only if the smooth components are also approximated exponentially. One such case occurs when f is a polynomial, as noted in the theorem below.

Theorem 3.79 *Consider the hp-version for problem* (3.4.24), $0 < d \le 1$, *where* $\Sigma = (\mathcal{T}, \boldsymbol{p})$, *with*

$$\boldsymbol{p} = \{p,p,p\} \quad \textit{and} \quad \mathcal{T} = \{-1, -1+\kappa p,\ 1-\kappa p, 1\}\ ,$$

if $\kappa\tilde{p}d < 2$ *and*

$$\boldsymbol{p} = \{p\},\ \mathcal{T} = \{-1, 1\}$$

otherwise.

Then there exist constants $\alpha < 1$ and $C > 0$ independent of p and d such that with respect to solution sets H^B_{d,Π_n} and error measure the energy norm, this version is robust with uniform order $g(p) = Cd^{1/2}\alpha^{\tilde{p}}$ for $p \geq n$.

Proof The theorem follows easily by (3.4.33), Corollary 3.52 and Theorem 3.74. □

Exercise 3.80 Frequently, in particular in the theory of plates and shells, the boundary layer functions are not given in the explicit form (3.4.7). One has rather (see also Theorem 3.81 ahead)

$$u_d(x) = \exp(-a(d)(1+x)),\ \overline{u}_d(x) = \exp(-a(d)(1-x)) \qquad (3.4.89)$$

where, for small d,

$$a(d) = a_{-1}d^{-1} + a_0 + a_1 d + \ldots,\ a_{-1} > 0\,, \qquad (3.4.90)$$

and the function $da(d)$ is analytic in $[0, d_0]$ for some $d_0 > 0$.

Prove, using Theorem 3.74, an exponential convergence result for the "three-element" *hp*-FEM mesh.

3.4.7 *Numerical results*

In this section, we present the results of numerical computations for the model problem (3.4.2)–(3.4.3) where

$$f(x) \equiv 1,\quad \alpha^+ = \alpha^- = 0,\quad a = 1\,. \qquad (3.4.91)$$

The exact solution is then given by

$$u_d(x) = 1 - \frac{\cosh(xd)}{\cosh(1d)}\,, \qquad (3.4.92)$$

so that

$$\|u_d\|_d^2 = B_d(u_d, u_d) = (1, u_d) = 2 - 2d\tanh(1d) = O(1)\,. \qquad (3.4.93)$$

Note that since $f(x)$ is a polynomial of degree 0, Corollary 3.52 applies. Noting (3.4.93), we conclude that the relative error in the energy norm,

$$E_R(d) = \|u_d - u_d^S\|_d / \|u_d\|_d$$

should behave like $\Phi(d, S)$ given by (3.4.35). All graphs shown in this section will depict $E_R(d)$ versus the number of degrees of freedom in the finite element method. The value of d (and, where applicable, of κ) will be stated with the figures. All computations were done in double precision on an SGI-2 workstation using MATLAB 4.1.

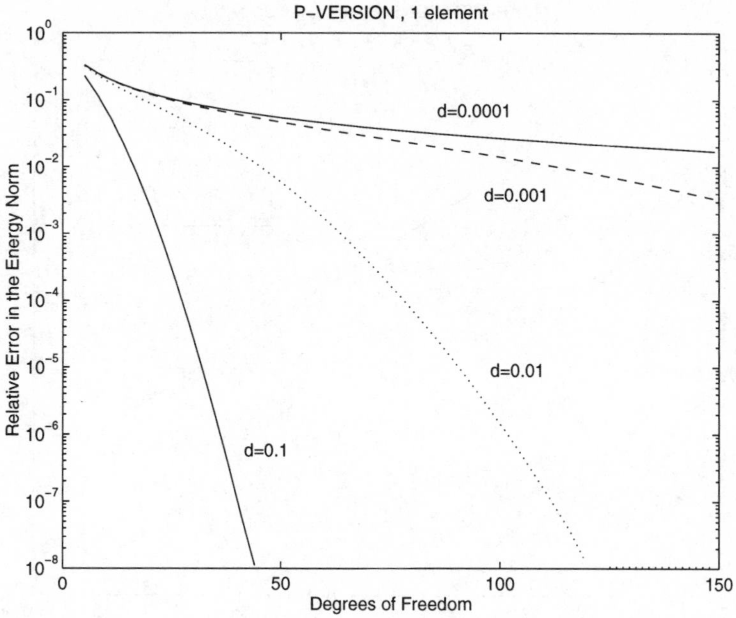

FIG. 3.3. The p-version with one element

We first consider the p version over a single element. Figure 3.3 shows $E_R(d)$ plotted versus the number of degrees of freedom $N = p - 1$, for various values of d, in a semi-log scale. By Corollary 3.65, for $\tilde{p} > e/2d$, the error will be in the asymptotic (superexponential) range. This is only reached, however, when $p \approx 13$ for $d = 0.1$, $p \approx 136$ for $d = 0.01$, $p \approx 1359$ for $d = 0.001$ and $p \approx 13\,591$ for $d = 0.0001$. We see that, except for the first value, none of the rest can be considered as within a practical range of p. For $d = 0.1$, however, the graph in the semi-log scale of Figure 3.3 **is** close to a straight line for p in this range, showing agreement with the theory.

Turning to the case $d = 0.01$, Corollary 3.68 predicts that for d small and $\sqrt{3/(4d)} \leq \tilde{p} \leq 2d$, i.e., $5 \leq p \leq 200$, $\log(E_R)$ should behave like $-\gamma\,\tilde{p}^2 d$ where $\gamma > 0$ is independent of d (the value of γ in Corollary 3.52 is $1/3$, but this may not be optimal). Hence, we should observe a parabolic curve for $d = 0.01$ when p is large enough. Again, the graph in Figure 3.3 is consistent with this bound.

As d becomes even smaller, the error in Figure 3.3 is seen to deteriorate further. For $1 \leq \tilde{p} \leq Kd^{-\frac{1}{2}}$, Theorem 3.70 predicts a convergence rate of only CKp^{-1}. This is precisely what is observed in Figures 3.7 ($d = 10^{-3}$) and 3.8 ($d = 10^{-6}$) ahead. The graphs are now in a log-log scale, and we observe straight lines with slope -1. The "doubling" over the rate of convergence with the uniform h-version is also clearly apparent from these figures.

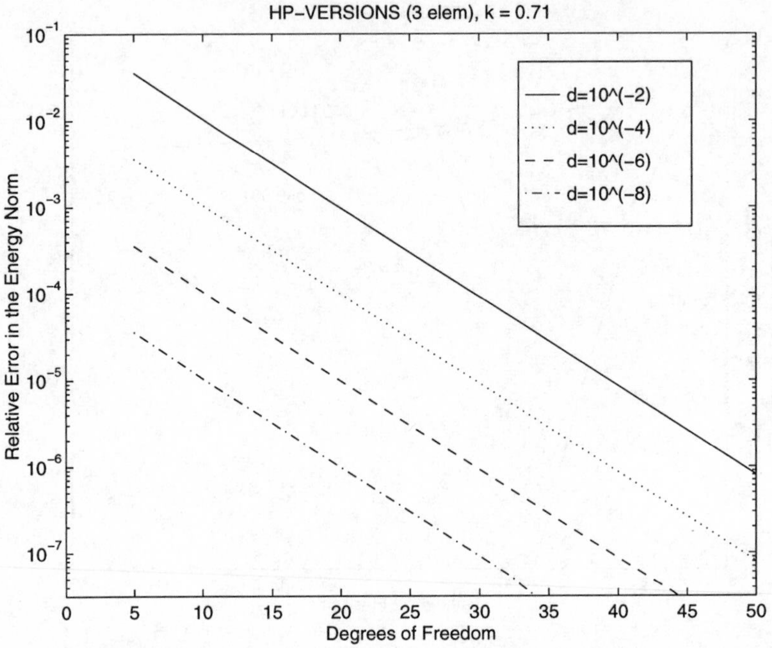

FIG. 3.4. The *hp* version for three elements with $\kappa = 0.71$

Let us now consider the *hp*-version, i.e., the *p*-version on a variable mesh. Since our model solution (3.4.92) has a boundary layer at each endpoint of the domain, the minimum number of elements suggested by Theorem 3.74 will now be 3, with the mesh given by

$$\mathcal{T} = \{-1, -1 + \kappa\tilde{p}d, 1 - \kappa\tilde{p}d, 1\},\ 0 < \kappa < 4/e\ .$$

Since f is a polynomial of degree 0, by Theorem 3.79, the minimal degree vector is now $\boldsymbol{p} = \{p, 1, p\}$. From Theorems 3.74, 3.79, we have the error estimate

$$E_R \leq C(\kappa)d^{1/2}\alpha^{N/2}, \quad N = \dim(S_0(\Sigma)) = 2p + 1\ . \tag{3.4.94}$$

with α given by (3.4.72). The experiments in Figure 3.4, obtained with $\kappa = \kappa^* = 0.71$, clearly show the uniform exponential convergence as well as the factor $d^{1/2}$, since $\log(E_R(d))$ plotted against N is a straight line, which translates downwards as d decreases. By Remark 3.75, for $\kappa = 0.71$, $\alpha \approx e^{-0.71}$ for p large – this is the same value that emerges by measuring the slopes in Figure 3.4. Note that the striking accuracy obtained for small d is not possible with a comparable number of degrees of freedom and methods based on a single element (see, e.g., the results in [42, 88]).

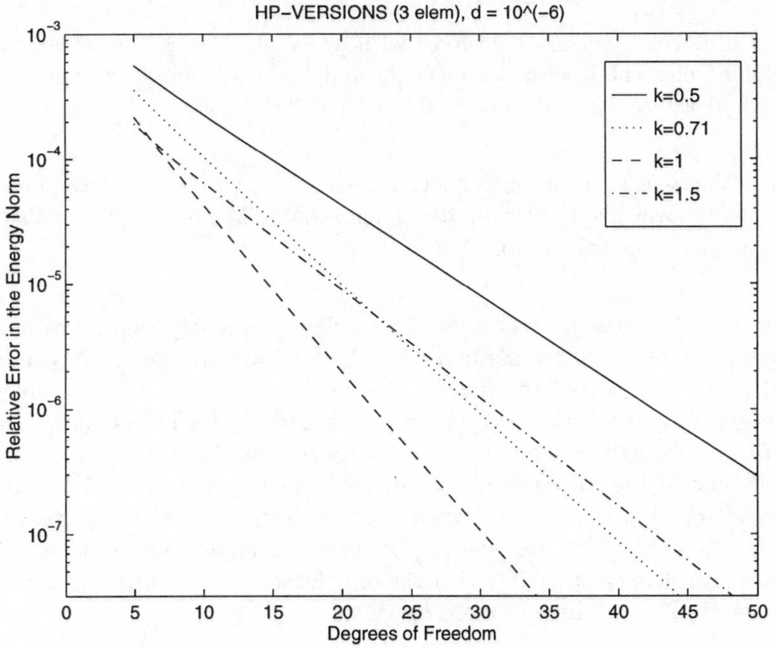

FIG. 3.5. The dependence on parameter κ

In Figure 3.5, we investigate the convergence of the three-element *hp*-version for different values of κ, when $d = 10^{-6}$ (other values of d show similar results). We observe that $\kappa = \kappa^* = 0.71$ is not quite optimal, since $\kappa = 1$ gives better results. Careful examination shows that the graph for $\kappa = 1$ consists of two linear pieces with different slopes. This is due to the fact that initially, the error in the central interval is dominant, so that the value of α in (3.4.74), (3.4.94) is close to $e^{-\kappa}$. As p increases, the size of this interval decreases and the error in the other two intervals eventually dominates, with α behaving like $\kappa e/4$. (Recall that we obtained κ^* by setting $e^{-\kappa}$ equal to $\kappa e/4$, so that only one straight line is observed in this case.)

Finally, in Figures 3.5–3.7, we show a performance comparison between the various methods for $d = 10^{-2}$, 10^{-3} and 10^{-6} respectively (smaller values of d up to 10^{-8} were tested, for which the behavior was similar to $d = 10^{-6}$). In these figures, we have shown the results with four methods: a) the p-version with one element, b) the h-version with $p = 1$, c) the hp-version with three elements taking $\boldsymbol{p} = (p, p, p)$ and $\kappa = 1$ and d) the h-version (taking $p = 1$) with the exponential mesh $\mathcal{T} = \{-1, x_1, \ldots, x_{m-1}, 1\}$ where, for m even,

$$x_{\frac{1}{2}\pm i} = \mp\, d\tilde{p} \ln\left(1 - c\,\frac{2i}{m}\right), \quad i = 0, \ldots \frac{1}{2}, \tag{3.4.95}$$

where $c = 1 - \exp(-1/(d\tilde{p}))$. The mesh (3.4.95) is derived in Exercise 3.42. We observe the following.

a) The uniform h version converges with order $O(N^{-1/2})$, the p-version on a single element with order $O(N^{-1})$ and the h-version with exponential mesh at the optimal algebraic rate of $O(N^{-1})$.

b) Both the h-version with exponential mesh and the hp-version have an error which behaves like $O(d^{1/2})$ in dependence on d. The other two versions do not display this translation as $d \to 0$.

c) For $d = 10^{-2}$, the p-version rapidly reaches a superexponential rate, and eventually becomes the method with the fastest convergence. Asymptotically, i.e., for $\kappa\tilde{p}d > 2$ and fixed d, the p-version with a single element will always have the best convergence rate according to Theorems 3.64 and 3.74. Accordingly, Theorem 3.74 indicates that at about $\kappa\tilde{p}d = 2$ one must switch from the hp-version to a single-element p-version. For $d = 10^{-2}$, this is apparent in Figure 3.6 where the one-element p-version becomes superior at some point. However, as is clearly visible in Figures 3.7 and 3.8, this point may occur so late that the only feasible method (in the practical range of p) is the three-element hp-version.

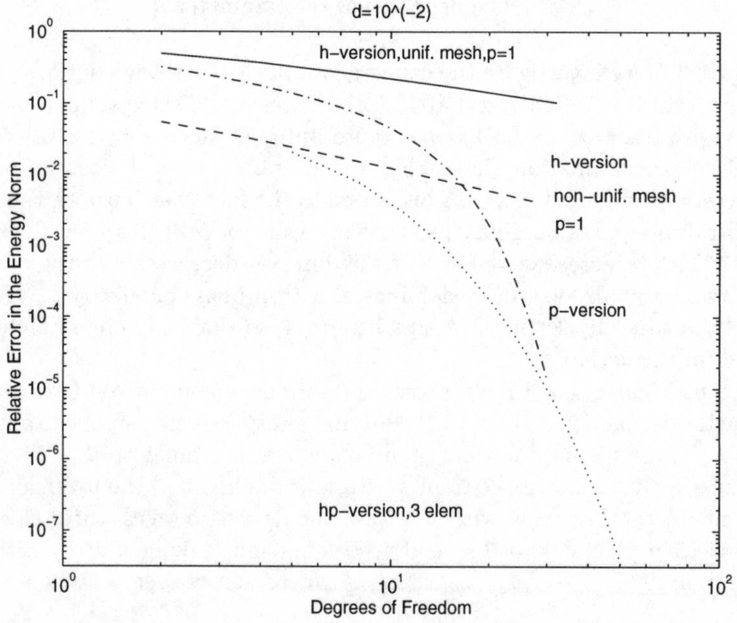

FIG. 3.6. Comparison of various methods, $d = 10^{-2}$

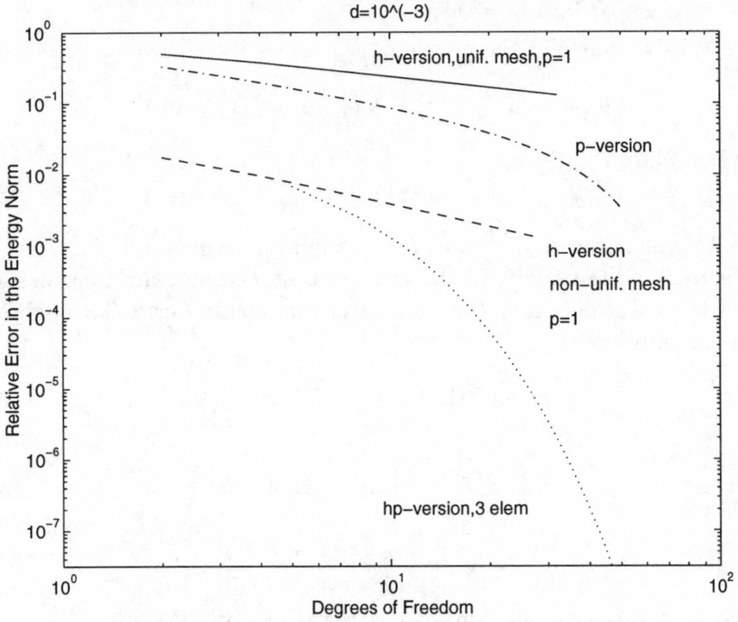

FIG. 3.7. Comparison of various methods, $d = 10^{-3}$

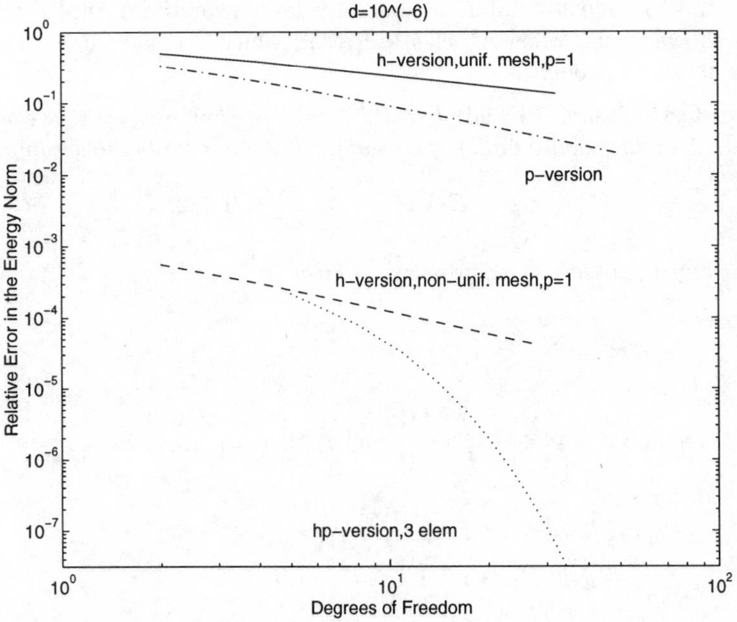

FIG. 3.8. Comparison of various methods, $d = 10^{-6}$

3.4.8 *Convection–diffusion problem*

In $\Omega = (-1, 1)$ we consider the problem

$$Lu := -du'' + a(x)u' + b(x)u = f(x) \quad \text{in } \Omega \tag{3.4.96}$$

with the boundary conditions

$$u(\pm 1) = 0. \tag{3.4.97}$$

Here $d > 0$ is the diffusivity, $u(x)$ is, for example, concentration of a transported substance, $a(x)$ is the velocity of the transporting medium, $b(x)$ specifies losses/sources of the substance and $f(x)$ is an external source term. Throughout, we make the assumptions

$$a(x) \in C^1([-1,1]), \quad a(x) \geq \underline{a} > 0, \tag{3.4.98}$$

$$b(x) \in C^0([-1,1]), \quad b(x) \geq \underline{b}, \tag{3.4.99}$$

$$\underline{a}^2 + 4d\underline{b} \geq \gamma > 0 \tag{3.4.100}$$

which ensure the unique solvability of (3.4.96), (3.4.97). The finite element approximation of (3.4.96), (3.4.97) for small d is nontrivial due to the singular perturbation character of the problem which manifests itself in two distinct phenomena.

First, the solution u exhibits a boundary layer near the outflow boundary $x = 1$ as shown in the following, classical result which is analogous to Theorem 3.50 (see [95] for a proof).

Theorem 3.81 *Assume* (3.4.98)–(3.4.100) *and that* $f(x)$, $a(x)$, $b(x)$ *are analytic in* $[-1, 1]$. *Then the solution* $u(x)$ *of* (3.4.96) *exists and admits the decomposition*

$$u(x) = w_M + C_M\, u_d^+ + r_M, \quad M = 0, 1, 2, \ldots \tag{3.4.101}$$

where the **asymptotic expansion** w_M *is given by*

$$w_M(x) := \sum_{j=0}^{M} d^j\, u_j(x) + \alpha^-\, e^{-\Lambda(x)} ,$$

$$u_{j+1}(x) := e^{-\Lambda(x)} \int_{-1}^{x} \frac{e^{\Lambda(t)}}{a(t)}\, u_j''(t)\, dt, \; j = 0, \ldots, M-1 ,$$

$$u_0(x) := e^{-\Lambda(x)} \int_{-1}^{x} \frac{e^{\Lambda(t)}}{a(t)}\, f(t)\, dt ,$$

$$\Lambda(x) := \int_{-1}^{x} \lambda(t)\, dt, \; \lambda(x) := \frac{b(x)}{a(x)} ,$$

and the **outflow boundary layer** $u_d^+(x)$ *solves the problem*

$$Lu_d^+ = 0 \text{ in } (-1,1), \; u_d^+(-1) = 0, \; u_d^+(1) = 1 \,,$$

and C_M *is given by*

$$C_M := \alpha^+ - w_M(1) \,.$$

The remainder $r_M(x)$ *satisfies*

$$Lr_M = \varepsilon^{M+1} u_M'' \text{ in } (-1,1), \; r_M(\pm 1) = 0 \,.$$

There exist constants $C, K > 0$ *independent of* d *such that for*

$$0 < \varepsilon M K \leq 1$$

holds

$$\|w_M^{(n)}\|_{L^\infty(-1,1)} \leq CK^n n! \,, \quad n \in \mathbb{N}_0 \,,$$
$$\|r_M^{(n)}\|_{L^\infty(-1,1)} \leq Cd^{1-n}(dMK)^M, \quad n = 0,1,2$$
$$|C_M| \leq C \,.$$

Notice that in the remainder estimate the dependence on d as well as on M are explicit – this is necessary for the convergence analysis of *hp*-FEM.

Exercise 3.82 Prove Theorem 3.81 (follow the proof of Theorem 3.50).

The expression for λ shows that (3.4.100) is sharp; if $a + 4db \leq 0$, the problem will have eigensolutions, in general. Splittings similar to (3.4.101) have been given in the literature, see for example [65]. The main point is that w_M is a smooth (analytic) function the derivatives of which are bounded independently of d and that the remainder estimates are explicit in the expansion order M.

The second difficulty arises from the fact that symmetric variational formulations of (3.4.96), (3.4.97) based on H_0^1 as trial and test space are not uniformly stable with respect to the parameter d. One possible remedy is the use of **streamline-diffusion techniques** which amount in effect to a nonconforming method (see, e.g., [119] and the references there). Crucial to its convergence analysis are certain elementwise inverse inequalities which allow to control the higher order terms introduced into the variational formulation through the streamline-diffusion term. While this idea is, in principle, also feasible for *p*- or spectral element methods, the convergence rates obtained that way will be suboptimal due to the higher loss of derivatives in inverse inequalities for polynomials. This is even more pronounced for the combined *hp*-type FEM, in particular in two and three dimensions, where the optimal approximation of boundary layers mandates elements of arbitrary high aspect ratio (see [131]) for which suitable inverse inequalities do not seem to be available. The *hp*-FEM is nevertheless very attractive for such problems, since *hp*-trial spaces can be shown to approximate boundary layers at a *robust exponential convergence rate* (see [128], [131]). This will also be true for any stable numerical method based on these trial spaces. We present here Galerkin–Petrov methods where test and trial spaces are distinct, an approach that has been followed by numerous authors in the finite difference and finite volume setting (see [98, 119] for an account of this work and many references). We base the *hp*-FEM on the variational framework from [94, 146] where the h-version FEM was analyzed and optimal convergence rates, uniform in d, were shown. Asymptotically exact h-version a posteriori error estimators for this variational formulation have also been developed in [146], and it was shown that the numerical solutions exhibit few spurious oscillations and good pointwise convergence. The crucial ingredient in [94, 146] was the construction of suitable, upwinded test functions by asymptotic analysis of the elemental adjoint problem. The generalization of this approach to high order elements and higher dimensions is not straightforward.

For *hp*-trial spaces with any mesh-degree combination *upwinded testfunctions* can be stably computed. The calculation of the test functions is completely localized to either a single element or a patch of elements and can be done by a least-squares like method (which is uniformly stable in d), see [95] for details. This can be simply performed as part of the usual element stiffness matrix generation in the *hp*-FEM.

3.4.9 *Variational formulation*

To motivate our variational formulation, we observe that multiplication of (3.4.96) by a test function v and twofold integration by parts gives a so-called *very weak variational formulation*: find $u \in L^2(\Omega)$ such that

$$B(u,v) := \int_\Omega uL^*v dx = \int_\Omega fv dx =: F(v) \quad \text{for all} \quad v \in H^2 \cap H_0^1(\Omega).$$

Here, L^* denotes the adjoint of L, i.e.,

$$L^*u = -du'' - a(x)u' + (b - a')(x)u \tag{3.4.102}$$

which is defined when $a \in C^1([-1,1])$. There are several drawbacks with FEM based on very weak variational formulations: first, a' is in general not globally continuous, but only elementwise smooth (if it stems, for example, from linearization of the nonlinear problem around a FE approximation of u), second, to obtain a good test-space for a given trial space of possibly discontinuous functions, a global adjoint problem must be solved for each basis function and third, the essential boundary conditions (3.4.97) are generally not satisfied by FE solutions. This leads us to a formulation which is situated "between" the weak one based on $H_0^1 \times H_0^1$ and the very weak one based on $L^2 \times H_0^1 \cap H^2$.

We present Sobolev **spaces with mesh-dependent norms** introduced in [146]. Let again $\mathcal{T} := \{-1 = x_0 < x_1 < ... < x_M = 1\}$ be any mesh in Ω and set $I_j := (x_{j-1}, x_j)$, $h_j := |I_j| = x_j - x_{j-1}$, $m_j = (x_{j-1} + x_j)/2$ for $j = 1, ..., M$. Let further $\{\rho_j\}_{j=1}^{N-1}$ be a sequence of positive numbers and set

$\rho := \rho_1 + \rho_2 + ... + \rho_{M-1}$, $h := \max\{h_j : j = 1, ..., M\}$. Then we define the trial space $H^0_{\mathcal{T}}$ as completion of $H^1_0(\Omega)$ with respect to the norm

$$\|u\|_{H^0_{\mathcal{T}}} := \Big(\int_{-1}^{1} |u|^2 \, dx + \sum_{j=1}^{M-1} \rho_j \, |u(x_j)|^2 \Big)^{\frac{1}{2}}. \tag{3.4.103}$$

The space $H^0_{\mathcal{T}}$ thus obtained is a Hilbert space and is isomorphic to $L^2(\Omega) \oplus \mathbb{R}^{M-1}$ so that every $u \in H^0_{\mathcal{T}}$ is of the form $u = (\tilde{u}, d_1, d_2, ..., d_{M-1})$ and

$$\|u\|_{H^0_{\mathcal{T}}} = \Big(\|\tilde{u}\|^2_{L^2} + \sum_{j=1}^{M-1} \rho_j \, |d_j|^2 \Big)^{\frac{1}{2}}. \tag{3.4.104}$$

If $u \in H^0_{\mathcal{T}} \cap H^1$ then $\tilde{u} \in H^1$ and $d_j = \tilde{u}(x_j)$.

Next, we introduce the test space

$$H^2_{\mathcal{T}} := \{ v \in H^1_0(\Omega) : v|_{I_j} \in H^2(I_j), j = 1, ..., M \} \tag{3.4.105}$$

and assume here and in the following that $a(x)$ satisfies (3.4.98). On the pair $H^0_{\mathcal{T}} \times H^2_{\mathcal{T}}$ we define the bilinear form $B_{\mathcal{T}}(\cdot, \cdot)$ by

$$B_{\mathcal{T}}(u, v) := \sum_{j=1}^{M} \int_{I_j} \tilde{u} L^* v dx - \sum_{j=1}^{M-1} d_j \, [dv'(x_j)] \tag{3.4.106}$$

where $[v'(x_j)]$ denotes the jump of v' at $x_j \in \mathcal{T}$. We equip the space $H^2_{\mathcal{T}}$ with the norm

$$\|v\|_{H^2_{\mathcal{T}}} := \Big(\sum_{j=1}^{M} \|L^* v\|^2_{L^2(I_j)} + \sum_{j=1}^{M-1} \frac{|[dv'(x_j)]|^2}{\rho_j} \Big)^{\frac{1}{2}}. \tag{3.4.107}$$

We remark in passing that so far we have used $a(x) \in C^0([-1, 1]) \cap C^1(\bar{I}_j)$, $j = 1, ..., M$, rather than (3.4.98). With these definitions we have the following result.

Proposition 3.83 *For any mesh $\mathcal{T}$ and any positive sequence $\{\rho_j\}_{j=1}^{N-1}$, the bilinear form $B_{\mathcal{T}}(\cdot, \cdot)$ satisfies*

$$|B_{\mathcal{T}}(u, v)| \le \|u\|_{H^0_{\mathcal{T}}} \|v\|_{H^2_{\mathcal{T}}} \quad \textit{for all} \;\; u \in H^0_{\mathcal{T}}, v \in H^2_{\mathcal{T}}, \tag{3.4.108}$$

$$\inf_{0 \ne v \in H^2_{\mathcal{T}}} \; \sup_{0 \ne u \in H^0_{\mathcal{T}}} \frac{B_{\mathcal{T}}(u, v)}{\|u\|_{H^0_{\mathcal{T}}} \|v\|_{H^2_{\mathcal{T}}}} \ge 1, \tag{3.4.109}$$

and

$$\textit{for all} \;\; 0 \ne u \in H^0_{\mathcal{T}} : \quad \sup_{v \in H^2_{\mathcal{T}}} B_{\mathcal{T}}(u, v) > 0. \tag{3.4.110}$$

Proof The bound (3.4.108) follows directly from the definition of the norms and the Schwarz inequality.

To show (3.4.109), for given $v \in H^2_{\mathcal{T}}$, we select $u_v = (\tilde{u}, d_1, ..., d_{M_1}) \in H^0_{\mathcal{T}}$ as follows:

$$\tilde{u}|_{I_j} = \mathrm{sgn}(L^* v|_{I_j})\big|L^* v|_{I_j}\big|, \qquad j = 1, ..., M,$$
$$d_j = -\rho_j^{-1}\mathrm{sgn}([v'(x_j)])|\,[dv'(x_j)]\,|, \quad j = 1, ..., M-1.$$

Then $\|u_v\|_{H^0_{\mathcal{T}}} \le \|v\|_{H^2_{\mathcal{T}}}$ and $B_{\mathcal{T}}(u_v, v) = \|v\|^2_{H^2_{\mathcal{T}}}$, whence for every $0 \neq v \in H^2_{\mathcal{T}}$

$$\sup_{0 \neq u \in H^0_{\mathcal{T}}} \frac{B_{\mathcal{T}}(u,v)}{\|u\|_{H^0_{\mathcal{T}}}\|v\|_{H^2_{\mathcal{T}}}} \ge \frac{B_{\mathcal{T}}(u_v,v)}{\|u_v\|_{H^0_{\mathcal{T}}}\|v\|_{H^2_{\mathcal{T}}}} \ge \frac{B_{\mathcal{T}}(u_v,v)}{\|v\|^2_{H^2_{\mathcal{T}}}} = 1$$

which proves (3.4.109) and (3.4.110). □

Proposition 3.83 allows to prove the existence of a solution $u \in H^0_{\mathcal{T}}$ of the problem:

$$u \in H^0_{\mathcal{T}}, \qquad B_{\mathcal{T}}(u, v) = F(v) \text{ for all } v \in H^2_{\mathcal{T}}. \tag{3.4.111}$$

Proposition 3.84 *Assume* (3.4.98)–(3.4.100). *Then for every* $f \in L^1(\Omega)$, *every* $0 < d \le 1$ *and every mesh* $\mathcal{T}$, *every positive sequence* $\{\rho_j\}_{j=1}^{M-1}$, *the problem* (3.4.111) *admits a unique solution* $u \in H^0_{\mathcal{T}}$.

Proof We proceed in several steps.

Step 1: We claim that for any mesh $\mathcal{T}$, any positive sequence $\{\rho_j\}_{j=1}^{N-1}$ and any $d \in (0, 1]$, assumptions (3.4.98)–(3.4.100) imply

$$\|v\|_{L^\infty(\Omega)} \le C_1\sqrt{\rho}\|v\|_{H^2_{\mathcal{T}}}, \tag{3.4.112}$$

where C_1 is independent of d and $\mathcal{T}$. To prove it, we note that (3.4.98)–(3.4.100) imply the existence of a Green's function $G(x, y)$ for the problem (3.4.96), (3.4.97) which is bounded uniformly with respect to x, y, d (see [146], Theorem 2.7), i.e.,

$$\max_{(x,y)\in[-1,1]^2} |G(x,y)| \le C_G.$$

For $v \in H^2_{\mathcal{T}}$, we can write

$$v(y) = \sum_{j=1}^{M} \int_{I_j} G(x,y)(L^*v)(x)dx - \sum_{j=1}^{M-1} [dv'(x_j)]\, G(x_j, y) \quad \text{for all } y \in [-1, 1].$$

Using the boundedness of $G(x,y)$, we then estimate

$$\begin{aligned}|v(y)| &\leq C_G\Big\{\sum_{j=1}^{M}\int_{I_j}|L^*v|\,dx+\sum_{j=1}^{M-1}\rho_j^{-1/2}\,|[dv'(x_j)]|\,\rho_j^{1/2}\Big\}\\ &\leq \sqrt{2}C_G\max\{\sqrt{2},\sqrt{\rho}\}\|v\|_{H^2_{\mathcal{T}}}\end{aligned}$$

which proves (3.4.112).

Step 2: For $f\in L^1(\Omega)$ and $v\in H^2_{\mathcal{T}}$, we therefore have

$$|F(v)|\leq\|f\|_{L^1(\Omega)}\|v\|_{L^\infty(\Omega)}\leq C_1\max\{1,\sqrt{\rho}\}\|f\|_{L^1(\Omega)}\|v\|_{H^2_{\mathcal{T}}}.$$

Hence, $F(\cdot)$ is a continuous, linear functional on $H^2_{\mathcal{T}}$ the norm of which is bounded uniformly with respect to d and $\mathcal{T}$. By Propositions 3.83 and 1.22, we have also

$$\inf_{0\neq u\in H^0_{\mathcal{T}}}\ \sup_{0\neq v\in H^2_{\mathcal{T}}}\frac{B_{\mathcal{T}}(u,v)}{\|u\|_{H^0_{\mathcal{T}}}\|v\|_{H^2_{\mathcal{T}}}}\geq 1\,,$$

$$\text{for all }\ 0\neq v\in H^2_{\mathcal{T}}:\qquad \sup_{u\in H^0_{\mathcal{T}}}B_{\mathcal{T}}(u,v)>0\,.$$

This implies with $F\in(H^2_{\mathcal{T}})'$ and Proposition 3.83 that (3.4.111) admits a unique solution and that the a priori estimate

$$\|u\|_{H^0_{\mathcal{T}}}\leq C_1\max\{1,\sqrt{\rho}\}\|f\|_{L^1(\Omega)} \tag{3.4.113}$$

holds. The variational formulation (3.4.111) is the basis of the FE discretization.

3.4.10 *hp-finite element discretization*

3.4.10.1 *The finite element spaces*

We associate with each element I_j a polynomial degree $p_j\geq 1$ and combine the p_j in the degree-vector $\boldsymbol{p}$. We also set $p:=\max\{p_j:1\leq j\leq M\}$. The trial space of our finite element method is the usual space of continuous, piecewise polynomials of degree p_j:

$$S^{\boldsymbol{p}}(\mathcal{T}):=\big\{u\in C^0_0(\Omega):u|_{I_j}\in\Pi_{p_i}(I_j),j=1,...,M,u(\pm1)=0\big\} \tag{3.4.114}$$

where $C^0_0(\Omega)$ denotes the continuous functions in Ω which vanish on $\partial\Omega$.

As test space we choose, following [146], the space of L-splines of degree $\boldsymbol{p}$ defined by

$$\begin{aligned}S^{\boldsymbol{p}}_L(\mathcal{T}):=\{v\in C^0_0(\Omega):\ &(L^*v)|_{I_j}=0\text{ if }p_j=1\,,\\ &(L^*v)|_{I_j}\in\Pi_{p_j-2}(I_j)\text{ if }p_j\geq 2\}\,.\end{aligned} \tag{3.4.115}$$

We omit the argument $\mathcal{T}$ when it is clear from the context which mesh is meant. Note that $S_L^{\mathbf{p}}$ is well defined if the coefficient $a(x)$ is piecewise C^1 due to (3.4.102). We also observe that

$$N = \dim(S^{\mathbf{p}}) = -1 + \sum_{j=1}^{N} p_j = \dim(S_L^{\mathbf{p}}). \tag{3.4.116}$$

The finite element approximation u_M is then obtained in the usual way:

$$u_N \in S^{\mathbf{p}}, \qquad B_{\mathcal{T}}(u_M, v) = F(v) \text{ for all } v \in S_L^{\mathbf{p}}. \tag{3.4.117}$$

Due to (3.4.116), problem (3.4.117) amounts to solving a linear system of N equations for the N unknown coefficients of u_N.

3.4.10.2 *Stability*

Our main result in this section is the following.

Theorem 3.85 *Select*

$$\rho_j := (h_j + h_{j+1})/2, \quad j = 1, ..., N-1. \tag{3.4.118}$$

Then for all $0 < d \leq 1$, $\mathcal{T}$ *and* $\mathbf{p}$ *it holds that*

$$\inf_{0 \neq v \in S_L^{\mathbf{p}}} \sup_{0 \neq u \in S^{\mathbf{p}}} \frac{B_{\mathcal{T}}(u,v)}{\|u\|_{H^0_{\mathcal{T}}} \|v\|_{H^2_{\mathcal{T}}}} \geq \frac{1}{\gamma_N} \tag{3.4.119}$$

with $\gamma_N = \max\{\sqrt{5}, \sqrt{p+3}\}$.

Proof We show that for every $v \in S^{\mathbf{p}}$ there exists $u_v \in S_L^{\mathbf{p}}$ such that

$$B_{\mathcal{T}}(u_v, v) \geq \|v\|^2_{H^2_{\mathcal{T}}}, \quad \|u_v\|_{H^0_{\mathcal{T}}} \leq \gamma_N \|v\|_{H^2_{\mathcal{T}}}.$$

To this end, we write

$$u_v|_{I_j} = \sum_{i=0}^{p_j} a_{ij} L_i\Big(2\,\frac{x - m_j}{h_j}\Big), \tag{3.4.120}$$

where L_i denotes the ith Legendre polynomial on $(-1,1)$ normalized such that $L_i(1) = 1$. □

A basis for $S_L^{\mathbf{p}}$ can be obtained as follows: First, we define **external, nodal upwinded shape functions** $\psi_{-1,j}$ by

$$\begin{aligned} L^*\psi_{-1,j} &= 0 \text{ in } I_{j-1} \cup I_j, j = 2, ..., M, \\ \psi_{-1,j}(x_k) &= \delta_{j,k+1}, k = 1, ..., M, \\ \psi_{-1,j} &= 0 \quad \text{elsewhere}. \end{aligned} \tag{3.4.121}$$

These nodal shape functions are augmented for $p_j \geq 2$ by **internal, upwinded shape functions** $\psi_{i,j} \in H^1(I_j)$ defined as solutions of

$$
\begin{aligned}
L^*\psi_{i,j} &= L_i\Big(2\,\frac{x-m_j}{h_j}\Big) \text{ in } I_j, j=1,...,M,\ i=0,...,p_j-2,\\
\psi_{i,j} &= 0 \text{ elsewhere.}
\end{aligned}
\tag{3.4.122}
$$

Then any $v \in S_L^{\vec{p}}$ can be written as

$$
v(x) = \sum_{j=2}^{M} v(x_{j-1})\psi_{-1,j}(x) + \sum_{j=1}^{M}\sum_{i=0}^{p_j-2} b_{ij}\psi_{i,j}(x) \tag{3.4.123}
$$

where b_{ij} are the Legendre coefficients of $L^*v|_{I_j}$. Further, from the definition (3.4.121) of the $\psi_{-1,j}$ we have

$$
L^*v|_{I_j} = \sum_{i=0}^{p_j-2} b_{ij}L_i\Big(2\,\frac{x-m_j}{h_j}\Big), \quad j=1,...,M
$$

which yields with the orthogonality properties of the Legendre polynomials and a scaling argument

$$
\sum_{j=1}^{M} \|L^*v\|^2_{L^2(I_j)} = \sum_{j=1}^{M} h_j \sum_{i=0}^{p_j-2} \frac{|b_{ij}|^2}{2i+1}. \tag{3.4.124}
$$

Combining (3.4.124) with (3.4.107), we obtain for $v \in S_L^{\mathbf{p}}$ an expression for $\|v\|_{H^2_{\mathcal{T}}}$ in terms of the b_{ij}:

$$
\begin{aligned}
\|v\|_{H^2_{\mathcal{T}}} &= \Big(\sum_{j=1}^{M} h_j \sum_{i=0}^{p_j-2} \frac{|b_{ij}|^2}{2i+1} + \sum_{j=1}^{M-1} \rho_j^{-1}\,|[dv'(x_j)]|^2\Big)^{\frac{1}{2}}\\
&\text{for all } v \in S_L^{\vec{p}}.
\end{aligned}
\tag{3.4.125}
$$

Writing $v(x)$ in the form (3.4.123) and $u_v(x)$ as in (3.4.120) and inserting into (3.4.106), we find in the same way

$$
B_{\mathcal{T}}(u_v, v) = \sum_{j=1}^{M-1}\Big\{h_j \sum_{i=0}^{p_j-2} \frac{a_{ij}b_{ij}}{2i+1}\Big\} - \sum_{j=1}^{M-1} u_v(x_j)\,[dv'(x_j)]\ . \tag{3.4.126}
$$

For given $v \in S_L^p$, i.e., given b_{ij}, we choose now a_{ij} as follows: first, we select

$$a_{ij} = b_{ij}\,, \quad i = 0, ..., p_j - 2 \tag{3.4.127}$$

which leaves $a_{p_j-1,j}, a_{p_j,j}$ to be determined, for each I_j. Since u_v must be continuous, two conditions per interval must be enforced. We prescribe u_v at each endpoint of I_j as follows:

$$u_v(x_{j-1}^+) = a_j^- := \begin{cases} -|[dv'(x_{j-1})]| \operatorname{sgn}([v'(x_{j-1})])/\rho_{j-1} & \text{if } j > 1, \\ 0 & \text{if } j = 1 \end{cases} \tag{3.4.128}$$

and

$$u_v(x_j^-) = a_j^+ := \begin{cases} -|[dv'(x_j)]| \operatorname{sgn}([v'(x_j)])/\rho_j & \text{if } j < N, \\ 0 & \text{if } j = N. \end{cases} \tag{3.4.129}$$

Conditions (3.4.128), (3.4.129) ensure continuity of u_v. Since $L_i(\pm 1) = (\pm 1)^i$ implies

$$u_v(x_{j-1}^+) = \sum_{i=0}^{p_j} (-1)^i a_{ij}, \quad u_v(x_j^-) = \sum_{i=0}^{p_j} (-1)^i a_{ij}$$

we get with (3.4.127) the linear system

$$\begin{bmatrix} (-1)^{p_j-1} & (-1)^{p_j} \\ 1 & 1 \end{bmatrix} \begin{bmatrix} a_{p_j-1,j} \\ a_{p_j,j} \end{bmatrix} = \begin{bmatrix} a_j^- - \sum_{i=0}^{p_j-2} (-1)^i b_{ij} \\ a_j^+ - \sum_{i=0}^{p_j-2} b_{ij} \end{bmatrix}. \tag{3.4.130}$$

Its determinant is nonzero for any p_j, therefore u_v is uniquely determined by (3.4.127) and (3.4.130).

From (3.4.126), (3.4.127) and (3.4.128), (3.4.129) we get

$$B_{\mathcal{T}}(u_v, v) = \|v\|_{H_{\mathcal{T}}^2}^2 .$$

It remains therefore to show

$$\|u_v\|_{H_{\mathcal{T}}^0} \leq \gamma_N \|v\|_{H_{\mathcal{T}}^2} \tag{3.4.131}$$

with γ_N as in (3.4.119).

Since u_v is continuous, we have

$$
\begin{aligned}
\|u_v\|^2_{H^0_{\mathcal{T}}} &= \sum_{j=1}^{M} \|u_v\|^2_{L^2(I_j)} + \sum_{j=1}^{M-1} \rho_j \,|u_v(x_j)|^2 \\
&= \sum_{j=1}^{M} h_j \sum_{i=0}^{p_j} \frac{|a_{ij}|^2}{2i+1} + \sum_{j=1}^{M-1} \rho_j \,|a_j^+|^2 \\
&= \sum_{j=1}^{M} \Big\{ h_j \sum_{i=0}^{p_j-2} \frac{|b_{ij}|^2}{2i+1} + h_j \sum_{i=p_j-1}^{p_j} \frac{|a_{ij}|^2}{2i+1} \Big\} \qquad (3.4.132) \\
&\quad + \sum_{j=1}^{M-1} \rho_j^{-1} \,|[dv'(x_j)]|^2 + \sum_{j=1}^{M-1} \rho_j^{-1} \,|[dv'(x_j)]|^2 \\
&= \|v\|^2_{H^2_{\mathcal{T}}} + \sum_{j=1}^{M} h_j \sum_{i=p_j-1}^{p_j} \frac{|a_{ij}|^2}{2i+1} \,.
\end{aligned}
$$

We estimate $|a_{ij}|^2$ for $i = p_j - 1, p_j$. From (3.4.130), we get

$$
\begin{bmatrix} a_{p_j-1,j} \\ a_{p_j,j} \end{bmatrix} = \frac{1}{2(-1)^{p_j}} \begin{pmatrix} 1 & (-1)^{p_j-1} \\ -1 & (-1)^{p_j-1} \end{pmatrix} \begin{pmatrix} a_j^- - \sum_{i=0}^{p_j-2} (-1)^i b_{ij} \\ a_j^+ - \sum_{i=0}^{p_j-2} b_{ij} \end{pmatrix}
$$

$$
= \frac{1}{2(-1)^{p_j}} \begin{pmatrix} a_j^- + (-1)^{p_j-1} a_j^+ + \sum_{i=0}^{p_j-2} b_{ij}((-1)^{p_j} - (-1)^i) \\ -a_j^- + (-1)^{p_j-1} a_j^+ + \sum_{i=0}^{p_j-2} b_{ij}((-1)^i + (-1)^{p_j}) \end{pmatrix}.
$$

We estimate

$$
\max\{|a_{ij}| : i = p_j - 1, p_j\} \le \frac{1}{2}\left(|a_j^-| + |a_j^+|\right) + \sum_{i=0}^{p_j-2} |b_{ij}|
$$

and get with (3.4.128), (3.4.129) that

$$
\max\{|a_{ij}|^2 : i = p_j - 1, p_j\} \le d^2 \Big(\frac{|[v'(x_{j-1})]|^2}{\rho_{j-1}^2} + \frac{|[v'(x_j)]|^2}{\rho_j^2} \Big) + 2 \Big(\sum_{i=0}^{p_j-2} |b_{ij}| \Big)^2.
$$

With the understanding that $[v'(x_0)] = [v'(x_N)] = 0$ and $\rho_0 = \rho_N = \infty$ we estimate further

$$\begin{aligned}
&\sum_{j=1}^{M} h_j \sum_{i=p_j-1}^{p_j} \frac{|a_{ij}|^2}{2i+1} \\
&\leq \sum_{j=1}^{M} \frac{h_j}{2p_j-1} \left\{ \frac{\|[dv'(x_{j-1})]\|^2}{\rho_{j-1}^2} + \frac{\|[dv'(x_j)]\|^2}{\rho_j^2} + 2\left(\sum_{i=0}^{p_j-2} |b_{ij}|\right)^2 \right\} \\
&\leq \sum_{j=1}^{M} \frac{h_j}{2p_j-1} \left\{ \frac{\|[dv'(x_{j-1})]\|^2}{\rho_{j-1}^2} + \frac{\|[dv'(x_j)]\|^2}{\rho_j^2} + 2\sum_{i=0}^{p_j-2} (2i+1) \sum_{i=0}^{p_j-2} \frac{|b_{ij}|^2}{2i+1} \right\} \\
&= \sum_{j=1}^{M} \frac{h_j}{2p_j-1} \left\{ \frac{\|[dv'(x_{j-1})]\|^2}{\rho_{j-1}^2} + \frac{\|[dv'(x_j)]\|^2}{\rho_j^2} + 2(p_j-1)^2 \sum_{i=0}^{p_j-2} \frac{|b_{ij}|^2}{2i+1} \right\}.
\end{aligned}$$

Now using $h_j/\rho_j \leq 2$, $h_j/\rho_{j-1} \leq 2$ and

$$\max\left\{ \frac{2(p_j-1)^2}{2p_j-1} : j = 1, ..., M \right\} \leq p - 1$$

we arrive at

$$\begin{aligned}
\sum_{j=1}^{M} h_j \sum_{i=p_j-1}^{p_j} \frac{|a_{ij}|^2}{2i+1} &\leq 4 \sum_{j=1}^{M-1} \frac{\|[dv'(x_j)]\|^2}{\rho_j} + (p+2) \sum_{j=1}^{M} h_j \sum_{i=0}^{p_j-2} \frac{|b_{ij}|^2}{2i+1} \\
&\leq \max\{4, p+2\} \|v\|_{H^2_{\mathcal{T}}}^2
\end{aligned}$$

where we used (3.4.107) and (3.4.124). Referring to (3.4.132) completes the proof. □

Remark 3.86 In Theorem 3.85, we selected a specific sequence $\{\rho_j\}$. Inspection of the proof shows, however, that any positive sequence is admissible. Then, however,

$$\gamma_N = C\sqrt{p} \max_{1 \leq j \leq M} \{h_j/\rho_j, h_j/\rho_{j-1}\}. \tag{3.4.133}$$

This shows that in order for γ_N to be independent of $\mathcal{T}$, the ρ_j must essentially be of the order of the local meshwidth.

3.4.10.3 *Consistency and convergence*

Theorem 3.85 implies with Theorem 2.21 and (3.4.108), (3.4.109) that

$$\inf_{0\neq u\in S^{\vec{p}}} \sup_{0\neq v\in S_L^{\vec{p}}} \frac{B_{\mathcal{T}}(u,v)}{\|u\|_{H^0_{\mathcal{T}}}\|v\|_{H^2_{\mathcal{T}}}} \geq \frac{1}{\gamma_N}, \tag{3.4.134}$$

$$\text{for all } 0\neq v\in S_L^{\vec{p}}: \quad \sup_{u\in S^{\vec{p}}} B(u,v) > 0. \tag{3.4.135}$$

Referring to the continuous inf–sup condition (3.4.109), we deduce from (3.4.134) and from Theorem 2.21 that for every mesh-degree combination $(\boldsymbol{p},\mathcal{T})$ there exists a unique FE solution u_M of (3.4.117). In particular, the $M\times M$ (generally nonsymmetric) stiffness matrix corresponding to (3.4.117) is nonsingular. Moreover, the FE solution u_M is quasioptimal, i.e.,

$$\|u-u_N\|_{H^0_{\mathcal{T}}} \leq (1+\gamma_N)\|u-w\|_{H^0_{\mathcal{T}}} \text{ for all } w\in S^{\vec{p}}. \tag{3.4.136}$$

The rate of convergence of the FEM (3.4.117) is therefore given by the approximability of the exact solution u from the trial space $S^{\boldsymbol{p}}$. We show that proper selection of $\mathcal{T}$ and $\boldsymbol{p}$ yield an *exponential rate of convergence*, uniform in d.

Theorem 3.87 *Assume $a(x)=a$, $b(x)=b$ are constant and that f in* (3.4.96) *is a polynomial of degree k. Define further $\boldsymbol{p}$ and $\mathcal{T}$ by*

$$\begin{aligned} \boldsymbol{p}&=\{p,p,p\}\ \mathcal{T}=\{-1,-1+(p+1/2)d/a,1\} && \text{if } \tilde{p}d/a<2,\\ \boldsymbol{p}&=\{p,p\}\quad \mathcal{T}=\{-1,1\} && \text{else}\end{aligned} \tag{3.4.137}$$

and let $\{\rho_j\}_{j=1}^{M-1}$ be any sequence of positive numbers. Then for every $0<d\leq 1$ there exists $v\in S^{\vec{p}}$ such that

$$\|u-v\|_{H^0_{\mathcal{T}}} \leq C\max\{1,\sqrt{\rho}\}\exp(-bN) \tag{3.4.138}$$

where $N=\dim(S^{\vec{p}})$ and $b,C>0$ are independent of $d,\mathcal{T}$, p and ρ.

The proof can be found in [95].

Results analogous to Theorem 3.87 hold also true when $f(x)$ is piecewise analytic on $[-1,1]$ (see [93]); then, however, additional internal layers arise at points of nonanalyticity of f which must be accounted for by adding further $O(dp)$ elements. Theorem 3.87 should also hold true for nonconstant, piecewise analytic coefficients $a(x),b(x)$. The regularity estimates necessary to establish Theorem 3.87 are given in [95].

Theorem 3.87 shows with (3.4.136) that exponential convergence can be achieved by the FE scheme (3.4.117) provided the space $S^{\boldsymbol{p}}$ is designed properly (i.e., with one element of size $O(pd/a)$ in the outflow boundary layer) and the corresponding stable test space S^{p}_{L} is available. This is, in general, not the case and corresponding approximate test spaces must be used; they still ensure stability and can be obtained computationally with moderate effort (see [95]).

3.5 A posteriori error estimate

The error estimates obtained so far in this chapter were all based on the abstract setting (3.3.1), i.e., they were determined by the approximability $Z(u_0, S, X)$ of the exact solution u_0. Much of Section 3 was devoted to estimates of Z for some specific classes of exact solutions u_0. The resulting error estimates were, as a rule, **asymptotic**, i.e. they pertain to the case when $N = \dim(S^{p,1}(\Omega, \mathcal{T})) \to \infty$, and **a priori**, i.e. they applied under some a-priori known regularity properties of the solution.

In computational practice, however, the following question arises: assume that, for a given, specific set of data, we have obtained a FE solution $u_S \in S^{p}(\Omega, \mathcal{T})$. Then we wish to know a computable bound on the error $\|u_0 - u_S\|_X$. It turns out that this can indeed be done and computable **a posteriori** (i.e., **after** calculating the FE solution) estimates can be derived. We present the basic ideas here and refer to the literature [149] for more.

3.5.1 *Abstract residual error estimate*

We are in the abstract setting of Chapter 1, Section 1.3 and Chapter 2. Let $B(\cdot,\cdot) : X \times Y \to \mathbb{R}$ be a stable bilinear form, i.e.,

$$\inf_{0 \neq u \in X} \sup_{0 \neq v \in Y} \frac{B(u,v)}{\|u\|_X \, \|v\|_Y} \geq \alpha > 0\,, \tag{3.5.1}$$

let $F \in Y'$ and $u_0 \in X$ solve the problem

$$u_0 \in X : \;\; B(u_0, v) = F(v) \quad \text{for all} \;\; v \in Y\,. \tag{3.5.2}$$

Let $u_S \in X$ be **any** approximation of u_0, assumed to be known explicitly. Then the **error**

$$e_S := u_0 - u_S \tag{3.5.3}$$

satisfies the **residual problem**

$$e_S \in X : \;\; B(e_S, v) = R(v) := F(v) - B(e_S, v) \quad \text{for all} \;\; v \in Y\,. \tag{3.5.4}$$

We observe that the residual is, in principle, known and (3.5.1) therefore implies

$$\alpha \, \|e_S\|_X \leq \sup_{0 \neq v \in Y} \frac{B(e_S, v)}{\|v\|_Y} = \sup_{0 \neq v \in Y} \frac{R(v)}{\|v\|_Y} = \|R\|_{Y'}\,,$$

i.e., we get the **residual a posteriori error estimate**

$$\|e_S\|_X \le \frac{\|R\|_{Y'}}{\alpha} . \tag{3.5.5}$$

The problem in practice is how to estimate $\|R\|_{Y'}$.

Let $S \subset X, V \subset Y$ and $u_S \in S$ be a FE approximation

$$u_S \in S : \; B(u_S, v) = F(v) \text{ for all } v \in Y . \tag{3.5.6}$$

Then by (3.5.4),

$$R(v) = F(v) - B(u_S, v) = B(e_S, v) = 0 \quad \text{for all } v \in V$$

and hence, for any $v \in Y, w \in V$ and $Z \supset Y$:

$$\begin{aligned} R(v) &= R(v - w) \le \|R\|_{Z'} \, \|v - w\|_Z \\ &\le \Phi(V, Y, Z) \, \|R\|_{Z'} \, \|v\|_Y \end{aligned} \tag{3.5.7}$$

where

$$\Phi(V, Y, Z) = \sup_{0 \neq v \in Y} \; \inf_{0 \neq w \in V} \; \frac{\|v - w\|_Z}{\|v\|_Y}$$

is the width of V in Y, i.e., with (3.5.5) we get

$$\|e_S\|_X \le \alpha^{-1} \, \Phi(V, Y, Z) \, \|R\|_{Z'} . \tag{3.5.8}$$

Typically, Z is selected such that $\|R\|_{Z'}$ is easy to compute or estimate and a sharp bound on $\Phi(V, Y, Z)$ is known. For example, in the FEM one often has

$$X = Y = H^1, \;\; Z \cong Z' = L^2 .$$

We now apply these ideas to a model problem, essentially following [29].

3.5.2 *Application to hp-FEM [6]*

We consider again the model problem (3.2.1) with $AE = 1$, $c = 0$, i.e.,

$$-u'' = f \text{ in } \Omega = (-1, 1) , \tag{3.5.9}$$

$$u(-1) = u(1) = 0 , \tag{3.5.10}$$

and the variational formulation (3.2.4) with FE discretization (3.2.5):

$$u_S \in S_0^{\mathbf{p},1}(\Omega, \mathcal{T}) : \quad B(u_S, v) = F(v) \text{ for all } v \in S_0^{\mathbf{p},1}(\Omega, \mathcal{T}) . \tag{3.5.11}$$

Here $\mathcal{T} = \{-1 =: x_0 < x_1 < \ldots < x_M := 1\}$ and $\boldsymbol{p} = \{p_1, \ldots, p_M\}$ are the FE mesh and the polynomial degree vector, respectively, and

$$B(u,v) = \int_{-1}^{1} u'v'dx \, .$$

We have, for any $w \in S_0^{\boldsymbol{p},1}(\Omega, \mathcal{T})$,

$$\begin{aligned} R(v) &= \sum_{j=1}^{M} R^{[j]}(v) = \sum_{j=1}^{M} F^{[j]}(v) - B^{[j]}(u_0 + u_S, v) \\ &= \sum_{j=1}^{M} \int_{x_{j-1}}^{x_j} (f + u_S'')\, v \, dx + \sum_{j=1}^{M-1} [u_S'](x_j)\; v(x_j) \end{aligned} \tag{3.5.12}$$

where $[u_S'](x_j)$ denotes the jumps of u_S' at x_j.

We now select

$$w = \pi_{\boldsymbol{p}}\, v \tag{3.5.13}$$

where $\pi_{\boldsymbol{p}}\, v \in S_0^{\boldsymbol{p},1}(\Omega, \mathcal{T})$ denotes the approximation of v constructed in Theorem 3.17. Then

$$(v - w)(x_j) = (v - \pi_{\boldsymbol{p}}\, v)(x_j) = 0, \quad j = 0, \ldots, M \tag{3.5.14}$$

by (3.3.13) and we get in (3.5.12), using $R(\pi_{\boldsymbol{p}}\, v) = 0$,

$$\begin{aligned} R(v) &= R(v - \pi_{\boldsymbol{p}}\, v) \\ &= \sum_{j=1}^{M} \int_{x_{j-1}}^{x_j} (f + u_S'')(v - \pi_{\boldsymbol{p}}\, v)\, dx + \int_{x_{j-1}}^{x_j} (f - \Pi_S f)\, \pi_{\boldsymbol{p}}\, v \, dx \end{aligned} \tag{3.5.15}$$

where $\Pi_S f$ denotes the L^2-projection of f onto $S^{\boldsymbol{p},0}(\Omega, \mathcal{T})$, i.e., the elementwise Legendre series of f (since $\pi_{\boldsymbol{p}}\, v \in S^{\boldsymbol{p},1}(\Omega, \mathcal{T}) \subset S^{\boldsymbol{p},0}(\Omega, \mathcal{T})$, the last term in (3.5.15) vanishes).

Regrouping terms, we get using the Schwarz inequality

$$\begin{aligned} R(v) &= \sum_{j=1}^{M} \int_{x_{j-1}}^{x_j} (u_S'' + \Pi_S f)(v - \pi_{\boldsymbol{p}}\, v)\, dx + \int_{x_{j-1}}^{x_j} (f - \Pi_S f)\, v \, dx \\ &\le \sum_{j=1}^{M} \Big(\int_{x_{j-1}}^{x_j} (u_S'' + \Pi_S f)^2\, \omega_j \, dx \Big)^{\frac{1}{2}} \Big(\int_{x_{j-1}}^{x_j} \frac{(v - \pi_{\boldsymbol{p}}\, v)^2}{\omega_j(x)}\, dx \Big)^{\frac{1}{2}} \\ &\quad + \Big(\int_{x_{j-1}}^{x_j} (f - \Pi_S f)^2\, \omega_j \, dx \Big)^{\frac{1}{2}} \Big(\int_{x_{j-1}}^{x_j} \frac{(v - \pi_{\boldsymbol{p}}\, v)^2}{\omega_j(x)}\, dx \Big)^{\frac{1}{2}} \end{aligned} \tag{3.5.16}$$

where the weight function $\omega_j(x)$ is given by

$$\omega_j(x) := (x_j - x)(x - x_{j-1}) .$$

From (3.3.17) with $t = 0$ and a scaling argument, we get

$$\int_{x_{j-1}}^{x_j} \frac{(v - \pi_{\mathbf{p}}\, v)^2}{\omega_j(x)}\, dx \le \frac{1}{p_j(p_j+1)} \int_{x_{j-1}}^{x_j} (v')^2\, dx .$$

Inserting this in (3.5.16) gives the bound

$$\begin{aligned} R(v) \le \Big\{ \sum_{j=1}^{M} \frac{1}{p_j(p_j+1)} \Big[\|(f - \Pi_S f)\omega_j^{\frac{1}{2}}\|^2_{L^2(\Omega_j)} \\ + \|r_j\, \omega_j^{1/2}\|^2_{L^2(\Omega_j)} \Big] \Big\}^{\frac{1}{2}} \|v'\|_{L^2(\Omega)} , \end{aligned} \tag{3.5.17}$$

where the **element residuals** $r_j(x)$ are defined by

$$r_j(x) := \Pi_S f + u_S'',\ x \in \Omega_j,\ j = 1, \dots, M .$$

We therefore obtain the following result.

Proposition 3.88 *For any mesh $\mathcal{T}$ and any degree distribution $\mathbf{p}$, the a posteriori error estimate*

$$\|e\|_E = \|u_0' - u_S'\|_{L^2(\Omega)} \le EST \tag{3.5.18}$$

holds where the **error estimator** *EST is defined by*

$$(EST)^2 = \sum_{j=1}^{M} \eta_j^2 + \frac{1}{p_j(p_j+1)} \|(f - \Pi_S f)\, \omega_j^{1/2}\|^2_{L^2(\Omega_j)} \tag{3.5.19}$$

and the **elemental error indicators** *η_j are given by*

$$\eta_j^2 := \frac{1}{p_j(p_j+1)} \|r_j\, \omega_j^{1/2}\|^2_{L^2(\Omega_j)} . \tag{3.5.20}$$

Proof

$$\begin{aligned} \|u_0' - u_S'\|_{L^2(\Omega)} &=: \|e\|_E = \sup_{0 \ne v \in H_0^1} \frac{B(e, v)}{\|v'\|_{L^2(\Omega)}} \\ &\le \sup_{0 \ne v \in H_0^1} \frac{R(v)}{\|v'\|_{L^2(\Omega)}} \\ &\overset{(3.5.17)}{\le} EST . \end{aligned}$$

□

Proposition 3.88 shows that EST is **reliable**, i.e., it always overestimates the error $\|e\|_E$. We show now that it is also **efficient**, i.e., it does not overestimate by too much.

Proposition 3.89 *There exists $c > 0$, independent of $\mathbf{p}$ and $\mathcal{T}$, such that*

$$\eta_j \le c\,\|u_0' - u_S'\|_{L^2(\Omega_j)} + \frac{1}{\sqrt{p_j(p_j+1)}}\,\|(f - \Pi_S f)\,\omega_j^{\frac{1}{2}}\|_{L^2(\Omega_j)}\,, \qquad j = 1, \ldots, M\,. \tag{3.5.21}$$

Proof With (3.5.12), we find for any $v \in H_0^1(\Omega)$ that

$$\begin{aligned} B(e,v) &= \int_{-1}^{1} (u_0' - u_S')\, v'\, dx = R(v) \\ &= \sum_{j=1}^{M} \int_{\Omega_j} (u_S'' + \Pi_S f)\, v\, dx + \sum_{j=1}^{M-1} [u_S'](x_j)\, v(x_j) \\ &\quad + \int_{\Omega} (f - \Pi_S f)\, v\, dx\,. \end{aligned} \tag{3.5.22}$$

We now select the function v in (3.5.22) as follows:

$$v|_{\Omega_j} = r_j\,\omega_j, \quad v|_{\Omega\setminus\Omega_j} = 0\,. \tag{3.5.23}$$

This gives

$$\begin{aligned} \int_{\Omega_j} (r_j(x))^2\,\omega_j(x) dx &= B(e,v) - \int_{\Omega} (f - \Pi_S f)\, v\, dx \\ &\le \|e'\|_{L^2(\Omega_j)} \|v'\|_{L^2(\Omega_j)} \\ &\quad + \|(f - \Pi_S f)\,\omega_j^{\frac{1}{2}}\,\|_{L^2(\Omega_j)} \|r_j\,\omega_j^{\frac{1}{2}}\,\|_{L^2(\Omega_j)}\,. \end{aligned} \tag{3.5.24}$$

We have

$$\|v'\|^2_{L^2(\Omega_j)} \le 2\int_{\Omega_j} (r_j')^2\,\omega_j^2\, dx + 2\int_{\Omega_j} (r_j)^2\,(\omega_j')^2\, dx\,.$$

Since $|\omega_j(x)| \le h_j$ for $x \in \Omega_j$, we get

$$\|v'\|^2_{L^2(\Omega_j)} \le 2\,\{\|r_j'\,\omega_j\|^2_{L^2(\Omega_j)} + h_j^2\|r_j\|^2_{L^2(\Omega_j)}\}\,.$$

Using the inverse inequalities (see Theorems 3.95 and 3.96 below)

$$\int_{-1}^{1} (\pi_p')^2\,(1-\xi^2)^2\,d\xi \le cp^2 \int_{-1}^{1} (\pi_p)^2\,(1-\xi^2)\,d\xi$$

$$\int_{-1}^{1} (\pi_p)^2\,d\xi \le cp^2 \int_{-1}^{1} (\pi_p)^2\,(1-\xi^2)\,d\xi$$

and a scaling argument, we deduce

$$\begin{aligned}\|v'\|^2_{L^2(\Omega_j)} &\le cp_j^2 \int_{\Omega_j} (r_j)^2\,\omega_j\,dx + cp_j^2\,h_j^2 \int_{\Omega_j} (r_j)^2\,\omega_j\,dx \\ &= cp_j^2\,(1+h_j^2) \int_{\Omega_j} (r_j)^2\,\omega_j\,dx\,.\end{aligned}$$

Inserting into (3.5.24) gives

$$\|r_j\,\omega_j^{1/2}\|_{L^2(\Omega_j)} \le cp_j\,(1+h_j^2)^{\frac{1}{2}}\,\|e'\|_{L^2(\Omega_j)} + \|(f-\Pi_S f)\,\omega_j^{\frac{1}{2}}\,\|_{L^2(\Omega_j)}$$

and, after dividing by $\sqrt{p_j(p_j+1)}$,

$$\eta_j \le c\,\|e'\|_{L^2(\Omega_j)} + \frac{1}{\sqrt{p_j(p_j+1)}}\,\|(f-\Pi_S f)\,\omega_j^{\frac{1}{2}}\,\|_{L^2(\Omega_j)}\,.$$

□

We observe that the local lower bound (3.5.21) gives immediately

$$\eta_j^2 \le 2c^2\,\|e'\|^2_{L^2(\Omega_j)} + \frac{2}{p_j(p_j+1)}\,\|(f-\Pi_S f)\,\omega_j^{\frac{1}{2}}\,\|^2_{L^2(\Omega_j)}\,.$$

Summing over all elements, it follows that

$$(EST)^2 \le 2c^2\,\|e\|_E^2 + 3\sum_{j=1}^{M} \frac{1}{p_j(p_j+1)}\,\|(f-\Pi_S f)\,\omega_j^{\frac{1}{2}}\,\|^2_{L^2(\Omega_j)}\,,$$

i.e., up to consistency errors in the approximation of the right hand side and a constant which is independent of $\mathcal{T}$ and $\boldsymbol{p}$ we also have a global lower bound.

We remark that residual error estimates and indicators are available for many linear as well as nonlinear problems with sufficient structure, see [149] and the references there. In all the reliability and efficiency results shown there, the constants still depend on p, however. More advantageous if robustness with respect to p is desired is the so-called equilibrated residual method which gives upper bounds for the error in energy norm independently of p, see e.g. [4], Section 5; there, however, some local auxiliary problems must be solved to obtain the estimate.

Exercise 3.90 Inspecting the proof of Proposition 3.89, estimate the constant c in (3.5.21).

3.6 Inverse inequalities

3.6.1 *Basic inverse inequality*

Theorem 3.91 *Let $I = (a, b)$ and $h = b - a$. Then for every polynomial $v \in S^p(I)$ it holds that*

$$\|v'\|_{L^2(I)} \leq 2\sqrt{3}\ \frac{p^2}{h}\ \|v\|_{L^2(I)}\ . \tag{3.6.1}$$

Proof We first assume that $\Omega = I$, i.e., $b = 1$, $a = -1$, $h = 2$.

Step 1: We have for every $n \geq 1$ that

$$\int_{-1}^{1} (L_n'(\xi))^2 d\xi = n(n+1)\ . \tag{3.6.2}$$

To see this, we use integration by parts and the orthogonality properties of the Legendre polynomials to get

$$\begin{aligned}\int_{-1}^{1} (L_n'(\xi))^2 d\xi &= -\int_{-1}^{1} L_n(\xi)\, L_n''(\xi) d\xi + (L_n L_n')\Big|_{-1}^{1} \\ &= L_n'(1) - (-1)^n\, L_n'(-1)\ .\end{aligned}$$

Since $L_n'(\pm 1) = \pm\, n(n+1)/2$, we find (3.6.2).

Step 2: Every $v \in S^p(I)$ may be expanded into a Legendre series (cf. Chapter 3, Section 3.3)

$$v(x) = \sum_{n=0}^{p} a_n\, L_n(x)$$

where

$$\|v\|^2_{L^2(I)} = \sum_{n=0}^{p} \frac{2}{2n+1}\, |a_n|^2\ .$$

By Step 1, we get

$$\|v'\|_{L^2(I)} \leq \sum_{n=0}^{p} |a_n|\ \|L_n'\|_{L^2(I)} \leq \sum_{n=0}^{p} |a_n|\ (n(n+1))^{1/2}\ .$$

By the Cauchy–Schwarz inequality we get

$$\begin{aligned}
\|v'\|^2_{L^2(I)} &\le \Big(\sum_{n=0}^{p} \frac{2}{2n+1}\,|a_n|^2\Big)\Big(\sum_{n=1}^{p} n(n+1)(2n+1)/2\Big) \\
&\le \|v\|^2_{L^2(I)}\, p \max_{1\le n\le p} n(n+1)(2n+1)/2 \\
&= \|v\|^2_{L^2(I)}\, p^2(p+1)(p+1/2)
\end{aligned}$$

whence it follows that

$$\|v'\|^2_{L^2(I)} \le 3p^4\, \|v\|^2_{L^2(I)}\,.$$

The proof is completed by a scaling argument. □

An analog of Theorem 3.91 can be obtained for the L^∞-norm. We have the following result.

Theorem 3.92 *Let $I = (a, b)$ be a bounded interval and $h = b - a$. Then for every $v \in S^p(I)$ it holds that*

$$\|v'\|_{L^\infty(I)} \le \frac{2p^2}{h}\, \|v\|_{L^\infty(I)}\,, \tag{3.6.3}$$

$$\|v\|_{L^\infty(I)} \le 2\Big(\frac{8}{b-a}\Big)^{1/q} p^{2/q}\, \|v\|_{L^q(I)}\,,\ \ 1 \le q \le \infty\,. \tag{3.6.4}$$

For a proof of this result, we refer to [114].

(3.6.3) is sometimes also referred to as Markov's inequality.

Remark 3.93 Based on (3.6.1) and (3.6.3), inverse inequalities in L^r for any $2 \le r \le \infty$ may be obtained by interpolation.

From Theorem 3.91 we may furthermore deduce by iteration

Corollary 3.94 *Let $0 \le s \le t$ be integers. Then for every $v \in S^p(I)$ holds with $I = (a, b)$, $h = b - a$*

$$\|v^{(t)}\|_{L^2(I)} \le (2\sqrt{3})^{t-s}\, (p^2/h)^{t-s}\, \|v^{(s)}\|_{L^2(I)} \tag{3.6.5}$$

By interpolation (see Appendix B), one can easily deduce (3.6.5) for noninteger s and t.

3.6.2 *Weighted inverse estimates*

The above inverse inequalities have the esthetic imperfection that **per differentiation order, two powers of** p but only **one power of** h is lost. The bounds are, nevertheless, optimal in standard Sobolev spaces. It turns out that in **weighted spaces**, inverse inequalities hold with the loss of only one power of p. We have the following result.

Theorem 3.95 *Let $I = (-1,1)$ and $v \in S^p(I)$. Then*

$$\int_I (v'(x))^2(1-x^2)\,dx \le p(p+1) \int_I (v(x))^2\,dx\,,$$

$$\int_I (v'(x))^2(1-x^2)^{\frac{1}{2}}\,dx \le c\,p^2 \int_I (v(x))^2(1-x^2)^{-\frac{1}{2}}\,dx\,,$$

where c is independent of v and p.

Proof We expand v in a Legendre series

$$v(x) = \sum_{k=0}^{p} a_k L_k(x)\,.$$

This gives

$$\int_{-1}^{1} (v'(x))^2\,(1-x^2)dx = \sum_{k=0}^{p}\sum_{\ell=0}^{p} a_k a_\ell \int_{-1}^{1} L_k'(x)\,L_\ell'(x)\,(1-x^2)dx\,.$$

Using (3.13), (3.14) of Appendix C, we find

$$\int_{-1}^{1} L_k'(x)\,L_\ell'(x)\,(1-x^2)dx = \frac{2}{2k+1}\,\frac{(k+1)!}{(k-1)!}\,\delta_{k\ell} = \frac{2k(k+1)}{2k+1}$$

and therefore

$$\begin{aligned}\int_{-1}^{1} (v'(x))^2\,(1-x^2)dx &= \sum_{k=0}^{p} \frac{2k(k+1)}{2k+1}\,(a_k)^2 \\ &\le p(p+1)\sum_{k=0}^{p} \frac{2}{2k+1}\,(a_k)^2 \\ &= p(p+1)\int_{-1}^{1} (v(x))^2\,dx\end{aligned}$$

which is the first assertion. The second estimate is proved in the same way, using expansions into Chebyčev polynomials $P_k(x;-1/2)$ rather than Legendre polynomials. □

Using expansions into the Jacobi polynomials from Appendix C, Theorem 3.95 can be generalized to other weights. We leave it to the reader.

Another class of weighted inverse estimates relates different weighted norms, without differentiating the polynomials.

Theorem 3.96 *For any $\alpha \geq 0$, there exists $C(\alpha)$ such that for any $v \in \mathcal{S}^p(I)$*

$$\int_{-1}^{1} (v(x))^2\, dx \leq C(\alpha)\, p^{2\alpha} \int_{-1}^{1} (1-x^2)^\alpha\, (v(x))^2\, dx\ .$$

Proof We start with $\alpha = 1/2$. Let $m \geq 1$ be an integer and set $\varepsilon_{mp} = 1 - \cos(1/mp)$. Then we estimate

$$\begin{aligned}\int_{-1}^{1} \frac{(v(x))^2}{\sqrt{1-x^2}}\, dx &\leq \frac{1}{\sin(1/mp)} \int_{-1+\varepsilon_{mp}}^{1-\varepsilon_{mp}} (v(x))^2\, dx \\ &\quad + 2\, \|v\|^2_{L^\infty(I)} \int_{1-\varepsilon_{mp}}^{1} (1-x^2)^{-\frac{1}{2}}\, dx \\ &\leq 2mp \int_{-1+\varepsilon_{mp}}^{1-\varepsilon_{mp}} (v(x))^2\, dx + \frac{2}{mp}\, \|v\|^2_{L^\infty(I)}\end{aligned} \tag{3.6.6}$$

since $mp \geq 1$ and $1/(\sin(x)) \leq 2/x$ if $0 < x \leq 1$.

Next, we claim that

$$\|v\|^2_{L^\infty(I)} \leq \frac{2(p+1)}{\pi} \int_{-1}^{1} \frac{(v(x))^2}{\sqrt{1-x^2}}\, dx\ . \tag{3.6.7}$$

To see it, expand $v(x)$ in a **Chebyčev series** $v(x) = \sum_{k=0}^{p} v_k\, T_k(x)$, where $T_0(x) = 1/\sqrt{2}$, $T_k(x) = \cos(k \arccos(x))$. Then

$$(v(x))^2 = \Big(\sum_{k=0}^{p} v_k\, T_k(x)\Big)^2 \leq (p+1)\Big(\sum_{k=0}^{p} (v_k)^2\Big)$$

since $T_k(x) \leq 1$ for $|x| \leq 1$. Since

$$\begin{aligned}\int_{-1}^{1} \frac{(v(x))^2}{\sqrt{1-x^2}}\, dx &= \sum_{k=0}^{p} \sum_{\ell=0}^{p} v_k v_\ell \int_{-1}^{1} \frac{T_k(x)\, T_\ell(x)}{\sqrt{1-x^2}}\, dx \\ &= \sum_{k,\ell=0}^{p} v_k v_\ell \int_{0}^{\pi} \cos(k\theta)\ \cos(\ell\theta)\, d\theta \\ &= \frac{\pi}{2} \sum_{k=0}^{p} (v_k)^2\end{aligned}$$

we find (3.6.7). Inserting (3.6.6), we get

$$\left(1 - \frac{4(p+1)}{\pi mp}\right) \int_{-1}^{1} \frac{(v(x))^2}{\sqrt{1-x^2}}\, dx \le 2mp \int_{-1}^{1} (v(x))^2\, dx\,.$$

Selecting m sufficiently large now completes the proof in the case $\alpha = 1/2$.

Now let $\alpha \ge 0$ be arbitrary. With ε_{mp} as above we get

$$\begin{aligned}\int_{-1}^{1} (1-x^2)^\alpha\, (v(x))^2\, dx &\ge \left(\sin\left(\frac{1}{mp}\right)\right)^{2\alpha} \int_{-1+\varepsilon_{mp}}^{1-\varepsilon_{mp}} (v(x))^2\, dx \\ &\ge \left(\sin\left(\frac{1}{mp}\right)\right)^{2\alpha} \int_{-1}^{1} (v(x))^2\, dx - 2\varepsilon_{mp} \left(\sin\left(\frac{1}{mp}\right)\right)^{2\alpha} \|v\|^2_{L^\infty(I)}\,.\end{aligned} \tag{3.6.8}$$

By (3.6.7), we have the useful estimate

$$\|v\|^2_{L^\infty(I)} \le Cp \int_{-1}^{1} \frac{(v(x))^2}{\sqrt{1-x^2}}\, dx \le Cp^2\, \|v\|^2_{L^2(I)}$$

(which is (3.6.4) for $q = 2$) where we used (3.6.6) for sufficiently large m.

Inserting in (3.6.8) yields

$$\int_{-1}^{1} (1-x^2)^\alpha\, (v(x))^2\, dx \ge \left(\sin\left(\frac{1}{mp}\right)\right)^{2\alpha} (1 - 2C\,\varepsilon_{mp}\, p^2)\, \|v\|^2_{L^\infty(I)}\,.$$

Since $\varepsilon_{mp} = m^{-2} p^{-2} + O(m^{-4} p^{-4})$, we get for m large enough (but independent of p) the assertion. □

Theorem 3.96 will be used in Chapter 5. The proof is taken from [107].

3.7 Bibliographical remarks

The FE spaces $S^{\mathbf{p},\ell}(\Omega, \mathcal{T})$ are generalizations of the classical spaces $S^{p,\ell}(\Omega, \mathcal{T})$ of (fixed) polynomial degree p and were introduced in [24]; see also [144]. The exposition and notation in Sections 3.1 -3.3 follow [24]. The results in Section 3.2.3 on the p-version approximation are from [129]. Section 3.3 is essentially a simplification of [71] to the one-dimensional case. Section 3.4 follows closely [132] while Section 4 is taken from [129].

We remark that in one dimension exponential convergence of an *hp*-version FEM for a special singular solution was first shown in [69]. There, one also finds a very detailed investigation of the optimality geometric meshes and linear degree vectors – if, for example, the polynomial degree is uniform, then a radical mesh yields somewhat better rates of convergence (in the sense that the constant b in (3.3.58) is unproved). In all cases, the values of b in (3.3.58) are between 1 and 2 if $u_0(x)$ is of the type $|x - x_0|^\alpha$. The *hp*-idea, however, is somewhat older, see, e.g., [123] and the references there.

We note further that all approximation results presented here were L^2-based; this allowed to prove them by rather explicit constructions of quasi-interpolants, namely, the truncated Legendre series. Analogous results are, however, valid in L^r-based spaces, $1 \leq r \leq \infty$. We refer to Section 3 of [57] for analogs of Theorem 3.36 in an L^r-setting.

4

HP-FINITE ELEMENTS IN TWO DIMENSIONS

In the present chapter, we analyze *hp*-FEM for scalar elliptic problems in two-dimensional domains with a piecewise analytic boundary, such as the curvilinear polygon introduced in Chapter 1. It is well known that the solution of, for example, Laplace's equation in such a domain is analytic except in the vertices A_k of the polygon Ω, provided the input data are analytic. The regularity of the solution will be discussed in more detail in Section 4.2, after introduction of the model problem in Section 4.1. In Section 4.3, a complete analysis of the rate of convergence of linear finite elements with mesh refinement toward the corners of the domain is given. We prove that linear FEM on a properly refined mesh exhibit the same rate of convergence as on a quasiuniform mesh if the solution were smooth. Although we consider only low order elements, the proof yield several weighted norm estimates that are used later on for the analysis of *hp*-FEM. Section 4.4 presents the *hp*-FE subspaces $S^{\mathbf{p},\ell}(\Omega, \mathcal{T}_\sigma^n)$ on geometric meshes $\mathcal{T}_\sigma^n$. Section 4.5 is devoted to the proof and discussion of exponential convergence of the *hp*-FEM on geometric meshes in polygons. Section 4.6 is a synopsis of various inverse ("Bernstein") inequalities and polynomial trace liftings that are used in Sections 4.5 and 4.7. Finally, in Section 4.7, we address the fast iterative solution and the preconditioning of the linear systems of equations.

4.1 Model problem

Let $\Omega \subset \mathbb{R}^2$ be a curvilinear polygon (see Definition 1.2 where the notation is defined). In Ω, we consider the **membrane problem**

$$Lu = f \ \text{ in } \Omega \tag{4.1.1}$$

where

$$Lu(x) = -\mathrm{div}\big(AE(x)\,\mathbf{grad}\,u(x)\big) + c(x)\,u(x)$$

with boundary conditions

$$\begin{aligned} \gamma_0\,u(x) &= g^0 \ \text{ on } \ \Gamma^{[0]}\,, \\ \gamma_1\,u(x) &= g^1 \ \text{ on } \ \Gamma^{[1]}\,, \end{aligned} \tag{4.1.2}$$

with γ_1 denoting the conormal derivative operator. We assume here in particular $\Gamma^{[2]} = \varnothing$ and, for convenience, that

$$AE(x) = \mathbf{1}\,, \quad c(x) = 1 \tag{4.1.3}$$

so that $Lu = -u + u$. All results to be proved in the present chapter will hold for the case when $AE(x)$ and $c(x)$ are sufficiently smooth (analytic) in $\overline{\Omega}$, however, provided AE is strongly elliptic, i.e., provided (1.3.22) holds. We assume further that $\Gamma^{[i]}$, $i = 0, 1$, is the union of a finite number of analytic boundary segments Γ_j.

For given data

$$f \in L^2(\Omega), \quad g^0 \in H^{\frac{1}{2}}(\Gamma^{[0]}), \quad g^1 \in L^2(\Gamma^{[1]}) \tag{4.1.4}$$

(actually, $g^1 \in [H_{00}^{1/2}(\Gamma^{[1]})]'$ would be sufficient. See Chapter 1, Section 1.4 for more on $H_{00}^{1/2}(\Gamma^{[1]})$): problem (4.1.1)–(4.1.3) admits the **variational formulation**: find $u \in \{g^0\} + H_0^1(\Omega)$ such that

$$B(u, v) = F(v) \text{ for all } v \in H_0^1(\Omega) \tag{4.1.5}$$

where $H_0^1(\Omega) = H^1(\Omega) \cap \{u : \gamma_0 u = 0 \text{ on } \Gamma^{[0]}\}$ and

$$B(u, v) = \int_\Omega (\nabla u \cdot \nabla v + uv)\, dx, \quad F(v) = \int_\Omega f v\, dx + \int_{\Gamma^{[1]}} g^1 \gamma_0\, v\, do\,.$$

Since $B(u, u) = \|u\|_{H^1}^2$ and $|F(v)| \le C\,\|v\|_{H^1}$, problem (4.1.5) admits a unique solution by the Lax-Milgram lemma, (Theorem 2.19).

Let $S \subset H^1(\Omega)$ be a finite-dimensional subspace and set $S_0 = S \cap H_0^1(\Omega)$. Assume that the Dirichlet data g^0 is such that

$$\text{there exists } s_0 \in S \text{ such that } \gamma_0 s_0 = g^0 \text{ on } \Gamma^{[0]}\,. \tag{4.1.6}$$

Then the FE approximation of (4.1.5) is: find $u_S \in S$ such that $u_S = g^0$ on $\Gamma^{[0]}$ and

$$B(u_S, v) = F(v) \ \text{ for all } \ v \in S_0\,. \tag{4.1.7}$$

Again by the Lax–Milgram lemma, there exists a unique solution $u_S \in S$ of (4.1.7). The error $e_S = u - u_S$ vanishes on $\Gamma^{[0]}$ by (4.1.6) and satisfies

$$B(e_S, v) = 0 \ \text{ for all } \ v \in S \cap H_0^1(\Omega)\,. \tag{4.1.8}$$

For any $v \in S$, $v = g^0$ on $\Gamma^{[0]}$, we have

$$\|e_S\|_{H^1}^2 = B(e_S, e_S) = B(e_S, u - v) \le \|e_S\|_{H^1}\, \|u - v\|_{H^1}$$

whence

$$\begin{aligned} \|e_S\|_{H^1} &\le \inf_{\substack{v \in S \\ v = g^0 \text{ on } \Gamma^{[0]}}} \|u - v\|_{H^1} \\ &= Z(u, S_0 + \{g^0\}, H^1)\,. \end{aligned} \tag{4.1.9}$$

Therefore, the FEM is stable and the FE approximation error is determined by the best approximation error of the solution. To estimate the consistency Z, we need two basic ingredients: the **regularity of the solution** and the **structure of the FE space** S. In general, $Z > 0$ in (4.1.9), i.e., S does not contain the exact solution. The FE space S should therefore provide the minimal error at a given dimension for u belonging to the anticipated solution class. In the following subsections, we will address these issues for the model problem (4.1.5).

4.2 Regularity of the solution - I

We collect, without proofs, some facts about the regularity of the solution u of (4.1.5). In the one-dimensional analog of (4.1.1), i.e.

$$-u'' + u = f \text{ in } (-1,1)\,,$$

the solution u is as smooth as the right-hand side allows: $f \in H^k \Longrightarrow u \in H^{k+2}$, $f \in C^k \Longrightarrow u \in C^{k+2}$ for any $k \geq 0$. This is the so-called "**shift theorem**". The shift theorem does not hold generally for (4.1.5) in two dimensions: Even if the data f, g^0, g^1 are smooth (C^∞ say), the solution u may fail to be in $H^2(\Omega)$ owing to singularities induced by the corner points A_j of the polygon Ω. It turns out, however, that u can be decomposed into a singular part and a regular remainder u^0 which is smooth as the right-hand side allows. More precisely, we have the following result.

Theorem 4.1 *Assume* (4.1.4) *and* $f \in H^{-1+s}(\Omega)$,

$$g^0 \in H^{\frac{1}{2}+s}(\Gamma_i) \text{ for } \Gamma_i \subset \Gamma^{[0]},\;\; g^1 \in H^{-\frac{1}{2}+s}(\Gamma_i) \textit{ for } \Gamma_i \subset \Gamma^{[1]},\; s > 0\,,$$

and that Ω is a straight polygon. Let (r_i, θ_i) be polar coordinates at the vertex A_i of Ω (see Figure 1.1.2*), with θ_i oriented counterclockwise. Define the* **singular functions**

$$S_{ij}(r_i,\theta_i) = \begin{cases} r_i^{\alpha_{ij}} \sin(\alpha_{ij}\theta_i) \textit{ if } \Gamma_{i+1} \subset \Gamma^{[0]} \\ r_i^{\alpha_{ij}} \cos(\alpha_{ij}\theta_i) \textit{ if } \Gamma_{i+1} \subset \Gamma^{[1]} \end{cases} \tag{4.2.1}$$

if $\alpha_{ij} \notin \mathbb{N}$ where the **singular exponents** α_{ij} *are*

$$\alpha_{ij} = \begin{cases} j\pi/\omega_i & \textit{if } A_i \textit{ is a Dirichlet or Neumann vertex}, \\ (j-1/2)\,\pi/\omega_i & \textit{else}\,. \end{cases} \tag{4.2.2}$$

If $\alpha_{ij} \in \mathbb{N}$, (4.2.1) *is modified to*

$$S_{ij}(r_i,\theta_i) = \begin{cases} r_i^{\alpha_{ij}} \left(\log(r_i) \sin(\alpha_{ij}\,\theta_i) + \theta_i \cos(\alpha_{ij}\,\theta_i)\right) \Gamma_{i+1} \subset \Gamma^{[0]} \\ r_i^{\alpha_{ij}} \left(\log(r_i) \cos(\alpha_{ij}\,\theta_i) + \theta_i \sin(\alpha_{ij}\,\theta_i)\right) \Gamma_{i+1} \subset \Gamma^{[1]}\,. \end{cases} \tag{4.2.3}$$

Then we have the decomposition

$$u = u^0 + \sum_{\alpha_{ij}<s} a_{ij}\, S_{ij}(r_i, \theta_i) \tag{4.2.4}$$

where $u^0 \in H^{1+s}(\Omega)$ *and the coefficients* $a_{ij} \in \mathbb{R}$.

Remark 4.2 Theorem 4.1 remains valid for slit domains used to model cracks. In this case $\omega_i = 2\pi$ and $\alpha_{i1} = 1/2$. The coefficients a_{i1} of the dominant singularities in (4.2.4) are called **stress intensity factors**. Results analogous to Theorem 4.1 hold for any strongly elliptic system with sufficiently smooth coefficients in a curvilinear polygon (see, e.g., [51], [68], [83], [84]).

An alternative to the decomposition (4.2.4) is to measure the regularity of u in terms of **weighted Sobolev spaces**. Let $A_i, i = 1, \ldots, M$ denote the vertices of Ω and $\beta = (\beta_1, \ldots, \beta_M)$, $0 \le \beta_i < 1$ be an M-tuple. For any integer k we define further $\beta \pm k := (\beta_1 \pm k, \beta_2 \pm k, \ldots, \beta_M \pm k)$ and the weight function

$$\Phi_\beta(x) = \prod_{i=1}^{M} (r_i^*(x))^{\beta_i},\ r_i^*(x) = \min(1, |x - A_i|)\ . \tag{4.2.5}$$

For integers $m \ge \ell \ge 0$, we define the seminorms

$$|u|^2_{H_\beta^{m,\ell}(\Omega)} := \sum_{k=\ell}^{m} \|\, |D^k u|\, \Phi_{\beta+k-\ell}\|^2_{L^2(\Omega)}\ . \tag{4.2.6}$$

By $H_\beta^{m,\ell}(\Omega)$, $m \ge \ell \ge 0$, we denote the completion of $C^\infty(\overline{\Omega})$ with respect to the norms

$$\begin{aligned} \|u\|^2_{H_\beta^{m,\ell}(\Omega)} &= \|u\|^2_{H^{\ell-1}(\Omega)} + |u|^2_{H_\beta^{m,\ell}(\Omega)}, && \ell \ge 1\,, \\ \|u\|^2_{H_\beta^{m}(\Omega)} &= \|u\|^2_{H_\beta^{m,0}(\Omega)} + \sum_{k=0}^{m} \|\, |D^k u|\, \Phi_{\beta+k}\|^2_{L^2(\Omega)}, && \ell = 0\,. \end{aligned} \tag{4.2.7}$$

If $m = \ell = 0$, we also write $H_\beta^{0,0} = H_\beta^0 = L_\beta(\Omega)$. Here $|D^k u|^2 = \sum_{|\alpha|=k} |D^\alpha u|^2$. this and the following section, we use mainly the weighted space $H_\beta^{2,2}$; the more general spaces $H_\beta^{m,\ell}(\Omega)$ will be required in Section 4.5 ahead in the convergence analysis of the *hp*-version.

Proposition 4.3

$$H_\beta^{2,2}(\Omega) \subset C^0(\overline{\Omega})\ . \tag{4.2.8}$$

For a proof, we refer to [19]. We also have from Hölder's inequality and the embedding $H^1(\Omega) \subset L^p(\Omega)$ for $1 \le p < \infty$ that

$$\|u\|^2_{H^0_{-\beta}} := \int_\Omega \Phi_\beta^{-2} u^2\, dx \le C_\beta\, \|u\|^2_{H^1(\Omega)}\ . \tag{4.2.9}$$

Therefore, if $f \in H^0_\beta(\Omega)$, we find that $F(v) = \int_\Omega f v\,dx$ defines a continuous linear functional on $H^1_0(\Omega)$. Then the following holds.

Theorem 4.4 *Assume* $0 \le \beta_i < 1$, $1 \le i \le M$, *and that* $1 - \alpha_{i1} < \beta_i < 1$ *if* $\alpha_{i1} < 1$. *Assume further* $g^0 = g^1 = 0$ *and* $f \in H^0_\beta(\Omega)$. *Then the solution* u *of* (4.1.5) *belongs to* $H^{2,2}_\beta(\Omega)$ *and the a priori estimate*

$$\|u\|_{H^{2,2}_\beta(\Omega)} \le C\, \|f\|_{H^0_\beta(\Omega)} \tag{4.2.10}$$

holds.

For a proof, we refer to [15].

We obtain for the pure Dirichlet or Neumann problem in a convex polygon that $\omega_i > \pi$, $\alpha_{i1} \ge 1$ and hence that

$$\|u\|_{H^2(\Omega)} \le C\, \|f\|_{L^2(\Omega)}\ .$$

4.3 Linear finite elements with mesh refinement

4.3.1 *Meshes consisting of triangles*

Let Ω be a polygon with straight sides and a mesh $\mathcal{T} = \{T_j\}_{j=1}^{M(\mathcal{T})}$ be given corresponding to a partition of Ω into M open, disjoint triangles T_j, i.e., $T_i \cap T_j = \emptyset$ if $i \ne j$, $\overline{\Omega} = \bigcup_{T\in\mathcal{T}} \overline{T}$.

Definition 4.5

i) *A mesh* $\mathcal{T} = \{T_j\}_{j=1}^M$ *is* **regular**, *if* $\overline{T}_i \cap \overline{T}_j$, $i \ne j$, *is either empty or it consists of a vertex or an entire side of* T_i.

ii) *For each* $T \in \mathcal{T}$, *define*

$$h_T = \mathrm{diam}(T)\ , \tag{4.3.1}$$

$$\rho_T = \sup\{\mathrm{diam}(B) : B \textit{ is a ball contained in } T\}\ . \tag{4.3.2}$$

The **meshwidth** h *of* $\mathcal{T}$ *is* $h(\mathcal{T}) = \max_{T\in\mathcal{T}}\{h_T\}$.

iii) *A family of triangular meshes* $\{\mathcal{T}_i\}_{i=1}^\infty$ *is* **shape regular** *if it is regular and if there exists a constant* κ, *independent of* i, *such that*

$$\sup_{T\in\mathcal{T}_i} \frac{h_T}{\rho_T} \le \kappa < \infty, \quad i = 1, 2, \dots\ . \tag{4.3.3}$$

iv) $\{\mathcal{T}_i\}_{i=1}^\infty$ *is a* **quasi uniform mesh** *sequence, if there exist constants* $\kappa_1, \kappa_2 > 0$ *independent of* i, *such that*

$$\kappa_1 h(\mathcal{T}) \le h_T \le \kappa_2 \rho_T,\ \textit{for all}\ \ T \in \mathcal{T}_i,\ \ i = 1, 2, \dots\ . \tag{4.3.4}$$

v) $\{\mathcal{T}_i\}_{i=1}^\infty$ *is a* **uniform mesh** *sequence if all triangles* $T_j \in \mathcal{T}_i$ *are congruent.*

We evidently have

$$\text{uniform} \Longrightarrow \text{quasi uniform} \Longrightarrow \text{shape regular} .$$

Example 4.6 Consider the ***L*-shaped polygon** in Figure 4.1. Continuing the refinement near the reentrant corner O yields a family $\{\mathcal{T}_i\}_{i=1}^{\infty}$ of meshes which are **geometrically refined** towards O. They are shape regular, but evidently not quasiuniform.

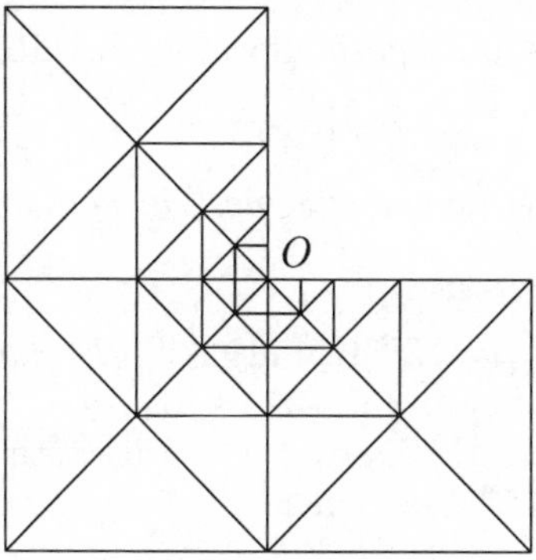

FIG. 4.1. L-shaped domain with a geometric mesh

For the approximation of solutions $u \in H^{2,\beta}(\Omega)$ of the model problem we use meshes which are systematically refined near the vertices A_j of Ω.

Definition 4.7 *Let $\gamma = (\gamma_1, \ldots, \gamma_N)$, $0 \le \gamma_i < 1$ and $\Phi_\gamma(x)$ be the corresponding weight function defined in (4.2.5). Then we call a regular triangulation $\mathcal{T}$ a* **triangulation of type** (h, γ, K), *if*

i) *$\mathcal{T}$ is shape regular (i.e., (4.4.18) holds),*
ii) *$\Phi_\gamma > 0$ on $\overline{T}$, then $\kappa^{-1} h \Phi_\gamma(x) \le h_T \le \kappa h \Phi_\gamma(x)$ for all $x \in \overline{T}$,*
iii) *$\Phi_\gamma(x) = 0$ at some $x \in \overline{T}$, then*

$$\kappa^{-1} \le \frac{h_T}{h \sup\limits_{x \in T} \Phi_\gamma(x)} \le \kappa .$$

Exercise 4.8 Show that if $\gamma_i = 0$, $i = 1, \ldots, M$, a family $\{\mathcal{T}_j\}$ of triangulations of type $(h, 0, K)$ is quasiuniform.

Lemma 4.9 *Let $\mathcal{T}$ be a triangulation of type (h, γ, k) and let $N(\mathcal{T})$ denote the number of vertices in $\mathcal{T}$. Then there exists a constant $C(\Omega, \gamma, K)$ independent of h, such that*

$$N \le C h^{-2} . \tag{4.3.5}$$

Proof We have $N \leq 3\sum_{T\in\mathcal{T}} 1$. If $\Phi_\gamma \neq 0$ in $\overline{T}$, we have

$$1 \leq C h_T^{-2} \int_T dx \leq C h^{-2} \int_T \Phi_\gamma^{-2}(x) dx \leq C h^{-2} .$$

Since there are only finitely many triangles in which Φ_γ vanishes, we have (4.3.5). □

Lemma 4.9 states that a mesh $\mathcal{T}$ of type (h, γ, k) has asymptotically, as $h \to 0$, the same number of nodes as a quasiuniform mesh with meshwidth h.

4.3.2 *The finite element spaces $S^{p,\ell}(\Omega, \mathcal{T})$*

Definition 4.10 *Let $\mathcal{T}$ be a triangular mesh in the (straight) polygon Ω, $p \geq 0$ and $\ell \geq 0$ integers. Then*

$$S^{p,\ell}(\Omega, \mathcal{T}) = \{u \in H^\ell(\Omega) : u|_T \in \mathcal{S}_1^p(T) \;\; for \;\; T \in \mathcal{T}\} . \tag{4.3.6}$$

Of greatest importance are the cases $\ell = 0$ (discontinuous functions), $\ell = 1$ (C^0-functions) and $\ell = 2$ (C^1-functions).

Example 4.11 For $\ell = p = 1$, we obtain the piecewise linear Courant triangles. Since a linear function on a triangle is determined by its values in the vertices, $u \in S^{1,1}$ is uniquely determined by the function values in the vertices of $\mathcal{T}$.

Definition 4.12 *A set of points in Ω such that $u \in S^{p,\ell}(\Omega, \mathcal{T})$ is (uniquely) determined by its values at these points are called (unisolvent)* **nodes**.

Functions $u \in S^{p,\ell}$ which are nonzero in exactly one node are called a **nodal basis** *of $S^{p,\ell}$.*

Example 4.13 Let Ω be a polygon and $\mathcal{T}$ a mesh consisting of triangles. A nodal basis of $S^{1,1}(\Omega, \mathcal{T})$ consists of piecewise linear functions which are equal to one in exactly one node and zero in all other nodes.

In particular,

$$N = \dim(S^{1,1}(\Omega, \mathcal{T})) = \# \text{ of nodes in } \mathcal{T} .$$

Remark 4.14 (on symbolic element notation). For fixed, uniform polynomial degree p and when all elements in $\mathcal{T}$ are of the same type (in particular for the h-version FEM), it is customary in the literature to express the subspace $S^{p,\ell}(\Omega, \mathcal{T})$ symbolically by a sketch of the reference element and the degrees of freedom corresponding to the nodal basis. For example, for $S^{1,1}(\Omega, \mathcal{T})$, the degrees of freedom are the function values at the vertices of the triangles $T \in \mathcal{T}$. This would be expressed symbolically as in Figure 4.2.

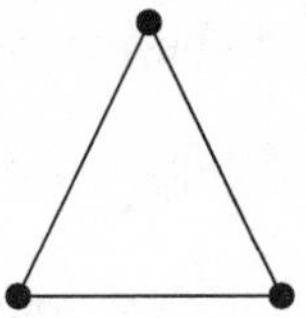

FIG. 4.2.

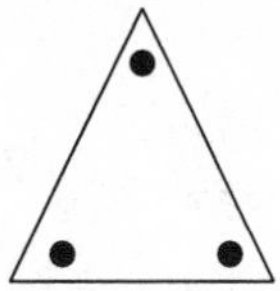

FIG. 4.3.

The bold dots correspond to the elemental degrees of freedom and their location on the element boundary means they are shared by adjacent elements due to the C^0-continuity of $S^{1,1}(\Omega, \mathcal{T})$. In this way, $S^{1,0}(\Omega, \mathcal{T})$ (the space of piecewise linear, **discontinuous** functions on $\mathcal{T}$) would be denoted as in Figure 4.3. The elemental degrees of freedom are still associated with the vertices, but are now depicted inside the element, since they are not linked to corresponding degrees of freedom on the adjacent elements.

This kind of symbolic notation and the corresponding definition of "an element" are closely related to the h-version FEM where all elements K are, in a sense, equivalent to a reference element $\widehat{K}$ with a canonical set of nodal degrees of freedom.

We emphasize, however, that this symbolic notation of finite elements does not convey any global information on the mesh, etc. Moreover, in the context of hp-FEM, this notation becomes meaningless. It is therefore more advisable to work with the subspace notation and to view the design of FEM as a method of designing subspaces with good stability or approximation properties.

4.3.3 *Approximation properties of* $S^{1,1}(\Omega, \mathcal{T})$

We will be interested in estimating

$$Z(u, S^{1,1}, H^1) = \inf_{v \in S^{1,1}(\Omega, \mathcal{T})} \|u - v\|_{H^1(\Omega)} \tag{4.3.7}$$

where $u \in H^{2,2}_\beta(\Omega)$ is the solution of our model problem.

Theorem 4.15 *Let $\{\mathcal{T}_i\}$ denote a family of triangulations of type (h_i, β, κ) with $h_i \to 0$ as $i \to \infty$. Then there is a constant C (depending only on β) such that*

$$\inf_{v \in S^{1,1}(\Omega, \mathcal{T}_i)} \|u - v\|_{H^1(\Omega)} \leq C\, \kappa h_i \|u\|_{H^{2,2}_\beta(\Omega)} \,. \tag{4.3.8}$$

If $u = 0$ on $\Gamma^{[0]}$, so is v. In particular, for $\beta = (0, \ldots, 0)$, $u \in H^2(\Omega)$, the meshes $\mathcal{T}_i$ are quasiuniform and we get the classical estimate

$$\inf_{v \in S^{1,1}(\Omega, \mathcal{T}_i)} \|u - v\|_{H^1(\Omega)} \leq C h(\mathcal{T}_i) \|u\|_{H^2(\Omega)} \,. \tag{4.3.9}$$

Using (4.3.5) we get an error estimate in terms of the number of degrees of freedom N_i

$$\inf_{v\in S^{1,1}(\Omega,\mathcal{T}_i)} \|u-v\|_{H^1(\Omega)} \le C\kappa\, N_i^{-1/2}\, \|u\|_{H^{2,2}_\beta(\Omega)} \,.$$

We observe that the use of refined triangulations yields a convergence rate of $N^{-1/2}$ independently of β, the strength of the singularities.

We will prove that (4.3.8) holds actually for $v = I_{\mathcal{T}}u$, the piecewise linear interpolant of u in the nodes of $\mathcal{T}$. The proof will be given in several steps.

We begin with the classical case (4.3.9). As in the one-dimensional setting, global error estimates are obtained by adding scaled local ones; we have for $v \in S^{1,1}(\Omega,\mathcal{T})$:

$$\|u-v\|^2_{H^1(\Omega)} = \sum_{T\in\mathcal{T}} \|u-v\|^2_{H^1(T)} \,. \tag{4.3.10}$$

We select $v = I_{\mathcal{T}}u$ to be the interpolant of u in the nodes of $\mathcal{T}$ (by Proposition 4.3, $u \in C^0(\overline{\Omega})$ and v is well defined). Therefore the crucial part of the proof is to estimate the element contributions in (4.3.10).

Lemma 4.16 *Let T be a triangle which is shape regular, i.e.,* (4.3.3) *holds, and with vertices $A_1 = (0,0)$, A_2, A_3. Let $u \in H^{2,2}_\beta(T)$ with $\Phi_\beta(x) = |x|^\beta$, $0 \le \beta < 1$. Then the linear interpolant $I_T u$ of u in the vertices of T satisfies*

$$\|u - I_T u\|_{H^1(T)} \le C h_T^{1-\beta}\, \|\, |x|^\beta\, D^2 u\|_{L^2(T)} \,, \tag{4.3.11}$$

$$\|u - I_T u\|_{H^{2,2}_\beta(T)} \le C\, \|\, |x|^\beta\, D^2 u\|_{L^2(T)} \,. \tag{4.3.12}$$

Let us show how Lemma 4.16 implies Theorem 4.15. Due to (4.3.10), we estimate

$$\begin{aligned} \|u - I_{\mathcal{T}}u\|^2_{H^1(\Omega)} &= \sum_{\substack{T\in\mathcal{T}\\ A_i\notin\overline{T}}} \|u - I_{\mathcal{T}}u\|^2_{H^1(T)} + \sum_{\substack{T\in\mathcal{T}\\ A_i\in\overline{T}}} \|u - I_{\mathcal{T}}u\|^2_{H^1(T)} \\ &\le C \sum_{\substack{T\in\mathcal{T}\\ A_i\notin\overline{T}}} \|h_T\, D^2u\|^2_{L^2(T)} + C \sum_{\substack{T\in\mathcal{T}\\ A_i\in\overline{T}}} h_T^{2-2\gamma_i}\, \|\Phi_\gamma\, D^2 u\|^2_{L^2(T)} \end{aligned}$$

where we applied (4.3.11) with $\beta = 0$ for the first sum and with $\beta = \gamma_i$ for the second one.

Since $\mathcal{T}$ is of type (h,γ,κ), we get that $h_T \le \kappa h \Phi_\gamma(x)$ if $A_i \notin \overline{T}$ and $h_T \le \kappa h \sup_{x\in T} \Phi_\gamma(x) \le \kappa h\, h_T^{\gamma_i}$ or $h_T^{1-\gamma_i} \le \kappa h$ if $A_i \in \overline{T}$. Hence

$$\|u - I_{\mathcal{T}}u\|^2_{H^1(\Omega)} \le C\kappa^2\, h^2 \sum_{T\in\mathcal{T}} \|\Phi_\gamma\, D^2u\|^2_{L^2(T)} + C\kappa^2\, h^2\, \|\Phi_\gamma\, D^2u\|^2_{L^2(\Omega)} \,,$$

which is (4.3.8). It remains therefore to prove (4.3.11) and (4.3.12).

We give the proof in several steps.

Lemma 4.17 *(Hardy inequality). Let $\alpha \neq 1$. Then*

$$\int_0^1 t^{\alpha-2}(z(t)-a)^2\,dt \le C(\alpha)\int_0^1 t^{\alpha}(z'(t))^2\,dt \tag{4.3.13}$$

where

$$a = \begin{cases} z(0) & \text{if } \alpha < 1\,, \\ z(1) & \text{if } \alpha > 1\,. \end{cases}$$

Proof If $\alpha < 1$, we have from the Hardy inequality ([75], Theorem 2.55)

$$\int_0^1 s^{-2}\,w(s)^2 ds \le C\int_0^1 (w'(x))^2\,ds + C(w(1))^2,\quad w(0)=0\,.$$

Since $(w(1))^2 \le \int_0^1 (w'(x))^2\,ds$, we change variables $s = t^{1-\alpha}$ and set $z(t) = w(t) - w(0)$ to get (4.3.13).

For $\alpha > 1$, we use [75], Theorem 2.55:

$$\int_0^\infty s^{-2}(w(s))^2\,ds \le 4\int_0^\infty (w'(s))^2\,ds,\quad w(0)=0\,.$$

Setting here $t = s^{1-\alpha}$, $v(t) = w(t) - w(1)$ for $t < 1$, $v(t) = 0$ else, (4.3.13) follows. □

Lemma 4.18 *Let T have vertex $A_1 = (0,0)$. Let $\alpha \neq 0$ and let u satisfy $\int_T |x|^\alpha\,|D^1u|^2\,dx < \infty$. Then there exists a constant a and $C = C(\alpha, T)$ such that*

$$\int_T |x|^{\alpha-2}\,|u-a|^2\,dx \le C\int_T |x|^\alpha\,|D^1u|^2\,dx \tag{4.3.14}$$

and

$$|a| \le C\int_T r^\alpha\,|D^1u|\,dx\,.$$

If $\alpha < 0$ and $u \in C^0(\overline{T})$, then $a = u(0)$.

Proof Let θ_0 be the angle at A_1 and S be the sector given by (r,θ), $0 \le \theta \le \theta_0$, $0 \le r \le 1$. Then T may be mapped into S by a smooth map and it is sufficient to prove (4.3.14) with T replaced by S. Define

$$\overline{u}(r) = \theta_0^{-1}\int_0^{\theta_0} u(r,\theta)\,d\theta\,.$$

Then, switching from Cartesian to polar coordinates,

$$\int_0^1 r^{\alpha+1} \left|\frac{d\bar{u}}{dr}\right|^2 dr \leq C \int_S r^\alpha \, |D^1 u|^2 \, dx < \infty \, . \tag{4.3.15}$$

We use (4.3.13) and (4.3.15) to obtain that there exists a such that

$$\int_0^1 r^{\alpha-1} \, [\overline{u}(r) - a]^2 dr \leq C \int_S r^{\alpha+1} \left|\frac{d\bar{u}}{dr}\right|^2 dr \leq C \int_S r^\alpha \, |D^1 u|^2 \, dx \, . \tag{4.3.16}$$

Note that $\overline{u}(r)$ is continuous on $[0,1]$ and we have for $\alpha < 0$ that $a = \overline{u}(0)$ in (4.3.16). If, moreover, $u(r,\theta) \in C^0(\overline{T})$ and $\alpha < 0$, then also $a = u(0,0)$. We integrate next (4.3.16) with respect to θ and get

$$\int_0^1 r^{\alpha-2} \, |\overline{u} - a|^2 dx \leq C \int_S r^\alpha \, |D^1 u|^2 \, dx \, . \tag{4.3.17}$$

Further,

$$u(r,\varphi) - u(r,\psi) = \int_\psi^\varphi \frac{\partial u}{\partial \theta} \, (r,\theta) \, d\theta \, .$$

Taking the average with respect to ψ on both sides gives

$$\begin{aligned} |u(r,\varphi) - \overline{u}(r)| &= \theta_0^{-1} \left| \int_0^{\theta_0} \int_\psi^\varphi \frac{\partial u}{\partial \theta} \, (r,\theta) \, d\theta d\psi \right| \\ &\leq C \left(\int_0^{\theta_0} \left| \frac{\partial u}{\partial \theta} \, (r,\theta) \right|^2 d\theta \right)^{\frac{1}{2}} . \end{aligned}$$

Squaring and integrating, we arrive at

$$\int_0^{\theta_0} |u(r,\varphi) - \overline{u}(r)|^2 \, d\varphi \leq C \int_0^{\theta_0} \left| \frac{\partial u}{\partial \theta} \, (r,\theta) \right|^2 d\theta \, . \tag{4.3.18}$$

We multiply both sides by $r^{\alpha-1}$ and integrate to get

$$\int_S r^{\alpha-2} \, |u - \overline{u}|^2 \, dx \leq C \int_S r^\alpha \, |D^1 u|^2 \, dx \, . \tag{4.3.19}$$

We combine (4.3.17) and (4.3.18) with the triangle inequality to conclude. □

Lemma 4.19 *Let T be a triangle with vertex $A_1 = (0,0)$, $\Phi_\beta(x) = |x|^\beta$, $0 \leq \beta < 1$. Then $H^{2,2}_\beta(T)$ is compactly embedded into $H^1(T)$.*

Proof If $\beta = 0$, $H^{2,2}_\beta(T) = H^2(T)$ and this is just Rellich's theorem. The idea is now to reduce the case $\beta > 0$ to the case $\beta = 0$. We assume first that $u(0) = 0$, and put $v = r^\beta u$ where (r, θ) are polar coordinates at A_1. We denote partial derivatives by subscripts. Then we verify

$$\begin{aligned}
v_{\theta^{\alpha_2}} &= r^\beta\, u_{\theta^{\alpha_2}}\,, && 0 \leq \alpha_2 \leq 2\,,\\
v_{r\theta^{\alpha_2}} &= r^\beta\, u_{r\theta^{\alpha_2}} + \beta r^{\beta-1} u_{\theta^{\alpha_2}}\,, && 0 \leq \alpha_2 \leq 1\,,\\
v_{rr} &= r^\beta u_{rr} + 2\beta r^{\beta-1}\, u_r + \beta(\beta-1)\, r^{\beta-2}\, u\,.
\end{aligned}$$

We have

$$\|r^{\beta-1}\, u_r\|_{L^2(T)} \leq C\, \|u\|_{H^{2,2}_\beta(T)}\,.$$

As $u \in C^0(\widetilde{T})$ by Proposition 4.3 and $u(0) = 0$, Lemma 4.18 with $\alpha = 2(\beta - 1)$ gives

$$\|r^{\beta-2}\, u\|_{L^2(T)} \leq C\, \|r^{\beta-1}\, D^1 u\|_{L^2(T)} \leq C\, \|u\|_{H^{2,2}_\beta(T)}\,.$$

Hence $v = r^\beta u \in H^2(T)$.

Let now $\{u_j\}_{j=1}^\infty$ be a bounded sequence in $H^{2,2}_\beta(T)$ with $u_j(0) = 0$. Then $v_j = r^\beta u_j$ is uniformly bounded in $H^2(T)$. By Rellich's theorem, the embedding $H^2(T) \subset W^{1,s}(T)$ is compact for $1 \leq s < \infty$. Hence there exists a convergent subsequence, again denoted by $\{v_j\}$, and $\widetilde{v} \in W^{1,s}(T)$, so that $\widetilde{v}(0) = 0$ and $\|v_j - \widetilde{v}\|_{W^{1,s}(T)} \to 0$, $1 \leq s < \infty$. Put $\widetilde{u} = r^{-\beta}\, \widetilde{v}$. Then

$$\begin{aligned}
\widetilde{u}_x &= r^{-\beta}\, \widetilde{v}_x - \beta r^{-\beta-1}\, \widetilde{v} x/r\,,\\
\widetilde{u}_y &= r^{-\beta}\, \widetilde{v}_y - \beta r^{-\beta-1}\, \widetilde{v} y/r\,.
\end{aligned}$$

Since $\widetilde{v}(0) = 0$ we have by Lemma 4.18

$$\|r^{-\beta-1}\, \widetilde{v}\|_{L^2(T)} \leq C\, \|r^{-\beta}\, D^1\, \widetilde{v}\|_{L^2(T)}\,,$$

and hence

$$\begin{aligned}
\|D^1\, \widetilde{u}\|_{L^2(T)} &\leq C\, \|r^{-\beta}\, D^1\, \widetilde{v}\|_{L^2(T)}\\
&\leq C\, \|r^{-\beta s}\|^{1/s}_{L^2(T)}\ \|(D^1\, \widetilde{v})^{s'}\|^{1/s'}_{L^2(T)}\\
&\leq C\, \|r^{-\beta s}\|^{1/s}_{L^2(T)}\ \|\widetilde{v}\|_{W^{1,2s'}(T)}
\end{aligned}$$

where $1/s + 1/s' = 1$ and $s > 1$, $s\beta < 1$. Therefore $\widetilde{u} \in H^1(T)$. In the same way we prove

$$\|u_j - \widetilde{u}\|_{H^1(T)} \leq C\, \|u_j - \widetilde{u}\|_{W^{1,2s'}(T)} = C\, \|v_j - \widetilde{v}\|_{H^2(T)} \to 0\,,$$

hence the sequence u_j converges in $H^1(T)$.

Assume now that $u_j(0) \neq 0$. By (4.2.8), $|u_j(0)| \leq C \ \|u\|_{H^{2,2}_\beta(T)}$, hence there exists a subsequence (still denoted $\{u_j\}$) so that $u_j(0) \to A$. Applying the first part of the proof, a subsequence of $\overline{u}_j(x) = u_j(x) - u_j(0)$ is Cauchy in $H^1(T)$. For this sequence

$$\|u_j - u_k\|_{H^1(T)} \leq \|\overline{u}_j - \overline{u}_k\|_{H^1(T)} + C\,|u_j(0) - u_k(0)|$$

so that $\{u_j\}$ is Cauchy in $H^1(T)$. □

Proof of Lemma 4.16 Assume first that $T = \widehat{T}$ has vertices $A_1 = (0,0)$, $A_2 = (1,0)$, $A_3(0,1)$. Let $U \in H^{2,2}_\beta(\widehat{T})$ with $\Phi_\beta = r^\beta$ and $V = I_{\widehat{T}}U$ ($I_{\widehat{T}}U$ is well defined by (4.2.8)). We claim

$$\|U\|^2_{H^{2,2}_\beta(\widehat{T})} \leq C\Big\{\|r^\beta\, D^2 U\|^2_{L^2(\widehat{T})} + \sum_{i=1}^{3} |U(A_i)|^2\Big\}\,. \tag{4.3.20}$$

with C independent of U. The proof is by contradiction. If (4.3.20) were false, there would exist a sequence $U_j \in H^{2,2}_\beta(\widehat{T})$, $j = 1, 2, \ldots$, such that $\|U_j\|_{H^{2,2}_\beta(\widehat{T})} = 1$ and

$$\|r^\beta\, D^2 U_j\|^2_{L^2(\widehat{T})} + \sum_{i=1}^{3} |U_j(A_i)|^2 \longrightarrow 0\,. \tag{4.3.21}$$

Since $H^{2,2}_\beta(\widehat{T}) \subset H^1(\widehat{T})$ compactly by Lemma 4.19, there exists a subsequence, again denoted by $\{U_j\}$, which converges in $H^1(\widehat{T})$. By (4.3.21), $\{U_j\}$ is Cauchy in $H^{2,2}_\beta(\widehat{T})$ and, arguing as in the proof of Lemma 4.19, another subsequence (still denoted $\{U_j\}$) can be found such that $\{r^\beta\, U_j\}_{j=1}^\infty$ is Cauchy in $H^2(\widehat{T})$, $r^\beta\, U_j \to V$ in $H^2(\widehat{T})$ and $U_j \to \overline{U} = r^{-\beta}\, V$ in $H^{2,2}_\beta(\widehat{T})$. Now $\|r^\beta\, D^2 U_j\|_{L^2(\widehat{T})} \to 0$, hence $|D^2\,\overline{U}| = 0$ and $\overline{U}$ is a linear function. By (4.3.21), $U_j(A_i) \to 0$ for $i = 1, 2, 3$ as $j \to \infty$, hence $\overline{U}(A_i) = 0$. Therefore $\overline{U} \equiv 0$ on $\widehat{T}$, a contradiction to $1 = \lim_{j\to\infty} \|U_j\|_{H^{2,2}_\beta(\widehat{T})} = \|\overline{U}\|_{H^{2,2}_\beta(\widehat{T})}$.

We apply (4.3.20) to $U - V$ which proves (4.3.11), (4.3.12) for the triangle $\widehat{T}$. The general case is done by affinely mapping T to $\widehat{T}$ (see Exercise 4.20). □

Exercise 4.20 Let T be a triangle with vertex $A_1 = (0,0)$ and $\widehat{T}$ the reference triangle in the proof of Lemma 4.16. Let $x = Q_T(\hat{x}) = B_T\hat{x} : \widehat{T} \to T$ denote affine mapping of $\widehat{T}$ into T. Assume that (4.3.11) holds for $\widehat{T}$. Deduce from this that (4.3.11) holds for T (it is not essential that $A_1 = (0,0)$).

Exercise 4.21 Let $\{\mathcal{T}_i\}$ be a quasiuniform mesh family and let $u \in H^1(\Omega)$ solve (4.1.1)–(4.1.3).

i) Using the decomposition Theorem 4.1, find the largest s such that $u \in H^{1+s}(\Omega)$ (s depends on the singularity exponents α_{ij}!).

ii) What is the best rate of convergence you can expect for quasiuniform meshes and linear elements in this case? How does this rate compare with the one obtained in Theorem 4.15?

(Hint: use (4.3.9) and interpolation).

4.3.4 *Extension to meshes containing quadrilaterals*

In the previous sections we considered meshes $\mathcal{T}$ consisting only of triangles. Here we show that the same results hold also for meshes consisting of quadrilaterals. On quadrilateral elements, the solution is approximated by a **bilinear function** that is, a function which belongs to

$$\{u(x) : \; u = a + bx_1 + cx_2 + dx_1x_2\} \,. \tag{4.3.22}$$

Exercise 4.22 Show that a bilinear function on a square is uniquely determined by its function values in the vertices. Find a nodal basis $\overset{0}{N_i}(\xi)$ of the bilinear functions in the unit square $\widehat{Q} = (-1,1)^2$.

An arbitrary quadrilateral element $\Omega_i \subset \mathbb{R}^2$ is, by Exercise 4.20, the image of $\widehat{Q}$ under a (componentwise) bilinear map $Q_i : \widehat{Q} \to \Omega_i$ which is completely determined by the location of its vertices.

In symbolic notation (cf. Remark 4.14), bilinear continuous elements would be depicted as in Figure 4.4.

FIG. 4.4. Bilinear continuous element–symbolic notation

The definition of various classes of quadrilateral meshes is completely analogous to Definition 4.5 for meshes consisting of triangles. The only difference is shape regularity.

Definition 4.23 *A family of meshes $\{\mathcal{T}_i\}_{i=1}^{\infty}$ consisting of quadrilaterals is* **shape regular** *if it is regular, if all elements are convex and if, moreover,*

$$\sup_i \max_{T \in \mathcal{T}_i} \max_{1 \le k \le 4} |\cos(\theta_k(T))| \le \gamma < 1 \,. \tag{4.3.23}$$

Here $\theta_k(T)$ denotes the interior angles of element T. Condition (4.3.23) *requires that these angles must approach neither zero nor π as the meshes $\mathcal{T}_i$ get finer.*

Analogously to Definition 4.7 we can also define quadrilateral meshes of type (h, γ, K). Then we have the following result.

Theorem 4.24 *Let $\{\mathcal{T}_i\}$ denote a shape regular family of triangulations consisting of quadrilaterals of type (h_i, γ, K) with $h_i \to 0$ as $i \to \infty$. Then, for $u \in H^{2,2}_{\gamma}(\Omega)$,*

$$\inf_{v \in S^{1,1}(\Omega, \mathcal{T}_i)} \|u - v\|_{H^1(\Omega)} \leq C\, K h_i \, \|u\|_{H^{2,2}_{\gamma}(\Omega)} \,. \tag{4.3.24}$$

Here γ is as in Definition 4.7.

The crucial ingredient in the proof of Theorem 4.24 is the following analog of Lemma 4.16.

Lemma 4.25 *Let Q be a convex quadrilateral which is shape regular, i.e., (4.3.23) holds, with vertices $A_1 = (0,0)$, A_2, A_3, A_4. Let $u \in H^{2,2}_{\beta}(T)$ with $\Phi_\beta(x) = |x|^\beta$, $0 \leq \beta < 1$. Then the bilinear interpolant of u in the vertices, $I_Q u$, satisfies*

$$\|u - I_Q u\|_{H^1(Q)} \ \leq C h_Q^{1-\beta} \, \| \, |x|^\beta \, D^2 u\|_{L^2(Q)} \,, \tag{4.3.25}$$

$$\|u - I_Q u\|_{H^{2,2}_{\beta}(Q)} \leq C \qquad \| \, |x|^\beta \, D^2 u\|_{L^2(Q)} \,. \tag{4.3.26}$$

Proof It is sufficient to prove the assertion for the unit square $\widehat{Q} = (-1,1)^2$, the general case follows by a scaling argument similar to that of Exercise 4.20.

We next observe that Lemmas 4.17 and 4.18 also hold when T is replaced by $\widehat{Q}$. Then, as in the proof of Lemma 4.16, we get by a contradiction argument that

$$\|U\|^2_{H^{2,2}_{\beta}(\widehat{Q})} \leq C \left\{ \|r^\beta D^2 U\, \|^2_{L^2(\widehat{Q})} + \sum_{k=1}^{4} |U(A_k)|^2 \right\} .$$

Now let W be the linear function which interpolates U at the vertices A_2, A_3, A_4 and let $\overset{0}{N}_1$ be the bilinear nodal function associated with the origin, i.e., $\overset{0}{N}_1$ $(A_k) = \delta_{1k}$. Put $Z = U - W$, then $I_{\widehat{Q}} U = W + Z(A_1)\, \overset{0}{N}_1$ and

$$\begin{aligned} \|U - I_{\widehat{Q}} U\|_{H^{2,2}_{\beta}(\widehat{Q})} &\leq \|U - W\|_{H^{2,2}_{\beta}(\widehat{Q})} + |Z(A_1)| \, \| \, \overset{0}{N}_1 \, \|_{H^{2,2}_{\beta}(\widehat{Q})} \\ &\leq C \, \|U - W\|_{H^{2,2}_{\beta}(\widehat{Q})} \end{aligned}$$

by the embedding (4.2.8). As in the proof of Lemma 4.16 we get now that

$$\|U - W\|_{H^{2,2}_{\beta}(\widehat{Q})} \leq C \| \, |x|^\beta D^2 U\|_{L^2(\widehat{Q})}$$

from where (4.3.25) and (4.3.26) follow. □

Exercise 4.26 Work out the proof of Lemma 4.25 in detail and prove Theorem 4.24.

Remark 4.27 Theorem 4.24 remains valid also for (h, γ, K)-meshes consisting of combinations of triangular and quadrilateral elements.

4.4 *hp*-Finite elements in two dimensions. Space $S^{\mathbf{p},\ell}(\Omega,\mathcal{T})$

In the previous sections we studied piecewise linear or bilinear finite elements in a (straight) polygon Ω with mesh refinement towards the corners A_k. We showed that proper mesh refinement can compensate the singular behavior of the solution at the vertices A_i. Specifically, the rate of convergence of the FEM with linear/bilinear elements and optimally refined meshes near the vertices of the polygon is given by

$$Z(u, S^{1,1}(\Omega,\mathcal{T}), H^1) = \inf_{v\in S^{1,1}(\Omega,\mathcal{T})} \|u-v\|_{H^1(\Omega)} \le C\kappa N^{-\frac{1}{2}}\|u\|_{H^{2,2}_\gamma(\Omega)} \tag{4.4.1}$$

where $N = \dim(S^{1,1}(\Omega,\mathcal{T}))$ and the constant C is independent of κ and N provided the mesh $\mathcal{T}$ was of type (h,γ,κ) (see Definition 4.7). This is the same **algebraic** rate achieved for sufficiently smooth (e.g. $H^2(\Omega)$) solutions using a quasi uniform mesh and linear finite elements (which follows by setting $\gamma = 0$ in (4.4.1), see also Theorem 4.15). The asymptotic rate (4.4.1) cannot be improved with piecewise linear/bilinear elements.

In the present section, we define the FE spaces $S^{\mathbf{p},\ell}(\Omega,\mathcal{T})$ underlying the *hp*-FEM in a straight polygon Ω. We will prove that by a judicious combination of mesh refinement towards the corners of the polygon and increase of the polynomial degree p used in the approximation **exponential convergence** can be achieved. We begin our analysis with the so-called **geometric meshes** $\mathcal{T}$ on Ω.

4.4.1 *The geometric mesh families*

The geometric meshes in two-dimensional domains are direct generalizations of those introduced in one dimension in Chapter 3. The main idea is now to refine the meshes geometrically in a radial direction towards those points of the domain where the r^α singularities of the solution are concentrated (cf. Section 4.2). These are in general **all** corner points A_k of the polygonal domain Ω (regardless of being reentrant or not) and also points on the boundary where the type of boundary condition changes. Elliptic regularity states that the solution u of (4.1.1) is, for analytic data, itself analytic in $\overline{\Omega}$ minus these singular points. Geometrically graded meshes therefore ensure the analyticity of the solution restricted to each element not abutting at a singular point and, most importantly, a domain of analyticity for these restrictions the size of which is proportional to the element diameter, uniform in the number of refinement steps.

Definition 4.28 *Let Ω be a polygon with N vertices A_k and $\{\mathcal{T}_i\}$ a family of partitions of Ω into triangular and/or quadrilateral elements K_{ij}. Then $\{\mathcal{T}_i\}$ is a family of geometric meshes of type $K > 0$, if*

i) *$\mathcal{T}_i$ is shape-regular with constant κ, i.e.,*

$$\sup_{K_{ij}\in\mathcal{T}_i} \frac{h_{K_{ij}}}{\rho_{K_{ij}}} \le \kappa < \infty \tag{4.4.2}$$

with κ independent of i and

ii) *K_{ij} does not contain a vertex A_k,*

$$\kappa^{-1} \leq \frac{h_{K_{ij}}}{\min_k(\mathrm{dist}(K_{ij}, A_k))} \leq K < \infty, \; K_{ij} \in \mathcal{T}_i, \; i = 1, 2, \ldots . \tag{4.4.3}$$

Example 4.29 (the basic geometric mesh in $(-1,1)^2$). By symmetry, it is sufficient to define the mesh only in a quarter of the domain Ω, say $(-1,0)^2$. Upon translation, this becomes $\widehat{\Omega} = (0,1)^2$ with one singular vertex $A = (0,0)$.

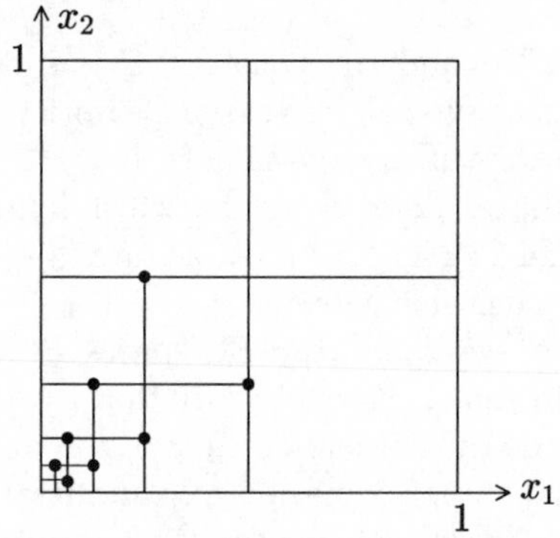

FIG. 4.5. Basic geometric mesh patch $\widehat{\mathcal{T}}_\sigma^n$ on the unit square $\widehat{\Omega}$ $(n = 5, \; \sigma = 0.5)$ with refinement towards O.

Definition 4.30 *As in the one-dimensional case, we define in $\widehat{\Omega}$ the* **basic geometric mesh family** *$\widehat{\mathcal{T}}_\sigma^n$ with $n+1$* **layers** *and* **grading factor** *$0 < \sigma < 1$ recursively as follows: if $n = 0$, $\widehat{\mathcal{T}}_\sigma^0 = \{\widehat{\Omega}\}$. Given $\widehat{\mathcal{T}}_\sigma^n$, $n \geq 0$, generate $\widehat{\mathcal{T}}_\sigma^{n+1}$ by subdividing the element Ω_{ij} abutting at the singular vertex $0 \in \overline{\Omega}_{ij}$ into four smaller rectangles by dividing its sides in a $\sigma : (1-\sigma)$ ratio.*

Figure 4.5 depicts a typical example with $\sigma = 0.5$ and $n = 5$. Another typical example of a geometric mesh consisting of triangles is depicted in Figure 4.1.

Remark 4.31 Note that the meshwidths $h_i = \max_{\Omega_{ij} \in \mathcal{T}_i}\{h_{\Omega_{ij}}\}$ of a family $\{\mathcal{T}_i\}_{i=1}^{\infty}$ of geometric meshes in general do not tend to zero as $i \to \infty$, unlike in the (h, γ, κ)-meshes introduced in Section 4.3. As in one dimension, convergence of the FE approximations on such mesh families must be achieved by increasing the polynomial degree in the large elements.

Example 4.32 **A geometric mesh in a polygon** Ω is obtained by mapping the basic mesh $\widehat{\mathcal{T}}_\sigma^n$ from $\widehat{\Omega}$ linearly to a vicinity of each convex corner of Ω. At reentrant corners, three copies of $\widehat{\mathcal{T}}_\sigma^n$, suitably scaled, are used (see Figure 4.6).

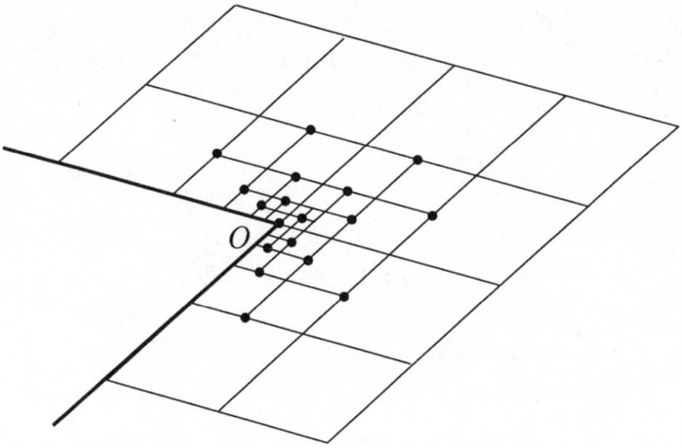

FIG. 4.6. Basic geometric mesh near a reentrant corner O

This defines the basic geometric mesh in the neighborhood of each corner. The remainder of Ω is then partitional into a fixed number of subdomains. In summary, the special geometric mesh in a polygon Ω consists of

a) patches of basic geometric meshes mapped from the unit square and
b) a quasi–uniform, fixed partition in the interior of Ω which is not covered by the mapped mesh patches.

Remark 4.33 The structure of the basic geometric meshes (as well as that of the proper geometric meshes to be introduced below) is such that **mesh refinement** takes place only in those elements K_{ij} which abut at a vertex A_k of Ω.

So far, we have discussed special geometric meshes which consist exclusively of quadrilateral elements. Such meshes necessarily contain **irregular nodes**.

Definition 4.34 *An element vertex is a* **regular node** *if it is a vertex to all adjacent elements. Vertices which are not regular nodes are called* **irregular nodes**. *The corresponding meshes are referred to* **regular** *or* **irregular** *(geometric)* **meshes**.

For example, the nodes marked with a bold dot in Figures 4.5 and 4.6 are irregular nodes and the corresponding meshes are hence irregular geometric meshes, whereas the mesh shown in Figure 4.1 is a regular geometric mesh.

Remark 4.35 Irregular nodes such as the ones in Figures 4.5 and 4.6 are sometimes also called **hanging nodes** in the literature.

Irregular nodes require special, constrained polynomial subspaces on the corresponding triangulations $\mathcal{T}_i$ in order to ensure the global continuity of the FE approximations, as we will see in the next section. In order to avoid this (and also to accommodate FE codes which cannot handle irregular nodes) a **regular geometric mesh** can be associated with every special, irregular geometric mesh as follows: one simply subdivides those elements which contain an irregular node as a mid-side node into triangles. Figure 4.7 illustrates this for the special geometric mesh in $\widehat{\Omega} = (0,1)^2$.

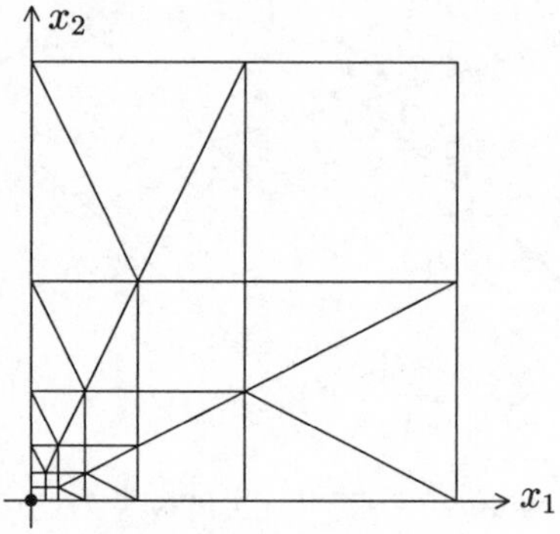

FIG. 4.7. Regular geometric mesh $\mathcal{T}_\sigma^n$ corresponding to the special geometric mesh $\widehat{\mathcal{T}}_\sigma^n$ from Figure 4.5

Definition 4.36 Regular geometric meshes $\mathcal{T}_\sigma^n$ with n layers in general polygons *are obtained by the same procedure as in the case of irregular meshes, i.e. by linear mapping and combination of several patches as in Figure* 4.7 *at each vertex. A possibly remaining gap between these patches is then meshed with finitely many triangles and/or quadrilaterals.*

4.4.2 *Construction of $S^{\mathbf{p},\ell}(\Omega, \mathcal{T})$*

4.4.2.1 *Reference elements*

The definition of $S^{\mathbf{p},\ell}(\Omega, \mathcal{T})$ is quite similar to that in the one-dimensional case in Chapter 3. Again, functions in $S^{\mathbf{p},\ell}(\Omega, \mathcal{T})$ are, in local coordinates, polynomials of the appropriate degrees. Since the proper geometric meshes $\mathcal{T}$ may contain both triangles and quadrilateral elements (see Figure 4.7), we need **two reference elements**, namely the **reference square**

$$\widehat{Q} = \{(\xi_1, \xi_2) \in \mathbb{R}^2 : \; |\xi_i| \le 1, \; i = 1, 2\} \tag{4.4.4}$$

and the (equilateral) **reference triangle**

$$\widehat{T} = \{(\xi_1, \xi_2) \in \mathbb{R}^2 : \begin{array}{ll} 0 \le \xi_2 \le (1+\xi_1)\sqrt{3} & -1 \le \xi_1 \le 0 \text{ or} \\ 0 \le \xi_2 \le (1-\xi_1)\sqrt{3} & 0 \le \xi_1 \le 1\} \; . \end{array} \tag{4.4.5}$$

We use the notation $\widehat{K}$ for the generic reference element. Let $\mathcal{T}$ be any (regular or irregular) mesh on Ω, $\mathcal{T} = \{K_j\}$. With each (triangular or quadrilateral) element Ω_j we associate an **element mapping**

$$Q_j : \; \widehat{K} \to K_j \, , \quad K_j \in \mathcal{T} \, . \tag{4.4.6}$$

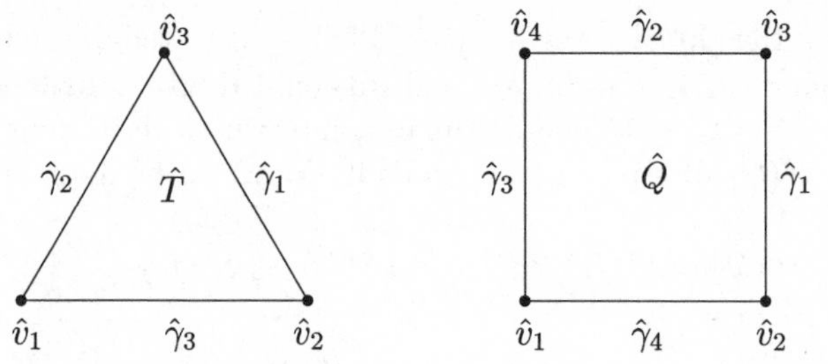

FIG. 4.8. Reference triangle $\widehat{T}$ and square $\widehat{Q}$ and the notation

If the quadrilateral elements $K_j \in \mathcal{T}$ are parallelograms (as is the case for the geometric meshes defined in Section 4.1), the element mappings Q_j can be chosen as affine mappings, i.e.,

$$x = Q_j(\xi) = B_j\,\xi + b_j\,, \quad \xi \in \widehat{Q}\,. \tag{4.4.7}$$

Note in particular that affine element mappings Q_j are vector functions and that their Jacobian is constant.

4.4.2.2 *Polynomial spaces and shape functions on $\widehat{K}$*

We define several sets $\mathcal{S}_i^p$ of polynomials on $\widehat{K} = \widehat{T}$ and $\widehat{K} = \widehat{Q}$. We begin with

$$\mathcal{S}_1^p(\widehat{K}) = \Big\{ u = \sum_{\substack{0 \le i,j \le p \\ i+j \le p}} c_{ij}\,\xi_1^i\,\xi_2^j\,,\ c_{ij} \in \mathbb{R},\ (\xi_1, \xi_2) \in \widehat{K} \Big\}\,, \tag{4.4.8}$$

the space of polynomials of (total) degree at most p on $\widehat{K}$. Evidently,

$$\dim(\mathcal{S}_1^p(\widehat{K})) = (p+1)(p+2)/2 = p^2/2 + O(p)\,.$$

On the reference square $\widehat{Q}$, we consider two kinds of spaces. First,

$$\mathcal{S}_2^p(\widehat{Q}) = \Big\{ u = \sum_{0 \le i,j \le p} c_{ij}\,\xi_1^i\,\xi_2^j\,,\ c_{ij} \in \mathbb{R},\ (\xi_1, \xi_2) \in \widehat{Q} \Big\}\,, \tag{4.4.9}$$

the **tensor product space** of one-dimensional polynomials of degree p with

$$\dim(\mathcal{S}_2^p(\widehat{Q})) = (p+1)^2 = p^2 + O(p)\,.$$

(i.e., the space of functions which are polynomials of degree $\le p$ **in each variable**). Notice that

$$\mathcal{S}_2^p(\widehat{Q}) \subseteq \mathcal{S}_1^{2p}(\widehat{Q})\,.$$

An alternative space in $\widehat{Q}$ of polynomials of degree p with a smaller dimension is the so-called **trunk** or **serendipity space** $\mathcal{S}_3^p(\widehat{Q})$. In order to define it and also to define a basis of the global space $S^{p,\ell}(\Omega, \mathcal{T})$ below, it is useful to partition the degrees of freedom on $\widehat{K}$ into **external** and **internal degrees of freedom**. To this end, let $\widehat{I}_i = \{\xi_i \lfloor\, |\xi_i| \leq 1\}$ denote the unit interval on the ξ_i-axis ($i = 1, 2$). Then the **space $\mathcal{E}^p(\widehat{Q})$ of external degrees of freedom of degree** p **on** $\widehat{Q}$ is

$$\mathcal{E}^p(\widehat{Q}) = \mathcal{S}^1(\widehat{I}_1) \otimes \mathcal{S}^p(\widehat{I}_2) \cup \mathcal{S}^p(\widehat{I}_1) \otimes \mathcal{S}^1(\widehat{I}_2) \,. \tag{4.4.10}$$

A basis for $\mathcal{E}^p(\widehat{Q})$ consists of four bilinear **nodal shape functions**

$$\begin{aligned} \overset{0}{N}_1(\xi_1, \xi_2) &= (1-\xi_1)(1-\xi_2)/4\,, \\ \overset{0}{N}_2(\xi_1, \xi_2) &= (1+\xi_1)(1-\xi_2)/4\,, \\ \overset{0}{N}_3(\xi_1, \xi_2) &= (1+\xi_1)(1+\xi_2)/4\,, \\ \overset{0}{N}_4(\xi_1, \xi_2) &= (1-\xi_1)(1+\xi_2)/4 \end{aligned} \tag{4.4.11}$$

and $p-1$ **side shape functions** $\hat{N}_i^{[j]}$ of degree $p \geq 2$ associated with each side $\hat{\gamma}_j$ of $\widehat{Q}$, see Figure 4.8, which are defined as

$$\begin{aligned} \overset{1}{N}{}_i^{[1]}(\xi_1, \xi_2) &= \frac{1}{2}\,(1+\xi_1)\psi_i(\xi_2), \qquad i = 1, 2, \ldots, p-1\,, \\ \overset{1}{N}{}_i^{[2]}(\xi_1, \xi_2) &= \frac{1}{2}\,(1+\xi_2)\psi_i(\xi_1), \qquad i = 1, 2, \ldots, p-1\,, \\ \overset{1}{N}{}_i^{[3]}(\xi_1, \xi_2) &= \frac{(-1)^i}{2}\,(1-\xi_1)\psi_i(\xi_2), \; i = 1, 2, \ldots, p-1\,, \\ \overset{1}{N}{}_i^{[4]}(\xi_1, \xi_2) &= \frac{(-1)^i}{2}\,(1-\xi_2)\psi_i(\xi_1), \; i = 1, 2, \ldots, p-1\,. \end{aligned} \tag{4.4.12}$$

Here $\psi_i(\eta)$ are the normalized antiderivatives of the Legendre polynomials introduced in Chapter 3, i.e.,

$$\psi_i(\eta) = \sqrt{\frac{2i+1}{2}} \int_{-1}^{\eta} L_i(t)dt\,, \;\; i = 1, 2, 3, \ldots\,. \tag{4.4.13}$$

The term $(-1)^i$ in (4.4.12) is needed to ensure invariance of the $\overset{1}{N}{}_i^{[j]}$ with respect to a rotation of coordinates in $\widehat{Q}$.

Now we define **two sets of internal shape functions** on $\widehat{Q}$. First, define the **basic bubble function** in $\widehat{Q}$

$$b_{\widehat{Q}}(\xi_1, \xi_2) = (1-\xi_1^2)(1-\xi_2^2) \;\;. \tag{4.4.14}$$

Then we set

$$\mathcal{I}^p(\widehat{Q}) := \{b_{\widehat{Q}}v \mid v \in \mathcal{S}_1^{p-4}(\widehat{Q})\} \tag{4.4.15}$$

and

$$\mathcal{J}^p(\widehat{Q}) := \{b_{\widehat{Q}}v \mid v \in \mathcal{S}_2^{p-2}(\widehat{Q})\} . \tag{4.4.16}$$

Then the **tensor product space** is

$$\mathcal{S}_2^p(\widehat{Q}) = \mathcal{E}^p(\widehat{Q}) \oplus \mathcal{J}^p(\widehat{Q})$$

and we define the **serendipity** or **trunk space** as follows:

$$\mathcal{S}_3^p(\widehat{Q}) := \mathcal{E}^p(\widehat{Q}) \oplus \mathcal{I}^p(\widehat{Q}) , \quad p \geq 4 . \tag{4.4.17}$$

A **basis for** $\mathcal{S}_3^p(\widehat{Q})$ consists of the external shape functions (4.4.11)–(4.4.13) plus, for $p \geq 4$, the $(p-2)(p-3)/2$ **internal shape functions**

$$\overset{2}{N}_{ij} (\xi_1, \xi_2) = b_{\widehat{Q}} (\xi_1 , \xi_2)\, L_i(\xi_1)\, L_j(\xi_2) , \quad 0 \leq i + j \leq p - 4 \tag{4.4.18}$$

(recall that $\overset{i}{N}$ means that N is associated with a geometric entity of dimension i). Notice that for large p we have

$$\dim(\mathcal{S}_3^p(\widehat{Q})) = p^2/2 + O(p) \text{ and } \dim(\mathcal{S}_2^p(\widehat{Q})) = p^2 + O(p) .$$

We will later see that $\mathcal{S}_3^p$ has the same asymptotic approximation properties as $\mathcal{S}_2^p$ in many cases.

A **basis for** $\mathcal{S}_2^p(\widehat{Q})$ consists of (4.4.11) - (4.4.13) plus (4.4.18) for $0 \leq i, j \leq p-2$, so that

$$\mathcal{S}_2^p(\widehat{Q}) = \mathcal{E}^p(\widehat{Q}) \oplus \mathcal{J}^p(\widehat{Q}) , \quad p \geq 2 . \tag{4.4.19}$$

Let us now turn to **shape functions for** $\widehat{T}$.

To define hierarchic shape functions on $\widehat{T}$, we introduce **barycentric coordinates** via

$$\lambda_1 := (1 - \xi_1 - \xi_2/\sqrt{3})/2, \quad \lambda_2 = (1 + \xi_1 - \xi_2/\sqrt{3})/2, \quad \lambda_3 = \xi_2/\sqrt{3} . \tag{4.4.20}$$

Evidently

$$\lambda_1 + \lambda_2 + \lambda_3 = 1 ,$$

and λ_i is identically equal to one at vertex $\hat{v}_i$ (see Figure 4.8) and vanishes on the opposite side $\hat{\gamma}_i$.

The **hierarchic shape functions on** $\widehat{T}$ consist again of three groups: 3 **nodal shape functions**

$$\overset{0}{N}_i\,(\xi_1,\xi_2) = \lambda_i, \quad i = 1,2,3\,, \tag{4.4.21}$$

and $p-1$ **side shape functions** $\overset{1}{N}{}_i^{[j]}\,(\xi_1,\xi_2)$, $i = 1,\ldots,p-1$, $j = 1,2,3$ associated with side $\hat{\gamma}_j$ of $\widehat{T}$ which vanish at the vertices. We construct them only for the side $\hat{\gamma}_1$, i.e. for $\overset{1}{N}{}_i^{[j]}$. The side shape functions $\overset{1}{N}{}_i^{[j]}$, $j = 2,3$, can then be obtained by permutation of indices.

To construct the side shape functions $\overset{1}{N}{}_i^{[j]}$, observe that $\psi_i(\eta)$ defined in (4.2.8) satisfies $\psi_i(\pm 1) = 0$. Hence we can write

$$\psi_i(\eta) = \frac{1}{4}\,(1-\eta^2)\,\varphi_i(\zeta), \quad i = 1,2,3,\ldots \tag{4.4.22}$$

with $\varphi_i(\eta)$ being a polynomial of degree $i-1$. Then the side modes are defined by

$$\begin{aligned} \overset{1}{N}{}_i^{[1]}(\xi_1,\xi_2) &= \lambda_2\lambda_3\,\varphi_i(\lambda_3-\lambda_2)\,, \\ \overset{1}{N}{}_i^{[2]}(\xi_1,\xi_2) &= \lambda_3\lambda_1\,\varphi_i(\lambda_1-\lambda_3)\,, \quad i = 1,\ldots,p-1\,, \\ \overset{1}{N}{}_i^{[3]}(\xi_1,\xi_2) &= \lambda_1\lambda_2\,\varphi_i(\lambda_2-\lambda_1)\,. \end{aligned} \tag{4.4.23}$$

Exercise 4.37 Show that the triangle side shape functions (4.4.23) exactly match the corresponding ones for the quadrilaterals defined in (4.4.11), (4.4.12). To this end, consider the element pair (T,Q) shown in Figure 4.9. Show that on the common side γ the triangle, quadrilateral side modes $\overset{1}{N}_i$ coincide. This shows that $\mathcal{E}^p(\widehat{T})$ and $\mathcal{E}^p(\widehat{Q})$ are compatible. By (4.3.2) and (4.3.4), triangular *p*-elements are compatible with the product as well as the serendipity space.

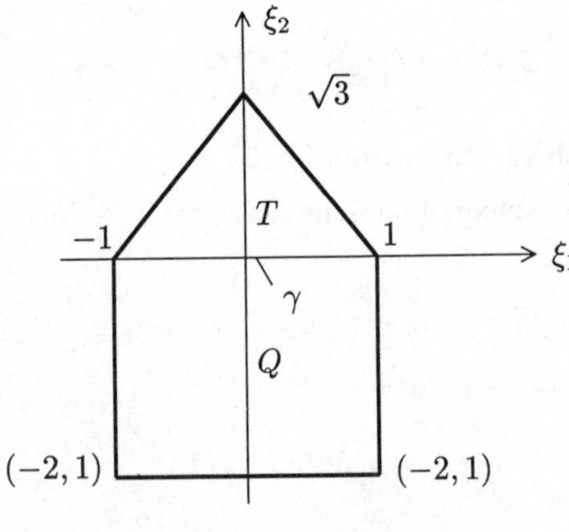

FIG. 4.9. Element pair T, Q with common edge γ

As in the quadrilateral case, we define the **set of external degrees of freedom on the triangle $\widehat{T}$**

$$\mathcal{E}^p(\widehat{T}) := \operatorname{span}\{\overset{0}{N}{}_i^{[j]}, \overset{1}{N}{}_i^{[j]} : 1 \le j \le 3, 1 \le i \le p-1\} \tag{4.4.24}$$

and the **internal shape functions**

$$\mathcal{I}^p(\widehat{T}) = \operatorname{span}\{\overset{2}{N}_i : \overset{2}{N}_i \in (\mathcal{S}_1^p \cap H_0^1)(\widehat{T})\} . \tag{4.4.25}$$

Evidently, dim $(\mathcal{I}^p)$ = $\dim(\mathcal{S}_1^p(\widehat{T}))$ − $\dim(\mathcal{E}^p(\widehat{T}))$. Since dim $(\mathcal{S}_1^p)$ = $(p+1)\,(p+2)/2$, dim $(\mathcal{I}^p(\widehat{T})) = (p+1)(p+2)/2-(3+3(p-1)) = (p-1)(p-2)/2$. Hence, nontrivial internal shape functions on $\widehat{T}$ exist only if $p \ge 3$. For $p = 3$, there is exactly one internal shape function, the **basic bubble function on $\widehat{T}$**

$$b_T(\xi_1, \xi_2) = \lambda_1\lambda_2\lambda_3 = \frac{\xi_2}{4\sqrt{3}}\left((1-\xi_2/\sqrt{3})^2 - \xi_1^2\right) . \tag{4.4.26}$$

Exercise 4.38 Prove the important characterization

$$\mathcal{I}^p(\widehat{T}) = \{b_{\widehat{T}}\, v | v \in \mathcal{S}_1^{p-3}(\widehat{T})\} = \{b_{\widehat{T}}\} \otimes \mathcal{S}_1^{p-3}(\widehat{T}), \;\; p \ge 3 . \tag{4.4.27}$$

(Hint: compare the dimensions of $\mathcal{I}^p(\widehat{T})$, $\mathcal{S}_1^p(\widehat{T})$).

Higher-order internal shape functions can therefore be constructed from any basis for $\mathcal{S}_1^p(\widehat{T})$.

Remark 4.39 (on static condensation). The actual implementation of the *hp*-FEM in two dimensions follows the algorithmic pattern outlined in Chapter 3, Section 3.2. The differences from the one-dimensional case are therefore only of a quantitative nature. Since now the number of degrees of freedom per element grows like $O(p^2)$, substantially more work is spent on the generation of the element stiffness matrices than in the one-dimensional case. It must be emphasized, though, that the p - and *hp*-FEM provide ample opportunity for parallelization with distributed memory typically available on workstation clusters: the numerical integration for the calculation of the element stiffness matrices parallelizes perfectly; even more important is the **static condensation of the internal degrees of freedom**: as we saw in Section 3.2, this **Schur complement calculation** can be performed completely in parallel, and in two dimensions, the number of internal degrees of freedom is $p^2 + O(p)$ for $\mathcal{S}_2^p$ or $p^2/2 + O(p)$ for $\mathcal{S}_1^p$ and $\mathcal{S}_3^p$, whereas the number of external ones is $O(p)$. It follows that the number of degrees of freedom to be assembled in the global stiffness matrix (and hence the size of that matrix, the interprocessor communication, etc.) after condensation is determined by the $O(p)$ external degrees of freedom. In this sense, *p*- and *hp*-FEM have a domain decomposition flavour; the assembly of the global system and its solution can be avoided if iterative schemes are used, but even then static condensation is advantageous (see Section 4.7 below).

Alternatively to the algorithmic pattern in Chapter 3, Section 3.2, one might also consider assembling first the global stiffness matrix **with** internal degrees of freedom and then applying an advanced direct solver based, e.g. on nested dissection. Such a solver will irrevocably "detect" the locality of the internal degrees of freedom for each element and will start the elimination process by condensing these out first at the expense of additional work, however.

4.4.2.3 *Definition of* $S^{\mathbf{p},\ell}(\Omega,\mathcal{T})$

To define the global FE space $S^{\mathbf{p},\ell}(\Omega,\mathcal{T})$, we associate with each $K \in \mathcal{T}$ a polynomial degree p_K and combine the elemental polynomial degrees in the **degree vector** $\boldsymbol{p}$. If $\ell = 0$, functions in $S^{\mathbf{p},0}(\Omega,\mathcal{T})$ are discontinuous, and no interelement continuity needs to be observed. If, however, $\ell = 1$ or $\ell = 2$, constraints are enforced to ensure interelement continuity or differentiability. These constraints are relatively straightforward if $p_K = p$ for all $k \in \mathcal{T}$, but some care is required if p_K is variable. In this case it is actually useful to associate with each edge e in the triangulation τ an edge-degree p_e. Let $e = \overline{K} \cap \overline{K}'$, i.e., e is common to elements K and K'. Then we require that

$$p_e = \min\{p_K, p_{K'}\} . \tag{4.4.28}$$

In this way, with each triangle $K \in \mathcal{T}$ (or quadrilateral $K \in \mathcal{T}$) three (or four) edge degrees $\{p_e : e \subset \partial K\}$ are associated which we assume to satisfy

$$\min\{p_e : e \subset \partial K\} \leq p_K = \max\{p_e : e \subset \partial K\} . \tag{4.4.29}$$

As above, we split the elemental polynomial space into internal and external modes. Conditions (4.4.28), (4.4.29) then imply that the internal modes of degree p_K are used whereas the external modes $\mathcal{E}^{p_K}(K)$ must be reduced so that on each edge $e \subset \partial K$ only the side shape functions up to edge degree p_e are in $\mathcal{E}^{p_K}(K)$. Degrees of freedom corresponding to side modes of order p satisfying $p_e < p \leq p_K$ are constrained to zero in the assembly process.

The definition of $S^{\mathbf{p},\ell}(\Omega,\mathcal{T})$ is then as follows.

Definition 4.40 *Let* $\ell = 0,1,2$, $\Omega \subset \mathbb{R}^2$ *be a polygon and* $\mathcal{T}$ *a possibly irregular mesh in* Ω. *Let further* $\mathbf{p} = \{p_K : K \in \mathcal{T}\}$ *denote a distribution of polynomial degrees on* $\mathcal{T}$. *Then*

$$u \in S^{\mathbf{p},\ell}(\Omega,\mathcal{T}) \text{ if } u \in H^\ell(\Omega) \text{ and } (u|_K \circ Q_K)(\xi) \in \mathcal{S}^{p_K}(\widehat{K}) \text{ for } K \in \mathcal{T} . \tag{4.4.30}$$

Here the element map Q_K *is linear if* K *is a triangle or a parallelogram and bilinear if* K *is a general quadrilateral.*

Further, $\mathcal{S}^{p_K}(\widehat{K}) = \mathcal{S}_1^{p_K}(\widehat{T})$ *if* $\widehat{K} = \widehat{T}$ *and if* $\widehat{K} = \widehat{Q}$, *we have either* $\mathcal{S}^{p_K}(\widehat{K}) = \mathcal{S}_2^{p_K}(\widehat{K})$ *or* $\mathcal{S}^{p_K}(\widehat{K}) = \mathcal{S}_3^{p_K}(\widehat{K})$ *(i.e., tensor product or trunk spaces on quadrilateral elements).*

Remark 4.41 If $\ell = 1$, functions $u \in S^{p,\ell}(\Omega, \mathcal{T})$ are necessarily continuous in $\bar{\Omega}$. If $\ell = 0$, discontinuities on element interfaces are possible.

4.5 The rate of convergence of the *hp*-FEM

We show that the rate of convergence of the *hp*-FEM for the membrane model problem (4.1.1)–(4.1.3) is **exponential** provided that the data f, g^0, g^1 of the problem are piecewise analytic and the spaces $S^{p,\ell}(\Omega, \mathcal{T})$ are properly designed. "Proper mesh design" means here that

a) the mesh is geometrically graded towards the singular points of the solution (for example, the vertices A_k of Ω or the points where the boundary conditions change), preferably with grading factor $\sigma = 0.15$,

b) the polynomial degree increases linearly away from the singular points from 2 to a maximal value proportional to the number of layers in the geometric mesh.

We will analyze both kinds of geometric meshes, the proper geometric mesh and the special geometric mesh which consists of mapped quadrilateral elements and has hanging nodes (cf. Figures 4.2–4.4 and Definition 4.34). The pattern of proof is quite similar to that for the piecewise linear finite elements on (h, γ, K)-meshes: since the variational formulation (4.1.5) is stable with stability constant 1, the rate of convergence of the *hp*-FEM in the H^1-norm is governed by that of the best approximation of the solution u_0. We therefore analyze here the approximability $Z(u_0, S^{p,\ell}(\Omega, \mathcal{T}), H^\ell)$ for $\ell = 0, 1$. For the proof of exponential convergence more refined regularity statements about the solution u_0 than the weighted a priori estimates used in Section 4.2 are necessary. Here we use, as in Chapter 3, that the solution u_0 belongs to an appropriate, **countably normed space** $\mathcal{B}^{\ell+1}_\beta(\Omega)$, i.e., the derivatives of u_0 arbitrarily high order are controlled here.

4.5.1 *Regularity of the solution*

We recall the spaces $H^{m,\ell}_\beta(\Omega)$ defined in Section 2, with the norms (4.2.6), (4.2.7). The countably normed spaces $\mathcal{B}^\ell_\beta(\Omega)$ are then defined as follows.

Definition 4.42 *Let Ω be a polygon in $\mathbb{R}^2$ with singular vertices $A_1, \ldots, A_N$ and let $\beta = (\beta_1, \ldots, \beta_N)$, $\beta_i \in [0,1)$ denote the vector in the definition* (4.2.5) *of the vertex weight function $\Phi_\beta(x)$. Let $\ell \geq 0$ be an integer. Then $u \in \mathcal{B}^\ell_\beta(\Omega)$ if:*

$$u \in H^{m,\ell}_\beta(\Omega)\,, \quad \text{for all } m \geq \ell \geq 0\,, \tag{4.5.1}$$

and

$$\|\Phi_{\beta+k-\ell}|D^k u|\|_{L^2(\Omega)} \leq C d^{(k-\ell)}(k-\ell)!\,, \quad \text{for } k = \ell, \ell+1, \ldots \tag{4.5.2}$$

for some constants $C > 0$, $d \geq 1$ independent of k.

We will also need the weighted spaces $H^{s,\ell}_\beta(\Omega)$ for noninteger $s = k+\theta$, $0 < \theta < 1$. They are defined by the K-method of interpolation (see Appendix B) as

$$H^{k+\theta,\ell}_\beta(\Omega) = \left(H^{k,\ell}_\beta(\Omega), H^{k+1,\ell}_\beta(\Omega)\right)_{\theta,\infty}, \quad 0 < \theta < 1. \tag{4.5.3}$$

If $u \in \mathcal{B}^\ell_\beta(\Omega)$, then for any $k \geq \ell$ it holds that

$$\|u\|_{H^{k+\theta,\ell}_\beta(\Omega)} \leq C d^{k+\theta}(k+\theta)^{1/2}\,\Gamma(k+1-\ell+\theta)\ . \tag{4.5.4}$$

We also need certain trace spaces of $\mathcal{B}^\ell_\beta(\Omega)$. To this end, let $\mathcal{M} \subset \{1, \ldots, N\}$ be an index set and define

$$\gamma = \bigcup_{j\in\mathcal{M}} \overline{\Gamma}_j \subset \Gamma\ . \tag{4.5.5}$$

Then $\mathcal{B}^{\ell-1/2}_\beta(\gamma)$ is the set of traces from $\mathcal{B}^\ell_\beta(\Omega)$ on γ. In the case $\ell = 2$, sufficient conditions for the boundary data g^i to belong to $\mathcal{B}^{3/2-i}_\beta(\Gamma^{[i]})$ are given in the following lemma.

Lemma 4.43 *Let $\Omega \subset \mathbb{R}^2$ be a polygon and $\partial\Omega = \overline{\Gamma}^{[0]} \cup \overline{\Gamma}^{[1]}$ be a partition of $\partial\Omega$ into Dirichlet and Neumann boundary. Assume further that $\partial\Omega$ is Lipschitz, i.e., that $0 < \omega_i < 2\pi$ for $1 \leq i \leq N$. Then the following statements hold:*

i) *If f is analytic in $\bar{\Omega}$, $f \in \mathcal{B}^0_\beta(\Omega)$.*

ii) *Let g^1 be analytic on every segment $\overline{\Gamma}_i \subset \overline{\Gamma}^{[1]}$. Then $g^1 \in \mathcal{B}^{1/2}_\beta(\Gamma^{[1]})$.*

iii) *Let g^0 be continuous on $\Gamma^{[0]}$ and analytic on every segment $\overline{\Gamma}_i \subset \Gamma^{[0]}$. Then $g^0 \in \mathcal{B}^{3/2}_\beta(\Gamma^{[0]})$.*

For a proof, we refer to part II of [15].

We can now state the regularity result needed in the *hp* error analysis.

Theorem 4.44 *Let $\Omega \subset \mathbb{R}^2$ be a polygon. Then there exist $0 < \beta_j < 1$, $j = 1, \ldots, N$, such that for $f \in \mathcal{B}^0_\beta(\Omega)$, $g^{[i]} \in \mathcal{B}^{3/2-i}_\beta(\Gamma^{[i]})$, $i = 0, 1$, in* (4.1.4), *the weak solution $u_0 \in H^1(\Omega)$ of* (4.1.5) *exists and belongs to $\mathcal{B}^2_\beta(\Omega)$.*

For a proof of this theorem, we refer to [15].

Exercise 4.45 Let Ω be a polygon and assume that $\Phi(x) = |x - x_0|$ for some $x_0 \in \mathbb{R}^2 \backslash \overline{\Omega}$ with $\delta := \text{dist}(x_0, \Omega) > 0$. Show that $u \in B^2_\beta(\Omega)$ implies that u is analytic in $\overline{\Omega}$.

4.5.2 *Basic approximation results for hp-FEM*

We have seen that under typical (analyticity) assumptions on the data the variational solution $u_0 \in H^1(\Omega)$ of the membrane problem (4.1.1)–(4.1.3) in a polygon $\Omega \subset \mathbb{R}^2$ exists and belongs in fact to $\mathcal{B}^2_\beta(\Omega)$ for a certain β. Since the FEM

(4.1.7) is stable with stability constant 1, the rate of convergence of the FEM is determined by the **approximability**

$$Z(u_0,\ S^{p,\ell}(\Omega,\mathcal{T}^n_\sigma),\ H^\ell) = \inf_{v\in S^{p,\ell}(\Omega,\mathcal{T})} \|u_0 - v\|_{H^\ell(\Omega)}\ ,\ \ \ell = 0,1\ , \tag{4.5.6}$$

of the solution $u_0 \in \mathcal{B}^{\ell+1}_\beta(\Omega)$. The purpose of the present section is to estimate Z in (4.5.6) for both, the regular and the irregular geometric meshes defined in Section 4.4. Since these meshes were defined in terms of (mapped) geometric meshes in the unit square, we analyze this case first. The analysis proceeds along the lines of the one-dimensional case (see Chapter 3, Section 3.3.6). We first obtain an approximation result on the unit square $Q = (-1,1)^2$. Then, by a scaling argument, a discontinuous piecewise polynomial approximation on the basic geometric mesh in Figure 4.5 is obtained and the exponential convergence is established. Finally, in order to obtain a continuous approximation in $H^1(\Omega)$, the interelement discontinuities in the approximation are removed without reducing the rate of convergence (this last part is new as compared to the one-dimensional case). This will be done separately for the irregular geometric mesh $\widehat{\mathcal{T}}^n_\sigma$ consisting of quadrilaterals (see Figure 4.6) and for the regular, proper geometric mesh $\mathcal{T}^n_\sigma$ consisting of triangles and quadrilaterals (see Figure 4.7). The asymptotic rate of convergence is in both cases proved to be bounded by

$$Z(u_0,\ S^{p,\ell}(\Omega,\mathcal{T}),H^\ell) \le C\ \exp(-b\sqrt[3]{N}),\ \ N = \dim(S^{p,\ell}(\Omega,\mathcal{T}))\ .$$

4.5.2.1 *Basic approximation result on a square and parallelogram*

Let $\widehat{Q} = (-1,1)^2$. Then we have the following result.

Theorem 4.46 *For any* $u \in H^{k+3}(\widehat{Q}),\ k = \max(k_1,k_2),\ k_1,k_2 \ge 2$, *there exists a polynomial*

$$\phi(\xi_1,\xi_2) = \sum_{\substack{0\le i\le k_1\\ 0\le j\le k_2}} d_{ij}\ \xi_1^i\ \xi_2^j$$

such that

$$\begin{aligned} \|D^m(u-\phi)\|^2_{L^2(\widehat{\Omega})} \le C\Big\{ &\frac{(k_1-s_1)!}{(k_1+s_1+2-2m)!} \sum_{\ell=0}^{2} \|\partial_1^{s_1+1}\,\partial_2^\ell u\|^2_{L^2(\widehat{\Omega})} \\ &+ \frac{(k_2-s_2)!}{(k_2+s_2+2-2m)!} \sum_{\ell=0}^{2} \|\partial_1^\ell \partial_2^{s_2+1}\,u\|^2_{L^2(\widehat{\Omega})}\Big\} \end{aligned} \tag{4.5.7}$$

where $0\le m\le 2,\ \ 1\le s_i\le k_i,\ \ i=1,2$ *and* C *is independent of* $s_i\,,k_i$. *Moreover,*

$$u = \phi \ \text{ at the vertices of } \widehat{Q}. \tag{4.5.8}$$

Proof The proof is elementary, but lengthy and broken into several steps. It may be skipped at first reading.

Step 1: By density, it suffices to prove (4.5.6) and (4.5.8) for $u \in C^\infty(\overline{\widehat{Q}})$. For such u, one verifies the formula

$$\begin{aligned} u(\xi_1,\xi_2) = & \int_{-1}^{\xi_1}\int_{-1}^{\xi_2}\int_{-1}^{\eta_1}\int_{-1}^{\eta_2} \partial_1^2\,\partial_2^2 u(\tilde\eta_1,\tilde\eta_2)d\tilde\eta_1 d\tilde\eta_2 d\eta_1 d\eta_2 \\ & + (1+\xi_1)\int_{-1}^{\xi_2} \partial_1\partial_2 u(-1,\eta_2)d\eta_2 + (1+\xi_2)\int_{-1}^{\xi_1} \partial_1\partial_2 u(\eta_1,-1)d\eta_1 \\ & + \int_{-1}^{\xi_1} \partial_1 u(\eta_1,-1)d\eta_1 + \int_{-1}^{\xi_1} \partial_1 u(\eta_1,-1)d\eta_1 \\ & + u(-1,-1) - (1+\xi_1)(1+\xi_2)(\partial_1\partial_2 u)(-1,-1)\,. \end{aligned} \tag{4.5.9}$$

We expand the following derivatives of u in (4.5.9) into Legendre series:

$$\begin{aligned} (\partial_1^2\partial_2 u)(\xi_1,\xi_2) &= \sum_{i=0}^{\infty} a_i L_i(\xi_1)\,, \quad (\partial_1\partial_2^2 u)(\xi_1,\xi_2) = \sum_{j=0}^{\infty} b_j L_j(\xi_2)\,, \\ (\partial_1^2\partial_2^2 u)(\xi_1,\xi_2) &= \sum_{i,j=0}^{\infty} c_{ij} L_i(\xi_1)L_j(\xi_2)\,, \\ (\partial_1^2 u)(\xi_1,-1) &= \sum_{i=0}^{\infty} d_i L_i(\xi_1)\,, \quad (\partial_2^2 u)(-1,\xi_2) = \sum_{j=0}^{\infty} e_j L_j(\xi_2)\,. \end{aligned}$$

Step 2: The approximation ϕ of u will be constructed by truncating these series. We set

$$\begin{aligned} \chi_1(\xi_1,\xi_2) &:= \sum_{0\le i\le k_1-2}\ \sum_{0\le j\le k_2-2} c_{ij}L_i(\xi_1)L_j(\xi_2)\,, \\ \chi_2(\xi_1) &:= \sum_{0\le i\le k_1-2} a_i L_i(\xi_1)\,, \quad \tilde\chi_2(\xi_1) := \int_{-1}^{\xi_1} \chi_2(\eta_1)d\eta_1 + (\partial_1\partial_2 u)(-1,1)\,, \\ \chi_3(\xi_2) &:= \sum_{0\le j\le k_2-2} b_j L_j(\xi_2)\,, \quad \tilde\chi_3(\xi_2) := \int_{-1}^{\xi_2} \chi_3(\eta_2)d\eta_2 + (\partial_1\partial_2 u)(-1,1)\,, \\ \chi_4(\xi_1) &:= \sum_{0\le i\le k_1-2} d_i L_i(\xi_1)\,, \quad \tilde\chi_4(\xi_1) := \int_{-1}^{\xi_1} \chi_4(\eta_1)d\eta_1 + (\partial_1 u)(-1,1)\,, \\ \chi_5(\xi_2) &:= \sum_{0\le j\le k_2-2} e_j L_j(\xi_2)\,, \quad \tilde\chi_5(\xi_2) := \int_{-1}^{\xi_2} \chi_5(\eta_2)d\eta_2 + (\partial_2 u)(-1,1)\,. \end{aligned}$$

By (4.5.9), the approximating polynomial is then $\phi = \phi_1 + \phi_2$ with

$$\begin{aligned}\phi_1 := &\int_{-1}^{\xi_1}\int_{-1}^{\xi_2}\int_{-1}^{\eta_1}\int_{-1}^{\eta_2} \chi_1(\tilde{\eta}_1,\tilde{\eta}_2)d\tilde{\eta}_1 d\tilde{\eta}_2 d\eta_1 d\eta_2 \\ &+ (1+\xi_2)\int_{-1}^{\xi_1}\tilde{\chi}_2(\eta_1)d\eta_1 + \int_{-1}^{\xi_1}\tilde{\chi}_4(\eta_1)d\eta_1 ,\end{aligned} \tag{4.5.10}$$

$$\begin{aligned}\phi_2 := &(1+\xi_1)\int_{-1}^{\xi_2}\tilde{\chi}_3(\eta_2)d\eta_2 + \int_{-1}^{\xi_2}\tilde{\chi}_5(\eta_2)d\eta_2 + u(-1,-1) \\ &+ (1+\xi_1)(1+\xi_2)(\partial_1\partial_2 u)(-1,-1) .\end{aligned} \tag{4.5.11}$$

Comparing (4.5.9) with (4.5.10), (4.5.11) we get

$$u - \phi = \sum_{\nu=1}^{6} F_\nu \tag{4.5.12}$$

with

$$F_1 = \int_{-1}^{\xi_1}\int_{-1}^{\xi_2}\int_{-1}^{\eta_1}\int_{-1}^{\eta_2} \sum_{i=k_1-1}^{\infty}\sum_{j=0}^{\infty} c_{ij}L_i(\tilde{\eta}_1)L_j(\tilde{\eta}_2)d\tilde{\eta}_1 d\tilde{\eta}_2 d\eta_1 d\eta_2 ,$$

$$F_2 = (1+\xi_2)\int_{-1}^{\xi_1}\int_{-1}^{\eta_1} \sum_{i=k_1-1}^{\infty} a_i L_i(\tilde{\eta}_1)d\tilde{\eta}_1 d\eta_1 ,$$

$$F_3 = \int_{-1}^{\xi_1}\int_{-1}^{\eta_1} \sum_{i=k_1-1}^{\infty} d_i L_i(\tilde{\eta}_1)d\tilde{\eta}_1 d\eta_1 ,$$

$$F_4 = \int_{-1}^{\xi_1}\int_{-1}^{\xi_2}\int_{-1}^{\eta_1}\int_{-1}^{\eta_2} \sum_{i=0}^{k_1-2}\sum_{j=k_2-1}^{\infty} c_{ij}L_i(\tilde{\eta}_1)L_j(\tilde{\eta}_2)d\tilde{\eta}_1 d\tilde{\eta}_2 d\eta_1 d\eta_2 ,$$

$$F_5 = (1+\xi_1)\int_{-1}^{\xi_2}\int_{-1}^{\eta_2} \sum_{j=k_2-1}^{\infty} b_j L_j(\tilde{\eta}_2)d\tilde{\eta}_2 d\eta_2 ,$$

$$F_6 = \int_{-1}^{\xi_2}\int_{-1}^{\eta_2} \sum_{j=k_2-1}^{\infty} e_j L_j(\tilde{\eta}_2)d\tilde{\eta}_2 d\eta_2 .$$

Step 3: It remains to estimate $\|D^m F_\nu\|_{L^2(\widehat{Q})}$. To this end we need

$$\int_{\widehat{Q}} |D^\alpha u|^2 (1-\xi_1^2)^{\alpha_1}(1-\xi_2^2)^{\alpha_2} d\xi = \sum_{\substack{i\geq\alpha_1\\ j\geq\alpha_2}} |a_{ij}|^2 \frac{2}{2i+1}\frac{2}{2j+1}\frac{(i+\alpha_1)!(j+\alpha_2)!}{(i-\alpha_1)!(j-\alpha_2)!}, \tag{4.5.13}$$

where $u(\xi_1,\xi_2)=\sum_{i,j=0}^{\infty} a_{ij}L_i(\xi_1)L_j(\xi_2)$ is the Legendre series of u.

Step 4: We estimate the F_ν. Obviously

$$\begin{aligned}\partial_1^2 F_1 &= \int_{-1}^{\xi_2}\int_{-1}^{\eta_2} \sum_{i=k_1-1}^{\infty}\sum_{j=0}^{\infty} c_{ij}L_i(\xi_1)L_j(\tilde\eta_2)d\tilde\eta_2 d\eta_2 \\ &= \sum_{i=k_1-1}^{\infty}\sum_{j=0}^{\infty} c_{ij}L_i(\xi_1)\Big\{\frac{L_{j+2}(\xi_2)-L_j(\xi_2)}{2j+3} - \frac{L_j(\xi_2)-L_{j-2}(\xi_2)}{2j-1}\Big\}.\end{aligned}$$

Then, from the orthogonality of the Legendre polynomials and $1\leq s_1\leq k_1$:

$$\begin{aligned}\|\partial_1^2 F_1\|^2_{L^2(\widehat{Q})} &\leq C \sum_{i=k_1-1}^{\infty}\sum_{j=0}^{\infty}|c_{ij}|^2\frac{2}{2i+1}\frac{2}{2j+1}(2j+1)^{-4} \\ &\leq C\frac{(k_1-s_1)!}{(k_1+s_1-2)!}\sum_{i=s_1-1}^{\infty}\sum_{j=0}^{\infty}|c_{ij}|^2\frac{2}{2i+1}\frac{2}{2j+1}\frac{(i+s_1-1)!}{(i-(s_1-1))!} \\ &\leq C\frac{(k_1-s_1)!}{(k_1+s_1-2)!}\,\|\partial_1^{s_1+1}\partial_2^2 u\|^2_{L^2(\widehat{Q})}\end{aligned}$$

by (4.5.13).

In exactly the same way we estimate the other derivatives of F_1. This gives for $0\leq m\leq 2$

$$\|D^m F_1\|^2_{L^2(\widehat{Q})} \leq C\frac{(k_1-s_1)!}{(k_1+s_1+2-2m)!}\,\|\partial_1^{s_1+3}\partial_2^2 u\|^2_{L^2(\widehat{Q})}, \tag{4.5.14}$$

$$\|D^m F_4\|^2_{L^2(\widehat{Q})} \leq C\frac{(k_2-s_2)!}{(k_2+s_2+2-2m)!}\,\|\partial_1^2\partial_2^{s_2+3} u\|^2_{L^2(\widehat{Q})}. \tag{4.5.15}$$

To estimate F_2, we observe

$$\partial_1^2 F_2 = (1+\xi_2) \sum_{i=k_1-1}^{\infty} a_i L_i(\xi_1) ,$$

and estimate

$$\begin{aligned}
\|\partial_1^2 F_2\|^2_{L^2(\widehat{Q})} &\le C \sum_{i=k_1-1}^{\infty} |a_i|^2 \frac{2}{2i+1} \\
&\le C \frac{(k_1-s_1)!}{(k_1+s_1-2)!} \sum_{i=s_1-1}^{\infty} |a_i|^2 \frac{2}{2i+1} \frac{(i+s_1-1)!}{(i-(s_1-1))!} \\
&\le C \frac{(k_1-s_1)!}{(k_1+s_1-2)!} \|\partial_1^{s_1+1}\partial_2 u\|^2_{L^2(\widehat{Q})} .
\end{aligned}$$

Analogously it follows that

$$\begin{aligned}
\|\partial_1 F_2\|^2_{L^2(\widehat{Q})} &\le C \frac{(k_1-s_1)!}{(k_1+s_1)!} \|\partial_1^{s_1+1}\partial_2 u\|^2_{L^2(\widehat{Q})} , \\
\|F_2\|^2_{L^2(\widehat{Q})} &\le C \frac{(k_1-s_1)!}{(k_1+s_1+2)!} \|\partial_1^{s_1+1}\partial_2 u\|^2_{L^2(\widehat{Q})} .
\end{aligned}$$

Since $\partial_2^2 F_2 = 0$ and also

$$\|\partial_2 F_2\|_{L^2(\widehat{Q})} \le C \|F_2\|_{L^2(\widehat{Q})} ,$$

we obtain for $0 \le m \le 2$

$$\|D^m F_2\|^2_{L^2(\widehat{Q})} \le C \frac{(k_1-s_1)!}{(k_1+s_1+2-2m)!} \|\partial_1^{s_1+1}\partial_2 u\|^2_{L^2(\widehat{Q})} . \tag{4.5.16}$$

In the same way we obtain for $0 \le m \le 2$ the estimates

$$\|D^m F_5\|^2_{L^2(\widehat{Q})} \le C \frac{(k_2-s_2)!}{(k_2+s_2+2-2m)!} \|\partial_1 \partial_2^{s_2+1} u\|^2_{L^2(\widehat{Q})} , \tag{4.5.17}$$

$$\|D^m F_3\|^2_{L^2(\widehat{Q})} \le C \frac{(k_1-s_1)!}{(k_1+s_1+2-2m)!} \|\partial_1^{s_1+1} u\|^2_{L^2(\widehat{Q})} , \tag{4.5.18}$$

$$\|D^m F_6\|^2_{L^2(\widehat{Q})} \le C \frac{(k_2-s_2)!}{(k_2+s_2+2-2m)!} \|\partial_2^{s_2+1} u\|^2_{L^2(\widehat{Q})} . \tag{4.5.19}$$

Combining (4.5.14)–(4.5.19) gives (4.5.6).

Step 5: We verify (4.5.8). From the orthogonality properties of the Legendre polynomials we have

$$0 = F_1(\pm 1, \xi_2) = F_4(\xi_1, \pm 1) = F_2(\pm 1, \xi_2) = F_5(\xi_1, \pm 1) = F_3(\pm 1) = F_6(\pm 1) .$$

Therefore (4.5.8) follows from (4.5.12). □

Scaling the estimate (4.5.6) yields immediately the following result.

Corollary 4.47 *Let $\Omega = (a,b) \times (c,d)$ with $h_1 = b-a$, $h_2 = d-c$. If $u \in H^{k+3}(\Omega)$, $k = \max(k_1,k_2)$, $k_1,k_2 \geq 2$, then there is a polynomial*

$$\phi = \sum_{i=0}^{k_1} \sum_{j=0}^{k_2} c_{ij}\, \xi_1^i\, \xi_2^j$$

such that

$$\|D^m(u-\phi)\|^2_{L^2(\Omega)}$$

$$\leq Ch^{-2m}\Big\{\frac{(k_1-s_1)!}{(k_1+s_1+2-2m)!}\left(\frac{h_1}{2}\right)^{2(s_1+1)} \sum_{\ell=0}^{2} h_2^{2\ell}\|\partial_1^{s_1+1}\partial_2^{\ell} u\|^2_{L^2(\Omega)} \tag{4.5.20}$$

$$+ \frac{(k_2-s_2)!}{(k_2+s_2+2-2m)!}\left(\frac{h_2}{2}\right)^{2(s_2+1)} \sum_{\ell=0}^{2} h_1^{2\ell}\|\partial_1^{\ell}\partial_2^{s_2+1} u\|^2_{L^2(\Omega)}\Big\} .$$

Here $h = \min(h_1,h_2)$, $1 \leq s_i \leq k_i$, $i = 1,2$ are integers $0 \leq m \leq 2$ and C is independent of s_i, k_i. Moreover, $u = \phi$ at the vertices of Ω.

We use Corollary 4.47 for the approximation of $u \in \mathcal{B}^{\ell}_{\beta}(\Omega)$ on a typical element Ω_j of the basic geometric mesh (cf. Figure 4.5), i.e., a mesh for which the property (4.4.2) holds.

Lemma 4.48 *Let $\Omega = (0,1)^2$ and $\Omega_1 = (a,b) \times (c,d) \subset \Omega$ such that $h_1 := b-a \leq \lambda r_0$, $h_2 = d-c \leq \lambda r_0$, $\lambda > 0$ and $r_0 = \operatorname{dist}(O,\Omega_1)$. Assume further that $\max(h_1,h_2)/\min(h_1,h_2) \leq \kappa$. Then for any $u \in H^{k+3,2}_{\beta}(\Omega)$ with $\Phi_\beta(x) = |x|^\beta, 0 < \beta < 1$, there exists a polynomial $\phi(x_1,x_2)$ of degree k in each variable such that*

$$\|D^m(u-\phi)\|^2_{L^2(\Omega_1)}$$

$$\leq Cr_0^{2(2-m-\beta)} \frac{\Gamma(k-s+1)}{\Gamma(k+s+3-2m)}\left(\frac{\lambda}{2}\right)^{2s} \|u\|^2_{H^{s+3,2}_{\beta}(\Omega)} , \quad m = 0,1,2 . \tag{4.5.21}$$

Here $1 \leq s \leq k$ and C is independent of s and k, but depends on λ and κ. Moreover, $u = \phi$ at the vertices of Ω_1.

Proof We apply Corollary 4.47 and have for any integers $\tilde{k}_i$ with $1 \leq \tilde{k}_i \leq k$, $i = 1, 2$, the error bound (4.5.20) with s_i replaced by $\tilde{k}_i$ and Ω replaced by Ω_1. The first term in the bound is, under our assumptions, bounded by

$$Cr_0^{2(2-m-\beta)} \frac{(k-\tilde{k}_1)!}{(k+\tilde{k}_1+2-2m)!} \left(\frac{\lambda}{2}\right)^{2(\tilde{k}_1+1)} \|u\|^2_{H_\beta^{s+3,2}(\Omega)} .$$

Analogously, the second term can be bounded and we obtain for $\tilde{k}_1 = \tilde{k}_2 = \tilde{k}$

$$\|D^m(u-\phi)\|^2_{L^2(\Omega_1)} \leq Cr_0^{2(2-m-\beta)} \frac{(k-\tilde{k})!}{(k+\tilde{k}+2-2m)!} \left(\frac{\lambda}{2}\right)^{2\tilde{k}} \|u\|^2_{H_\beta^{\tilde{k}+3,2}(\Omega)} . \tag{4.5.22}$$

Let $T : u \to D^m(u-\phi)$ denote the error operator for the approximation ϕ. By (4.5.22), T is linear and bounded from $H_\beta^{\tilde{k}+3,2}(\Omega)$ to $L^2(\Omega_1)$ and its norm is bounded by

$$|||T|||^2 \leq Cr_0^{2(2-m-\beta)} \frac{(k-\tilde{k})!}{(k+\tilde{k}+2-2m)!} \left(\frac{\lambda}{2}\right)^{2\tilde{k}} , \quad 1 \leq \tilde{k} \leq k .$$

Now let $s = \tilde{k} - 1 + \theta$ for some $0 < \theta < 1$. Since $H_\beta^{s+3,2}(\Omega)$ is the interpolation space (4.5.3), T is also linear and continuous from $H_\beta^{s+3,2}(\Omega) \to L^2(\Omega_1)$ and its norm is bounded by

$$|||T|||^2 \leq Cr_0^{2(2-m-\beta)} \left(\frac{\lambda}{2}\right)^{2s} \frac{(k-\tilde{k})!}{(k+\tilde{k}+2-2m)!} \frac{(k-\tilde{k}+1)^{1-\theta}}{(k+\tilde{k}+2-2m)^{\theta}} .$$

Since

$$0 < C_1 \leq \frac{n!\,(n+1)^{\theta}}{\Gamma(n+1+\theta)} \leq C_2 < \infty , \quad n \in N , \quad 0 < \theta < 1$$

with C_1, C_2 independent of n and θ, we have (see also (3.3.51))

$$|||T|||^2 \leq Cr_0^{2(2-m-\beta)} \left(\frac{\lambda}{2}\right)^{2s} \frac{\Gamma(k-s+1)}{\Gamma(k+s+3-2m)}$$

and (4.5.21) follows. □

So far, all local approximation results have been valid for a square or rectangle. In order to obtain a global approximation, for example for the geometric mesh shown in Figure 4.6, we need local approximation results on a parallelogram. Such a result can be proved as Lemma 4.48. First, however, one needs an analog of Corollary 4.47 which can be deduced from it by the linear coordinate transformation (see also Figure 4.10)

$$x_1 = \tilde{x}_1 - \tilde{x}_2 \cot\omega , \quad x_2 = \tilde{x}_2 / \sin\omega , \quad 0 < \omega < \pi/2 . \tag{4.5.23}$$

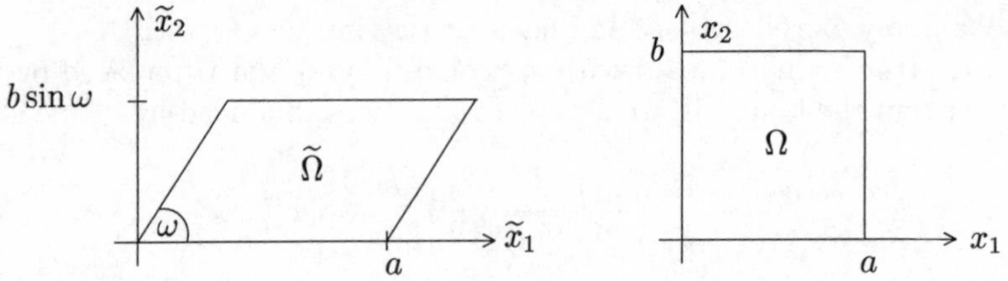

FIG. 4.10. Parallelogram $\widetilde{\Omega}$ and the variable transformation $(x_1, x_2) \leftrightarrow (\widetilde{x}_1, \widetilde{x}_2)$

Corollary 4.49 *Let $\Omega_1 \subset \Omega = (0,1)^2$ denote a rectangle as in Corollary 4.47, and let $\widetilde{\Omega}_1$ denote its image under the map (4.5.23). If $u \in H^{k+3}(\widetilde{\Omega}_1)$, there exists a polynomial $\phi(\widetilde{x}_1, \widetilde{x}_2) = \sum_{\substack{0\le i\le k_1 \\ 0\le j\le k_2}} c_{ij} x_1^i x_2^j$, with $1 \le k_1, k_2 \le k$ such that for $0 \le m \le 2$*

$$\begin{aligned} \|D^m(u-\phi)\|^2_{L^2(\widetilde{\Omega}_1)} \le Ch^{-2m}\Big\{ & \frac{\Gamma(k_1 - s_1 + 1)}{\Gamma(k_1 + s_1 + 3 - 2m)!}\left(\frac{\tilde{h}_1}{2}\right)^{2(s_1+1)} \\ & \sum_{\ell=0}^{2}\sum_{\ell'=0}^{\ell} h_2^{2\ell}\,\|\partial_1^{s_1+1+\ell'}\partial_2^{\ell-\ell'} u\|^2_{L^2(\widetilde{\Omega}_1)} \\ + & \frac{\Gamma(k_2 - s_2 + 1)}{\Gamma(k_2 + s_2 + 3 - 2m)!}\left(\frac{\tilde{h}_2}{2}\right)^{2(s_2+1)} \\ & \sum_{\ell=0}^{2}\sum_{\ell'=0}^{\ell} h_1^{2\ell}\,\|\partial_1^{\ell-\ell'}\partial_2^{s_2+1+\ell'} u\|^2_{L^2(\widetilde{\Omega}_1)}\Big\} \end{aligned} \tag{4.5.24}$$

where $\tilde{h}_1, \tilde{h}_2$ are the edge lengths of $\widetilde{\Omega}_1$ and $1 \le s_i \le k_i$ are arbitrary real numbers. The constant C in (4.5.24) depends on ω, but is independent of s_i, k_i. Moreover, $u = \phi$ at the vertices of $\widetilde{\Omega}_1$.

Analogous to Lemma 4.48 we can now prove the following result.

Lemma 4.50 *Let $\widetilde{\Omega}_1 \subset \widetilde{\Omega}$ be the images of Ω_1, Ω in Lemma 4.48 under the affine map (4.5.23) and let $\tilde{h}_1, \tilde{h}_2$ be the edge lengths of $\widetilde{\Omega}_1$, with $\tilde{h}_1 \le \lambda r_0$, $\tilde{h}_2 \le \lambda r_0$, where $r_0 = \mathrm{dist}(O, \widetilde{\Omega}_1)$. If $u \in H_\beta^{k+3,2}(\widetilde{\Omega})$ with $\Phi_\beta(x) = |x|^\beta$, $0 < \beta < 1$, the polynomial $\phi(\widetilde{x}_1, \widetilde{x}_2)$ of Corollary 4.49 satisfies the estimate*

$$\begin{aligned} & \|D^m(u-\phi)\|^2_{L^2(\widetilde{\Omega}_1)} \\ & \le C r_0^{2(2-m-\beta)} \frac{\Gamma(k - s + 1)}{\Gamma(k + s + 3 - 2m)}\left(\frac{\lambda}{2}\right)^{2s} \|u\|^2_{H_\beta^{s+3,2}(\Omega)}\,. \end{aligned} \tag{4.5.25}$$

Here $1 \le s \le k$ is any real number and $C > 0$ is independent of s and k, but depends on m, ω, λ and the ratio of $\tilde{h}_1$ and $\tilde{h}_2$.

4.5.2.2 *hp-approximation of* $u \in \mathcal{B}^2_\beta$ *on a square*

We prove now the approximation properties of $S^{\mathbf{p},\ell}(\widehat{\Omega}, \mathcal{T}^n_\sigma)$ on a square $\widehat{Q}$ for functions $u \in \mathcal{B}^2_\beta(\widehat{\Omega})$ with weight $\Phi_\beta(x) = |x|^\beta$, i.e., which are analytic in $[0,1]^2$ except at the origin. Here $\mathcal{T}^n_\sigma$ are regular geometric meshes with $n+1$ layers as defined in Section 4.1 (see also Figure 4.7) with grading factor $\sigma \in (0,1)$. In particular, $\mathcal{T}^n_\sigma$ consists of triangles and quadrilaterals and $S^{\mathbf{p},\ell}(\widehat{\Omega}, \mathcal{T}^n_\sigma)$ is as in Definition 4.40 with the tensor product polynomial spaces $\mathcal{S}^{p_i}_2$ on quadrilateral elements. The polynomial degree vector $\boldsymbol{p}$ increases linearly away from the origin. More precisely, let $\mathcal{T}^n_\sigma = \{\Omega_{ij}, \Omega^k_{ij} : 1 \le i \le 3,\ 1 \le j \le n+1,\ 1 \le k \le 3\}$ be as in Figure 4.11.

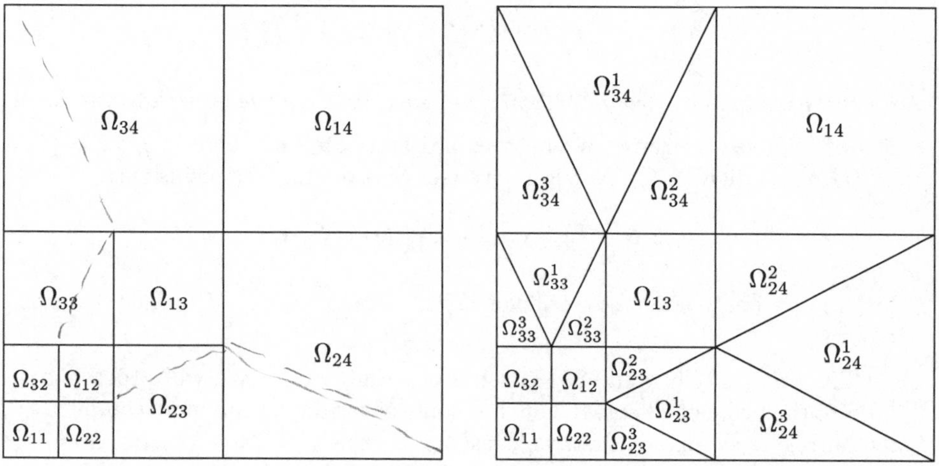

FIG. 4.11. Meshes $\widehat{\mathcal{T}}^n_\sigma$ and $\mathcal{T}^n_\sigma$ for $\sigma = 0.5$, $n = 3$

The layer j of elements is defined as $\{\Omega_{ij},\ \Omega^k_{ij}\colon 1 \le i,\ k \le 3\}$. Then we call a polynomial degree distribution **linear with slope** $\mu > 0$ if

$$\boldsymbol{p} = \{p_{ij} = p^k_{ij} = p_j,\ \ 1 \le i,k \le 3,\ \ p_j = \max(2, \lfloor \mu j \rfloor)\}\,. \tag{4.5.26}$$

Now we can state the main result of this section.

Theorem 4.51 *Let* $\widehat{\Omega} = (0,1)^2$, $0 < \sigma < 1$ *and let* $\mathcal{T}^n_\sigma$ *be the regular geometric mesh with* $n+1$ *layers in* $\widehat{\Omega}$. *Let further* $\mathbf{p}$ *be the linear degree vector in* (4.5.26) *and* $u \in \mathcal{B}^2_\beta(\widehat{\Omega})$ *with some* $0 < \beta < 1$ *and* $\Phi_\beta(x) = |x|^\beta$.

Define the hp-FE space as in (4.4.30)*:*

$$S^{\mathbf{p},1}(\widehat{\Omega}, \mathcal{T}^n_\sigma) = \{u \in H^1(\widehat{\Omega}) : (u|_K \circ Q_K)(\xi) \in \mathcal{S}^{p_K}_2(\widehat{K}) \quad \textit{for all } K \in \mathcal{T}^n_\sigma\}\,.$$

Then there exists some $\mu > 0$ *in* (4.5.26) *such that*

$$\inf_{v \in S^{\mathbf{p},1}(\widehat{\Omega}, \mathcal{T}_\sigma^n)} \|u - v\|_{H^1(\widehat{\Omega})} \leq C \exp(-b\sqrt[3]{N}) \tag{4.5.27}$$

where $C, b > 0$ *do not depend on* $N = \dim(S^{\mathbf{p},1}(\widehat{\Omega}, \mathcal{T}_\sigma^n))$.

We emphasize that the elementwise approximations belong to the full tensor product space $\mathcal{S}^{p_K}(K)$, even if K is a triangle. The general case, however, follows easily.

Corollary 4.52 *Under the assumptions of Theorem* 4.51, *with*

$$\begin{aligned} S^{\mathbf{p},1}(\widehat{\Omega}, \mathcal{T}_\sigma^n) = \{u \in H^1(\widehat{\Omega}) : (u|_K \circ Q_K)(\xi) \in \mathcal{S}_1^{p_K}(T) \\ \text{or } \in \mathcal{S}_3^{p_K}(Q) \text{ for } T, Q \in \mathcal{T}_\sigma^n\} \end{aligned}$$

we still have the error bound (4.5.27), *however, with* μ *replaced by* 2μ *in* (4.5.26).

Proof We use the result of Theorem 4.51 and note that there $(u|_K \circ Q_K)(\xi) \in \mathcal{S}_2^{p_K}(K)$ for **all** $K \in \mathcal{T}_\sigma^n$. Now we have (see Section 4.2) the inclusions

$$\mathcal{S}_1^{2p}(T) \supseteq \mathcal{S}_2^p(T), \; \mathcal{S}_3^{2p}(Q) \supseteq \mathcal{S}_2^p(Q)$$

for any (affinely mapped) triangle $T \in \mathcal{T}_\sigma^n$ or $Q \in \mathcal{T}_\sigma^n$. The assertion follows. □

The proof of Theorem 4.51 consists of several steps. We will proceed similarly to the proof of Theorem 3.36, but some of the details are more complicated. An outline is as follows: first, we construct on the irregular geometric mesh $\widehat{\mathcal{T}_\sigma^n}$ corresponding to $\mathcal{T}_\sigma^n$ (see Figure 4.11) elementwise a polynomial approximation using Lemma 4.48. This approximation will, however, generally not be continuous across element interfaces. We therefore subdivide some quadrilaterals into triangles (resulting in the mesh $\mathcal{T}_\sigma^n$ and removing the hanging nodes in $\widehat{\mathcal{T}_\sigma^n}$) and then adjust the interelement jumps without downgrading the approximation order on the geometric mesh $\mathcal{T}_\sigma^n$.

We begin with Step 1. From Lemma 4.48 we obtain the following result.

Lemma 4.53 *On* $\Omega_{ij} \in \widehat{\mathcal{T}_\sigma^n}$ *(cf. Figure* 4.11*), there exists* ϕ_{ij} *of degree* p_j *in each variable such that*

$$\begin{aligned} &\|D^m(u - \phi_{ij})\|^2_{L^2(\Omega_{ij})} \\ &\leq C\, \sigma^{2(n+2-j)(2-m-\beta)} \frac{\Gamma(p_j - s_j + 1)}{\Gamma(p_j + s_j + 3 - 2m)} \left(\frac{\rho}{2}\right)^{2s_j} \|u\|^2_{H_\beta^{s_j+3,2}(\Omega_{ij})} \end{aligned} \tag{4.5.28}$$

for $0 \leq m \leq 2$ *and* $2 \leq j \leq n+1$, $1 \leq i \leq 3$ *and any* $1 \leq s_j \leq p_j$ *where*

$$\rho = \max\left(1, \frac{1-\sigma}{\sigma}\right). \tag{4.5.29}$$

Exercise 4.54 Verify (4.5.28).

The global approximation $\phi \in S^{p,1}(\Omega, \widehat{\mathcal{T}}_\sigma^n)$ will be obtained by joining the elementwise local approximations ϕ_{ij} together. This yields at first only a discontinuous global approximation. For Step 2, i.e., to adjust the ϕ_{ij} for their discontinuities across interelement boundaries, we need the following.

Lemma 4.55 *By K we denote a triangle (or quadrilateral) and by $\gamma \subseteq \partial K$ the boundary or some sides of K. Further, we define*

$$H^k(\gamma) = \prod_{i=1}^{m} H^k(\gamma_i), \; k = 0, 1, \text{ and } m \leq 3 \text{ (or } m \leq 4\text{)},$$

where γ_i are the sides of K and assume that one vertex of K coincides with the origin. We have

i) *if $u \in H^2(K)$ and $u = 0$ at the vertices of K, then $u \in H^1(\gamma)$ and*

$$\|u\|^2_{H^1(\gamma)} \leq C h_K \, |u|^2_{H^2(K)} \, . \tag{4.5.30}$$

ii) *if $u \in H^{2,2}_\beta(K)$ with $\Phi_\beta(x) = |x|^\beta$, $0 < \beta < 1$ and $u = 0$ at the vertices of K, then*

$$\|u\|^2_{H^1(\gamma)} \leq C h_K^{1-2\beta} \, |u|^2_{H^{2,2}_\beta(K)} \tag{4.5.31}$$

if $\overline{\gamma}$ does not contain the origin.

iii) *Assume that γ_1 lies on the x_1-axis and that $v(x_1) \in \mathcal{S}^p(\gamma_1)$ such that $v(x_1)$ vanishes in the endpoints of γ_1. Then there exists $V(x_1, x_2)$ of degree p ($V \in \mathcal{S}^1_p$ if K is a triangle and $V \in \mathcal{S}^2_p$ if K is quadrilateral) such that $V \equiv 0$ on γ_i, $i \neq 1$, $V = v$ on γ_1 and*

$$\begin{aligned} \|V\|^2_{H^1(K)} &\leq C h_K \, \|v\|^2_{H^1(\gamma_1)} \, , \\ \|V\|^2_{H^1(K)} &\leq C \, h_K^{-1} \, \|v\|^2_{L^2(\gamma_1)} \leq C h_K \, \|v\|^2_{H^1(\gamma_1)} \, . \end{aligned} \tag{4.5.32}$$

Proof

i) Let K be the unit square $\widehat{Q} = (0,1)^2$. By the trace theorem

$$\begin{aligned} \|u\|^2_{L^2(\gamma)} &\leq C \, \|u\|^2_{H^1(\widehat{Q})} = C \left(|u|^2_{H^1(\widehat{Q})} + \|u\|^2_{L^2(\widehat{Q})} \right), \\ |u|^2_{H^1(\gamma)} &\leq C \left(|u|^2_{H^2(\widehat{Q})} + |u|^2_{H^1(\widehat{Q})} \right). \end{aligned}$$

Since $u = 0$ in the vertices of $\widehat{Q}$ (see [13]), we have

$$\|u\|_{L^2(\widehat{Q})} \leq C_1 \, |u|_{H^2(\widehat{Q})}, \quad |u|_{H^1(\widehat{Q})} \leq C_1 \, |u|_{H^2(\widehat{Q})} \, .$$

Hence

$$\|u\|_{H^1(\gamma)} \le C_2 \, |u|_{H^2(\widehat{Q})}$$

and (4.5.30) follows by a scaling argument. The proof for a triangle is analogous.

ii) Define $\widehat{Q}_1 = (0,1)\times(1/2,1)$ and let $\gamma = \{(x_1,1) : 0 < x_1 < 1\}$. By applying (4.5.30) on $\widehat{Q}_1$, we get

$$\|u\|_{H^1(\gamma)} \le C \, |u|_{H^2(\widehat{Q}_1)} \le C \, |u|_{H^{2,2}_\beta(\widehat{Q})} \, .$$

A scaling yields (4.5.31).

iii) Let $\widehat{Q} = (0,1)^2$, $\gamma_1 = \{(x_1,0) : 0 < x_1 < 1\}$ and set $V(x_1,x_2) = v(x_1)(1 - x_2)$. Then $V(x_1,1) = V(0,x_2) = V(1,x_2) = 0$ and $V(x_1,0) = v(x_1)$. Moreover,

$$\|V\|_{H^1(\widehat{Q})} \le C \, \|v\|_{L^2(\gamma_1)} \le C \, \|v\|_{H^1(\gamma_1)} \, .$$

Since $v(0) = v(1) = 0$, it follows from the first Poincaré inequality in an interval that

$$\|v\|_{L^2(\gamma)} \le C \, |v|_{H^1(\gamma_1)} \, ,$$

i.e.,

$$\|V\|_{H^1(\widehat{Q})} \le C \, |v|_{H^1(\gamma_1)} \, .$$

A scaling argument and an affine mapping yield (4.5.32) for a rectangle K.

The proof for a triangle is more complicated and can be found in [14], Section 7 (see also Section 4.6.3 below).

□

Remark 4.56 The trace lifting constructed in part iii) of Lemma 4.53 is a polynomial of degree p only if K is an axiparallel rectangle. The construction does not work any more if K is a triangle or if $V(x_1,x_2)$ must be in $\mathcal{S}^p_3$ rather than $\mathcal{S}^2_p$. These cases are addressed in [14], Section 7, and in Section 4.6.3 below.

We will now remove the interelement discontinuities between the ϕ_{ij} constructed in Lemma 4.53.

Lemma 4.57 *Assumptions and notation as in Theorem 4.51. Then there exists* $\psi \in S^{p,1}(\widehat{\Omega}, \mathcal{T}^n_\sigma)$ *such that*

$$\|u - \psi\|^2_{H^1(\widehat{\Omega})} \le C \Big\{ \sigma^{2(n+1)(1-\beta)} \|u\|^2_{H^{2,2}_\beta(\Omega_{11})}$$

$$+ \sum_{\substack{1\le i\le 3 \\ 2\le j\le n+1}} \sigma^{2(n+2-j)(1-\beta)} \, \frac{(p_j - s_j)!}{(p_j + s_j - 2)!} \left(\frac{\rho}{2}\right)^{2s_j} \|u\|^2_{H^{s_j+3,2}_\beta(\Omega_{ij})} \Big\}$$

where ρ *is as in* (4.5.29).

Proof We denote $\hat{x}_0 = 0$ and $\hat{x}_j = \sigma^{n+1-j}$, $1 \le j \le n+1$. Then the nodes of the meshes $\widehat{\mathcal{T}}_\sigma^n$, $\mathcal{T}_\sigma^n$ shown in Figure 4.11 are given by $A_{ij} = (\hat{x}_i, \hat{x}_j)$, $0 \le i, j \le n+1$. By construction, the polynomial approximations ϕ_{ij} in Lemmas 4.48 and 4.53 coincide with u in the regular nodes A_{ij}, but (possibly) not in the irregular nodes A_{ij}. We adjust ϕ_{ij} so that it coincides in irregular nodes with u. Consider $\Omega_{2j} \in \widehat{\mathcal{T}}_\sigma^n$ (cf. Figure 4.12) with irregular node $A_{j-1,j-2}$. Here and in what follows, we denote $\phi_{2j}|_{\Omega_{2j}^k}$ by ϕ_{2j}^k and by h_{ij} the diameter of $\Omega_{ij} \in \widehat{\mathcal{T}}_\sigma^n$. Notice that $\phi_{2j}^k \in \mathcal{S}_{2_j}^k \in \mathcal{S}_2^{p_j}(\Omega_{2j}^k)$, $k = 1, 2, 3$.

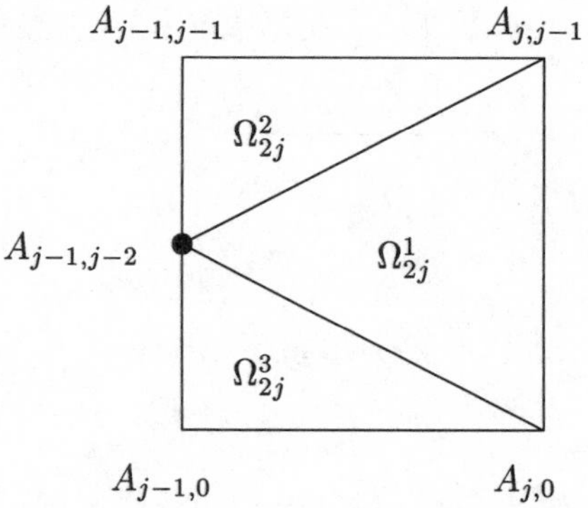

FIG. 4.12. Ω_{2j} and the scheme of the nodes

Let $v_{2j}^k(x)$ be linear on Ω_{2j}^k such that for $k = 1, 2, 3$

$$v_{2j}^k(A_{j-1,j-2}) = u(A_{j-1,j-2}) - \Phi_{2j}^k(A_{j-1,j-2}),$$
$$v_{2j}^k = 0 \text{ in the other nodes of } \Omega_{2j}^k.$$

Then $\widetilde{\phi}_{2j}^k := \phi_{2j}^k + v_{2j}^k$ is equal to u on all vertices of Ω_{2j}^k. Applying Lemma 4.16 with $\beta = 0$ we get for $i \ge 2, j > 2$

$$\|u - \widetilde{\phi}_{ij}^k\|_{H^2(\Omega_{ij}^k)}^2 = \|u - \phi_{ij}^k - v_{ij}^k\|_{H^2(\Omega_{ij}^k)}^2 \le C\,|u - \phi_{ij}^k|_{H^2(\Omega_{ij}^k)}^2 \quad (4.5.33)$$

$$\|u - \widetilde{\phi}_{ij}^k\|_{H^1(\Omega_{ij}^k)}^2 \le C\, h_{ij}^2\, |u - \phi_{ij}^k|_{H^2(\Omega_{ij}^k)}^2 \,. \quad (4.5.34)$$

Now ϕ_{ij} for $i = 1$ or $j < 2$ and $\widetilde{\phi}_{ij}^k$ for $i \ge 2$, $j > 2$ must be adjusted so as to be continuous across interelement boundaries.

There are six basic cases, shown in Figure 4.13.

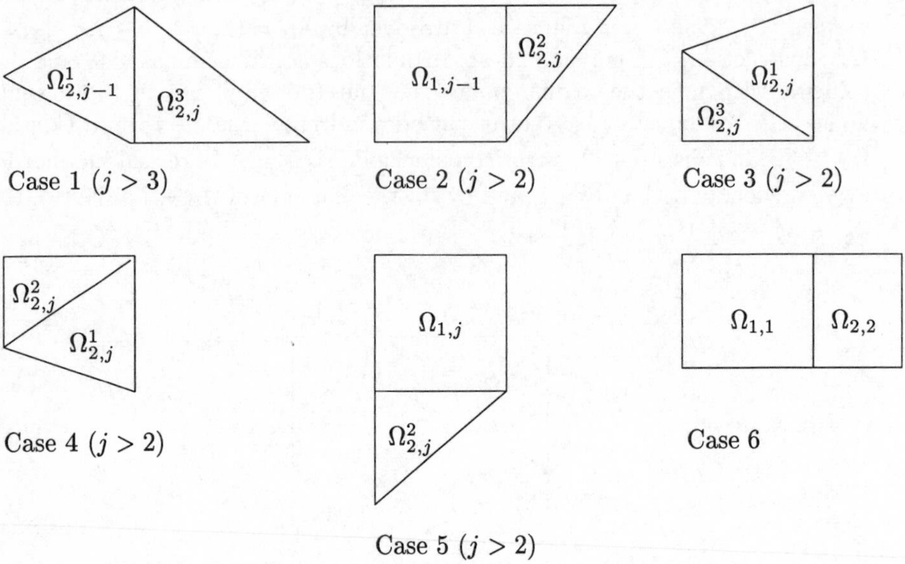

FIG. 4.13.

We show the details for Cases 2, 3, and 6: the remaining ones are completely analogous.

Case 2: Here $\phi_{1,j-1}$ and $\widetilde{\phi}^2_{2j}$ coincide with u at the vertices of $\Omega_{1,j-1}$ and Ω^2_{2j}, respectively. Hence $w := (\phi_{1,j-1} - \widetilde{\phi}^2_{2j})|_\gamma = 0$ at the endpoints of $\gamma := \overline{\Omega_{1,j-1}} \cap \overline{\Omega^2_{2j}}$. By construction, w is a polynomial of degree $\leq p_j$ in x_2 on γ. By Lemma 4.55, iii), there exists $\hat{w}(x_1, x_2)$ of degree $\leq p_j$ in x_2 and of degree 1 in x_1 such that for $\hat{x}_{j-2} < x_2 < \hat{x}_{j-1}$, $\hat{w}(\hat{x}_{j-1}, x_2) = w(x_2)$, $\hat{w} = 0$ on the other two sides of Ω^2_{2j} and, by Lemma 4.55, i).

$$
\begin{aligned}
\|\hat{w}\|^2_{H^1(\Omega^2_{2j})} &\leq C\, h_{2j}\, \|w\|^2_{H^1(\gamma)} \\
&\leq C\, h_{2j}\, (\|\widetilde{\phi}^2_{2j} - u\|^2_{H^1(\gamma)} + \|\phi_{1,j-1} - u\|^2_{H^1(\gamma)}) \\
&\leq C\, \{h^2_{2j}\, \|\widetilde{\phi}^2_{2j} - u\|^2_{H^2(\Omega^2_{2j})} \\
&\qquad + h_{2j}\, h_{1,j-1} \|\phi_{1,j-1} - u\|^2_{H^2(\Omega_{1,j-1})}\} \\
&\leq C\, \{h^2_{2j}\, \|\widetilde{\phi}^2_{2j} - u\|^2_{H^2(\Omega_{2j})} \\
&\qquad + h^2_{1,j-1}\, \|\phi_{1,j-1} - u\|^2_{H^2(\Omega_{1,j-1})}\}\,.
\end{aligned}
\tag{4.5.35}
$$

We define

$$\psi_{2j}^2 = \widetilde{\phi}_{2j}^2 + \hat{w} \text{ in } \Omega_{2j}^2, \; \psi_{1,j-1} := \phi_{1,j-1} \text{ in } \Omega_{1,j-1}$$

and get

$$\begin{aligned}\|u - \psi_{2j}^2\|_{H^1(\Omega_{2j}^2)}^2 &\le 2\{\|u - \widetilde{\phi}_{2j}^2\|_{H^1(\Omega_{2j}^2)}^2 + \|\hat{w}\|_{H^1(\Omega_{2j}^2)}^2\} \\ &\le C\{\|u - \widetilde{\phi}_{2j}^2\|_{H^1(\Omega_{2j}^2)}^2 + h_{2j}^2 \|u - \widetilde{\phi}_{2j}^2\|_{H^2(\Omega_{2j})}^2 \\ &\quad + h_{1,j-1}^2 \|u - \phi_{1,j-1}\|_{H^2(\Omega_{1,j-1})}^2\} \\ &\le C\{h_{2j}^2 |u - \phi_{2j}|_{H^2(\Omega_{2j})}^2 \\ &\quad + h_{1,j-1}^2 \|u - \phi_{i,j-1}\|_{H^2(\Omega_{1,j-1})}^2\}\end{aligned} \tag{4.5.36}$$

by (4.5.32) and (4.5.33) and also

$$\|u - \psi_{1,j-1}\|_{H^1(\Omega_{1,j-1})}^2 = \|u - \phi_{1,j-1}\|_{H^1(\Omega_{1,j-1})}^2 . \tag{4.5.37}$$

Case 3: Here $\widetilde{\phi}_{2j}^1 = \widetilde{\phi}_{2j}^3 = u$ at the endpoints of $\gamma := \overline{\Omega_{2j}^1} \cap \overline{\Omega_{2j}^3}$. Put $w := \widetilde{\phi}_{2j}^1 - \widetilde{\phi}_{2j}^3$. Then $w = v_{2j}^1 - v_{2j}^2$ is linear, $w \equiv 0$ on γ and the piecewise polynomial $\widetilde{\phi}_{ij}^k$ is continuous across γ.

Case 6: ($j = 2$). Here ϕ_{11} is the bilinear interpolant of u at the vertices of Ω_{11}. Then $\gamma := \overline{\Omega}_{11} \cap \overline{\Omega}_{22}$ is separated from the origin A_{00} and $w := (\phi_{22} - \phi_{11})|_\gamma = 0$ at the endpoints of γ. By Lemma 4.55, iii), there exists a polynomial of degree 2 in x_2 and of degree 1 in x_1 such that $\hat{w}|_\gamma = w$ and $\hat{w} = 0$ on the other edges of Ω_{22}. Moreover, $\|\hat{w}\|_{H^1(\Omega_{22})}^2 \le C\, h_{22}\, \|w\|_{H^1(\gamma)}^2$ and by Lemma 4.55, i) and ii),

$$\begin{aligned}\|\hat{w}\|_{H^1(\Omega_{22})}^2 &\le C\, h_{22} \{h_{22} \|u - \phi_{22}\|_{H^2(\Omega_{22})}^2 + h_{11}^{1-2\beta} \|u - \phi_{11}\|_{H_\beta^{2,2}(\Omega_{11})}^2\} \\ &\le C\{h_{22}^2 \|u - \phi_{22}\|_{H^2(\Omega_{22})}^2 + h_{11}^{2(1-\beta)} \|u - \phi_{11}\|_{H_\beta^{2,2}(\Omega_{11})}^2\} .\end{aligned}$$

Setting

$$\psi_{22} = \phi_{22} - \hat{w} \text{ in } \Omega_{22}, \; \psi_{11} = \phi_{11} \text{ in } \Omega_{11}$$

we find

$$\begin{aligned}\|u - \psi_{22}\|_{H^1(\Omega_{22})}^2 &\le \|u - \phi_{22}\|_{H^1(\Omega_{22})}^2 + \|\hat{w}\|_{H^1(\Omega_{22})}^2 \\ &\le C\{\|u - \phi_{22}\|_{H^1(\Omega_{22})}^2 + h_{22}^2 \|u - \phi_{22}\|_{H^2(\Omega_{22})}^2 \\ &\quad + h_{11}^{2(1-\beta)} \|u - \phi_{11}\|_{H_\beta^{2,2}(\Omega_{11})}^2\} .\end{aligned} \tag{4.5.38}$$

Since $\psi_{11} = \phi_{11}$ is the bilinear interpolant of u at the vertices of Ω_{11}, we have by Lemma 4.25,

$$\|u - \psi_{11}\|^2_{H^1(\Omega_{11})} \leq C\, h_{11}^{2(1-\beta)} \, \|u\|^2_{H^{2,2}_\beta(\Omega_{11})} \,. \tag{4.5.39}$$

After treating the remaining cases, we obtain $\psi^k_{ij}(i > 1, j > 2)$ and $\psi_{i2}(i = 1, 2)$ and set

$$\psi = \begin{cases} \psi^k_{ij} \text{ on } \Omega^k_{ij}\ 1 \leq k \leq 3,\ 2 \leq i \leq 3,\ 3 \leq j \leq n+1\,, \\ \psi_{ij} \text{ on } \Omega_{ij}\ i = 1,\ 1 \leq j \leq n+1 \text{ and } 1 \leq i \leq 3,\ j = 2\,. \end{cases}$$

Then, by construction, $\psi \in S^{\mathbf{p},1}(\widehat{\Omega}, \mathcal{T}^n_\sigma)$ and

$$\begin{aligned} \|u - \psi\|^2_{H^1(\widehat{\Omega})} \leq C \Big\{ & h_{11}^{2(1-\beta)} \, \|u - \phi_{11}\|^2_{H^{2,2}_\beta(\Omega_{11})} \\ & + \sum_{\substack{1 \leq i \leq 3 \\ 1 \leq j \leq n+1}} \|u - \phi_{ij}\|^2_{H^1(\Omega_{ij})} + h^2_{ij} \, \|u - \phi_{ij}\|^2_{H^2(\Omega_{ij})} \Big\} \,. \end{aligned} \tag{4.5.40}$$

The assertion follows with (4.3.26), (4.5.28) and (4.5.39). □

Exercise 4.58 So far, we have established Lemma 4.57 for a general $u \in \mathcal{B}^2_\beta(\widehat{\Omega})$. Assume now that $u \in H^1_0(\widehat{\Omega}) \cap \mathcal{B}^2_\beta(\widehat{\Omega})$. Show, using Lemma 4.55, that there exists $\psi \in S^{\mathbf{p},1}_0(\widehat{\Omega}, \mathcal{T}^n_\sigma) := S^{\mathbf{p},1}(\widehat{\Omega}, \mathcal{T}^n_\sigma) \cap H^1_0(\widehat{\Omega})$ such that Lemma 4.57 remains valid. Hint: Use the techniques in the proof of Lemma 4.57 to also adjust the (in general) nonzero boundary values of ψ.

Exercise 4.59 Discuss cases 1, 4 and 5 in the proof of Lemma 4.57.

We can now give the following proof.

Proof of Theorem 4.51: The argument is based on analyzing the general error bound in Lemma 4.55 for the particular linear degree vector (4.5.26). The technical details are quite analogous to the proof of Theorem 3.36, and we use some notation introduced there.

Since $u \in \mathcal{B}^2_\beta(\Omega)$, we have (4.5.4), which implies with Lemma 4.55 that there exists $\psi \in S^{\mathbf{p},1}(\widehat{\Omega}, \mathcal{T}^n_\sigma)$ such that

$$\begin{aligned} &\|u - \psi\|^2_{H^1(\widehat{\Omega})} \\ &\leq C \Big\{ \sigma^{2(n+1)(1-\beta)} + \sum_{j=2}^{n+1} \sigma^{2(n+2-j)(1-\beta)} \, \frac{\Gamma(p_j - s_j + 1)}{\Gamma(p_j + s_j - 1)} \, s_j (\Gamma(s_j + 2))^2 \left(\frac{\rho d}{2}\right)^{2s_j} \Big\} \\ &\leq C \sigma^{2(n+1)(1-\beta)} \\ &\quad \times \Big\{ 1 + \sum_{j=2}^{n+1} \sigma^{2(1-j)(1-\beta)} \frac{\Gamma(p_j - s_j + 1)}{\Gamma(p_j + s_j - 1)} \, s_j (\Gamma(s_j + 2))^2 \left(\frac{\rho d}{2}\right)^{2s_j} \Big\} \,. \end{aligned}$$

We show that the curved parentheses are bounded independently of n, p_j, s_j. To this end, we select $s_j = \alpha_j p_j$ with a parameter $0 < \alpha_j < 1$ still at our disposal. Using (3.3.77), we find

$$\|u - \psi\|^2_{H^1(\widehat{\Omega})} \leq C\sigma^{2(n+1)(1-\beta)} \Big\{ 1 + \sum_{j=2}^{n+1} \sigma^{2(1-j)(1-\beta)} \alpha_j^3 \, p_j^4 (F(\rho d, \alpha_j))^{p_j} \Big\}$$

where the function F is defined in (3.3.74). Now the remainder of the proof is completely analogous to that of Theorem 3.36, in particular the slope μ of the degree vector $\boldsymbol{p}$ satisfies (3.3.77). □

Exercise 4.60 Consider a mesh $\mathcal{T}_\sigma^n$ as in Figure 4.11 which is geometrically refined towards the origin with n layers. Prove that for linear degree vectors p there holds $N \sim np^2 \sim p^3 \sim n^3$. Show that for a constant polynomial degree p in all elements, we also have $n \sim p^3$. How do the leading constants in the asymptotics compare?

Remark 4.61

i) Inspection of the proof of Theorem 4.51 shows that for a **constant polynomial degree**, i.e., $p_{ij} = p = n + 1$ in (4.5.26), we also have exponential convergence. Moreover, in this case (4.5.27) holds and the condition (3.3.77) on the slope μ is not necessary.

ii) If $u_0 \in \mathcal{B}^2_\beta(\Omega)$ with weight $\Phi(x) = |x - x_0|$ for some $x_0 \in \mathbb{R}^2 \backslash \overline{\Omega}$, then u is analytic in $\overline{\Omega}$ (cf. Exercise 4.45) and that we have on any fixed mesh exponential convergence for increasing p, i.e., there exist constants $C, b > 0$ depending only on the mesh, x_0, and on u_0 such that

$$Z(u_0, S^{\boldsymbol{p},1}, H^1(\Omega)) \leq Ce^{-bp} \leq C \exp(-b\sqrt{N})$$

as $p \to \infty$.

We turn now to **geometric meshes with hanging nodes**. Here the technique used in the proof of Lemma 4.57 can be used also to adjust the discontinuous approximation obtained by applying Lemma 4.53 to each element Ω_{ij} of an irregular geometric mesh $\widehat{\mathcal{T}}_\sigma^n$.

Exercise 4.62 Let $\widehat{\mathcal{T}}_\sigma^n$ be the irregular geometric mesh shown in Figure 4.11 and let ϕ_{ij} be the piecewise polynomial approximation to $u \in \mathcal{B}^2_\beta(\Omega)$ constructed in Lemma 4.53. Prove, using Lemma 4.55, that the interelement discontinuities can be removed without introducing triangular elements, i.e., construct a C^0 approximation in $S^{\boldsymbol{p},1}(\Omega, \widehat{\mathcal{T}}_\sigma^n)$ which satisfies the error bound of Lemma 4.57. Deduce the analog of Theorem 4.51 for meshes with hanging nodes. Repeat Exercise 4.59 for this case, too.

4.5.3 *hp-Approximation of* $u \in \mathcal{B}^2_\beta$ *on a polygon*

We will now establish an exponential convergence result like Theorem 4.51 for the *hp*-FEM in a (linear) polygon Ω. The basic idea here is to localize the approximation problem at each corner A_i and to use the above results.

Recall from Definition 4.36 that a proper geometric mesh $\mathcal{T}^n_\sigma$ in a polygon Ω is obtained by mapping linearly up to three regular (or irregular) geometric mesh patches like the ones in Figures 4.5 or 4.7 to a neighborhood of each vertex A_k of Ω; see, e.g., Figure 4.14.

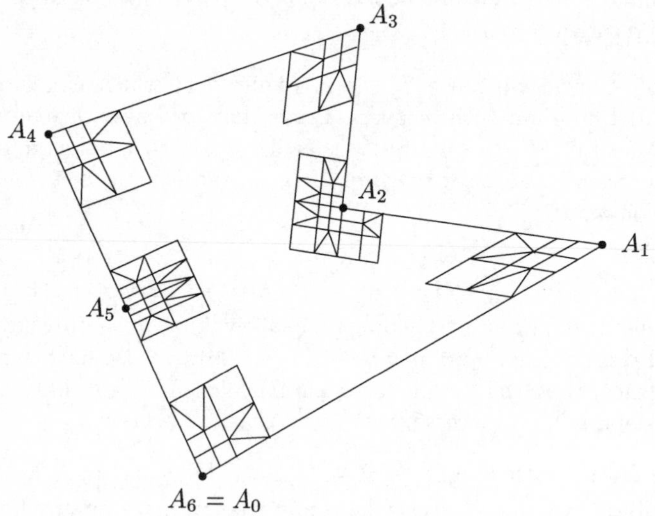

FIG. 4.14. Regular geometric mesh patches in a polygon Ω ($n = 2$ layers)

Here convex vertices (like A_0, A_1, A_3, A_4) require one patch, Dirichlet Neumann vertices (like A_5) require two and reentrant corners (like A_2) require three patches. On the remaining domain $\widetilde{\Omega} := \Omega \setminus \bigcup$ patches, $u \in \mathcal{B}^\ell_\beta(\Omega)$ is analytic (see Exercise 4.45). As in Definition 4.36, we mesh $\widetilde{\Omega}$ with a fixed mesh which is regularly connected with the geometric patches in Figure 4.14.

To construct a piecewise polynomial approximation v in $S^{\mathbf{p},1}(\Omega, \mathcal{T})$ of $u \in \mathcal{B}^2_\beta(\Omega)$, we proceed as follows: first we construct v in each parallelogram patch, and join these approximations continuously together as in the proof of Lemma 4.57. Since u is analytic in $\widetilde{\Omega}$ (cf. Exercise 4.45), we increase the polynomial degree on the fixed mesh in $\widetilde{\Omega}$ consistently with the largest degree in each patch (we assume that the degree vectors in each geometric mesh patch are identical), yielding exponential convergence also in $\widetilde{\Omega}$. Summing up the local error estimates from the subregions yields the following result.

Theorem 4.63 *Let $\Omega \subset \mathbb{R}^2$ be a polygon and $u_0 \in \mathcal{B}^2_\beta(\Omega)$. Then*

$$\inf_{v \in S^{\mathbf{p},1}(\Omega, \mathcal{T}^n_\sigma)} \|u_0 - v\|_{H^1(\Omega)} \leq C \exp(-b \sqrt[3]{N}) \tag{4.5.41}$$

where $C, b > 0$ are independent of $N = \dim(S^{\mathbf{p},1}(\Omega\Delta^n_\sigma))$. Here $\mathcal{T}^n_\sigma$ is the proper geometric mesh from Definition 4.36, $\mathbf{p}$ is the linear degree vector (4.5.26) with slope $\mu > 0$ as in (3.3.80) assumed identical in each geometric mesh patch and $n \sim |\mathbf{p}|$, i.e., the number of layers is proportional to the polynomial degree.

The same result holds for irregular geometric mesh patches $\widehat{\mathcal{T}}^n_\sigma$ with hanging nodes.

In order to prove Theorem 4.63, a generalization of Theorem 4.51 to parallelograms $\widetilde{\Omega}$ in Figure 4.10 is required. This can be established analogously to Theorem 4.51.

Exercise 4.64 Based on Corollary 4.49, prove a generalization of Theorem 4.51 for the parallelogram $\widetilde{\Omega}$ shown in Figure 4.10; use a geometric mesh which is an image of $\mathcal{T}^n_\sigma$ in Figure 4.11 under the linear map (4.5.23). Proceed along the lines of Lemmas 4.53, 4.55, 4.57.

So far, we have considered only the case $\ell = 1$, i.e. the design and approximation properties of $S^{\mathbf{p},1}(\Omega, \mathcal{T}) \subset H^1(\Omega)$ for $u \in \mathcal{B}^2_\beta(\Omega)$. For $H^\ell(\Omega)$ and $u \in \mathcal{B}^{\ell+1}_\beta(\Omega)$ with $\ell > 1$ analogous results hold, see [70]. Later on, for pressure approximation in the Stokes problem, we also require the case $\ell = 0$.

Exercise 4.65 Let $\Omega \subset \mathbb{R}^2$ be a polygon and assume that $u \in \mathcal{B}^1_\beta(\Omega)$. Show that

$$\inf_{v \in S^{\mathbf{p},0}(\Omega, \mathcal{T})} \|u - v\|_{L^2(\Omega)} \leq C \exp(-b N^{1/3}) \tag{4.5.42}$$

where the degree distribution $\boldsymbol{p}$ is as in Theorem 4.63 and the mesh $\mathcal{T}$ is as in Figure 4.14. Proceed as follows: first prove an analog of Theorem 4.46 by double Legendre expansions; note that due to $v \in S^{\mathbf{p},0}(\Omega, \mathcal{T})$ being allowed to be discontinuous ($\ell = 0$), (4.5.8) is not needed. Thus the result and its proof are considerably simplified. (Alternatively, use Theorem 3.11 and a tensor product argument. Deduce analogs of Corollary 4.47 and Lemma 4.48, Corollary 4.49 and Lemma 4.50. Then follow the discussion in Section 4.5.2.2 (note that Lemmas 4.55 and 4.57 are **not** needed in this case).

Remark 4.66 The following generalization of the previous exercise holds: for $u_0 \in \mathcal{B}^1_\beta(\Omega)$,

$$\inf_{v \in S^{\mathbf{p},1}(\Omega, \mathcal{T})} \|u_0 - v\|_{L^2(\Omega)} \leq C \exp(-bN^{1/3}) \,.$$

This is useful, since continuous pressure approximations are employed in Stokes problems. The proof, however, is more delicate than Exercise 4.65.

4.5.4 *hp-FEM on quasiuniform meshes*

So far, we have investigated the *hp*-FEM with linear degree vector on geometric meshes and proved exponential convergence of the *hp*-FEM if the solution u belongs to the class $\mathcal{B}^2_\beta(\Omega)$. It is, however, of interest also to consider the approximability of solutions u that just belong to $H^k(\Omega)$ for some $k \geq 1$.

We investigate this case by construction of a particular interpolation operator Π_k on the unit square $\widehat{Q}$ in Figure 4.8. For an analysis with a somewhat different operator Π_k, see [20].

Lemma 4.67 *Let $\pi_p : H^1(-1,1) \longmapsto S^p$ denote the interpolation operator in Theorem* 3.14 *and denote by $\Pi_p = \pi_p^{(x)} \otimes \pi_p^{(y)}$ its tensor product analog in two dimensions. Then, with the notation of Figure* 4.8, *we have*

$$\Pi_p \textit{ is well defined on } H^{1,1}(\widehat{Q}) \,, \tag{4.5.43}$$

$$(\Pi_p u)(\hat{v}_i) = u(\hat{v}_i) \textit{ in the vertices } \hat{v}_i \textit{ of } \widehat{Q} \,, \tag{4.5.44}$$

$$(\Pi_p u)|_{\hat{\gamma}_i} = \begin{cases} \pi_p^{(x)}(u|_{\hat{\gamma}_i}) \,, \text{ if } i = 2,4 \,, \\ \pi_p^{(y)}(u|_{\hat{\gamma}_i}) \,, \text{ if } i = 1,3 \,. \end{cases} \tag{4.5.45}$$

If, moreover, $u \in H^{k+1}(\widehat{Q})$ for some $k \geq 1$, then

$$\begin{aligned} |u - \Pi_p\, u|^2_{H^1(\widehat{Q})} &\\ \leq 2\, \frac{(p-s)!}{(p+s)!} \int_{\widehat{Q}} &\left\{(1-x^2)^s\, (\partial_x^{s+1} u)^2 + (1-y^2)^s\, (\partial_y^{s+1} u)\right\} dx\, dy \\ &\quad + \frac{2}{p(p+1)}\, \frac{(p-s+1)!}{(p+s-1)!} \\ \int_{\widehat{Q}} &\left\{(1-x^2)^{s-1}\, (\partial_y \partial_x^s\, u)^2 + (1-y^2)^{s-1}\, (\partial_x\, \partial_y^s\, u)^2\right\} dx\, dy \end{aligned} \tag{4.5.46}$$

for any $1 \leq s \leq \min(p,k)$.

Proof

i) Since $H^{1,1}(\widehat{Q}) = H^1(I, H^1(I))$, we see that $(\pi_p^{(y)}\, u)(x,y)$ is a polynomial of degree p in y with x-dependent coefficients that belong to $H^1(I)$, hence $\Pi_p\, u = \pi_p^{(x)}\, (\pi_p^{(y)}\, u)$ is well defined and also $\Pi_p\, u = \pi_p^{(y)}\, \pi_p^{(x)}\, u$.

ii) From (3.3.13) it follows that

$$(\pi_p^{(y)}\, u)(x, \pm 1) = u(x, \pm 1)$$

from where we get

$$(\Pi_p\, u)(x, \pm 1) = (\pi_p(u(\cdot, \pm 1))$$

which is (4.5.45) for $i = 2,4$. The proof for $i = 1,3$ is analogous.

iii) We observe that $H^{1,1}(\widehat{Q}) \subset C^0(\overline{\widehat{Q}})$, hence the point value $u(\hat{v}_i)$ is well defined for the vertices $\hat{v}_i$ of $\widehat{Q}$.

Using (4.5.45) and again (3.3.13), we get

$$(\Pi_p\, u)(x,1)|_{x=1} = \pi_p^{(x)}(u(\cdot,1))|_{x=1} = u(1,1)\ .$$

The assertion for the remaining vertices is analogous.

iv) To prove the error estimates (4.5.46), we use (3.3.16), (3.3.17):

$$||\partial_x(u - \Pi_p\, u)||_{L^2(\widehat{Q})} \leq ||\partial_x(u - \pi_p^{(x)}\, u)||_{L^2(\widehat{Q})} + ||\partial_x\, \pi_p^{(x)}(u - \pi_p^{(y)}\, u)||_{L^2(\widehat{Q})}\ .$$

Estimating the first term with (3.3.16) gives

$$||\partial_x(u - \pi_p^{(x)}\, u)||^2_{L^2(\widehat{Q})} \leq \frac{(p-s)!}{(p+s)!} \int_{\widehat{Q}} (1-x^2)^s \left(\partial_x^{s+1}\, u(x,y)\right)^2 dx\, dy$$

for any $0 \leq s \leq \min(p,k)$.

To estimate the second term, we observe that the interpolation operator π_p is bounded in H^1, i.e.,

$$||(\pi_p\, v)'||_{L^2(-1,1)} \leq ||v'||_{L^2(-1,1)}\ . \tag{4.5.47}$$

This follows from the definition

$$(\pi_p\, v)'(x) = \sum_{i=0}^{p-1} b_i\, L_i(x)$$

where b_i are the Legendre coefficients of $v'(x)$, whence

$$||(\pi_p\, v)'||^2_{L^2(-1,1)} = \sum_{i=0}^{p-1} \frac{2}{2i+1}\, |b_i|^2 \leq \sum_{i=0}^{\infty} \frac{2}{2i+1}\, |b_i|^2 = ||v'||^2_{L^2(-1,1)}$$

by (3.3.6). Therefore we have

$$\begin{aligned} ||\partial_x\, \pi_p^{(x)}\, (u - \pi_p^{(y)}\, u)||_{L^2(\widehat{Q})} &\leq ||\partial_x\, (u - \pi_p^{(y)}\, u)||_{L^2(\widehat{Q})} \\ &= ||\partial_x\, u - \pi_p^{(y)}\, (\partial_x\, u)||_{L^2(\widehat{Q})} \end{aligned}$$

since ∂_x and $\pi_p^{(y)}$ commute. Applying (3.3.17), we find for any $0 \le t \le \min(p,k)$ that

$$\begin{aligned}
&\|\partial_x \pi_p^{(x)} (u - \pi_p^{(y)} u)\|^2_{L^2(\widehat{Q})} \\
&\quad \le \|\partial_x u - \pi_p^{(y)} (\partial_x u)\|^2_{L^2(\widehat{Q})} \\
&\quad \le \frac{1}{p(p+1)} \frac{(p-t)!}{(p+t)!} \int_{\widehat{Q}} (1-y^2)^t \left(\partial_x \partial_y^{t+1} u(x,y)\right)^2 dx\, dy\,.
\end{aligned}$$

In the same way we estimate for any $0 \le \hat{t} \le \min(p,k)$

$$\begin{aligned}
&\|\partial_y (u - \Pi_p u)\|^2_{L^2(\widehat{Q})} \\
&\quad \le 2\,\|\partial_y (u - \pi_p^{(y)} u)\|^2_{L^2(\widehat{Q})} + 2\,\|\partial_y \pi_p^{(y)} (u - \pi_p^{(x)} u)\|^2_{L^2(\widehat{Q})} \\
&\quad \le 2\,\frac{(p-\hat{s})!}{(p+\hat{s})!} \int_{\widehat{Q}} (1-y^2)^{\hat{s}} \left(\partial_y^{\hat{s}+1} u(x,y)\right)^2 dx\, dy \\
&\quad + \frac{2}{p(p+1)} \frac{(p-\hat{t})!}{(p+\hat{t})!} \int_{\widehat{Q}} (1-x^2)^{\hat{t}} \left(\partial_y \partial_x^{\hat{t}+1} u(x,y)\right)^2 dx\, dy\,.
\end{aligned}$$

Selecting $\hat{s} = s$, $\hat{t} = t$ and $t = s - 1$, we get (4.5.46). □

We note some immediate consequences of Lemma 4.67.

Corollary 4.68 *The projector Π_p in Lemma 4.67 satisfies*

$$\begin{aligned}
|u - \Pi_p u|^2_{H^1(\widehat{Q})} \le \frac{2}{(2p)!} \Big\{ &\|\partial_x^{p+1} u\|^2_{L^2(\widehat{Q})} + \|\partial_y^{p+1} u\|^2_{L^2(\widehat{Q})} \\
&+ \|\partial_x \partial_y^p u\|^2_{L^2(\widehat{Q})} + \|\partial_x \partial_y^p u\|^2_{L^2(\widehat{Q})} \Big\}
\end{aligned} \tag{4.5.48}$$

if u is smooth in $\overline{\widehat{Q}}$ and

$$|u - \Pi_p u|^2_{H^1(\widehat{Q})} \le C(k)\, p^{-2k} \sum_{|\alpha|=k+1} \|D^\alpha u\|^2_{L^2(\widehat{Q})} \tag{4.5.49}$$

as $p \to \infty$ for any fixed $k \ge 1$ provided $u \in H^{k+1}(\widehat{Q})$.

Proof (4.5.48) follows from (4.5.46) with $s = p$ and $p(p+1) \ge 2p$ for $p \ge 1$. (4.5.49) follows from (4.5.46) with $s = k$ as $p \to \infty$, using Stirling's formula. □

On occasion, we will also need to bound the L^2-error of Π_p.

Corollary 4.69 *The projector* Π_p *in Lemma* 4.67 *satisfies*

$$\|u - \Pi_p u\|^2_{L^2(\widehat{Q})} \le \frac{2}{p(p+1)} \left\{ \frac{(p-s)!}{(p+s)!} \int_{\widehat{Q}} (1-x^2)^s \, (\partial_x^{s+1} u)^2 \, dx \, dy + \frac{(p-s+1)!}{(p+s-1)!} \int_{\widehat{Q}} (1-y^2)^{s-1} \, (\partial_x \partial_y^s u)^2 \, dx \, dy \right\} \tag{4.5.50}$$

for $1 \le s \le p$ *and, in particular, if* u *is smooth in* $\widehat{Q}$, *we have for any* p

$$\|u - \Pi_p u\|^2_{L^2(\widehat{Q})} \le \frac{2}{(2p)!} \left\{ \|\partial_x^{p+1} u\|^2_{L^2(\widehat{Q})} + \|\partial_x \partial_y^p u\|^2_{L^2(\widehat{Q})} \right\} \tag{4.5.51}$$

and, if $u \in H^{k+1}(\widehat{Q})$ *for some fixed* $k \ge 1$, *as* $p \to \infty$:

$$\|u - \Pi_p u\|^2_{L^2(\widehat{Q})} \le C(k) \, p^{-2k} \sum_{|\alpha| = k+1} \|D^\alpha u\|^2_{L^2(\widehat{Q})} \, . \tag{4.5.52}$$

Proof If (4.5.50) is established, (4.5.51) and (4.5.52) follow as in the proof of Corollary 4.68. To prove (4.5.50), we write, using (4.5.47) and (3.3.17),

$$\begin{aligned}
\|u - \Pi_p u\|^2_{L^2(\widehat{Q})} &\le 2 \, \|u - \pi_p^{(x)} u\|^2_{L^2(\widehat{Q})} + 2 \, \|\pi_p^{(x)} (u - \pi_p^{(y)} u)\|^2_{L^2(\widehat{Q})} \\
&\le 2 \, \|u - \pi_p^{(x)} u\|^2_{L^2(\widehat{Q})} + 2 \, \|\partial_x (u - \pi_p^{(y)} u)\|^2_{L^2(\widehat{Q})} \\
&= 2 \, \|u - \pi_p^{(x)} u\|^2_{L^2(\widehat{Q})} + 2 \, \|\partial_x u - \pi_p^{(y)} (\partial_x u)\|^2_{L^2(\widehat{Q})} \\
&\le \frac{2}{p(p+1)} \, \frac{(p-s)!}{(p+s)!} \int_{\widehat{Q}} (1-x^2)^s \, (\partial_x^{s+1} u)^2 \, dx \, dy \\
&\quad + \frac{2}{p(p+1)} \, \frac{(p-t)!}{(p+t)!} \int_{\widehat{Q}} (1-y^2)^t \, (\partial_x \, \partial_y^{t+1} u)^2 \, dx \, dy
\end{aligned}$$

for any $0 \le s, t \le p$. Selecting $t = s - 1$ gives (4.5.50). Selecting then $s = p$ implies (4.5.51) and (4.5.52) follows with $s = k$ and $t = k - 1$ by Stirling's formula. □

Remark 4.70 The L^2-bound (4.5.50) or (4.5.52) is suboptimal since the optimal rate of convergence as $p \to \infty$ is p^{-2k-2} rather than p^{-2k}.

Remark 4.71 The tensor product construction in Lemma 4.67 allows also for different polynomial degrees in x and y and for high aspect-ratio rectangles.

We can now formulate a result on *hp*-FE approximation on a parallelogram mesh with hanging nodes.

Theorem 4.72 *Let $\Omega \subset \mathbb{R}^2$ be a polygon and let $\mathcal{T}$ be a parallelogram mesh, with at most 1 hanging node per edge. Then, for any $u \in H^2(\Omega)$, there exists $\Pi u \in S^{p,1}(\Omega, \mathcal{T})$ such that*

$$\|u - \Pi u\|^2_{H^1(\Omega)} \leq C \sum_{K \in \mathcal{T}} \max\left\{\frac{(p-s_K)!}{(p+s_K)!}, \frac{(p-s_K+1)!}{p(p+1)(p+s_K-1)!}\right\} \times h_K^{2s_K} |u|^2_{H^{s_K+1}(K)} \tag{4.5.53}$$

for $1 \leq s_K \leq p$ such that the right-hand side in (4.5.53) is finite. Here C is independent of p, s_K and h_K.

If, moreover, $u = 0$ on $\Gamma \subseteq \partial\Omega$ and Γ is a union of edges, then also $\Pi u = 0$ on Γ. If, in particular, $u \in H^{s_K+1}(K)$ for some $s_K \geq 1$ and $K \in \mathcal{T}$, then as $p \to \infty$

$$\|u - \Pi u\|^2_{H^1(\Omega)} \leq C(k) \sum_{K \in \mathcal{T}} \left(\frac{h_K}{p}\right)^{2s_K} |u|^2_{H^{s_K+1}(K)} \, . \tag{4.5.54}$$

Proof We define a global interpolant $\widetilde{\Pi} u$ by setting

$$(\widetilde{\Pi} u)|_K \circ A_K = \Pi_{p_K}(u|_K \circ A_K), \quad K \in \mathcal{T}$$

where Π_p is the projector in Lemma 4.67 and $A_K : \widehat{Q} \longmapsto K$ is the affine element mapping (which preserves the local polynomial space). Summing up the local error estimates (4.5.46) and (4.5.50), properly scaled, gives the bound (4.5.53) with $C = 2$. If $\mathcal{T}$ does not contain hanging nodes, $\widetilde{\Pi} u$ is continuous on Ω. Also, if u vanishes on an edge, so does $\widetilde{\Pi} u$, and setting $\Pi = \widetilde{\Pi}$ the proof is complete in the case when $\mathcal{T}$ does not contain hanging nodes.

If $\mathcal{T}$ contains hanging nodes, however, the elementwise approximation $\widetilde{\Pi} u$ will in general be discontinuous across an edge with hanging nodes, cf. Figure 4.15.

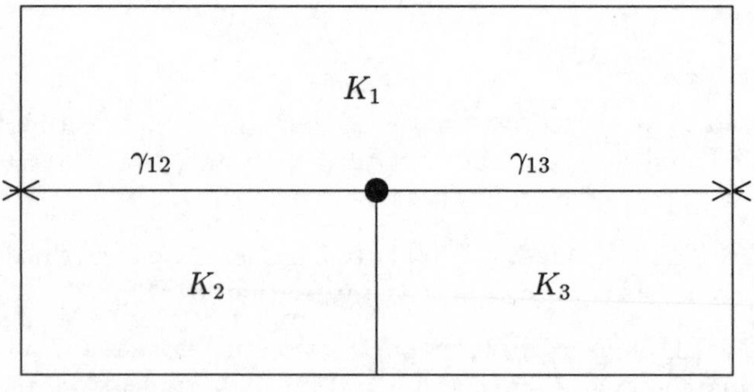

FIG. 4.15. Hanging node • and adjacent elements

Let $\gamma = \overline{\gamma_{12} \cup \gamma_{13}}$ with γ_{ij} as in Figure 4.15. By construction, the interpolant $\tilde{\Pi}u$ is continuous from K_2 to K_3 and the jump $[\widetilde{\Pi}u]$ across γ belongs to $H_0^1(\gamma) \cap (P_p(\gamma_{12} \cap P_p(\gamma_{13}))$.

As in Lemma 4.55 iii), we may construct a trace lifting V of $[\widetilde{\Pi}u]$ to $K_2 \cup K_3$ which is continuous in $\overline{K_2 \cup K_3}$ and belongs to $Q_p(K_2) \cup Q_p(K_3)$ and which vanishes on $\partial(\overline{K_2 \cup K_3})\backslash\gamma$. It satisfies

$$||V||^2_{H^1(K_2 \cup K_3)} \leq C\, |\gamma|^{-1} ||\,[\widetilde{\Pi}u]||^2_{L^2(\gamma)}$$

where $|\gamma|$ denotes the length of γ, and

$$||V||^2_{H^1(K_2 \cup K_3)} \leq C\, |\gamma|\, |\,[\widetilde{\Pi}u]|^2_{H^1(\gamma)}\,,$$

which yields, upon interpolation (see Appendix B), the bound

$$||V||^2_{H^1(K_2 \cup K_3)} \leq C\, ||[\widetilde{\Pi}u]||^2_{H^{1/2}(\gamma)}\,, \tag{4.5.55}$$

where C is independent of p and h_{K_2}.

Since u is continuous in $\overline{\Omega}$, $[u]_{12} = 0$ and, since the trace operator is bounded from $H^1(K_2 \cup K_3) \to H^{1/2}(\gamma)$, we estimate

$$\begin{aligned}||[\widetilde{\Pi}u]||^2_{H^{1/2}(\gamma)} &= ||[u - \widetilde{\Pi}u]||^2_{H^{1/2}(\gamma)} = ||(u - \widetilde{\Pi}u)_+ - (u - \widetilde{\Pi}u)_-||^2_{H^{1/2}(\gamma)} \\ &\leq 2||(u - \widetilde{\Pi}u)_+||^2_{H^{1/2}(\gamma)} + 2||(u - \widetilde{\Pi}u)_-||^2_{H^{1/2}(\gamma)} \\ &\leq C\, \textstyle\sum_{i=1}^3 ||u - \tilde{u}||^2_{H^1(K_i)}\end{aligned}$$

where C is independent of the size of K_i. This yields with (4.5.55) the bound

$$||V||^2_{H^1(K_2 \cup K_3)} \leq C\, ||u - \tilde{u}||^2_{H^1(K_i)}\,. \tag{4.5.56}$$

We set

$$\Pi u|_{K_1} = \widetilde{\Pi}u, \quad \Pi u|_{K_i} = \widetilde{\Pi}u - V \quad i = 2, 3.$$

Using the approximation error estimates for $\widetilde{\Pi}u$ yields (4.5.53). □

Remark 4.73 We assumed for convenience here $p_K = p$ for all $K \in \mathcal{T}$. If the polynomial degrees are not constant, the above construction still applies if different polynomial degrees are allowed on opposite sides of an element $K \in \mathcal{T}$. Estimate (4.5.53) generalizes various classical error estimates for the h-version FEM, the p-version FEM and the spectral element method, since it is explicit in h_K, p, s_K.

Remark 4.74 (p-FEM approximation on triangles). At present it is an open problem to derive an approximation result analogous to Lemma 4.67 on triangles;

a slightly weaker result corresponding to (4.5.49) can be derived as follows. Let $\widehat{T}$ be the reference triangle and assume that $u \in H^{k+1}(\widehat{T})$. Inscribe $\widehat{T}$ into the square $Q = (-1,1) \times (0,2)$. Then there exists a regularity preserving extension $\widetilde{u}$ of u to Q ([135], Chapter 6) such that

$$\widetilde{u} \in H^{k+1}(Q), \ \ \widetilde{u}|_{\widehat{T}} = u, \ \|\widetilde{u}\|_{H^{k+1}(Q)} \leq C(k)\, \|u\|_{H^{k+1}(\widehat{T})} \,.$$

We apply (4.5.49) to $\widetilde{u}$ and get $\widetilde{u}_p = (\Pi_p \widetilde{u})|_{\widehat{T}} \in \mathcal{S}_2^p(\widehat{T}) \subset \mathcal{S}_1^{2p}(\widehat{T})$ such that for every $p \geq 1$ we have from (4.5.49)

$$\begin{aligned} |\widetilde{u} - \widetilde{u}_p|^2_{H^1(\widehat{T})} \leq |\widetilde{u} - \Pi_p \widetilde{u}|^2_{H^1(Q)} &\leq C(k)\, p^{-2k}\, \|\widetilde{u}\|^2_{H^{k+1}(Q)} \\ &\leq C(k)\, p^{-2k}\, \|u\|^2_{H^{k+1}(\widehat{T})} \end{aligned}$$

and using (4.5.52),

$$\|\widetilde{u} - \widetilde{u}_p\|^2_{L^2(\widehat{T})} \leq C(k)\, p^{-2k}\, \|u\|^2_{H^{k+1}(\widehat{T})} \,.$$

By the trace theorem we get in particular also

$$\|\widetilde{u} - \widetilde{u}_p\|^2_{H^{1/2}(\partial\widehat{T})} \leq C(k)\, p^{-2k}\, \|u\|^2_{H^{k+1}(\widehat{T})} \,.$$

These estimates can be used to obtain spectral convergence estimates analogous to (4.5.53) on quasiuniform meshes consisting of triangles.

Exercise 4.75 Generalize Lemma 4.67 to the unit cube in $\mathbb{R}^3$ and prove a *p*-FEM error estimate for the unit tetrahedron.

We finally address the suboptimal L^2-convergence of $u - \Pi_p u$ in Remark 4.70. To this end, let $\mathbb{P}u \in S_0^{p,1}(\Omega, \mathcal{T})$ denote the H_0^1 projection of u onto $S_0^{p,1}(\Omega, \mathcal{T})$:

$$\int_\Omega \nabla \mathbb{P}u \cdot \nabla v \, dx = \int_\Omega \nabla u \cdot \nabla v \, dx \quad \text{for all } \ v \in S_0^{p,1}(\Omega, \mathcal{T}) \,. \tag{4.5.57}$$

Then we have with Π as in Theorem 4.72

$$|u - \mathbb{P}u|_{H^1(\Omega)} \leq |u - \Pi_p u|_{H^1(\Omega)}$$

and

$$\|u - \mathbb{P}u\|_{L^2(\Omega)} = \sup_{v \in L^2(\Omega)} \frac{\int_\Omega (u - \mathbb{P}u) v \, dx}{\|v\|_{L^2(\Omega)}} \,.$$

Given $v \in L^2(\Omega)$, we let w_v be the solution of the Dirichlet problem

$$w_v \in H_0^1(\Omega) : \int_\Omega \nabla w_v \cdot \nabla \varphi \, dx = \int_\Omega v \varphi \, dx \quad \text{for all } \ \varphi \in H_0^1(\Omega) \,. \tag{4.5.58}$$

We now assume H^2-regularity of the Dirichlet problem:

$$\text{for } v \in L^2(\Omega), \quad w_v \in (H_0^1 \cap H^2)(\Omega), \quad \|w_v\|_{H^2} \le C \|v\|_{L^2} . \tag{4.5.59}$$

This is satisfied if Ω is a convex polygon, for example. We estimate, using $\varphi = u - \mathbb{P}u$ in (4.5.58), for any $\psi \in S^{p,1}(\Omega, \mathcal{T})$:

$$\begin{aligned}\left|\int_\Omega (u - \mathbb{P}u) v \, dx\right| &= \left|\int_\Omega \nabla w_v \cdot \nabla (u - \mathbb{P}u) dx\right| \\ &= \left|\int_\Omega \nabla (w_v - \psi) \cdot \nabla (u - \mathbb{P}u) \, dx\right| \\ &\le |w_v - \psi|_{H^1(\Omega)} \, |u - \mathbb{P}u|_{H^1(\Omega)} \\ &\quad \text{for all } \psi \in S_0^{p,1}(\Omega, \mathcal{T})\end{aligned} \tag{4.5.60}$$

where we used the Galerkin orthogonality (4.5.57). Using (4.5.59)

$$\begin{aligned}\|u - \mathbb{P}u\|_{L^2(\Omega)} &\le C \, |u - \mathbb{P}u|_{H^1(\Omega)} \inf_{\psi \in S_0^{p,1}(\Omega,\mathcal{T})} \sup_{v \in L^2(\Omega)} \frac{|w_v - \psi|_{H^1(\Omega)}}{\|w_v\|_{H^2(\Omega)}} \\ &\le C \, |u - \mathbb{P}u|_{H^1(\Omega)} \sup_{w \in (H_0^1 \cap H^2)(\Omega)} \frac{|w - \Pi_p w|_{H^1(\Omega)}}{\|w\|_{H^2(\Omega)}} \\ &\le C \, \frac{\max_{K \in \mathcal{T}} h_K}{\min_{K \in \mathcal{T}} p_K} \, |u - \mathbb{P}u|_{H^1(\Omega)} .\end{aligned} \tag{4.5.61}$$

We see that we have gained one power of p as compared to the H^1-error bound (4.5.54), however, under a H^2-regularity assumption on the solution of the auxiliary Dirichlet problem (4.5.58) in Ω. The argument above is called **Aubin–Nitsche duality argument**. It can be used more generally to estimate the rate of convergence of linear functionals of the solution. In domains which do not provide H^2 regularity of the solution of the adjoint problem, the above argument has to be refined, see [74] for details.

4.6 Inverse inequalities and trace liftings

In the present section, we collect and prove some inverse inequalities for the *hp*-FEM. Throughout, $\widehat{T}, \widehat{Q}$ are the unit triangle and unit square defined in Section 4.2.1. On these domains, we consider the sets $\mathcal{S}_1^p(\widehat{T})$, $\mathcal{S}_2^p(\widehat{Q})$, $\mathcal{S}_3^p(\widehat{Q})$ of polynomials of degree p which were defined in Section 4.2.2.

4.6.1 *Basic inverse inequalities*

Theorem 4.76 *Let $\widehat{K} = \widehat{T}$ (or $\widehat{Q}$) and $\hat{\gamma}_i$ any one of its sides (cf. Figure 4.8). Then we have the following inverse estimates for $v \in \mathcal{S}_1^p(\widehat{T})$ or $v \in \mathcal{S}_2^p(\widehat{Q})$:*

$$\|v\|_{L^\infty(\widehat{K})} \leq C\, p^2 \|v\|_{L^2(\widehat{K})}\,, \tag{4.6.1}$$

$$\|D^1\, v\|_{L^\infty(\widehat{K})} \leq C\, p^2 \|v\|_{L^\infty(\widehat{K})}\,, \tag{4.6.2}$$

$$\|v\|_{L^\infty(\widehat{K})} \leq C\, \sqrt{\ell n(1+p)}\, \|v\|_{H^1(\widehat{K})}\,, \tag{4.6.3}$$

$$\|v\|_{L^2(\hat{\gamma}_i)} \leq C\, p \|v\|_{L^2(\widehat{K})}\,, \tag{4.6.4}$$

$$\|D^1\, v\|_{L^2(\widehat{K})} \leq C\, p^2 \|v\|_{L^2(\widehat{K})}\,. \tag{4.6.5}$$

Proof We prove the theorem only for $\widehat{K} = \widehat{T}$, the proofs for $\widehat{K} = \widehat{Q}$ are completely analogous.

We start with (4.6.2). To this end, it is sufficient to prove

$$\|\partial_1\, v\|_{L^\infty(\widehat{T})} \leq c p^2\, \|v\|_{L^\infty(\widehat{T})}\,. \tag{4.6.6}$$

To prove (4.6.6) we use the one-dimensional result (3.6.3) and define $\bar{x} = (\bar{x}_1, \bar{x}_2) \in \overline{\widehat{T}}$ by

$$\|\partial_1\, v\|_{L^\infty(\widehat{T})} = (\partial_1\, v)(\bar{x})\,.$$

Then we distinguish three cases.

Case 1: Assume that $\bar{x}_2 \leq 1/2$ and define

$$\varphi(x_1) := v(x_1, \bar{x}_2),\ x_1 \in I(\bar{x}_2) := \{x_1 : |x_1| \leq 1 - \bar{x}_2/\sqrt{3}\}\,.$$

Then, by (3.6.3) applied to $I(\bar{x}_2)$ and $0 \leq \bar{x}_2 \leq 1/2$,

$$\|\partial_1\, v\|_{L^\infty(\widehat{T})} = \|\varphi'\|_{L^\infty(I(\bar{x}_2))} \leq C\, p^2\, \|\varphi\|_{L^\infty(I(\bar{x}_2))}$$

which implies (4.6.6) in this case.

Case 2: Assume that $1/2 < \bar{x}_2 < \sqrt{3}$ and that

$$|(\partial_2\, v)(\bar{x})| \geq |(\partial_1\, v)(\bar{x})|/2\,.$$

Put $\varphi(x_2) = v(\bar{x}_1, x_2)$, $x_2 \in J(\bar{x}_1) := \{x_2 : 0 \leq x_2 \leq \sqrt{3}(1 - \bar{x}_1)\}$.

Then, using now (3.6.3) on $J(\bar{x}_1)$, we get

$$\begin{aligned}\|\partial_1 v\|_{L^\infty(\widehat{T})} &\le 2\,|(\partial_2 v)(\bar{x})| = 2\,|\varphi'(\bar{x}_2)| = 2\,\|\varphi'\|_{L^\infty(J(\bar{x}_1))} \\ &\le C\,p^2\,\|\varphi\|_{L^\infty(J(\bar{x}_1))} = C\,p^2\,\|v\|_{L^\infty(\widehat{T})}\,.\end{aligned}$$

Case 3: Assume again $1/2 < \bar{x}_2 < \sqrt{3}$ and

$$|(\partial_2 v)(\bar{x})| \le |(\partial_1 v)(\bar{x})|/2\,.$$

Put $\varphi(t) := \pi(t, \bar{x}_1 + \bar{x}_2 - t)$, $0 \le t \le \bar{x}_1 + \bar{x}_2$. Then

$$\|\partial_1 v\|_{L^\infty(\widehat{T})} \le 2\,|\varphi'(\bar{x}_1)| \le 2\,\|\varphi'\|_{L^\infty(0,\bar{x}_1+\bar{x}_2)} \le C\,p^2\,\|v\|_{L^2(\widehat{T})}\,.$$

The proof of (4.6.6) and hence of (4.6.2) is complete.

We prove (4.6.1). Pick again $\bar{x} \in \overline{\widehat{T}}$ such that $|v(\bar{x})| = \|v\|_{L^\infty(\widehat{T})}$. By continuity, there exists $B_\varepsilon(\bar{x}) = \{x \in \widehat{T} : |x - \bar{x}| \le \varepsilon\}$, $\varepsilon > 0$, such that

$$|v(x)| > \|v\|_{L^\infty(\widehat{K})}/2 \quad \text{for all } x \in B_\varepsilon(\bar{x})\,. \tag{4.6.7}$$

By (4.6.1), ε may be chosen as $\varepsilon = C_0/p^2$ where $C_0 > 0$ is independent of π and p. Then

$$\begin{aligned}\|v\|^2_{L^2(\widehat{T})} &= \int_{\widehat{T}} (v(x))^2\,dx \ge \frac{1}{4}\,|B_\varepsilon|\,\|v\|^2_{L^\infty(\widehat{T})} \\ &\ge C_1\,p^{-4}\,\|v\|^2_{L^\infty(\widehat{T})}\,.\end{aligned}$$

This proves (4.6.1).

To show (4.6.3), we use the following inequality: There exists a constant $C_3 < \infty$ such that for every $v \in H^1(\widehat{T})$

$$\int_{\widehat{T}} \left(\exp\left[\left(\frac{v(x)}{C_3\,\|v\|_{H^1(\widehat{T})}}\right)^2\right] - 1\right) dx \le 1\,. \tag{4.6.8}$$

(see, e.g., [63]). Using here (4.6.7) with $\varepsilon = C_0\,p^{-2}$, we get

$$|B_\varepsilon| \left(\exp\left[\left(\frac{\|v\|_{L^\infty(\widehat{T})}}{2C_3\|v\|_{H^1(\widehat{T})}}\right)^2\right] - 1\right) \le 1\,.$$

This implies (4.6.3).

The proof of (4.6.4) and (4.6.2) is left to the reader as an exercise. □

Exercise 4.77 Let $K = h\widehat{K}$ be an element of size h. In each of the estimates (4.6.1)–(4.6.9), exhibit the dependence on h by a scaling argument.

Exercise 4.78 Deduce analogs of the inverse estimates in Theorem 4.76 for higher norms, similar to Corollary 3.94.

4.6.2 *Inequalities in* $H^{1/2}(-1,1)$

Frequently in the analysis of *hp*-FEM for two-dimensional problems, we need to estimate norms of FE functions over boundaries (e.g. edges or interfaces) in terms of norms over the elements. This requires norm estimates of the trace operator for FE functions, in particular in the $H^{1/2}$-norm on the boundary.

We put $I = (a,b)$ and recall from Chapter 1, Section 1.4 the spaces $H^{1/2}(I)$, $H_{00}^{1/2}(I)$ that are equipped with the norms

$$\|u\|_{H^{1/2}(I)}^2 = \int_a^b \int_a^b \left(\frac{u(x)-u(y)}{x-y}\right)^2 dx\, dy + \|u\|_{L^2(I)}^2 , \tag{4.6.9}$$

$$\|u\|_{H_{00}^{1/2}(I)}^2 = \|u\|_{H^{1/2}(I)}^2 + \int_a^b \frac{(u(x))^2}{(x-a)(b-x)}\, dx . \tag{4.6.10}$$

Let $K \subset \mathbb{R}^2$ be a triangle or a quadrilateral with sides γ_i, $i = 1,\dots,L$, $L = 3$ or 4, respectively, and let v_i be a vertex common to $\gamma_{j(i)}$ and $\gamma_{\ell(i)}$. Then

$$\|u\|_{H^{1/2}(\partial K)}^2 \approx \sum_{i=1}^{L} \|u\|_{H^{1/2}(\gamma_i)}^2 + \int_0^c \frac{|u_{j(i)}(t) - u_{\ell(i)}(t)|^2}{|t|}\, dt \tag{4.6.11}$$

where t denotes the distance to vertex v_i and $c > 0$ is smaller than $\min_{1\le i\le L} |\gamma_i|$.

Let now $I = (-1,1)$ and $I^* = (0,\pi)$. Then the substitution $x = \cos\varphi$, $\varphi \in I^*$ transforms $v(x) \in H^{1/2}(I)$ into $H^{1/2}(I^*)$ with equivalence of norms, i.e.,

$$\begin{gathered} v(x) \in H^{1/2}(I) \iff v^*(\varphi) := v(\cos\varphi) \in H^{1/2}(I^*) , \\ 0 < C \le \|v\|_{H^{1/2}(I)} / \|v^*\|_{H^{1/2}(I^*)} < C^{-1} . \end{gathered} \tag{4.6.12}$$

We also have

$$v(x) \in H_{00}^{1/2}(I) \iff v^*(\varphi) \in H^{1/2}(I^*) . \tag{4.6.13}$$

By (4.6.12), we need to analyze only the weighted L^2-norm to prove this. It holds that

$$\begin{aligned} \int_{-1}^{1} \frac{(u(x))^2}{1-x^2}\, dx &= \int_0^\pi \frac{|u(\cos\varphi)|^2}{1-(\cos\varphi)^2} \sin\varphi\, d\varphi \\ &= \int_0^\pi \frac{|u^*(\varphi)|^2}{\sin\varphi}\, d\varphi \approx \int_0^\pi \frac{|u^*(\varphi)|^2}{\varphi(\pi-\varphi)}\, d\varphi . \end{aligned}$$

In particular, we have from (4.6.12) also

$$\text{for all } v \in H_{00}^{1/2}(I):\ 0 < C \leq \|v\|_{H_{00}^{1/2}(I)} / \|v^*\|_{H_{00}^{1/2}(I^*)} < C^{-1} .$$

We prove now the one-dimensional analog of (4.6.3), in a sense.

Theorem 4.79 *Let $I = (-1, 1)$ and $u \in \mathcal{S}^p(I)$. Then*

$$\|u\|_{L^\infty(I)} \leq C\sqrt{\ell n(1+p)}\ \|u\|_{H^{1/2}(I)},\ p \geq 0 . \tag{4.6.14}$$

Proof By (4.6.12), we consider

$$u^*(\varphi) = \sum_{k=0}^{p} b_k \cos(k\varphi),\ \|u^*\|^2_{H^{1/2}(I^*)} = \sum_{k=0}^{p} |b_k|^2 (k+1) .$$

Then, for any $x \in [-1, 1]$,

$$\begin{aligned} |u(x)| = |u(\cos\varphi)| = |u^*(\varphi)| &\leq \sum_{k=0}^{p} |b_k|\, (k+1)^{1/2}(k+1)^{-1/2} \\ &\leq \Big(\sum_{k=0}^{p} |b_k|^2\,(k+1)\Big)^{1/2} \Big(\sum_{k=0}^{p} (k+1)^{-1}\Big)^{1/2} \\ &\leq C\,\|u^*\|_{H^{1/2}(I^*)}\sqrt{\ln(p+1)} \end{aligned}$$

which implies (4.6.14) due to (4.6.12). □

We get the following corollary.

Corollary 4.80 *Let $u \in \mathcal{S}_i^p(\widehat{K})$, $i = 1, 2, 3$. Then*

$$\|u\|_{L^\infty(\partial\widehat{K})} \leq C\sqrt{\ln(p+1)}\ \|u\|_{H^1(\widehat{K})} . \tag{4.6.15}$$

This follows either from (4.6.3) or from (4.6.14).

Theorem 4.81 *Let $I = (-1, 1)$ and $u \in \mathcal{S}^p(I) \cap H_0^1(I)$. Then*

$$\|u\|_{H_{00}^{1/2}(I)} \leq C\ \ln(p+1)\, \|u\|_{H^{1/2}(I)} . \tag{4.6.16}$$

Proof We consider first

$$u^*(\varphi) = \sum_{k=0}^{p} b_k \cos(k\varphi),\ u^*(0) = \sum_{k=0}^{p} b_k = 0 .$$

Then

$$\int_0^\pi \frac{|u^*(\varphi)|^2}{\varphi}\,d\varphi = \int_0^\pi \frac{1}{\varphi}\Big(\sum_{k=1}^p b_k\,(\cos(k\varphi)-1)^2\Big)\,d\varphi$$
$$\le \int_0^\pi \frac{1}{\varphi}\Big(\sum_{k=1}^p |b_k|^2\,(k+1)\Big)\Big(\sum_{k=1}^p \frac{(\cos(k\varphi)-1)^2}{k+1}\Big)\,d\varphi$$
$$\le C\,\|u^*\|^2_{H^{1/2}(I^*)}\sum_{k=1}^p (k+1)^{-1}\int_0^\pi \varphi^{-1}\,(\cos(k\varphi)-1)^2\,d\varphi\,.$$

Since

$$\int_0^\pi \frac{(\cos(k\varphi)-1)^2}{\varphi}\,d\varphi = \int_0^{k\pi}\frac{(\cos\psi-1)^2}{\psi}\,d\psi \le C\,\log(k+1)\,,$$

we find

$$\int_0^\pi \frac{|u^*(\varphi)|^2}{\varphi}\,d\varphi \le C\,\|u^*\|^2_{H^{1/2}(I^*)}\sum_{k=1}^p\frac{\log k}{k+1} \le C\,\|u^*\|^2_{H^{1/2}(I^*)}(\log(p+1))^2\,.$$

The integral

$$\int_0^\pi \frac{|u^*(\varphi)|^2}{\pi-\varphi}\,d\varphi,\quad u^*(\pi)=0$$

is estimated analogously. □

Theorem 4.82 *Let $u \in \mathcal{S}^p(I)\cap H_0^1(I)$. Then*

$$\|u\|^2_{H_{00}^{1/2}(I)} \le \|u\|^2_{H^{1/2}(I)} + C\,\log(1+p)\,\|u\|^2_{L^\infty(I)}\,. \tag{4.6.17}$$

Proof Once again, we bound

$$\|u\|^2_{H_{00}^{1/2}(I)} - \|u\|^2_{H^{1/2}(I)} = \int_{-1}^1 \frac{(u(x))^2}{1-x^2}\,dx$$
$$\le C\,\Big\{\int_{-1}^1\frac{(u(x))^2}{1-x}\,dx + \int_{-1}^1\frac{(u(x))^2}{1+x}\,dx\Big\}\,.$$

Consider the second term. Since, by the inverse inequality (3.6.3),

$$|u(x)| = \Big|u(-1)+\int_{-1}^x u'(\xi)d\xi\Big| \le (1+x)\,\|u'\|_{L^\infty} \le C(1+x)\,p^2\,\|u\|_{L^\infty(I)}\,,$$

we obtain

$$\int_{-1}^{1} \frac{(u(x))^2}{1+x}\,dx \le C\,\|u\|^2_{L^\infty(I)} \Big\{ \int_{-1}^{-1+1/p^2} (1+x)\,p^4\,dx + \int_{-1+1/p^2}^{1} (1+x)^{-1}\,dx \Big\}$$
$$\le C\,(1+\log p)\,\|u\|^2_{L^\infty(I)}\,.$$

The first term is estimated in the same way. □

4.6.3 *Polynomial trace liftings*

Frequently in the construction of continuous, piecewise polynomial approximations of solutions we constructed elementwise polynomial approximations with the desired approximation properties which were, however, generally discontinuous between elements. Since the function to be approximated is continuous between elements, the jump in the approximation error across edges equals the jump of the approximation and this quantity is a polynomial across the edge. This jump is removed by extending the polynomial from the edge polynomially to the element, i.e., by a polynomial lifting. We present here some results on the existence of such polynomial trace liftings which were used in the *hp*-approximation theorems in Section 4.5.2. We do not elaborate on the (to some extent technical) proofs – for this, we refer to [14]. These results will also be used in the context of preconditioning in Section 4.7. Another application of such liftings appears in the context of inhomogeneous Dirichlet data addressed in Chapter 1, Section 1.4. See [18] for an analysis of the discrete trace liftings.

We let $\widehat{T}$, $\widehat{Q}$ be the reference elements shown in Figure 4.8 and use that notation in what follows. The first result addresses the lifting from one side of $\widehat{T}$.

Lemma 4.83 *Let $f \in C^0(\partial\widehat{T})$ and $f_i := f|_{\hat\gamma_i}$ and $f_1 \in P_p(\hat\gamma_1)$, $f_2 = f_3 = 0$. Then there exists $U \in S_1^p(\widehat{T})$ such that*

$$\|U\|_{H^1(\widehat{T})} \le C\,\|f_1\|_{H^1(\hat\gamma_1)},\quad U = f \text{ on } \partial\widehat{T} \tag{4.6.18}$$

where C is independent of f and p.

Next, we consider the lifting from all sides of $\widehat{T}$.

Theorem 4.84 *Let $f \in C^0(\partial\widehat{T})$, $f_i := f|_{\hat\gamma_i} \in P_p(\hat\gamma_i)$, $i = 1,2,3$. Then there exists $U \in S_1^p(\widehat{T})$ such that*

$$\|U\|_{H^1(\widehat{T})} \le C\,\|f\|_{H^{1/2}(\partial\widehat{T})},\quad U = f \text{ on } \partial\widehat{T} \tag{4.6.19}$$

where C is independent of p.

In the case of the unit square $\widehat{Q} = (-1,1)^2$, the result depends on whether the product space $S_2^p(\widehat{Q})$ or the trunk space $S_3^p(\widehat{Q})$ (see Section 4.2.2 for their definition) is used.

Theorem 4.85 *Let* $f \in C^0(\partial\widehat{Q})$ *and* $f_i := f|_{\hat\gamma_i} \in P_p(\hat\gamma_i)$, $i = 1, \dots, 4$.

Then for every $p > 0$ *there exists a lifting* $U \in \mathcal{S}_2^p(\widehat{Q})$ *such that*

$$\|U\|_{H^1(\widehat{Q})} \leq C\,\|f\|_{H^{1/2}(\partial\widehat{Q})}, \quad U = f \text{ on } \partial\widehat{Q}\,, \tag{4.6.20}$$

and there exists a lifting $\widetilde{U} \in \mathcal{S}_3^p(\widehat{Q})$ *such that*

$$\|\widetilde{U}\|_{H^1(\widehat{Q})} \leq Cp\,\|f\|_{H^{1/2}(\partial\widehat{Q})}, \quad \widetilde{U} = f \text{ on } \partial\widehat{Q}\,. \tag{4.6.21}$$

Finally, we consider the case that a corner value of a function must be lifted.

Theorem 4.86 *Let* $\widehat{Q}$ *be as in Figure 4.8. There exists* $U \in \mathcal{S}_3^p(\widehat{Q})$ *such that*

$$U(\hat{v}_1) = 1, \quad U|_{\hat\gamma_1} = U|_{\hat\gamma_2} = 0$$

and

$$\|U\|_{H^1(\widehat{Q})} \leq C\,(1 + \log p)^{1/2}\,.$$

4.7 *hp*-Preconditioning

4.7.1 *Preliminary remarks*

We consider again the model Poisson problem from Section 1 and its *hp*-FE approximation u_S (4.1.7) based on the *hp*-subspace $S^{p,1}(\Omega, \mathcal{T}_\sigma^n)$ introduced in Section 4.

If the stiffness matrix corresponding to the bilinear form $B(\cdot,\cdot)$ is ill conditioned, this limits the efficiency of iterative solution methods like conjugate gradients (CG) or GMRES. Ill conditioning for stiffness matrices corresponding to *h*-version FEM is well known. The phenomenon also arises in connection with *hp*-FEM for two reasons: a) the geometric mesh $\mathcal{T}_\sigma^n$ grading induces larger differences in the relative element sizes than those occurring in triangulations $\mathcal{T}$ of type (h, γ, K) in Definition 4.7 and b) the selection of the shape functions on the elements $k \in \mathcal{T}$ determines the condition number in dependence on the polynomial degree p. For example, it is well known that the mass matrix corresponding to the monomial basis $\{1, \xi, \xi^2, \dots, \xi^p\}$ of $S^p(-1,1)$ has a condition number which grows like $p!$ whereas the Legendre basis $\{c_k L_k(\xi)\}_{k=0}^p$, $c_k^2 = (2p+1)/2$, leads to a diagonal mass matrix with bounded condition number. Thus, **preconditioning** the stiffness and the mass matrix becomes essential for the efficient iterative solution.

We consider again the model problem (4.1.1)–(4.1.3) in the weak formulation (4.1.5) and its *hp*-FE discretization

$$u_S \in S_0^{p,1}(\Omega, \mathcal{T}) : B(u_S, v) = F(v) \quad \text{for all } v \in S_0^{p,1}(\Omega, \mathcal{T})\,, \tag{4.7.1}$$

where $B(u,v) = \int_\Omega \nabla u \cdot \nabla v dx$ or, in matrix form,

$$\boldsymbol{A}\boldsymbol{x} = \boldsymbol{b} \tag{4.7.2}$$

where $\boldsymbol{A}$ is the stiffness matrix corresponding to the form $B(\cdot,\cdot)$, and $\mathcal{T}$ is any shape-regular mesh. If, for example, $\mathcal{T} = \mathcal{T}_\sigma^n$ and if $N = \dim(S_0^{\boldsymbol{p},1}(\Omega, \mathcal{T}_\sigma^n))$, we have seen in Section 5 that the rate of convergence of the *hp*-FEM is exponential,

$$\|u - u_S\|_{H^1(\Omega)} \leq C \exp(-b\sqrt[3]{N}) \,. \tag{4.7.3}$$

This approximation-theoretic bound may be meaningless if the **cost** to compute u_S, or to solve (4.7.2), is excessive, since for the traditional h-version FEM rather efficient multigrid solvers exist. Assuming (pessimistically) that $\boldsymbol{A}$ in (4.7.2) is dense, we know that direct solution by Gaussian elimination takes $W = O(N^3)$ work, so that we get from (4.7.3) the following asymptotic **bound** for **convergence in terms of work**:

$$\|u - u_S\|_{H^1(\Omega)} \leq C \exp(-b\sqrt[9]{W}) \tag{4.7.4}$$

(this bound is somewhat pessimistic since the sparsity of $\boldsymbol{A}$ was not taken into account). This is better than the algebraic convergence rates that can in general be achieved with h-version FEM and fast solvers, such as multigrid (multigrid, on the other hand, is based on weaker regularity of the solution in Ω).

Here we are concerned with the preconditioning of (4.7.2) such that the condition number grows only moderately with respect to the polynomial degrees $\boldsymbol{p}$. We will present a preconditioner from [14] which supposes that the linear system corresponding to linear/bilinear elements on $\mathcal{T}_\sigma^n$ can be solved efficiently. Numerous iterative methods are available for this and yield, when combined with elementwise p-preconditioner, algorithms that are suitable for *hp*-FEM in very large scale applications. For example, in [2], [14], the elementwise p-preconditioner is combined with domain decomposition ideas for the $p = 1$ problem on the geometric mesh $\mathcal{T}$.

4.7.2 *Preconditioning hp-FEM by low-order elements*

We consider the problem (4.7.1) for an arbitrary, shape-regular mesh and a general degree distribution $\boldsymbol{p}$.

We say that a bilinear form $C(u,v)$ is a **preconditioning form** to $B(\cdot,\cdot)$ on the subspace S of our function space, if $C(\cdot,\cdot)$ is spectrally equivalent to $B(\cdot,\cdot)$, i.e.,

$$\text{for all } \; u \in S_0^{\boldsymbol{p},1}(\Omega, \mathcal{T}) : m_1 \, B(u,u) \leq C(u,u) \leq m_2 \, B(u,u) \,, \tag{4.7.5}$$

to the preconditioning form $C(\cdot,\cdot)$ corresponds the **preconditioning matrix** $\boldsymbol{C}$. In the course of an iterative solution of (4.7.2), we must solve repeatedly

$$\boldsymbol{C}\,\boldsymbol{r} = \boldsymbol{x} \,, \tag{4.7.6}$$

so that we have the following **requirements on the preconditioner**:

i) (4.7.6) should be inexpensive to solve,

ii) the constants m_1, m_2 in (4.7.5) should either be uniformly bounded with respect to $\mathcal{T}$ and $\boldsymbol{p}$ or grow slowly with $|\boldsymbol{p}| := \max\{p_K : K \in \mathcal{T}\}$.

To construct the preconditioner, we recall that for a typical element $K \in \mathcal{T}$ which is affine equivalent to $\widehat{K}$ we have for the polynomial space $\mathcal{S}^{p_K}(\widehat{K})$ two groups of shape functions (see Section 4.5):

external shape functions: $\quad \mathcal{E}^{p_K}(\widehat{K})$

internal shape functions:

$$
\begin{aligned}
\mathcal{I}^{p_K}(\widehat{T}) &:= b_{\widehat{T}} \otimes \mathcal{S}_1^{p_K-3}(\widehat{T}) \text{ if } \widehat{K} = \widehat{T}\,,\\
\mathcal{I}^{p_K}(\widehat{Q}) &:= b_{\widehat{Q}} \otimes \mathcal{S}_1^{p_K-4}(\widehat{Q}) \text{ if } \widehat{K} = \widehat{Q}\,,\\
\mathcal{J}^{p_K}(\widehat{Q}) &:= b_{\widehat{Q}} \otimes \mathcal{S}_2^{p_K-2}(\widehat{Q}) \text{ if } \widehat{K} = \widehat{Q}\,.
\end{aligned}
$$

We decompose $\mathcal{E}^{p_K}(\widehat{K})$ further:

$$
\mathcal{E}^{p_K}(\widehat{K}) = \bigoplus_{i=1}^{n} (\mathcal{N}_i \oplus \Gamma_i^{p_K})
$$

where $\mathcal{N}_i$ are the (linear or bilinear) nodal shape functions and Γ_i are the side shape functions associated with side γ_i of $\widehat{K}$ (see Figure 4.8), and $n = 3$ for triangles, $n = 4$ for quadrilaterals.

By a tilde, we denote the span of the respective set of shape functions, i.e., we write $\widetilde{\mathcal{I}}^{p_K}(K)$, $\widetilde{\Gamma}_i^{p_K}$, $\widetilde{\mathcal{N}}_i$.

We will focus first only on the spaces $\mathcal{S}_1^{p_K}(\widehat{T})$, $\mathcal{S}_2^{p_K}(\widehat{Q})$ and address the case of the serendipity space $\mathcal{S}_3^{p_K}(\widehat{Q})$ later. For $K \in \mathcal{T}$ and $u \in S^{\boldsymbol{p},1}(\Omega, \mathcal{T})$, we denote $u^K := u|_K$ and write

$$
B(u, v) = \sum_{K \in \mathcal{T}} B_K(u^K, v^K)\,, \tag{4.7.7}
$$

where $B_K(u, v)$ is the restriction of $B(\cdot,\cdot)$ to $K \in \mathcal{T}$. Since K is affine equivalent to the reference element $\widehat{K}$, we have for every $u, v \in S^{\boldsymbol{p},1}(\Omega, \mathcal{T})$ and every $K \in \mathcal{T}$,

$$
c_1^{\mathcal{T}}\, \hat{B}(\hat{u}^K, \hat{v}^K) \le B_K(u^K, v^K) \le c_2^{\mathcal{T}}\, \hat{B}(\hat{u}^K, \hat{v}^K) \tag{4.7.8}
$$

where

$$
\hat{B}(\hat{u}, \hat{v}) = \int_{\widehat{K}} \nabla \hat{u} \cdot \nabla \hat{v}\, d\hat{x} \tag{4.7.9}
$$

and where $c_1^{\mathcal{T}}, c_2^{\mathcal{T}}$ are independent of p_K and h_K (they depend only on the shape regularity of K).

Let $u \in S^{\mathbf{p},1}(\Omega, \mathcal{T})$ and $K \in \mathcal{T}$. Then

$$u^K = \sum_{i=1}^{n} u^K_{\boldsymbol{v},i} + \sum_{i=1}^{n} u^K_{\boldsymbol{s},i} + u^K_{\boldsymbol{i}} \tag{4.7.10}$$

where $u^K_{\boldsymbol{v},i} \in \widetilde{\mathcal{N}}_i(K)$, $u^K_{\boldsymbol{s},i} \in \widetilde{\Gamma}^{p_K}_i(K)$ and $u^K_{\boldsymbol{i}} \in \widetilde{\mathcal{I}}^{p_K}(K)$.

We define, for $u, v \in S^{\mathbf{p},1}(\Omega, \mathcal{T})$, the preconditioning form

$$C(u,v) = \sum_K C_K(u^K, v^K) \tag{4.7.11}$$

where

$$\begin{aligned} C_K(u^K, v^K) := B_K\Big(\sum_{i=1}^{n} u^K_{\boldsymbol{v},i}, \sum_{j=1}^{n} v^K_{\boldsymbol{v},j}\Big) + \sum_{j=1}^{n} B_K\left(u^K_{\boldsymbol{s},j}, v^K_{\boldsymbol{s},j}\right) \\ + B_K(u^K_{\boldsymbol{i}}, v^K_{\boldsymbol{i}}) \,. \end{aligned} \tag{4.7.12}$$

Notice that the stiffness matrix corresponding to $C(u,v)$ is block diagonal and can hence be easily inverted. The form $\hat{C}(\cdot,\cdot)$ is defined analogously in terms of $\hat{B}(\cdot,\cdot)$. We now show that it is sufficient to prove the preconditioning inequality (4.7.5) elementwise.

Lemma 4.87 *Assume that $u \in S^{\mathbf{p},1}(\Omega, K)$ and that*

$$m_1^{\mathcal{T}}\, B_K(u^K, u^K) \le C_K(u^K, u^K) \le m_2^{\mathcal{T}}\, B_K(u^K, u^K) \tag{4.7.13}$$

for all $K \in \mathcal{T}$ with $m_1^{\mathcal{T}}, m_2^{\mathcal{T}}$ independent of K. Then (4.7.5) *holds with $m_1 = m_1^{\mathcal{T}}$, $m_2 = m_2^{\mathcal{T}}$.*

The proof is by summation of (4.7.12) over all $K \in \mathcal{T}$.

Thus, in the light of (4.7.8), it is sufficient to precondition the form $\hat{B}(u,v)$ on $\widehat{K}$ by a form $\hat{C}$. The local forms C_K in (4.7.12) are then obtained by transporting the form $\hat{C}$ back to $K \in \mathcal{T}$.

We henceforth consider therefore $\hat{B}(\cdot,\cdot)$ on $\mathcal{S}^p \times \mathcal{S}^p$ where $\mathcal{S}^p$ is generic notation for any of the reference spaces $\mathcal{S}^p_i(\widehat{K})$ and $p = p_K$. We have the following result.

Lemma 4.88 *There holds, for any $\hat{u} \in \mathcal{S}^p(\widehat{K})$,*

$$\hat{m}_1\, \hat{B}(\hat{u},\hat{u}) \le \hat{C}(\hat{u},\hat{u}) \le \hat{m}_2\, \hat{B}(\hat{u},\hat{u}) \ \text{ with} \tag{4.7.14}$$

with

$$\hat{m}_2/\hat{m}_1 \le (n+2)(b_1 + nb_2 + b_3)\,, \tag{4.7.15}$$

provided that in the decomposition

$$\hat{u} = \sum_{i=1}^{n} \hat{u}_{\boldsymbol{v},i} + \sum_{i=1}^{n} \hat{u}_{\boldsymbol{s},i} + \hat{u}_{\boldsymbol{i}}, \ \hat{u} \in \mathcal{S}^p(\widehat{K}) \tag{4.7.16}$$

the component energies are bounded as follows:

$$\left| \sum_{j=1}^{n} \hat{u}_{\boldsymbol{v},j} \right|^2_{H^1(\widehat{K})} \le b_1 \left| \hat{u} \right|^2_{H^1(\widehat{K})} , \tag{4.7.17}$$

$$\left| \hat{u}_{\boldsymbol{i},j} \right|^2_{H^1(\widehat{K})} \le b_2 \left| \hat{u} \right|^2_{H^1(\widehat{K})} , \quad j = 1, \dots, n \tag{4.7.18}$$

$$\left| \hat{u}_{\boldsymbol{i}} \right|^2_{H^1(\widehat{K})} \le b_3 \left| \hat{u} \right|^2_{H^1(\widehat{K})} . \tag{4.7.19}$$

Proof We have from the definition (4.7.11), (4.7.12) of $C(\cdot, \cdot)$ that

$$\begin{aligned}
\hat{B}(\hat{u}, \hat{u}) &= \hat{B}\left(\sum_{i=1}^{n} \hat{u}_{\boldsymbol{v},i} + \sum_{i=1}^{n} \hat{u}_{\boldsymbol{i},i} + \hat{u}_{\boldsymbol{i}}, \sum_{i=1}^{n} \hat{u}_{\boldsymbol{v},i} + \sum_{i=1}^{n} \hat{u}_{\boldsymbol{s},i} + \hat{u}_{\boldsymbol{i}} \right) \\
&\le (n+2)\left\{ \hat{B}\left(\sum_{i=1}^{n} \hat{u}_{\boldsymbol{v},i}, \sum_{i=1}^{n} \hat{u}_{\boldsymbol{v},i} \right) + \sum_{i=1}^{n} \hat{B}(\hat{u}_{\boldsymbol{s},i}, \hat{u}_{\boldsymbol{s},i}) + \hat{B}\left(\hat{u}_{\boldsymbol{i}}, \hat{u}_{\boldsymbol{i}}\right) \right\} \\
&= (n+2)\, \hat{C}\left(\hat{u}, \hat{u}\right) ,
\end{aligned}$$

which implies that $\hat{m}_1 \ge 1/(n+2)$.

Using the energy bound (4.7.17)–(4.7.19), we get

$$\hat{C}\left(\hat{u}, \hat{u}\right) \le (b_1 + n\, b_2 + b_3) \left| \hat{u} \right|^2_{H^1(\widehat{K})} = \hat{m}_2\, \hat{B}\left(\hat{u}, \hat{u}\right)$$

whence (4.7.14) follows. □

It remains therefore to estimate the b_i in (4.7.17) - (4.7.19) in dependence on the polynomial degree p. While this dependence can be very unfavourable for certain sets of shape functions, there are good sets for which the b_i are almost bounded.

Lemma 4.89 *Let* $\mathcal{S}^p(\widehat{K}) = \mathcal{S}_1^p(\widehat{T})$ *or* $\mathcal{S}_2^p(\widehat{Q})$ *and assume orthogonality of side and internal shape functions (or that the side shape functions are discretely harmonic):*

$$\hat{B}\left(\hat{u}, \hat{v}\right) = 0 \quad \text{for all } \ \hat{u} \in \widetilde{\Gamma}_i, \quad \text{for all } \ \hat{v} \in \widetilde{\mathcal{I}}, \ \ i = 1, \dots, n \, . \tag{4.7.20}$$

Then (4.7.17)–(4.7.19) *hold with*

$$b_i \le C\left(1 + \log p\right)^2 . \tag{4.7.21}$$

Proof For $\hat{u} \in \mathcal{S}^p(\widehat{K})$, we write $\hat{u} = \overline{u} + \lambda$ for some $\lambda \in \mathbb{R}$. Then

$$\overline{u} = \sum_{j=1}^{n} \overline{u}_{\boldsymbol{v},j} + \sum_{j=1}^{n} \overline{u}_{\boldsymbol{s},j} + \overline{u}_{\boldsymbol{i}}$$

where

$$\overline{u}_{\boldsymbol{i}} = \hat{u}_{\boldsymbol{i}},\ \overline{u}_{\boldsymbol{s},j} = \hat{u}_{\boldsymbol{s},j} \ \text{ and } \sum_{j=1}^{n} \overline{u}_{\boldsymbol{v},j} = \lambda + \sum_{j=1}^{n} \hat{u}_{\boldsymbol{v},j}\,.$$

In view of the fact that only seminorms appear on the left-hand sides of (4.6.16)–(4.7.19), it suffices to obtain the desired bounds for $\overline{u}$ instead of $\hat{u}$. Selecting λ suitably, we find from Poincaré's inequality on $\widehat{K}$ that

$$\|\hat{u}\|_{H^1(\widehat{K})} \leq C\,|\hat{u}|_{H^1(\widehat{K})}\,.$$

We then apply the discrete Sobolev inequality (4.6.3) to get

$$\|\hat{u}_{\boldsymbol{v},j}\|^2_{H^1(\widehat{K})} \leq C\,\|\hat{u}_{\boldsymbol{v},j}\|^2_{L^\infty(\widehat{K})} \leq C(1+\log p)\,|\hat{u}|^2_{H^1(\widehat{K})}\,,$$

which gives (4.7.17) with $b_1 \leq C(1+\log p)$. We next see

$$\hat{u}_1 = \hat{u} - \sum_{j=1}^{n} \hat{u}_{\boldsymbol{v},j}\,.$$

Then $\hat{u}_1$ vanishes at the vertices of $\widehat{K}$ and

$$\|\hat{u}_1\|^2_{H^1(\widehat{K})} \leq C\,(1+\log p)\,\|\hat{u}\|^2_{H^1(\widehat{K})}\,.$$

By (4.6.3), we get

$$\|\hat{u}_1\|^2_{L^\infty(\widehat{K})} \leq C\|\hat{u}\|^2_{L^\infty(K)} \leq C(1+\ln p)\|\hat{u}\|^2_{H^1(K)}$$

and, from Theorem 4.82, since $\hat{u}_1|_{\hat{\gamma}_i} \in \mathcal{S}^p(\hat{\gamma}_i)$,

$$\begin{aligned}\|\hat{u}_1\|^2_{H^{1/2}_{00}(\hat{\gamma}_i)} &\leq \|\hat{u}_1\|^2_{H^{1/2}(\hat{\gamma}_i)} + C\log(1+p)\,\|\hat{u}_1\|^2_{L^\infty(\hat{\gamma}_i)} \\ &\leq C(1+\log p)^2\,\|\hat{u}\|^2_{H^1(\widehat{K})}\,,\quad i=1,\ldots,n\,.\end{aligned}$$

By Theorems 4.84 and 4.85, there exists a trace lifting $\widetilde{u}_{\boldsymbol{s},j} \in \widetilde{\Gamma}_j$, $j=1,\ldots,n$, with $\widetilde{u}_{\boldsymbol{s},j}|_{\hat{\gamma}_j} = \hat{u}_1$ and

$$\|\widetilde{u}_{\boldsymbol{s},j}\|^2_{H^1(\widehat{K})} \leq C\,\|\hat{u}_1\|^2_{H^{1/2}_{00}(\hat{\gamma}_i)} \leq C(1+\log p)^2\,|\hat{u}|^2_{H^1(\widehat{K})},\ j=1,\ldots,n\,.$$

We have $\hat{u}_{\boldsymbol{s},j} \in \widetilde{\Gamma}_j$, $\hat{u}_{\boldsymbol{s},j} = \widetilde{u}_{\boldsymbol{s},j}$ on $\partial\widehat{K}$. Then $\hat{u}_{\boldsymbol{s},j} - \widetilde{u}_{\boldsymbol{s},j} \in \widetilde{\mathcal{I}}$ and (4.7.20) implies

$$\hat{B}(\hat{u}_{\boldsymbol{s},j}, \hat{u}_{\boldsymbol{s},j}) \leq \hat{B}\left(\widetilde{u}_{\boldsymbol{s},j}, \widetilde{u}_{\boldsymbol{s},j}\right).$$

This gives

$$\|\hat{u}_{\boldsymbol{s},j}\|^2_{H^1(\widehat{K})} \leq C\,\|\widetilde{u}_{\boldsymbol{s},j}\|^2_{H^1(\widehat{K})} \leq C(1+\log p)^2\,|\hat{u}|^2_{H^1(\widehat{K})},$$

i.e., $b_2 \leq C(1+\log p)^2$.

Since $\hat{u}_{\boldsymbol{i}} = \hat{u}_1 - \sum_{j=1}^n \hat{u}_{\boldsymbol{s},j}$, the triangle inequality implies also $b_3 \leq C(1+\log p)^2$. □

In exactly the same way we deduce the bound for the serendipity space $\mathcal{S}_3^p(\widehat{Q})$ from (4.6.21) instead of (4.6.20).

Lemma 4.90 *If* $\widehat{K} = \widehat{Q}$ *and* $\mathcal{S}^p(\widehat{K}) = \mathcal{S}_3^p(\widehat{Q})$, *we have* (4.7.17)–(4.7.19) *with*

$$b_i \leq Cp^2(1+\log p)^2. \tag{4.7.22}$$

We can now prove the following result.

Theorem 4.91 *Let* $\mathcal{T}$ *be any affine, shape-regular mesh and assume that* $\mathcal{S}^p(\widehat{K}) = \mathcal{S}_1^p(\widehat{T})$ *or* $\mathcal{S}_2^p(\widehat{Q})$. *Then if for every element* (4.7.19) *holds, we have for* $C(\cdot,\cdot)$ *in* (4.7.11) *the equivalence* (4.7.5) *with*

$$m_2/m_1 \leq C(1+\log|\boldsymbol{p}|)^2. \tag{4.7.23}$$

If some quadrilaterals are serendipity elements, we have instead

$$m_2/m_1 \leq C\,|\boldsymbol{p}|^2\,(1+\log|\boldsymbol{p}|)^2. \tag{4.7.24}$$

Here C *depends only on the shape regularity of the elements in* $\mathcal{T}$, *not on their size nor on* $\mathbf{p}$ *the degree distribution* $\mathbf{p} := \{p_K : k \in \tau\}$.

Remark 4.92 Notice that the form $C(\cdot,\cdot)$ is symmetric and positive definite, hence the problem (4.7.6) has a unique solution.

Remark 4.93 The orthogonality (4.7.20) can be achieved elementwise by forming the **Schur Complement** of $\mathcal{E}^{p_K}(K)$ with respect to $\mathcal{I}^{p_k}(K)$ (cf. Remark 4.39); this step can be performed completely in parallel. The resulting condensed global stiffness matrix has, by Theorem 4.91, a moderate condition number.

Remark 4.94 The solution of (4.7.6) requires the solution of a global system for the nodal unknowns (cf. (4.7.11)) and of some local systems; for the global, low order system, any of the known fast material can be applied. The local systems can be solved independently in parallel.

4.8 Bibliographical remarks and further results

Section 4.2: The basic result on $\mathcal{B}^2_\beta(\Omega)$-regularity for the solution of the Poisson equation with analytic right hand side in a straight polygon $\Omega \subset \mathbb{R}^2$ is due to Babuška and Guo [15]. Regularity results in the closely related spaces $\mathcal{C}^2_\beta$ were proved, using the deep results of Morrey on interior analyticity of solutions of strongly elliptic systems [97], in [16] for curvilinear polygons with piecewise analytic boundary. There also the corresponding FE-approximations and the exponential rates of convergence are studied.

Section 4.3: Results and proofs are from [19]. Analogous results can also be achieved using a characterization of the exact solution in different weighted spaces, see [7]. The extension to quadrilateral meshes in Section 4.3.4 does not involve any essential new ideas. The compactness result Lemma 4.19 and the resulting proof of Lemma 4.16 are generalizations of the classical **Bramble–Hilbert lemma** to the context of weighted Sobolev spaces (see, e.g., [38] or [45] for more on the Bramble–Hilbert lemma). Rather than resorting to weighted spaces one can also use the explicit decomposition (4.4.9) of the solution. Then all that is required is the Finite Element approximation of the singular functions S_{ij} in (4.4.6), (4.4.8).

Section 4.4: The construction of *hp*-FE subspaces and the first implementation of a *p*-version FEM for engineering (elasticity) problems go back to Szabo and his group in the mid 1970s [109].

For more on the shape functions and implementational aspects of *p*- and *hp*-FEM we refer to [144].

Section 4.5: The convergence analysis of *hp*-FEM presented here follows [71], where the exponential convergence of the *hp*-FEM in a two-dimensional setting was proved for the first time. All results proved were obtained for straight polygonal domains. Analogous results hold, however, also for curvilinear polygons; see [16] for details on this. It is interesting to note that for the convergence result on curved elements a slightly stronger regularity requirement on the solution has to be imposed, namely that $u \in \mathcal{C}^2_\beta(\Omega)$; here $\mathcal{C}^2_\beta$ is also a countably normed space defined in terms of sup- rather than L^2-norms (see [16]). In this context it is important that the curved elements can be mapped exactly onto the reference element. The construction of such element maps, based only on analytic expressions for the boundary curves of the element, is possible by transfinite blending, see [144] for details.

In addition, for curved domains the entries in the stiffness matrix must be approximated by numerical quadrature. This is addressed in the context of *hp*-FEM in part I of [94]. Roughly speaking, quadratures of exactness order $O(p_j)$ in element $K_j \in \Delta$ are sufficient to preserve the exponential convergence rates. Of particular interest is the case of quadrilateral elements with local tensor product polynomials of degree p, i.e. $\mathcal{S}^p_2$. Choosing here a $p+1$-point Gauss–Legendre or Gauss–Gauss–Lebatto rule in each direction, it is possible to select **Lagrange-type** shape functions which vanish in all but one quadrature point, thereby reducing the quadrature work for the element stiffness matrix. This is particularly advantageous for complicated element mappings with expensive evaluations of the Jacobian. If, in addition, the element mappings are affine, i.e., the Jacobians are constant, further gains can be achieved by analogs to the fast Fourier transform (we refer to [30] and to [43] for more on such spectral element techniques).

Section 4.6: Theorem 4.76 is classical, except for (4.6.4) which was obtained by [33]. Sections 6.2 and 6.3 follow [14].

Section 4.7: Theorem 4.91 was obtained in [14]. Further results, in particular on the combination of domain decomposition with the above p-preconditioning, can be found [2, 73].

5

FINITE ELEMENT ANALYSIS OF SADDLE POINT PROBLEMS. MIXED *HP*-FEM IN INCOMPRESSIBLE FLUID FLOW

5.1 Introduction. Navier–Stokes equations

The Navier–Stokes Equations (NSE) which describe the d-dimensional motion of an incompressible fluid with constant density $\rho > 0$ in a domain $\Omega \subset \mathbb{R}^d$ are as follows.

a) **Balance of momentum:**

$$\rho \left(\partial_t u_i + \sum_{j=1}^{d} u_j \partial_j u_i \right) - \sum_{j=1}^{d} \partial_j \sigma_{ij} = \rho f_i \ , \quad 1 \le i \le d \tag{5.1.1}$$

b) **Incompressibility constraint:**

$$\operatorname{div} \boldsymbol{u} = \sum_{i=1}^{d} \epsilon_{ii}(\boldsymbol{u}) = 0 \tag{5.1.2}$$

where the **strain** (or **deformation rate**) **tensor** $\{\epsilon_{ij}\}$ is given by

$$\epsilon_{ij}(\boldsymbol{u}) = \frac{1}{2} \left(\partial_j u_i + \partial_i u_j \right) , \quad 1 \le i, j \le d \, . \tag{5.1.3}$$

The **hydrodynamic stresses** σ_{ij} are related to the strains by the **constitutive equation**

$$\sigma_{ij} = -P \, \delta_{ij} + 2\mu \, \epsilon_{ij}(\boldsymbol{u}) \, . \tag{5.1.4}$$

Here $\boldsymbol{u} = (u_1, \ldots, u_d)$ is the velocity of the fluid and $\mu > 0$ its viscosity; P is the pressure[6] and $\boldsymbol{f} = (f_1, \ldots, f_d)$ are body forces per unit mass. In this chapter, $d = 2$.

We nondimensionalize the NSE. There are two canonical ways of doing this which lead to the same mathematical problem (the different nondimensionalization procedures become important when formulating multiphase flow or free boundary problems which we do not consider here).

[6]For the polynomial degree in p- or hp-FEM in fluid flow problems we shall use in this chapter to avoid confusion with the kinematic pressure the variable k rather than p.

a) **First Nondimensionalization (kinematic variables)**

Define **kinematic pressure** p and **kinematic viscosity** ν by

$$p = P/\rho\,, \quad \nu = \mu/\rho\,. \tag{5.1.5}$$

Then (5.1.1) becomes

$$\partial_t u_i + \sum_{j=1}^{d} u_j \partial_j u_i - 2\nu \sum_{j=1}^{d} \partial_j\, \epsilon_{ij}(\boldsymbol{u}) + \partial_i\, p = f_i\,, \quad 1 \le i \le d \tag{5.1.6}$$

$$\operatorname{div} \boldsymbol{u} = 0\,. \tag{5.1.7}$$

Note that (5.1.7) implies

$$\sum_{j=1}^{d} \partial_j\, \epsilon_{ij}(\boldsymbol{u}) = \frac{1}{2} \sum_{j=1}^{d} (\partial_j^2 u_i + \partial_i \partial_j u_j) = \frac{1}{2}\, \nabla^2 u_i\,, \tag{5.1.8}$$

hence (5.1.6) becomes

$$\begin{gathered} \partial_t \boldsymbol{u} + \sum_{j=1}^{d} u_j \partial_j \boldsymbol{u} - \nu \nabla^2 \boldsymbol{u} + \mathbf{grad}\, p = \boldsymbol{f}\,, \\ \operatorname{div} \boldsymbol{u} = 0\,. \end{gathered} \tag{5.1.9}$$

b) **Second nondimensionalization (Reynolds number)**

Let L and U be a characteristic length and velocity, respectively, of a given problem. This determines the characteristic time $T = L/U$ and we define nondimensional quantities

$$x' = x/L, \;\; \boldsymbol{u}' = \boldsymbol{u}/U, \;\; t' = t/T\,. \tag{5.1.10}$$

Defining

$$p' = P/(\rho U^2), \;\; \boldsymbol{f}' = (L/U^2)\, \boldsymbol{f}\,, \tag{5.1.11}$$

the NSE become

$$\begin{gathered} \partial_t \boldsymbol{u}' + \sum_{j=1}^{d} u_j' \partial_j' \boldsymbol{u}' - \frac{\mu}{LU}\, \nabla_{x'}^2\, \boldsymbol{u}' + \mathbf{grad}_{x'}\, p' = \boldsymbol{f}'\,, \\ \operatorname{div}_{x'} \boldsymbol{u}' = 0\,. \end{gathered}$$

If we define the **Reynolds number** $Re = LU/\mu$, the NSE become

$$\begin{gathered} \partial_t \boldsymbol{u}' + \sum_{j=1}^{d} u_j \partial_j \boldsymbol{u}' - \frac{1}{Re}\, \nabla^2\, \boldsymbol{u}' + \mathbf{grad}\, p' = \boldsymbol{f}\,, \\ \operatorname{div} \boldsymbol{u} = 0 \end{gathered} \tag{5.1.12}$$

i.e., we get again (5.1.9) with ν replaced by $1/Re$.

We omit the primes and introduce two basic simplifications into (5.1.12):

$$\text{steady state assumption:} \quad \partial_t \boldsymbol{u} = 0\,, \tag{5.1.13}$$

$$\text{linearization:} \quad \sum_{j=1}^{d} u_j \partial_j \boldsymbol{u} = 0\,. \tag{5.1.14}$$

Then (5.1.12) becomes the **Stokes equations**:

$$-\nu\nabla^2 \boldsymbol{u} + \mathbf{grad}\, p = \boldsymbol{f}, \qquad \operatorname{div} \boldsymbol{u} = 0 \text{ in } \Omega\,. \tag{5.1.15}$$

The equations (5.1.15) must be amended by suitable boundary conditions. The simplest among several possible choices are Dirichlet conditions:

$$\boldsymbol{u} = \boldsymbol{h}^0 \quad \text{on} \quad \Gamma = \partial\Omega\,. \tag{5.1.16}$$

If $\boldsymbol{h}^0 = 0$, (5.1.16) is also referred to as **no-slip conditions**.

Due to the incompressibility constraint $\operatorname{div} \boldsymbol{u} = 0$ in Ω, the prescribed velocity must satisfy a **compatibility condition** by Green's formula

$$0 = (\operatorname{div} \boldsymbol{u}, v) = -(\boldsymbol{u}, \mathbf{grad}\, v) + \langle\, (\gamma_0\, \boldsymbol{u}) \cdot \boldsymbol{n},\ \gamma_0\, v \rangle\,.$$

Selecting here $v \equiv 1$ results in

$$\int_\Gamma \boldsymbol{h}^0 \cdot \boldsymbol{n}\; do = 0\,. \tag{5.1.17}$$

Throughout this chapter, the exterior unit normal vector to Ω at x will be denoted by $\boldsymbol{n}(x)$ to distinguish it from the viscosity ν.

To cast the Stokes problem (5.1.15), (5.1.16) into variational form, assume for now that $\boldsymbol{h}^0 = 0$. Multiplying (5.1.15) by a test function $\boldsymbol{v}$ and integrating by parts gives

$$\left.\begin{array}{l} \text{Find } (\boldsymbol{u}, p) \in \boldsymbol{H}_0^1(\Omega) \times L_0^2(\Omega) \text{ such that} \\ \quad a(\boldsymbol{u}, \boldsymbol{v}) + b(\boldsymbol{v}, p) = (\boldsymbol{f}, \boldsymbol{v}) \quad \boldsymbol{v} \in \boldsymbol{H}_0^1(\Omega) \end{array}\right\}\,. \tag{5.1.18}$$

Here

$$\begin{aligned} a(\boldsymbol{u}, \boldsymbol{v}) &= \nu \sum_{i,j=1}^{d} (\partial_j u_i,\ \partial_j v_i) = \nu(\mathbf{grad}\, \boldsymbol{u},\ \mathbf{grad}\, \boldsymbol{v})\,, \\ b(\boldsymbol{v}, p) &= -(p, \operatorname{div} \boldsymbol{v}) \end{aligned}$$

and

$$L_0^2(\Omega) = \Big\{ p \in L^2(\Omega) : \int_\Omega p\, dx = 0 \Big\}\,.$$

The $d+1$ variables u_i, p are not determined by the d equations (5.1.18). We therefore augment (5.1.18) by the **incompressibility condition** (5.1.7) **in variational form**

$$b(\boldsymbol{u}, q) = 0\,, \qquad q \in L_0^2(\Omega)\,. \tag{5.1.19}$$

(5.1.18) and (5.1.19) constitute a saddle point problem which is the variational principle underlying so-called **mixed FEM**.

Exercise 5.1 Carry out the conversion of the Stokes problem (5.1.15) into its variational form (5.1.18) in detail. What are the homogeneous natural boundary conditions of the Stokes equations?

5.2 Abstract Saddle point problems. Inf–sup conditions

5.2.1 *Minimization problems with constraints*

Let X, M be Hilbert spaces with norms $\|\circ\|_X$ and $\|\circ\|_M$, respectively (all spaces will be over $\mathbb{R}$ unless stated otherwise), and duality pairings $\langle\cdot,\cdot\rangle_{X\times X'}$, $\langle\cdot,\cdot\rangle_{M\times M'}$, respectively (the subscripts at the duality brackets are omitted when their meaning is clear from the context). Let $a(\cdot,\cdot)$: $X\times X \to \mathbb{R}$ and $b: X\times M \to \mathbb{R}$ be continuous bilinear forms, $a(u,v) = a(v,u)$ for every $u, v \in X$. Given $f \in X'$ and $g \in M'$, we consider the **minimization problem**

$$\min_{v\in X}\{J(v)\},\quad J(v) := \frac{1}{2}\,a(v,v) - f(v) \tag{5.2.1}$$

subject to the **linear constraint**

$$b(v,q) = g(q) \text{ for all } q \in M\,. \tag{5.2.2}$$

Let $v \in X$ satisfy (5.2.2). Then

$$J(v) = \mathcal{L}(v,\lambda) := J(v) + b(v,\lambda) - g(\lambda) \text{ for all } \lambda \in M\,, \tag{5.2.3}$$

where $\mathcal{L}$ denotes the Lagrange function associated with (5.2.1), (5.2.2). Any solution of (5.2.1), (5.2.2) should correspond to a critical point of $\mathcal{L}$. Proceeding as in the proof of Corollary 1.25, we arrive at the **first-order necessary conditions**: Find

$$\left.\begin{array}{l} (u,\lambda) \in X \times M \text{ such that} \\ \qquad a(u,v) + b(v,\lambda) = f(v) \text{ for all } v \in X\,, \\ \qquad b(u,\mu) \qquad\qquad = g(\mu) \text{ for all } \mu \in M \end{array}\right\}. \tag{5.2.4}$$

If $(u,\lambda) \in X \times M$ solves the saddle point problem (5.2.4), u is a solution of (5.2.1), (5.2.2). To prove that u and the Lagrange multiplier λ exist requires certain assumptions on $a(\cdot,\cdot)$ and $b(\cdot,\cdot)$. In the remainder of this section we will be concerned with abstract saddle point problems of the type (5.2.4) which are, however, not necessarily related to a constrained minimization problem (5.2.1), (5.2.2). We adopt a more general setting than necessary for the Stokes problem, since numerous other FEM (such as mixed and hybrid FEM in solid mechanics) can be formulated in such a general setting, see [41].

5.2.2 *Abstract saddle point problems*

We consider now abstract versions of the saddle point problem (5.2.4). Let X_i, M_i, $i = 1, 2$, be reflexive Banach spaces and denote by $\langle \cdot, \cdot \rangle$ the corresponding duality pairings (it will be indicated at each pairing which spaces are meant). We consider the following type of saddle point problem. Given continuous bilinear forms

$$a : X_1 \times X_2 \to \mathbb{R}, \quad b_i : X_{3-i} \times M_i \to \mathbb{R}, \quad i = 1, 2\,, \tag{5.2.5}$$

the general, nonsymmetric variant of (5.2.4) reads as follows: find $(u, p) \in X_1 \times M_1$ such that

$$a(u, v) + b_1(v, p) = \langle f, v \rangle_{X_2' \times X_2} \text{ for all } v \in X_2\,, \tag{5.2.6a}$$

$$b_2(u, q) \qquad = \langle g, q \rangle_{M_2' \times M_2} \text{ for all } q \in M_2\,. \tag{5.2.6b}$$

By the Riesz representation theorem the bilinear forms a, b_i induce corresponding operators $A \in \mathcal{L}(X_1, X_2')$, $B_i \in \mathcal{L}(X_{3-i}, M_i')$, $i = 1, 2$, and their adjoints A^*, B_i^* such that

$$\begin{aligned} a(u, v) &= \langle Au, v \rangle_{X_2' \times X_2} = \langle u, A^* v \rangle_{X_1 \times X_1'} && \text{for all } u \in X_1,\ v \in X_2 \\ b_i(u, q) &= \langle B_i u, q \rangle_{M_i' \times M_i} = \langle u, B_i^* q \rangle_{X_{3-i} \times X_{3-i}'} && \text{for all } u \in X_{3-i},\ q \in M_i,\ i = 1, 2. \end{aligned}$$

This allows to write (5.2.6) also in the operator notation

$$Au + B_1^* p = f \text{ in } X_2' \tag{5.2.7a}$$

$$B_2 u \qquad = g \text{ in } M_2'\,. \tag{5.2.7b}$$

We are interested in **existence** and **a priori estimates** for solutions (u, p) of (5.2.6) and (5.2.7). Since (5.2.7) is a triangular system, this can be done in several steps:

1. particular solution u_g of $B_2 u_g = g$,
2. solution of (5.2.6a) and (5.2.7a) for u,
3. solution p for given u and u_g.

In each step, (a suitable version of) Theorem 1.15 is used. A general solvability theorem for (5.2.6), (5.2.7), due to Brezzi [40], will be obtained by combining estimates and assumptions of each step.

We begin our analysis by observing that for $i = 1, 2$

$$\begin{aligned} \ker B_i &= \{u \in X_{3-i} : b_i(u, q) = 0 \text{ for all } q \in M_i\} \subseteq X_{3-i} \\ \ker B_i^* &= \{q \in M_i : \quad b_i(u, q) = 0 \text{ for all } u \in X_{3-i}\} \subseteq M_i \end{aligned} \tag{5.2.8}$$

are closed, linear subspaces. By Proposition 1.18, the **factor spaces** $X_{3-i}/\ker B_i$, $M_i/\ker B_i^*$, $i = 1,2$, are Banach spaces equipped with the norms

$$\begin{aligned} \|u\|_{X_{3-i}/\ker B_i} &= \inf_{v \in \ker B_i} \|u + v\|_{X_{3-i}}, \\ \|p\|_{M_i/\ker B_i^*} &= \inf_{q \in \ker B_i^*} \|p + q\|_{M_i} . \end{aligned} \tag{5.2.9}$$

The following result from functional analysis, the **closed range theorem**, (see, e.g. [152]) will be used repeatedly.

Proposition 5.2 *Let $b : X \times M \to \mathbb{R}$ be a continuous bilinear form and $B \in \mathcal{L}(X, M')$, $B^* \in \mathcal{L}(M, X')$ be corresponding operators with $b(u,p) = \langle Bu, p\rangle_{M' \times M} = \langle u, B^*p\rangle_{X \times X'}$. Then the following are equivalent.*

i) *$R(B) \subset M'$ is closed*

ii) *$R(B^*) \subset X'$ is closed*

iii) $(\ker B)^\circ = R(B^*)$

iv) $(\ker B^*)^\circ = R(B)$

v) *there exists $\beta > 0$ such that for every $u \in X/\ker B$*

$$\sup_{p \in M} \frac{b(u,p)}{\|p\|_M} \geq \beta \, \|u\|_{X/\ker B} .$$

vi) *for the same $\beta > 0$ as in v) and every $p \in M/\ker B^*$*

$$\sup_{u \in X} \frac{b(u,p)}{\|u\|_X} \geq \beta \, \|p\|_{M/\ker B^*} .$$

We begin now with Step 1 of the analysis of the triangular system (5.2.7).

Proposition 5.3 *Assume that for any $q \in M_2/\ker B_2^*$*

$$\sup_{u \in X_1} \frac{b_2(u,q)}{\|u\|_{X_1}} \geq \beta_2 \, \|q\|_{M_2/\ker B_2^*} \tag{5.2.10}$$

with $\beta_2 > 0$ independent of q.

Then for every $g \in M_2'$ with $\langle g, q\rangle_{M_2' \times M_2} = 0$ for $q \in \ker B_2^*$ the problem: find $u \in X_1$ such that

$$b_2(u,q) = \langle g, q\rangle_{M_2' \times M_2} \quad \textit{for all} \quad q \in M_2 \tag{5.2.11}$$

admits a solution $u_g \in X_1$ which is unique modulo $\ker B_2$ and which satisfies the **a priori estimate**

$$\|u_g\|_{X_1/\ker B_2} \leq \frac{1}{\beta_2} \, \|g\|_{M_2'} . \tag{5.2.12}$$

Proof Put $X = X_1/\ker B_2$, $Y = M_2/\ker B_2^*$. (5.2.10) implies that for every $0 \neq q \in Y$, $\sup_{u\in X} b_2(u,q) > 0$. By v) and vi) of Proposition 5.2, (5.2.10) implies also

$$\sup_{q\in M_2} \frac{b_2(u,q)}{\|q\|_{M_2}} \geq \beta_2 \|u\|_X , \quad u \in X ,$$

or, equivalently

$$\inf_{u\in X} \sup_{q\in Y} \frac{b_2(u,q)}{\|u\|_X \, \|q\|_Y} \geq \beta_2 > 0 .$$

We may therefore apply Theorem 1.20 and deduce that for every $g \in Y'$ there exists a unique u_g solution of (5.2.11). Now $Y' = (M_2/\ker B_2^*)' = (\ker B_2^*)^\circ$, i.e. $g \in Y'$ satisfies $\langle g, q\rangle_{M_2'\times M_2} = 0$ for every $q \in \ker B_2^*$. □

We proceed to Step 2, i.e., the solution of (5.2.6a) or (5.2.7a) for $u \in X_1$ which also satisfies (5.2.6b), (5.2.7b). To this end we use the nonuniqueness of u_g: any $u = u_0 + u_g$ with $u_0 \in \ker B_2$ also solves (5.2.6b). We therefore seek $u_0 \in \ker B_2$ such that

$$a(u_0, v) = \langle f, v\rangle_{X_2'\times X_2} - \langle B_1 v, p\rangle_{M_1'\times M_1} - a(u_g, v) \text{ for all } v \in X_2 .$$

To obtain an equation for u_0 alone, we restrict the test function v to $\ker B_1$. This yields the problem: find $u_0 \in \ker B_2$ such that

$$a(u_0, v) = L(v) := \langle f, v\rangle_{X_2'\times X_2} - a(u_g, v) \text{ for all } v \in \ker B_1 . \tag{5.2.13}$$

Proposition 5.4 *Assume that*

$$\inf_{u\in\ker B_2} \sup_{v\in\ker B_1} \frac{a(u,v)}{\|u\|_{X_1} \, \|v\|_{X_2}} \geq \alpha > 0 \tag{5.2.14}$$

and that for every $0 \neq v \in \ker B_1$

$$\sup_{u\in\ker B_2} a(u,v) > 0 . \tag{5.2.15}$$

Then (5.2.13) *admits a unique solution* $u_0 \in \ker B_2$ *and*

$$\|u_0\|_{X_1/\ker B_2} \leq \frac{1}{\alpha} \left\{ \|f\|_{X_2'} + \frac{\|a\|}{\beta_2} \|g\|_{M_2'} \right\} . \tag{5.2.16}$$

Proof Since $\ker B_i \subset X_{3-i}$ are closed subspaces, they are themselves reflexive Banach spaces. Therefore Theorem 1.20 yields the assertion and the a priori estimate

$$\|u_0\|_{X_1/\ker B_2} \leq \alpha^{-1} \sup_{v\in\ker B_1} |L(v)|/\|v\|_{X_2} \leq \alpha^{-1} \|L\|_{X_2'} .$$

From (5.2.13), $\|L\|_{X_2'} \le \|f\|_{X_2'} + \|a\|\ \|u_g\|_{X_1/\ker B_2}$ and referring to (5.2.12) completes the proof. □

We also have from (5.2.12) and (5.2.16) that

$$\begin{aligned}\|u\|_{X_1/\mathrm{kerB}_2} &\le \|u_0\|_{X_1/\mathrm{kerB}_2} + \|u_g\|_{X_1/\ker B_2} \\ &\le \frac{1}{\alpha}\|f\|_{X_2'} + \frac{1}{\beta_2}\left(1+\frac{\|a\|}{\alpha}\right)\|g\|_{M_2'}\,.\end{aligned} \tag{5.2.17}$$

Having obtained $u \in X_1$, we finally determine p from (5.2.6a), (5.2.7a). To this end, we need a result which is similar to Proposition 5.3.

Proposition 5.5 *Assume that for any $p \in M_1/\ker B_1^*$*

$$\sup_{v\in X_2} \frac{b_1(v,p)}{\|v\|_{X_2}} \ge \beta_1 \|p\|_{M_1/\ker B_1^*} \tag{5.2.18}$$

with $\beta_1 > 0$ independent of p. Then there exists a unique solution of the problem: find $p \in M_1/\ker B_1^*$ such that

$$b_1(v,p) = L(v) := \langle f, v\rangle_{X_2'\times X_2} - a(u,v) \text{ for all } v \in X_2 \tag{5.2.19}$$

and we have the a priori estimate

$$\|p\|_{M_1/\ker B_1^*} \le \frac{1}{\beta_1}\{\|f\|_{X_2'} + \|a\|\ \|u\|_{X_1}\}\,. \tag{5.2.20}$$

Proof Put $X = M_1/\ker B_1^*$, $Y = X_2/\ker B_1$ and $B(p,v) = b_1(v,p)$. Then (5.2.18) says that $B(\cdot,\cdot)$ satisfies the inf-sup condition (1.3.12). From v) and vi) of Proposition 5.2, we also see that for every $v \in Y$

$$\sup_{p\in M_1} \frac{b_1(v,p)}{\|p\|_{M_1}} \ge \beta_1 \|v\|_{X_2/\ker B_1}$$

which implies that for every $0 \ne v \in Y$,

$$\sup_{p\in X} B(p,v) > 0\,.$$

Hence Theorem 1.20 applies and (5.2.19) admits a unique solution $p \in X$ for every $L \in Y' = (X_2/\ker \mathrm{B}_1)' = (\ker B_1)^\circ$. Let us verify that L in (5.2.19) belongs to $(\ker B_1)^\circ$. By (5.2.13),

$$L(v) = \langle f, v\rangle_{X_2'\times X_2} - a(u_0,v) - a(u_g,v) = 0 \text{ for all } v \in \ker B_1$$

i.e., L in (5.2.19) belongs to $(\ker B_1)^\circ$. The proof is complete. □

We combine Propositions 5.3–5.5 into the following result.

Theorem 5.6 *Consider the problem* (5.2.6), (5.2.7). *Assume that*

$$\sup_{v \in X_2} \frac{b_1(v,p)}{\|v\|_{X_2}} \geq \beta_1 \|p\|_{M_1/\ker B_1^*} \tag{5.2.21}$$

for every $p \in M_1/\ker B_1^*$ *and that*

$$\sup_{u \in X_1} \frac{b_2(u,q)}{\|u\|_{X_1}} \geq \beta_2 \|q\|_{M_2/\ker B_2^*} \tag{5.2.22}$$

for every $q \in M_2/\ker B_2^*$. *Assume further that*

$$\inf_{u \in \ker B_2} \sup_{v \in \ker B_1} \frac{a(u,v)}{\|u\|_{X_1} \|v\|_{X_2}} \geq \alpha > 0 \tag{5.2.23}$$

and that for every $0 \neq v \in \ker B_1$

$$\sup_{u \in \ker B_2} a(u,v) > 0\,. \tag{5.2.24}$$

Then, for every $f \in X_2'$ *and* $g \in (\ker B_2^*)^\circ \subset M_2'$, *there exists a solution* $(u,p) \in X_1 \times M_1$ *of* (5.2.6), (5.2.7), *with* u *unique and* p *unique up to* $\ker B_1^*$. *Moreover, the following a priori estimates hold:*

$$\|u\|_{X_1} \leq \frac{1}{\alpha} \|f\|_{X_2'} + \frac{1}{\beta_2}\left(1 + \frac{\|a\|}{\alpha}\right)\|g\|_{M_2'} \tag{5.2.25}$$

and

$$\|p\|_{M_1/\ker B_1^*} \leq \frac{1}{\beta_1}\left(1 + \frac{\|a\|}{\alpha}\right)\left(\|f\|_{X_2'} + \frac{\|a\|}{\beta_2}\|g\|_{M_2'}\right). \tag{5.2.26}$$

We list an important special case, namely when $X_1 = X_2 = X$, $M_1 = M_2 = M$ are Hilbert spaces and $b_1 = b_2 = b$, separately for future reference.

Corollary 5.7 *Let* X, M *be Hilbert spaces and* $a : X \times X \to \mathbb{R}$, $b : X \times M \to \mathbb{R}$ *be continuous bilinear forms with corresponding operators* $A \in \mathcal{L}(X,X')$, $B \in \mathcal{L}(X,M')$. *Assume further that*

$$\inf_{u \in \ker B} \sup_{v \in \ker B} \frac{a(u,v)}{\|u\|_X \|v\|_X} \geq \alpha > 0 \tag{5.2.27}$$

and, for every $0 \neq v \in \ker B$

$$\sup_{u \in \ker B} a(u,v) > 0\,. \tag{5.2.28}$$

Finally, assume that $b(\cdot,\cdot)$ *is stable, i.e.,*

$$\sup_{u\in X} \frac{b(u,p)}{\|u\|_X} \geq \beta \|p\|_{M/\ker B^*} \text{ for all } p \in M . \tag{5.2.29}$$

Then, for every $f \in X'$, $g \in R(B) = (\ker B^*)^\circ \subset M'$ *there exists a solution* $(u,p) \in X \times M$ *of*

$$\begin{aligned} a(u,v) + b(v,p) &= \langle f, v\rangle_{X'\times X} \text{ for all } v \in X , \\ b(u,q) &= \langle g, q\rangle_{M'\times M} \text{ for all } v \in M \end{aligned} \tag{5.2.30}$$

and the a priori estimates

$$\|u\|_X \leq \frac{1}{\alpha}\|f\|_{X'} + \frac{1}{\beta}\left(1 + \frac{\|a\|}{\alpha}\right)\|g\|_{M'} \tag{5.2.31}$$

$$\|p\|_{M/\ker B^*} \leq \frac{1}{\beta}\left(1 + \frac{\|a\|}{\alpha}\right)\|f\|_{X'} + \frac{\|a\|}{\beta^2}\left(1 + \frac{\|a\|}{\alpha}\right)\|g\|_{M'} \tag{5.2.32}$$

hold. p *is unique modulo* $\ker B^*$.

Remark 5.8 If $a(\cdot,\cdot)$ is coercive on $\ker B$, i.e., there is $\alpha > 0$ such that

$$a(u,u) \geq \alpha \|u\|_X^2 , \quad u \in \ker B , \tag{5.2.33}$$

both (5.2.27) and (5.2.28) follow.

Remark 5.9 The problem (5.2.6) could have been reduced to the setting of Chapter 1 by introducing the bilinear form

$$B(U,V) := a(\boldsymbol{u},\boldsymbol{v}) + b_1(\boldsymbol{v},p) + b_2(\boldsymbol{u},q)$$

with $U = (\boldsymbol{u},p)$, $V = (\boldsymbol{v},q)$ defined on the pair $X \times Y$ given by

$$X = (X_1, M_1),\ Y = (X_2, M_2) \ .$$

The analysis in the present section then constitutes essentially a proof of the inf–sup condition for $B(\cdot,\cdot)$ on $X \times Y$. Nevertheless, the detailed analysis presented here yields the estimates (5.2.25), (5.2.26) which are sharper than what could be obtained from a direct analysis of $B(\cdot,\cdot)$ with $\|U\|_X = \|\boldsymbol{u}\|_{X_1} + \|p\|_{M_1}$ etc.

The analysis of the FE approximation of (5.2.4) will follow the same pattern.

5.3 Existence for the Stokes problem

We apply Corollary 5.7 to the Stokes problem (5.1.15) in the variational form (5.1.18), (5.1.19) which we recall here for convenience: find $(\boldsymbol{u}, p) \in \boldsymbol{H}_0^1(\Omega) \times L_0^2(\Omega)$ such that

$$\begin{aligned} a(\boldsymbol{u}, \boldsymbol{v}) + b(\boldsymbol{v}, p) &= (\boldsymbol{f}, \boldsymbol{v}), \quad \boldsymbol{v} \in \boldsymbol{H}_0^1(\Omega), \\ b(\boldsymbol{u}, q) &= 0, \qquad\quad q \in L_0^2(\Omega) \end{aligned} \tag{5.3.1}$$

where $a(\boldsymbol{u}, \boldsymbol{v}) = \nu\,(\mathbf{grad}\,\boldsymbol{u}, \ \mathbf{grad}\,\boldsymbol{v})$, $b(\boldsymbol{v}, q) = -(q, \operatorname{div} \boldsymbol{v})$, and $L_0^2 = \{p \in L^2(\Omega) : (p, 1) = 0\}$. As usual, $(\cdot,\cdot)$ denotes the $L^2(\Omega)$-inner product (taken componentwise for vector and matrix functions).

To prove existence of a solution of (5.3.1), we must verify (5.2.27), (5.2.29) (condition (5.2.28) follows from (5.2.27) since $a(\cdot,\cdot)$ is symmetric). We have $X = \boldsymbol{H}_0^1(\Omega)$, $M = L^2(\Omega)$. Then $B = -\text{div}$ and $B^* = \mathbf{grad}$. Hence

$$\begin{aligned} \{\boldsymbol{u} \in \boldsymbol{H}_0^1(\Omega) : \operatorname{div} \boldsymbol{u} = 0 \ \text{ in } \ L^2(\Omega)\} &= \ker B \\ \{p \in L^2(\Omega) : \ p = \text{const}\} &= \ker B^* . \end{aligned} \tag{5.3.2}$$

Therefore $M/\ker B^* = L_0^2(\Omega)$. Since for $\boldsymbol{u} \in X = \boldsymbol{H}_0^1(\Omega)$

$$a(\boldsymbol{u}, \boldsymbol{u}) = \nu(\mathbf{grad}\,\boldsymbol{u}, \mathbf{grad}\,\boldsymbol{u}) = \nu\,|\boldsymbol{u}|_{H^1(\Omega)}^2 \geq \nu C(\Omega)\,||\boldsymbol{u}||_{H^1(\Omega)}^2$$

by the first Poincaré inequality, $a(\cdot,\cdot)$ is in particular coercive on $\ker B$ and (5.2.27) holds with $\alpha = \nu C(\Omega)$. Thus, according to Remark 5.8 (5.2.27) and (5.2.28) are verified and the crucial condition to verify is (5.2.29) which is equivalent to

$$\sup_{0 \neq \boldsymbol{v} \in \boldsymbol{H}_0^1(\Omega)} \frac{(q, \operatorname{div} \boldsymbol{v})}{||\boldsymbol{v}||_{H_0^1(\Omega)}} \geq \beta\,||q||_{L^2(\Omega)}, \quad q \in L_0^2(\Omega) \tag{5.3.3}$$

for some $\beta > 0$ independent of q. (5.3.3) is also referred to as **divergence stability** of the pair $\boldsymbol{H}_0^1(\Omega) \times L_0^2(\Omega)$.

To prove (5.3.3), we use the following result.

Theorem 5.10 *Let $\Omega \subset \mathbb{R}^d$ be bounded and connected and let $\Gamma = \partial\Omega$ be Lipschitz. Then there holds*

i) *the range of* $\mathbf{grad}$*:* $L^2(\Omega) \to \boldsymbol{H}^{-1}(\Omega)$ *is closed in* $\boldsymbol{H}^{-1}(\Omega)$.
ii) *there exists* $c(\Omega)$ *such that*

$$||p||_{L^2(\Omega)} \leq c(\Omega)(||\mathbf{grad}\,p||_{H^{-1}(\Omega)} + ||p||_{H^{-1}(\Omega)}) \ \textit{for all} \ \ p \in L^2(\Omega), \tag{5.3.4}$$

$$||p||_{L^2(\Omega)} \leq c(\Omega)||\mathbf{grad}\,p||_{H^{-1}(\Omega)} \qquad \textit{for all} \ \ p \in L_0^2(\Omega). \tag{5.3.5}$$

The proof of Theorem 5.10 is technical and can be found, for example, in [99].

We now show how (5.3.5) implies (5.3.3). To this end we use the following general scheme outlined in Remark 1.16.

Given $q \in L_0^2(\Omega)$, find $\boldsymbol{v}_q \in H_0^1(\Omega)$ such that $\|\boldsymbol{v}_q\|_{H^1(\Omega)} \leq \alpha_1 \|q\|_{L^2(\Omega)}$ and such that $b(\boldsymbol{v}_q) = -(q, \operatorname{div} \boldsymbol{v}_q) \geq \alpha_2 \|q\|^2_{L^2(\Omega)}$.

Let $q \in L_0^2(\Omega)$. Then, by (5.3.5), $\|\mathbf{grad}\, q\|_{\mathbf{H}^{-1}(\Omega)} \geq C\, \|q\|_{L^2(\Omega)}$ and by the definition of the $\boldsymbol{H}^{-1}(\Omega)$-norm, there exists $\boldsymbol{v}_q \in \boldsymbol{H}_0^1(\Omega)$, $\|\boldsymbol{v}_q\|_{H^1(\Omega)} = 1$, such that

$$
\begin{aligned}
C\, \|q\|_{L^2(\Omega)} \leq \|\mathbf{grad}\, q\|_{H^{-1}(\Omega)} &= \langle \boldsymbol{v}_q, \mathbf{grad}\, q \rangle_{H_0^1 \times H^{-1}} \\
&= -\int_\Omega q \operatorname{div} \boldsymbol{v}_q \, dx \\
&= b(q, \boldsymbol{v}_q)
\end{aligned}
$$

which implies (5.3.3).

5.4 Finite element discretization of abstract saddle point problems

The FE discretization of the general saddle point problem (5.2.30) proceeds in the usual way. Given finite dimensional subspaces $X_N \subset X$ and $M_N \subset M$ of the Hilbert spaces X and M: **find** $(u_N, p_N) \in X_N \times M_N$ **such that**

$$
a(u_N, v) + b(v, p_N) = \langle f, v \rangle_{X' \times X}, \quad v \in X_N, \tag{5.4.1a}
$$

$$
b(u_N, q) = \langle g, q \rangle_{M' \times M}, \quad q \in M_N. \tag{5.4.1b}
$$

We will be interested in **existence** and **uniqueness** of u_N, p_N and in **error estimates** for $\|u - u_N\|_X$, $\|p - p_N\|_M$. Since (5.4.1) is a problem of the type (5.2.30), we expect the existence theory from the continuous case to apply here as well. This is essentially the case, although care must be taken to identify analogs of $\ker B$, $\ker B^*$.

By the Riesz representation theorem, there exist $A_N \in \mathcal{L}(X_N, X_N')$, $B_N \in \mathcal{L}(X_N, M_N')$ such that (5.4.1) can formally be written as

$$
\begin{aligned}
A_N u_N + B_N^* p &= P_{X_N} f, \\
B_N u_N &= P_{M_N} g,
\end{aligned} \tag{5.4.2}
$$

where $P_{X_N} : X \to X_N$, $P_{M_N} : M \to M_N$ denote orthogonal projections.

A word must be said about X_N', M_N'. The dual M_N' can be identified with a subspace of M' as follows: let $d(\cdot,\cdot)$ be a bilinear form on $M_N \times M_N$. Then $d(\cdot,\cdot)$ can be extended to a bilinear form on $M \times M$ by putting $d(u, v) = 0$ if either u or v are in $M_N^\perp \subseteq M$. Therefore, for $g_N' \in M_N'$:

$$
\langle g_N', q \rangle_{M' \times M} := \langle g_N', P_{M_N} q \rangle.
$$

For $g \in M'$, the natural definition of $P_{M'_N}\, g$ is also by duality:

$$\langle P_{M'_N}\, g, q\rangle_{M'\times M} = \langle g, P_{M_N}\, q\rangle_{M'\times M} = \langle P^*_{M_N}\, g, q\rangle_{M'\times M}\;. \tag{5.4.3}$$

Similar remarks hold for X'_N.

The operator B_N acts therefore from X_N into M'. However, B_N **does generally not coincide with the restriction of** B **to** X_N since for $v_N \in X_N$ and $q \in M'$

$$\begin{aligned} b(v_N, q) &= \langle B_N\, v_N, q\rangle_{M'\times M} = \langle B_N\, v_N, P_{M_N}\, q\rangle_{M'_N\times M_N} \\ &= b(v_N, P_{M_N}\, q) \qquad = \langle B\, v_N, P_{M_N}\, q\rangle_{M'_N\times M}\;. \end{aligned} \tag{5.4.4}$$

Since $P_{M'_N} = P^*_{M_N}$ by (5.4.3), we have by (5.4.4)

$$B_N\, v_N = P^*_{M_N}\, Bv_N = P_{M'_N}\, Bv_N \text{ for all } \; v_N \in X_N\;. \tag{5.4.5}$$

Hence B_N is the restriction of B only if $BX_N \subset M'_N$. This is generally not the case, however.

With this in mind, we define

$$Z_N(g) := \{v_N \in X_N : b(v_N, q) = \langle g, q\rangle_{M'\times M} \text{ for all } \; q \in M_N\}\;. \tag{5.4.6}$$

Then

$$Z_N(0) = \ker B_N \tag{5.4.7}$$

and we also have

$$\ker B^*_N = \{q \in M_N : b(v_N, q) = 0 \text{ for all } \; v_N \in X_N\}\;. \tag{5.4.8}$$

For the existence of FE solutions u_N, p_N we now reason as in the proof of Propositions 5.3–5.5.

For a solution of (5.4.1b) to exist, we must have $Z_N(g) \neq \varnothing$. Since the problem (5.4.1b) is finite dimensional, (5.2.22) holds then with a constant $\beta_2(N) > 0$ (since $R(B_N)$ is always closed in the finite-dimensional case, this follows from Proposition 5.2). Since $b_1 = b_2 = b$ here, (5.2.21) is satisfied as well. The only conditions to be checked are therefore (5.2.23), (5.2.24). An application of Theorem 5.6 now settles the question of existence of FE solutions (u_N, p_N).

Theorem 5.11 *Consider the problem* (5.4.1) *and assume* $Z_N(g) \neq \varnothing$. *Assume further that*

$$\inf_{u\in\ker B_N}\; \sup_{v\in\ker B_N} \frac{a(u,v)}{\|u\|_X\,\|v\|_X} \geq \alpha(N) > 0 \tag{5.4.9}$$

and that for every $0 \neq v \in \ker B_N$

$$\sup_{u\in\ker B_N} \; a(u,v) > 0\;. \tag{5.4.10}$$

Then for every $f \in X'$, $g \in M'$ *there exists a solution* (u_N, p_N) *of* (5.4.1), *with* u_N *unique in* X_N, p_N *unique modulo* $\ker B_N^*$.

Remark 5.12 In practice, (5.4.9) and (5.4.10) will follow from coercivity of $a(\cdot,\cdot)$ on $\ker B_N$, i.e.,

$$a(u,u) \geq \alpha(N)\, \|u\|_X^2 \quad \text{for all} \quad u \in \ker B_N \,. \tag{5.4.11}$$

Since in general $\ker B_N$ is not contained in $\ker B$, (5.4.11) cannot be deduced from (5.2.33), but must be checked independently, even if $a(\cdot,\cdot)$ is symmetric.

Remark 5.13 An important special case where (5.4.11) (and hence also (5.4.9), (5.4.10)) is satisfied for any selection of X_N, M_N is the **Stokes problem** (see Section 5.3). Here $X = \boldsymbol{H}_0^1(\Omega)$ and

$$a(\boldsymbol{u},\boldsymbol{u}) = \nu\, (\mathbf{grad}\, \boldsymbol{u}, \mathbf{grad}\, \boldsymbol{u}) \geq \nu C(\Omega) \|\boldsymbol{u}\|_{H^1(\Omega)}^2$$

for all $\boldsymbol{u} \in \boldsymbol{H}_0^1(\Omega)$. Thus (5.4.11) holds with $\alpha(N) = \nu C(\Omega)$ for any $\ker B_N \subset \boldsymbol{H}_0^1(\Omega)$. Note also that here $\alpha(N)$ is in fact independent of N.

Exercise 5.14 Let $Z_N(g)$ be as in (5.4.6). Prove that the following statements are equivalent:

i) for all $g \in R(B) : \ Z_N(g) \neq \varnothing$.

ii) For every $u \in X$ there exists $u_N \in X_N$ such that

$$b(u - u_N, q) = 0 \quad \text{for all} \quad q \in M_N \,.$$

iii) $\ker B_N^* = \ker B^* \cap M_N$.

Let us now turn to **error estimates** for u_N and p_N. We derive them in several steps, similar to what we did in Propositions 5.3–5.5. We begin with $\|u - u_N\|_X$.

Lemma 5.15 *Let* (u,p) *be a solution of* (5.2.30). *Assume* $Z_N(g) \neq \varnothing$ *and* (5.4.9), (5.4.10). *Let* (u_N, p_N) *solve* (5.4.1). *Then there holds*

$$\begin{aligned} \|u - u_N\|_X \leq \left(1 + \frac{\|a\|}{\alpha(N)}\right) & \inf_{v \in Z_N(g)} \|u - v\|_X \\ + \frac{\|b\|}{\alpha(N)} & \inf_{q \in M_N} \|p - q\|_M \,. \end{aligned} \tag{5.4.12}$$

If moreover $\ker B_N \subset \ker B$,

$$\|u - u_N\|_X \leq \left(1 + \frac{\|a\|}{\alpha(N)}\right) \inf_{v \in Z_N(g)} \|u - v\|_X \,. \tag{5.4.13}$$

Proof Let $v \in Z_N(g)$ be arbitrary. Then $u_N - v \in \ker B_N$, hence (5.4.9) implies

$$
\begin{aligned}
\alpha(N)\,\|u_N - v\|_X &\le \sup_{v_N \in \ker B_N} \frac{a(u_N - v, v_N)}{\|v_N\|_X} \\
&= \sup_{v_N \in \ker B_N} \frac{a(u_N - u, v_N) + a(u - v, v_N)}{\|v_N\|_X} \\
&= \sup_{v_N \in \ker B_N} \frac{a(u - v, v_N) + b(v_N, p - p_N)}{\|v_N\|_X}
\end{aligned}
$$

where we used that $a(u - u_N, v) + b(v, p - p_N) = 0$ for every $v \in X_N$.

If $\ker B_N \subset \ker B$, we have $b(v_N, p - p_N) = 0$ since $v_N \in \ker B$ and $\alpha(N)\|u_N - v\|_X \le \|a\|\,\|u - v\|_X$ and (5.4.13) follows.

If $\ker B_N \not\subset \ker B$ and $b(v_N, q) = 0$ for any $q \in M_N$, we get

$$|b(v_N, p - p_N)| = |b(v_N, p - q)| \le \|b\|\,\|v_N\|_X\,\|p - q\|_M\,.$$

Altogether, for any $v \in Z_N(g)$, $q \in M_N$ we have

$$\alpha(N)\,\|u_N - v\|_X \le \|a\|\,\|u - v\|_X + \|b\|\,\|p - q\|_M\,.$$

This implies (5.4.12). □

To obtain error estimates, it remains to relate $\inf_{v \in Z_N(g)} \|u - v\|_X$ to $\inf_{v \in X_N} \|u - v\|_X$. This will be achieved using the discrete inf-sup condition for $b(\cdot,\cdot)$.

Lemma 5.16 *In the setting of Lemma* 5.15, *assume in addition that for every* $q \in M_N$

$$\sup_{v \in X_N} \frac{b(v, q)}{\|v\|_X} \ge \beta(N)\,\|q\|_{M/\ker B_N^*}\,. \tag{5.4.14}$$

Then

$$
\begin{aligned}
\inf_{v \in Z_N(g)} \|u - v\|_X &\le \inf_{v \in X_N} \left\{ \|u - v\|_X + \frac{1}{\beta(N)} \sup_{q \in M_N} \frac{b(u - v, q)}{\|q\|_M} \right\} \\
&\le \left(1 + \frac{\|b\|}{\beta(N)}\right) \inf_{v \in X_N} \|u - v\|_X\,.
\end{aligned}
\tag{5.4.15}
$$

Proof Pick any $v_N \in X_N$ and find $r_N \in X_N$ such that

$$b(r_N, q) = b(u - v_N, q) \text{ for all } q \in M_N\,.$$

As $Z_N(g) \neq \varnothing$ by assumption, r_N exists and, by (5.4.14) and the continuity of $b(\cdot,\cdot)$, we have the a-priori estimate

$$\|r_N\|_X \le \frac{1}{\beta(N)} \sup_{q \in M_N} \frac{b(u - v_N, q)}{\|q\|_M} \le \frac{1}{\beta(N)}\,\|b\|\,\|u - v_N\|_X\,.$$

Moreover, we verify

$$b(r_N + v_N, q) = b(u, q) = \langle g, q\rangle_{M'\times M} \text{ for all } q \in M_N .$$

Hence $r_N + v_N \in Z_N(g)$. Therefore

$$\inf_{v\in Z_N(g)} \|u - v\|_X \le \|u - (r_N + v_N)\|_X \le \|u - v_N\|_X + \|r_N\|_X$$

which implies (5.4.15), since $v_N \in X_N$ was arbitrary. □

Remark 5.17 In the context of the Stokes problem, condition (5.4.14) with $\beta(N)$ independent of N reads

$$\sup_{\boldsymbol{v}\in X_N} \frac{\int_\Omega q \operatorname{div} \boldsymbol{v}\, dx}{\|\boldsymbol{v}\|_{H^1(\Omega)}} \ge \beta \|q\|_{L^2(\Omega)/\ker B_N^*} \tag{5.4.16}$$

for every $q \in M_N$. (5.4.16) is sometimes called **Babuska–Brezzi condition** or also **divergence stability** of the pair (X_N, M_N).

The a priori error estimates (5.4.12) and (5.4.15) give quasioptimality for $\|u-u_N\|_X$ in terms of the approximability of u and p from X_N and M_N, respectively. We turn now to a priori estimates for $\|p - p_N\|_M$.

Lemma 5.18 *Assumptions as in Lemmas* 5.15 *and* 5.16. *Then*

$$\|p - p_N\|_{M/\ker B_N^*} \le \left(1 + \frac{\|b\|}{\beta(N)}\right) \inf_{q\in M_N} \|p - q\|_M + \frac{\|a\|}{\beta(N)} \|u - u_N\|_X . \tag{5.4.17}$$

Proof We have for every $v \in X_N$

$$a(u - u_N, v) + b(v, p - p_N) = 0$$

hence for $q \in M_N$

$$b(v, q - p_N) = -a(u - u_N, v) - b(v, p - q) .$$

By (5.4.14), we get

$$\begin{aligned}
\|p_N - q\|_{M/\ker B_N^*} &\le \frac{1}{\beta(N)} \sup_{v\in X_N} \frac{b(v, p_N - q)}{\|v\|_X} \\
&= \frac{1}{\beta(N)} \sup_{v\in X_N} \frac{b(v, p - q) + a(u - u_N, v)}{\|v\|_X} \\
&\le \frac{1}{\beta(N)} (\|b\| \, \|p - q\| + \|a\| \, \|u - u_N\|) .
\end{aligned}$$

□

We combine the Lemmas 5.15, 5.16 and 5.18 into the following result.

Theorem 5.19 *Let (u,p) be a solution of* (5.2.30) *and assume $Z_N(g) \neq \varnothing$,* (5.4.9), (5.4.10) *and* (5.4.14). *Let (u_N, p_N) solve* (5.4.1). *Then the following a priori error estimates hold:*

$$\begin{aligned} \|u - u_N\|_X \leq & \left(1 + \frac{\|a\|}{\alpha(N)}\right)\left(1 + \frac{\|b\|}{\beta(N)}\right) \inf_{v \in X_N} \|u - v\|_X \\ & + \frac{\|b\|}{\alpha(N)} \inf_{q \in M_N} \|p - q\|_M \end{aligned} \tag{5.4.18}$$

$$\|p - p_N\|_{M/\ker B_N^*} \leq \left(1 + \frac{\|b\|}{\beta(N)}\right) \inf_{q \in M_N} \|p - q\|_M + \frac{\|a\|}{\beta(N)} \|u - u_N\|_X \, . \tag{5.4.19}$$

Remark 5.20 Notice that the bound (5.4.18) behaves as $1/\beta(N)$ whereas (5.4.19) behaves as $1/(\beta(N))^2$. A small stability constant $\beta(N)$ may therefore affect the accuracy of p_N more than that of u_N, which is indeed observed in many practical situations.

Theorem 5.19 addressed only the case (5.2.30), i.e., when $b_1 = b_2$ in (5.2.6), (5.2.7). A similar reasoning can, however, be used in the analysis of the general case also. The corresponding result is stated in the next theorem, the proof of which (along the lines of Lemma 5.15, 5.16 and 5.18) is left as an exercise.

Theorem 5.21 *Let (u,p) be a solution of* (5.2.6) *and assume* (5.2.21)–(5.2.24). *Let further $X_{i,N} \subset X_i$, $M_{i,N} \subset M_i$, $i = 1,2$, be families of finite dimensional subspaces, A_N and $B_{i,N}$ the corresponding finite-dimensional operators and define the kernels*

$$\begin{aligned} Z_{i,N} &= \{u_N \in X_{3-i,N} : b_i(u_N, q) = 0 \text{ for all } q \in M_{i,N}\},\ i = 1,2\,, \\ \ker B_{i,N}^* &= \{q_N \in M_{i,N} : \quad b_i(v, q_N) = 0 \text{ for all } v \in X_{3-i,N}\},\ i = 1,2\,. \end{aligned} \tag{5.4.20}$$

Assume the spaces $X_{i,N}$, $M_{i,N}$ satisfy the discrete inf–sup conditions

$$\inf_{u \in Z_{2,N}} \sup_{v \in Z_{1,N}} \frac{a(u,v)}{\|u\|_{X_1} \|v\|_{X_2}} \geq \alpha(N) > 0\,, \tag{5.4.21}$$

for every $0 \neq v \in Z_{1,N}$

$$\sup_{u \in Z_{1,N}} a(u,v) > 0\,, \tag{5.4.22}$$

and, for $i = 1,2$,

$$\inf_{q \in M_{i,N}} \sup_{v \in X_{3-i,N}} \frac{b_i(v,q)}{\|v\|_{X_{3-i}} \|q\|_{M_i}} \geq \beta_i(N) > 0\,. \tag{5.4.23}$$

Define further

$$Z_{2,N}(g) = \{u \in X_{1,N} \mid b_2(u,q) = \langle g,q \rangle \quad \textit{for all} \;\; q \in M_{2,N}\}$$

and assume that $Z_{2,N} \neq \varnothing$.

Then the approximate problem: find $(u_N, p_N) \in X_{1,N} \times M_{1,N}$ *such that*

$$\begin{aligned} a(u_N, v) + b_1(v, p_N) &= \langle f, v \rangle_{X_2' \times X_2} \;\; \textit{for all} \;\; v \in X_{2,N} \,, \\ b_2(u_N, q) &= \langle g, q \rangle_{M_2' \times M_2} \;\; \textit{for all} \;\; q \in M_{2,N} \end{aligned} \tag{5.4.24}$$

admits a solution (u_N, p_N), p_N *unique up to* $\ker B^*_{i,N}$, *and we have the error estimates*

$$\begin{aligned} \|u - u_N\|_{X_1} \le &\left(1 + \frac{\|a\|}{\alpha(N)}\right)\left(1 + \frac{\|b_2\|}{\beta_2(N)}\right) \inf_{v \in X_{1,N}} \|u - v\|_{X_1} \\ &+ \frac{\|b_1\|}{\alpha(N)} \inf_{q \in M_{1,N}} \|p - q\|_{M_1} \,, \end{aligned} \tag{5.4.25}$$

$$\begin{aligned} \|p - p_N\|_{M_1/\ker B_N^*} \le &\left(1 + \frac{\|b_1\|}{\beta_1(N)}\right) \inf_{q \in M_{1,N}} \|p - q\|_{M_1} \\ &+ \frac{\|a\|}{\beta_1(N)} \|u - u_N\|_{X_1} \,. \end{aligned} \tag{5.4.26}$$

5.5 Finite elements for the Stokes problem

5.5.1 *General results*

The general error analysis in Section 5.4 simplifies considerably in the case of the Stokes problem (5.3.1). This is due to $b_1 = b_2$, and the fact that we always have (5.4.9), (5.4.10) (see Remark 5.13) with $\alpha(N) = \nu\, C_1(\Omega)$ where $C_1(\Omega)$ is (essentially) the constant in the Poincaré inequality $|\boldsymbol{u}|^2_{H^1(\Omega)} \ge C_1(\Omega)\, \|\boldsymbol{u}\|^2_{H^1(\Omega)}$ for $\boldsymbol{u} \in H^1_0(\Omega)$. We can therefore specialize the general result Theorem 5.19 to

$$a(\boldsymbol{u}, \boldsymbol{v}) = \nu(\mathbf{grad}\, \boldsymbol{u}, \mathbf{grad}\, \boldsymbol{v}), \; b(\boldsymbol{v}, p) = -\int_\Omega p \operatorname{div} \boldsymbol{v} \, dx \,.$$

Theorem 5.22 *Let* $(\boldsymbol{u}, p) \in \boldsymbol{H}^1_0(\Omega) \times L^2(\Omega)$ *be a solution of the Stokes problem* (5.3.1). *Let a family* $\{(X_N, M_N)\}_N$ *of velocity and pressure subspaces be given which satisfy*

$$\begin{aligned} \ker B_N^* &= \{p \in M_N : \int_\Omega p \operatorname{div} \boldsymbol{v} \, dx = 0 \quad \textit{for all} \;\; \boldsymbol{v} \in X_N\} \\ &\subseteq \ker B^* = \operatorname{span}\{1\} \,, \end{aligned} \tag{5.5.1}$$

$$\sup_{\boldsymbol{v} \in X_N} \frac{\int_\Omega q \operatorname{div} \boldsymbol{v} \, dx}{\|\boldsymbol{v}\|_{H^1(\Omega)}} \ge \beta(N)\, \|q\|_{L^2(\Omega)/\ker B_N^*} \,, \quad \beta(N) > 0 \tag{5.5.2}$$

for every $q \in M_N/\ker B_N^*$. Then the following **error estimates** *hold for the FE-solutions $(\boldsymbol{u}_N, p_N) \in X_N \times M_N$ defined in* (5.4.1)*:*

$$\begin{aligned} \|\boldsymbol{u} - \boldsymbol{u}_N\|_{H^1(\Omega)} \le \Big(1 + \frac{1}{C_1}\Big)\Big(1 + \frac{c}{\beta(N)}\Big) \inf_{\boldsymbol{v} \in X_N} \|\boldsymbol{u} - \boldsymbol{v}\|_{H^1(\Omega)} \\ + \frac{c}{\nu C_1} \inf_{q \in M_N} \|p - q\|_{L^2(\Omega)} \end{aligned} \tag{5.5.3}$$

$$\begin{aligned} \|p - p_N\|_{L^2(\Omega)/\ker B_N^*} \le \Big(1 + \frac{1}{C_1}\Big)\Big(1 + \frac{c}{\beta(N)}\Big)\frac{\nu}{\beta(N)} \inf_{\boldsymbol{v} \in X_N} \|\boldsymbol{u} - \boldsymbol{v}\|_{H^1(\Omega)} \\ \times \Big(1 + \frac{c}{\beta(N)}\Big)\Big(1 + \frac{1}{C_1}\Big) \cdot \inf_{q \in M_N} \|p - q\|_{L^2(\Omega)} \end{aligned} \tag{5.5.4}$$

where c is an absolute constant (i.e., independent of N, ν and Ω), and

$$|\boldsymbol{u}|^2_{H^1(\Omega)} \ge C_1(\Omega)\, \|\boldsymbol{u}\|^2_{H^1(\Omega)} \quad \text{for} \quad \boldsymbol{u} \in H^1_0(\Omega)\,.$$

Remark 5.23 Assumption (5.5.1), i.e., that $\ker B_N^*$ is contained in the space of constant functions, is often satisfied and, in the light of Exercise 5.14, almost essential. It means that computed pressures p_N are unique up to constants. If (5.5.1) is violated, **parasitic pressure modes** from $\ker B_N^*/\{1\}$ may pollute p_N. Therefore (5.5.1) is essential for a reasonable approximation p_N. On the other hand, the inclusion

$$\ker B_N \subseteq \ker B = \{\boldsymbol{v} \in H^1_0(\Omega) : \operatorname{div} \boldsymbol{v} = 0\} \tag{5.5.5}$$

is rarely satisfied in practice, since the design of divergence-free velocity spaces X_N increases their complexity (see also Section 5.5.2 below for more on parasitic or spurious pressures).

Remark 5.24 The constants in the error bounds (5.5.3), (5.5.4) depend on the viscosity ν and on the inf–sup constant $\beta(N)$. Since the spaces X_N, M_N are finite dimensional, we always have (5.5.2) with some $\beta(N)$. The family (X_N, M_N) is called (divergence)-**stable** if $\beta(N) \ge \beta_0$ for some $\beta_0 > 0$ independent of N. Notice also that $\beta(N)$ enters in the pressure error estimate (5.5.4) as $1/(\beta(N))^2$, but in the velocity estimate only as $1/\beta(N)$. A small inf–sup constant $\beta(N)$ might be more visible in the pressure error $\|p - p_N\|_{L^2_0}$ than in the velocity error $\|\boldsymbol{u} - \boldsymbol{u}_N\|_{H^1}$. See also Remark 5.20.

We will now discuss several choices X_N, M_N of FE spaces and verify the inf–sup condition (5.5.2) for them.

The **setting** will always be the following: $\Omega \subset \mathbb{R}^2$ is a (straight) polygon (notation as in Chapter 1), and we consider homogeneous Dirichlet (i.e., no-slip) boundary conditions on all of $\Gamma = \partial\Omega$. In Ω, a mesh $\mathcal{T}$ is given which consists of triangles or of parallelogram elements Ω_j which are images of the reference elements $\widehat{K} = \widehat{T}$ or $\widehat{K} = \widehat{Q}$ under affine maps $F_j = B_j\,\hat{x} + c_j$:

$$\Omega_j = F_j(\widehat{K}),\ \ \widehat{K} = \widehat{T} \text{ or } \widehat{K} = \widehat{Q}\,. \tag{5.5.6}$$

We recall the definition of the *hp* FE spaces $S^{p,\ell}(\Omega, \mathcal{T})$ from Chapter 4 (Definition 4.40), $\ell = 0, 1$. Since the variables $\boldsymbol{u}$ and p belong to $\boldsymbol{H}_0^1(\Omega)$ and $L_0^2(\Omega)$, respectively, one is inclined to select subspaces X_N and M_N of different polynomial degrees and conformity. We consider first the simplest choice of this type, the so-called $Q_1 - P_0$ element, where the family (X_N, M_N) is obtained by h-refinement.

This element is unstable, as is evident by the existence of spurious checkerboard pressure modes on uniform meshes. We then show general techniques for verifying the discrete inf-sup condition (5.5.2), most notably Fortin's trick (Sections 5.5.3–5.5.5) and Stenberg's macroelement technique [136], [137] (Section 5.6). We illustrate them by showing stability of the $Q_1 - P_0$ element on special, distorted meshes, and also stability for other, low-order FE families on quite general meshes. We turn then to *hp*-finite elements on geometric meshes (cf., e.g., Figure 4.14). The basic strategy for proving (5.5.2) still applies, but the details are quite different here. In particular, it will not be possible to obtain $\beta(N) \geq \beta_0$ here (however, $\beta(N) \to 0$ "slowly" as $N \to \infty$).

5.5.2 *The $Q_1 - P_0$ element and spurious pressure modes*

We assume that the mesh $\mathcal{T}$ consists of quadrilaterals Ω_{ij}. The $Q_1 - P_0$ element is obtained for

$$X_N = S_0^{1,1}(\Omega, \mathcal{T})^2,\ M_N = S_0^{0,0}(\Omega, \mathcal{T})\,, \tag{5.5.7}$$

where $S_0^{1,1} := S^{1,1} \cap H_0^1$, $S_0^{0,0} := S^{0,0} \cap L_0^2$, i.e. we have bilinear, continuous velocities and constant, discontinuous pressures. The elemental degrees of freedom can be schematically represented as in Figure 5.1. They are velocity values at the vertices of Ω_{ij} (for each component of $\boldsymbol{v}$) and pressure values at the barycenter of Ω_{ij}. Since the element mappings F_{ij} are linear, constant pressures belong to $S^{0,0}(\Omega, \mathcal{T})$. These have been factored out in the definition (5.5.7) of M_N. Nevertheless, the pair (X_N, M_N) is not stable.

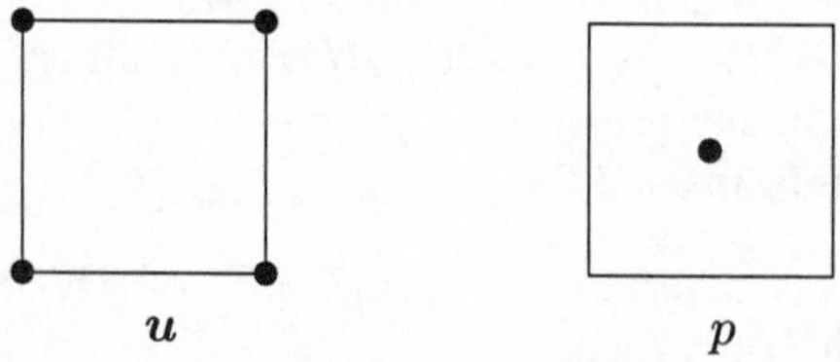

FIG. 5.1. Elemental degrees of freedom for $Q_1 - P_0$

Proposition 5.25 *Assume that Ω is a connected union of rectangles such that $\mathcal{T}$ is a uniform mesh consisting of axiparallel squares. Then*

$$\ker B_N^* = \operatorname{span}\{1, p^*\} \subset S^{0,0}(\Omega, \mathcal{T})$$

where p^* is the checkerboard pressure in Figure 5.2.

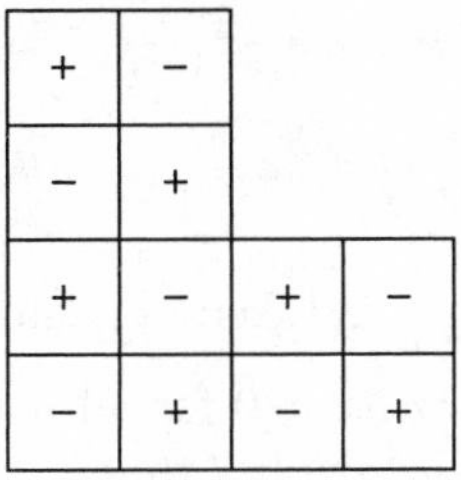

FIG. 5.2. Checkerboard pressure mode

Proof Let us number the vertices and barycenters of the elements Ω_{ij} as in Figure 5.3.

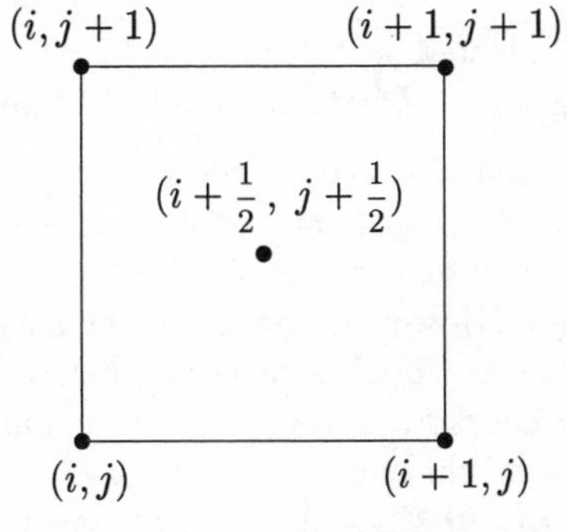

FIG. 5.3. Numbering of degrees of freedom in Ω_{ij}

Let $\boldsymbol{v}, q \in X_N \times M_N$. Then q is constant and div $\boldsymbol{v}$ linear on Ω_{ij}. Therefore, using the midpoint rule,

$$\begin{aligned} \int_{\Omega_{ij}} q \operatorname{div} \boldsymbol{v}\, dx &= h^2\, q_{i+\frac{1}{2}, j+\frac{1}{2}} (\operatorname{div} \boldsymbol{v})_{i+\frac{1}{2}, j+\frac{1}{2}} \\ &= h^2\, q_{i+\frac{1}{2}, j+\frac{1}{2}} \frac{1}{2h} [u_{i+1,j+1} + u_{i+1,j} - u_{i,j+1} - u_{i,j} \\ &\qquad + w_{i+1,j+1} + w_{i,j+1} - w_{i+1,j} - w_{i,j}] \end{aligned} \tag{5.5.8}$$

where $\boldsymbol{v} = (u, w)^\top$. Summing over all elements and rearranging the sums gives

$$\int_\Omega q \operatorname{div} \boldsymbol{v}\, dx = h^2 \sum_{i,j} \left[u_{i,j} (\nabla_1 q)_{i,j} + w_{i,j} (\nabla_2 q)_{i,j}\right] \tag{5.5.9}$$

where $(\nabla_k q)_{i,j}$ is given by

$$(\nabla_1 q)_{i,j} = \frac{1}{2h} \left[q_{i+\frac12,j+\frac12} + q_{i+\frac12,j-\frac12} - q_{i-\frac12,j+\frac12} - q_{i-\frac12,j-\frac12}\right],$$
$$(\nabla_2 q)_{i,j} = \frac{1}{2h} \left[q_{i+\frac12,j+\frac12} + q_{i-\frac12,j+\frac12} - q_{i+\frac12,j-\frac12} - q_{i-\frac12,j-\frac12}\right].$$

Since $\boldsymbol{v} \in H_0^1(\Omega)^2$, the sum $\sum_{i,j}$ is taken over all interior nodes.

Now $q \in \ker B_N^*$ if $\int_\Omega q \operatorname{div} \boldsymbol{v}\, dx = 0$ for all $\boldsymbol{v} \in X_N$, i.e. $\nabla_1 q$ and $\nabla_2 q$ must vanish at all interior nodes. This implies that

$$q_{i+\frac12,j+\frac12} = q_{i-\frac12,j-\frac12},\ q_{i+\frac12,j-\frac12} = q_{i-\frac12,j+\frac12}\,.$$

This is satisfied for $q = \text{const}$, but also for

$$q_{i+\frac12,j+\frac12} = \begin{cases} \alpha & \text{if } i+j \text{ is even}, \\ \beta & \text{otherwise}. \end{cases}$$

Here α and β must be selected such that $\int_\Omega p\, dx = 0$. This implies in particular that $\alpha = -\beta$, and we get $p^* \in L_0^2(\Omega)$ depicted in Figure 5.2. □

We emphasize that the above construction of p^* strongly depends on the regularity of the mesh. If $\mathcal{T}$ is slightly distorted, $\ker B_N^* = \{1\}$ usually. It is therefore sometimes recommended to avoid checkerboard pressures by distorting an "overly regular" mesh. However, analyzing the singular values of the matrix B_N^*, one has always one very small positive singular value with a singular vector which corresponds to a checkerboard pressure (see Chapter 2 for the connection of the inf–sup constant and the singular values of B_N, B_N^*). In general, therefore, this **stabilization of mixed FEM by mesh distortion** is not recommended, since for realistic geometries it is nonobvious what "good" meshes are.

The checkerboard pressure p^* was observed very early in FE calculations. In order to retain the simplicity of the $Q_1 - P_0$ calculations, one could try and restrict the space M_N further, e.g. to $M_N^* = M_N/\ker B_N^*$ or, equivalently,

$$M_N^* = \left\{p \in S^{0,0}(\Omega, \mathcal{T}) : \int_\Omega p\, dx = \int_\Omega p p^*\, dx = 0\right\}. \tag{5.5.10}$$

Then $B_N^* : M_N^* \to X_N'$ is injective, i.e. $\ker B_N^* = \{0\}$ and (5.5.2) holds with $\beta(N) > 0$. This idea underlies many stabilization and pressure filtering methods for the $Q_1^* - P_0$ element.

One might hope that the pair (X_N, M_N^*) obtained after filtering out the checkerboard modes is stable, i.e., that $\beta(N) \geq \beta_0 > 0$. Unfortunately, this is generally not so ([37], pp. 142–143).

Proposition 5.26 *Under the assumptions of Proposition* 5.25 *and with* M_N^* *given by* (5.5.10), (5.5.2) *holds with* $\beta(N) > 0$. *There exists a constant* $C(\Omega)$ *independent of* h *such that* $\beta(N) \leq C(\Omega)h$. *This result cannot be improved.*

5.5.3 *Verifying the inf–sup conditions: Fortin's lemma*

We present two techniques for checking the inf–sup condition (5.5.2) for a given pair $(X_N, M_N) \subset \boldsymbol{H}_0^1(\Omega) \times L_0^2(\Omega)$. Both are abstract in the sense that they are general strategies to be adapted in each special case. Their application will be illustrated in the next section. They can both be understood as special cases of the strategy in Remark 1.16.

Proposition 5.27 *Assume that the continuous inf–sup condition* (5.3.3) *holds with constant* $\beta > 0$. *Assume further that there exists a family of projectors:* $\Pi_N : X \to X_N$ *such that*

$$b(\boldsymbol{v} - \Pi_N \boldsymbol{v}, q) = 0 \ \text{for all} \ q \in M_N \,, \tag{5.5.11}$$

$$\|\Pi_N \boldsymbol{v}\|_X \leq C_N \, \|\boldsymbol{v}\|_X \tag{5.5.12}$$

where $b(\boldsymbol{v}, q) = \int_\Omega q \operatorname{div} \boldsymbol{v} \, dx$. *Then* (X_N, M_N) *satisfies* (5.5.1), *i.e.* $\ker B_N^* \subseteq \ker B^* = \{1\}$ *and there are no spurious pressure modes. Moreover, the discrete inf–sup condition holds with* $\beta(N) = \beta / C_N$.

Proof By (5.3.3), we have for every $q \in M_N \subset M$

$$\begin{aligned}
\beta \|q\|_M &\leq \sup_{0 \neq \boldsymbol{v} \in X} \frac{b(\boldsymbol{v}, q)}{\|\boldsymbol{v}\|_X} \\
&\overset{(5.5.11)}{=} \sup_{0 \neq \boldsymbol{v} \in X} \frac{b(\Pi_N \boldsymbol{v}, q)}{\|\boldsymbol{v}\|_X} \\
&= \sup_{\substack{\boldsymbol{v} \in X \\ \Pi_N \boldsymbol{v} \neq 0}} \frac{b(\Pi_N \boldsymbol{v}, q)}{\|\boldsymbol{v}\|_X} \\
&\overset{(5.5.12)}{\leq} C_N \sup_{\substack{\boldsymbol{v} \in X \\ \Pi_N \boldsymbol{v} \neq 0}} \frac{b(\Pi_N \boldsymbol{v}, q)}{\|\Pi_N \boldsymbol{v}\|_X} \\
&= C_N \sup_{0 \neq \boldsymbol{v} \in X_N} \frac{b(\boldsymbol{v}, q)}{\|\boldsymbol{v}\|_X}
\end{aligned}$$

□

Remark 5.28 By Exercise 5.14, (5.5.11) is equivalent to $\ker B_N^* \subseteq \ker B^*$, i.e., (5.5.1). Such elements will have (unlike the $Q_1 - P_0$ element) no spurious pressures. To see it, let $q \in \ker B_N^*$, i.e., $b(\boldsymbol{v}, q) = 0$ for every $\boldsymbol{v} \in X_N$. Then also $b(\boldsymbol{v}, q) = 0$ for every $\boldsymbol{v} \in X$ since q can be replaced by $\Pi_N \boldsymbol{v}$ by (5.5.11), hence $q \in \ker B^*$.

The problem in applying (5.5.11), (5.5.12) lies, of course, in the construction of Π_N. A useful technique is as follows.

Proposition 5.29 *Let $\Pi_{1N}\colon X \to X_N$ and $\Pi_{2N}\colon X \to X_N$ be such that*

$$\|\Pi_{iN}\boldsymbol{v}\|_X \le c_1 \|\boldsymbol{v}\|_X ,\ i = 1, 2 , \tag{5.5.13}$$

$$b(\boldsymbol{v} - \Pi_{2N}\boldsymbol{v}, q) = 0 \quad \textit{for all} \ \ q \in M_N , \tag{5.5.14}$$

$$\|\Pi_{2N}(I - \Pi_{1N})\boldsymbol{v}\|_X \le c_2 \|\boldsymbol{v}\|_X \tag{5.5.15}$$

and assume that the continuous inf–sup condition (5.3.3) *holds with $\beta > 0$. Then the discrete inf–sup condition* (5.5.2) *holds with $\beta(N) = \beta/(c_1 + c_2)$ and* $\ker B_N^* \subseteq \ker B^*$.

Proof Put $\Pi_N \boldsymbol{v} = \Pi_{2N}(\boldsymbol{v} - \Pi_{1N}\boldsymbol{v}) + \Pi_{1N}\boldsymbol{v}$. Then, for $q \in M_N$,

$$\begin{aligned} b(\Pi_N \boldsymbol{v}, q) &= b(\Pi_{2N}(v - \Pi_{1N}\boldsymbol{v}), q) + b(\Pi_{1N}\boldsymbol{v}, q) \\ &= b(\boldsymbol{v} - \Pi_{1N}\boldsymbol{v}, q) + b(\Pi_{1N}\boldsymbol{v}, q) \\ &= b(\boldsymbol{v}, q) \end{aligned}$$

hence we get (5.5.11).

One also checks that $\|\Pi_N \boldsymbol{v}\|_X \le (c_1 + c_2)\|\boldsymbol{v}\|_X$, which is (5.5.12). □

5.5.4 *An approximation result: Clément interpolant*

Typically, the operator $\Pi_{1N} : H^1(\Omega) \to X_N$ in Proposition 5.29 will be an interpolation operator constructed locally, i.e., elementwise. Since, however, functions in $H^1(\Omega)$ are not necessarily continuous in $\overline{\Omega}$, Π_{1N} can **not** be the nodal interpolant of $u \in H^1(\Omega)$.

An interpolation operator which works only with averages of the function u has been given by Clément [50].

Proposition 5.30 *Let $\Omega \subset \mathbb{R}^2$ be a polygon and $\mathcal{T}$ a regular mesh in Ω consisting of (triangular and/or quadrilateral) elements K which are affine images of a reference (triangular or quadrilateral) element $\widehat{K}$ (i.e., $K = F_K(\widehat{K})$ with $F_K(\hat{x}) = B_K\hat{x} + b_K$). For any $K \in \mathcal{T}$, define the patch δK of elements adjacent to K by $\delta K = \{K' \in \mathcal{T} : \overline{K} \cap \overline{K}' \neq \varnothing\}$ and put $h_{\delta K} = \max_{K' \in \delta K} h_{K'}$, $\rho_{\delta K} = \max_{K' \in \delta K}\{\rho_{K'}\}$ and $\sigma_{\delta K} = \max_{K' \in \delta K} \{h_{K'}/\rho_{K'}\}$, where $h_{K'} = \operatorname{diam}(K')$ and $\rho_{K'}$ denotes the radius of largest circle inscribed in K'.*

Then there is an interpolation operator

$$I_p^\ell : W^{1,s}(\Omega) \to S^{p,\ell}(\Omega, \mathcal{T}) ,$$

such that for every $K \in \mathcal{T}$, $1 \le s \le \infty$, $\ell \le m \le p+1$,

$$|v - I_p^\ell v|_{W^{\ell,s}(K)} \le C\,\sigma_{\delta K}\, h_{\delta K}^{m-\ell}\, |v|_{W^{m,s}(\delta K)} \,. \tag{5.5.16}$$

We note the following global estimate for $s = 2$ which we will need below.

Corollary 5.31 *If* $\{\mathcal{T}_i\}$ *is a shape regular family of affine meshes with*

$$\sup_i \max_{K\in\mathcal{T}_i} \{h_K/\rho_K\} \le \sigma < \infty$$

then, for $p \ge 1$ *and* $1 \le m \le p+1$, $\ell = 0, 1, \ldots, m$,

$$|v - I_p^\ell v|_{H^\ell(K)} \le C\sigma\, h_{\delta K}^{m-\ell}\, |v|_{H^m(\delta K)} \text{ for all } K \in \mathcal{T}_i \tag{5.5.17}$$

and we have the global estimate

$$\sum_{K\in\mathcal{T}_i} h_K^{2\ell-2}\, |v - I_p^\ell v|^2_{H^\ell(K)} \le C\, \|v\|^2_{H^1(\Omega)} \,,\ \ell = 0, 1 \,. \tag{5.5.18}$$

For $\ell = 1$ *in addition*

$$\sum_{e\in\mathcal{E}_i} h_e^{-1}\, \|v - I_p^\ell v\|^2_{L^2(e)} \le C\, \|v\|^2_{H^1(\Omega)} \tag{5.5.19}$$

where $\mathcal{E}_i$ *denotes the set of all interelement edges of the mesh* $\mathcal{T}_i$.

5.5.5 *Some stable low-order elements*

In applications of Proposition 5.29, in many cases Π_{1N} can be chosen to be Clément's interpolant while Π_{2N} has to be constructed specifically for the elements in question. We illustrate this for some h-version discretizations X_N, M_N.

Throughout, $\Omega \subset \mathbb{R}^2$ will be a polygon and $\mathcal{T}$ a regular mesh consisting of triangles unless explicitly stated otherwise.

Remark 5.32 No quasiuniformity of the mesh $\mathcal{T}$ is used and $\mathcal{T}$ can be chosen in particular to be a triangulation of type (h, γ, K) with mesh refinement towards the corners of Ω to achieve optimal rates of convergence despite the singularity of $\boldsymbol{u}$ and p.

5.5.5.1 $\boldsymbol{X_N = S_0^{1,1}(\Omega, \mathcal{T}),\ M_N = S^{1,1}(\Omega, \mathcal{T}) \cap L_0^2(\Omega)}$

Note that we have continuous pressures here. The pair $X_N \times M_N$ as defined does not satisfy an inf-sup condition with $\beta(N) \geq \beta_0 > 0$; however, we can **enrich the velocity space** so that this is true. To this end, let

$$B_k = \{b \in H_0^1(\Omega) : b(x)\,|_T \in P_k(T) \cap H_0^1(T) \quad \text{for all } \ T \in \mathcal{T}\},\ k \geq 3 \quad (5.5.20)$$

the set of **bubble functions** of degree k on T and put

$$X_N = (S^{1,1}(\Omega, \mathcal{T}) \otimes B_3)^2 \cap H_0^1(\Omega)^2 \,. \quad (5.5.21)$$

We claim that (X_N, M_N) is stable. To prove it, we use Proposition 5.29 with $\Pi_1 = I_1^1$, i.e. the Clément interpolant as defined in Proposition 5.30. This yields for $T \in \mathcal{T}$:

$$|\boldsymbol{v} - \Pi_1 \boldsymbol{v}|_{H^r(T)} \leq c h_T^{1-r}\, |\boldsymbol{v}|_{H^1(\delta K)},\ \ r = 0, 1 \,, \quad (5.5.22)$$

or, summing over all $T \in \mathcal{T}$, we get since $\mathcal{T}$ is regular that

$$\begin{aligned} |\Pi_1 \boldsymbol{v}|_{H^1(\Omega)} - |\boldsymbol{v}|_{H^1(\Omega)} &\leq \big|\, |\Pi_1 \boldsymbol{v}|_{H^1(\Omega)} - |\boldsymbol{v}|_{H^1(\Omega)} \big| \\ &\leq |\Pi_1 \boldsymbol{v} - \boldsymbol{v}|_{H^1(\Omega)} \leq c\, |\boldsymbol{v}|_{H^1(\Omega)} \end{aligned}$$

or

$$|\Pi_1 \boldsymbol{v}|_{H^1(\Omega)} \leq (c+1)\, |\boldsymbol{v}|_{H^1(\Omega)}$$

which is (5.5.13) for $i = 1$.

Define $\Pi_2 : H_0^1(\Omega) \to B_3$ via

$$\int_\Omega q \operatorname{div}(\Pi_2 \boldsymbol{v} - \boldsymbol{v})\, dx = \int_\Omega (\boldsymbol{v} - \Pi_2 \boldsymbol{v}) \cdot \mathbf{grad}\, q\, dx = 0 \ \text{ for all } \ q \in M_N \,. \quad (5.5.23)$$

Since $\mathbf{grad}\, q \in S^{0,0}(\Omega, \mathcal{T})$, (5.5.23) can be satisfied by choosing on each $T \in \mathcal{T}$ the projection $\Pi_2 \boldsymbol{v}$ to be

$$(\Pi_2 \boldsymbol{v})_i = \alpha_i b_T(x), \quad \alpha_i = \int_T v_i\, b_T\, dx \Big/ \int_T b_T\, dx, \ \ i = 1, 2 \,,$$

where $b_T \in B_3$ denotes the bubble function on T. We claim that

$$\|\Pi_2 \boldsymbol{v}\|_{H^1(T)} \leq c\, (h_T^{-1}\, \|\boldsymbol{v}\|_{L^2(T)} + |\boldsymbol{v}|_{H^1(T)}) \,. \quad (5.5.24)$$

This is evidently so in the reference element $\widehat{T}$ and follows in the general case by scaling.

Now (5.5.23) implies (5.5.14). Further, by (5.5.24) and (5.5.22) with $r = 0$ and 1 we get

$$\begin{aligned}\|\Pi_2(I - \Pi_1)\boldsymbol{v}\|_{H^1(K)} &\leq c h_T^{-1} \|\boldsymbol{v} - \Pi_1 \boldsymbol{v}\|_{L^2(K)} \\ &\leq c\,|\boldsymbol{v}|_{H^1(\delta K)} \,.\end{aligned}$$

Squaring and summing over all $K \in \mathcal{T}$, we have (5.5.15), i.e., the discrete inf–sup condition with $\beta(N) \geq \beta_0 > 0$. We get from Theorem 5.22.

Proposition 5.33 *Let $\mathcal{T}$ be a shape-regular triangulation of the polygon $\Omega \subset \mathbb{R}^2$. Then the pair*

$$X_N = (S_0^{1,1} \otimes B_3)^2, \ M_N = S^{1,1} \cap L_0^2(\Omega) \tag{5.5.25}$$

is divergence stable. Moreover, the FE solution (u_N, p_N) is quasioptimal, i.e., it holds that

$$\begin{aligned}\|\boldsymbol{u} - \boldsymbol{u}_N\|_{H^1(\Omega)} + \|p - p_N\|_{L_0^2(\Omega)} \leq c\Big(&\inf_{\boldsymbol{u} \in X_N} \|\boldsymbol{u} - \boldsymbol{v}\|_{H^1(\Omega)} \\ + &\inf_{q \in M_N} \|p - q\|_{L_0^2(\Omega)}\Big)\end{aligned} \tag{5.5.26}$$

with $c = c(\Omega)$ independent of N (it depends only on the parameters K in the definition (4.3.3) *of shape regularity).*

5.5.5.2 $\boldsymbol{X_N = S_0^{2,1}}$, $\boldsymbol{M_N = S_0^{0,0}}$ ($\boldsymbol{P_2 - P_0}$ *element*)

Note that we now have discontinuous approximate pressures p_N. To prove the inf–sup condition (5.5.2) with $\beta(N)$ independent of N, we use Proposition 5.29. Let Π_1 be the Clément interpolant introduced in Section 5.4.1 above. Then (5.5.13) holds for Π_1. Define Π_2 by

$$\Pi_2 \boldsymbol{v} = 0 \text{ at vertices of } \mathcal{T}, \ \int_e (\Pi_2 \boldsymbol{v} - \boldsymbol{v}) ds = 0 \tag{5.5.27}$$

for all interelement edges e of $\mathcal{T}$. Then for every $T \in \mathcal{T}$

$$\int_T \operatorname{div}(\boldsymbol{v} - \Pi_2 \boldsymbol{v}) dx = \int_{\partial T} (\Pi_2 \boldsymbol{v} - \boldsymbol{v}) \cdot \boldsymbol{n}\, ds = 0\,,$$

hence

$$\int_\Omega \operatorname{div}(\boldsymbol{v} - \Pi_2 v) q\, dx = 0 \quad \text{for all } \ q \in S^{0,0}(\Omega, \mathcal{T})\,, \tag{5.5.28}$$

which is (5.5.14).

Finally we verify

$$\|\Pi_2 \boldsymbol{v}\|_{H^1(T)} \leq c h_T^{-1} \|\boldsymbol{v}\|_{L^2(T)} \quad \text{for all } \ T \in \mathcal{T}\,, \tag{5.5.29}$$

from where (5.5.15) follows.

To prove (5.5.29), observe that on $\widehat{T}$ we have for any $v \in H^1(\widehat{T})^2$

$$|\hat{\Pi}_2 v|_{H^1(\hat{T})} \leq C \, \|v\|_{L^2(\partial\hat{T})} \leq C \, \|v\|_{H^1(\hat{T})}$$

by the trace theorem. A scaling argument yields

$$|\Pi_2 v|_{H^1(T)} \leq C \, (h_T^{-1} \, \|v\|_{L^2(T)} + |v|_{H^1(T)})$$

for every $T \in \mathcal{T}$. This implies (5.5.15). Referring to Propositions 5.27 and 5.29, we have proved the following result.

Proposition 5.34 *Let $\mathcal{T}$ be a regular triangulation of the polygon $\Omega \subset \mathbb{R}^2$. Then the pair*

$$X_N = S_0^{2,1}(\Omega, \mathcal{T})^2, \; M_N = S_0^{0,0}(\Omega, \mathcal{T}) \tag{5.5.30}$$

is stable and (u_N, p_N) is quasioptimal, i.e., (5.5.26) *holds.*

In exactly the same way the $Q_2' - P_0$ element on affine quadrilateral meshes can be shown to be stable.

Proposition 5.35 *Let $\mathcal{T}$ be a regular partition of Ω into quadrilaterals K, affine equivalent to $\widehat{K} = (-1,1)^2$. Then the pairing $S_0^{2,1}(\Omega, \mathcal{T}) = \{v \in H^1(\Omega) : v|_K \circ Q_K \in \mathcal{S}_3^2(\widehat{K})\}$ and $S_0^{0,0}(\Omega, \mathcal{T})$ as above is stable.*

Exercise 5.36 Mimicking the proof of Proposition 5.34, prove Proposition 5.35 and also a generalization to regular meshes $\mathcal{T}$ which contain any combination of shape-regular triangles and (affine) quadrilaterals.

In the proof of Proposition 5.29, the Clément operator Π_1 and the operator Π_2 satisfying (5.5.27) were combined into the projector

$$\widetilde{\Pi}_N := \Pi_1 + \Pi_2(Id - \Pi_1) \tag{5.5.31}$$

which satisfies (5.5.11), (5.5.12). This projector $\widetilde{\Pi}_N$ will be used in Proposition 5.39 below, where we give a general bubble stabilization result for discontinuous pressure elements.

5.5.5.3 $X_N = S_0^{2,1} \otimes B_3$, $M_N = S_0^{1,0}$ *(Crouzeix-Raviart element)*

We again apply Proposition 5.29. We select $\widetilde{\Pi}_1 := \Pi_1 + \Pi_2(Id - \Pi_1)$ with Π_1, Π_2 as for the $P_2 - P_0$ element in Section 5.2. Then $\widetilde{\Pi}_1$ satisfies (5.5.11) and (5.5.12), i.e.,

$$\|\widetilde{\Pi}_1 v\|_{H^1(\Omega)} \leq C \, \|v\|_{H^1(\Omega)} \text{ for all } v \in H_0^1(\Omega) \, , \tag{5.5.32}$$

$$\int_T \operatorname{div}(v - \widetilde{\Pi}_1 v) dx = 0 \qquad \text{for all } v \in H_0^1(\Omega), \text{ for all } T \in \mathcal{T} \, . \tag{5.5.33}$$

Next, we select $\Pi_2 : H_0^1(\Omega) \to B_3$. Since we later on use only $\Pi_2(Id - \widetilde{\Pi}_1)$, we may assume by (5.5.33) that $\int_T \operatorname{div} \boldsymbol{v}\, dx = 0$. Then we define $\Pi_2 \boldsymbol{v}$ by

$$\Pi_2 \boldsymbol{v} \in B_3(T) : \int_T q \operatorname{div}(\boldsymbol{v} - \Pi_2 \boldsymbol{v}) dx = 0 \text{ for all } q \in P_1(T) . \tag{5.5.34}$$

This is a linear system of 3 equations in 2 unknowns. It is, however, compatible, since

$$\int_T \operatorname{div} b\, dx = 0 \text{ for all } b \in B_3(T) . \tag{5.5.35}$$

Therefore $\Pi_2 \boldsymbol{v}$ is well defined and we must prove its boundedness, i.e.,

$$\|\Pi_2 \boldsymbol{v}\|_{H^1(T)} \le C\, \|\boldsymbol{v}\|_{H^1(T)} \quad \text{for all } \boldsymbol{v} \in \{\boldsymbol{v} \in \boldsymbol{H}_0^1(T) : \operatorname{div} \boldsymbol{v} = 0\} . \tag{5.5.36}$$

To prove (5.5.36), observe that due to $\int_T \operatorname{div} \boldsymbol{v}\, dx = 0$ we have from (5.5.34)

$$\int_T \mathbf{grad}\, q \cdot \Pi_2 \boldsymbol{v}\, dx = \int_T (q - \widetilde{q}) \operatorname{div} \boldsymbol{v}\, dx \text{ for all } \widetilde{q} \in P_0(T),\ q \in P_1(T) .$$

Changing variables in the integrals, we get

$$\int_{\widehat{T}} \widehat{\mathbf{grad}}\, q \cdot \hat{\Pi}_2 \boldsymbol{v}\, d\hat{x} = \int_{\widehat{T}} (q - \widetilde{q})\, \widehat{\operatorname{div}}\, \boldsymbol{v}\, d\hat{x} \quad \text{for all } \widetilde{q} \in P_0(\widehat{T}) .$$

Selecting $\widetilde{q} = [q] = \int_{\widehat{T}} q\, d\hat{x} / |\widehat{T}|$, we get

$$\begin{aligned} \left| \int_{\widehat{T}} (q - [q])\, \widehat{\operatorname{div}}\, \boldsymbol{v}\, d\hat{x} \right| &\le C\, |\boldsymbol{v}|_{H^1(\widehat{T})} \|q - [q]\|_{L^2(\widehat{T})} \\ &\le C\, |\boldsymbol{v}|_{H^1(\widehat{T})}\, |q|_{H^1(\widehat{T})} \end{aligned}$$

by the second Poincaré inequality. Hence, denoting by $\widetilde{H}^1(\widehat{T}) = \{u \in H^1(\widehat{T}) : \int_{\widehat{T}} u\, dx = 0\}$,

$$\|\hat{\Pi}_2 \boldsymbol{v}\|_{L^2(\widehat{T})} \le \sup_{q \in \widetilde{H}^1(\widehat{T})} \frac{\int_{\widehat{T}} \hat{\Pi}_2 \boldsymbol{v} \cdot \widehat{\operatorname{grad}}\, q\, d\hat{x}}{|q|_{H^1(\widehat{T})}} \le C\, |\boldsymbol{v}|_{H^1(\widehat{T})} .$$

Scaling back to $T \in \mathcal{T}$, we get

$$\|\Pi_2 \boldsymbol{v}\|_{L^2(T)} = C h_T\, |\boldsymbol{v}|_{H^1(T)} .$$

Since also $|\boldsymbol{w}|_{H^1(T)} \le C h_T^{-1}\, \|\boldsymbol{w}\|_{L^2(T)}$ for any $\boldsymbol{w} \in B_3(T)$, we get (5.5.36) and can apply Proposition 5.29. We have proved the following result.

Proposition 5.37 *Let $\mathcal{T}$ be a regular triangulation of the polygon $\Omega \subset \mathbb{R}^2$. Then the pair*

$$X_N = (S_0^{2,1} \otimes B_3)^2, \quad M_N = S_0^{1,0} \tag{5.5.37}$$

is stable and $(\boldsymbol{u}_N, p_N)$ is quasioptimal, i.e., (5.5.26) holds.

5.5.6 *Bubble stabilization of Stokes elements*

The stability proofs in Sections 5.5.5.1 and 5.5.5.3 had in common that the velocity space X_N was $S^{k,1} \otimes B_3$ where the bubble function served only stability purposes, but **not** approximability purposes. This is in fact a general principle: almost any element can be stabilized using bubble functions.

Let $M_N \subset L_0^2(\Omega)$ be any pressure space. Then we define

$$\begin{aligned} B(\mathbf{grad}\, M_N) := \{\boldsymbol{v} \in [H_0^1(\Omega)]^2 : \boldsymbol{v}\,|_K = b_K(\mathbf{grad}\, q)|_K \\ \text{for } q \in M_N\}\,. \end{aligned} \tag{5.5.38}$$

We have the following result.

Proposition 5.38 *(continuous pressures). Let $X_N \subset H_0^1(\Omega)^2$, $M_N \subset [H^1(\Omega) \cap L_0^2(\Omega)]$ be such that there exists $\Pi_1 : H_0^1(\Omega) \to X_N$, satisfying*

$$\sum_{K \in \mathcal{T}} h_K^{2r-2}\, |\boldsymbol{v} - \Pi_1 \boldsymbol{v}|^2_{H^r(K)} \leq C\, \|\boldsymbol{v}\|^2_{H^1(\Omega)}, \quad r = 0, 1\,, \tag{5.5.39}$$

and that

$$B(\mathbf{grad}\, M_N) \subseteq X_N\,. \tag{5.5.40}$$

Then (X_N, M_N) is stable, i.e., (5.5.2) holds with $\beta(N) \geq \beta_0 > 0$.

Proof We apply Proposition 5.29 with Π_1 in (5.5.39). Selecting $r = 1$, we get

$$\|\Pi_1 \boldsymbol{v}\|_{H^1(\Omega)} \leq \|\boldsymbol{v} - \Pi_1 \boldsymbol{v}\|_{H^1(\Omega)} + \|v\|_{H^1(\Omega)} \leq c_1\, \|\boldsymbol{v}\|_{H^1(\Omega)}\,,$$

i.e., (5.5.13). Now define $\Pi_2 : V \to B(\mathbf{grad}\, M_N)$ elementwise by

$$\begin{aligned} &\Pi_2 \boldsymbol{v}|_K \in B(\mathbf{grad}\, M_N)|_K = b_K\, \mathbf{grad}\, M_N|_K\,, \\ &\int_K (\Pi_2 \boldsymbol{v} - \boldsymbol{v}) \cdot \mathbf{grad}\, q\, dx = 0 \quad \text{for all } q \in M_N|_K\,. \end{aligned} \tag{5.5.41}$$

$\Pi_2 \boldsymbol{v}$ is well defined by (5.5.41). Moreover, since $M_N \subset H^1(\Omega)$,

$$\begin{aligned} \int_\Omega q \operatorname{div}(\Pi_2 \boldsymbol{v} - \boldsymbol{v}) dx &= -\int_\Omega (\Pi_2 \boldsymbol{v} - \boldsymbol{v}) \cdot \mathbf{grad}\, q\, dx \\ &= -\sum_K \int_K (\Pi_2 \boldsymbol{v} - \boldsymbol{v}) \cdot \mathbf{grad}\, q\, dx = 0\,, \end{aligned}$$

i.e., we have (5.5.14).

Finally, to prove (5.5.15), it is sufficient to establish

$$\|\Pi_2(Id - \Pi_1)\,\boldsymbol{v}\|_{H^1(K)} \le C_2\,\|\boldsymbol{v}\|_{H^1(K)}, \quad K \in \mathcal{T} \tag{5.5.42}$$

with C_2 independent of K. Then we may square and sum over all $K \in \mathcal{T}$ to get the result. Assume for now that

$$\|\Pi_2\boldsymbol{v}\|_{H^1(K)} \le C\,(h_K^{-1}\,\|\boldsymbol{v}\|_{L^2(K)} + |\boldsymbol{v}|_{H^1(K)}), \quad K \in \mathcal{T}\,. \tag{5.5.43}$$

This clearly implies (5.5.42) since with $\Pi_1 = I_1^1$ it follows from (5.5.18) that

$$\begin{aligned}\sum_{K\in\mathcal{T}} \|\Pi_2(Id-\Pi_1)\,\boldsymbol{v}\|_{H^1(K)}^2 &\le C \sum_{K\in\mathcal{T}} h_K^{-2}\,\|(Id-I_1^1)\,\boldsymbol{v}\|_{L^2(K)}^2 \\ &\qquad + |(Id-I_1^1)\,\boldsymbol{v}|_{H^1(K)}^2 \\ &\le C \sum_{K\in\mathcal{T}} |\boldsymbol{v}|_{H^1(\delta K)}^2 \\ &\le C\,\|\boldsymbol{v}\|_{H^1(\Omega)}^2\end{aligned}$$

provided the triangulation is shape regular.

To prove (5.5.43), we consider (5.5.41) on $\widehat{K}$. There, we have due to $\hat{\Pi}_2\boldsymbol{v} \in H_0^1(\widehat{K})$ that

$$\|\hat{\Pi}_2\boldsymbol{v}\|_{L^2(\widehat{K})} \le C\,|\hat{\Pi}_2\boldsymbol{v}|_{H^1(\widehat{K})} \le C\,\|\boldsymbol{v}\|_{H^1(\widehat{K})}\,.$$

Then on $K \in \mathcal{T}$ we have by a change of variables

$$h_K^{-1}\,\|\Pi_2\boldsymbol{v}\|_{L^2(K)} \le C\,|\Pi_2\boldsymbol{v}|_{H^1(K)} \le C\,(h_K^{-1}\,\|\boldsymbol{v}\|_{L^2(K)} + |\boldsymbol{v}|_{H^1(K)})$$

which implies (5.5.43). □

Proposition 5.39 *(discontinuous pressures). Let $X_N \subset H_0^1(\Omega)$, $M_N \subset L_0^2(\Omega)$ be such that there exists $\widetilde{\Pi}_1 : H_0^1(\Omega) \to X_N$ satisfying*

$$\sum_{K\in\mathcal{T}} h_K^{2r-2}\,|\boldsymbol{v} - \widetilde{\Pi}_1\boldsymbol{v}|_{H^r(K)}^2 \le C\,\|\boldsymbol{v}\|_{H^1(\Omega)}^2, \quad r = 0, 1\,, \tag{5.5.44}$$

and that

$$B(\mathbf{grad}\,M_N) \subseteq X_N\,. \tag{5.5.45}$$

Then (X_N, M_N) is stable.

Proof We apply Proposition 5.29 to construct Π_N. We first select $\Pi_1 = \widetilde{\Pi}_N$ in (5.5.31), i.e. the operator constructed in the stability proof of the $P_2 - P_0$

element. To apply Proposition 5.29, we must construct Π_2. We need to define Π_2 only on

$$\boldsymbol{V} = \left\{ \boldsymbol{v} \in H_0^1(\Omega) : \int_K \operatorname{div} \boldsymbol{v}\, dx = 0,\ K \in \mathcal{T} \right\}. \tag{5.5.46}$$

For $\boldsymbol{v} \in \boldsymbol{V}$, we construct $\Pi_2 \boldsymbol{v} \in B(\mathbf{grad}\, M_N)$ as follows:

$$\begin{aligned} &\Pi_2 \boldsymbol{v}|_K \in B(\mathbf{grad}\, M_N) = b_K\, (\mathbf{grad}\, M_N)|_K\,, \\ &\int_K q \operatorname{div}(\Pi_2 \boldsymbol{v} - \boldsymbol{v})\, dx = 0 \text{ for all } q \in M_N\,. \end{aligned} \tag{5.5.47}$$

The reasoning is now very similar to that of Section 5.5.5.2. Since $\int_K \operatorname{div} \boldsymbol{b}_T\, dx = 0$, the rank of the matrix arising in (5.5.47) in the determination of $\Pi_2 \boldsymbol{v}$ is $\dim(M_N|_K) - 1$ and the matrix is nonsingular. Hence $\Pi_2 \boldsymbol{v}$ is defined on each element and belongs to $H_0^1(\Omega)$. The proof that $\|\Pi_2 \boldsymbol{v}\|_{H^1} \leq C\, \|\boldsymbol{v}\|_{H^1}$ is done exactly as in Sections 5.5.5.2. We then apply Proposition 5.29 with Π_1 and Π_2. □

5.5.7 *Macroelement techniques*

In Section 5.2 we showed that the $Q_1 - P_0$ element is unstable and that it actually admits spurious pressure modes on uniform square grids. This seems to rule out this element as a useful tool for the discretization of the Stokes problem. Nevertheless, in many practical applications one observes that $Q_1 - P_0$ yields quite reasonable velocity and, in many cases, even good pressure approximations are achieved. The observation is that apparently irregular meshes can stabilize an otherwise unstable mixed method. A partial analysis of this empirical observation is the purpose of this section. The main tool in the analysis is the macroelement-technique to establish divergence stability which is in certain cases more powerful than the techniques based on Fortin's lemma.

5.5.7.1 *The macroelement technique*

We assume that $\Omega \subset \mathbb{R}^2$ is a polygon and $\{\mathcal{T}_i\}$ a family of shape-regular meshes consisting of triangular and/or quadrilateral elements $K \in \mathcal{T}_i$, affinely equivalent to the reference triangle $\widehat{T}$ or the reference square $\widehat{Q}$. Generically, the reference element is denoted by $\widehat{K}$ and $K = Q_K(\widehat{K})$. Let $\widehat{V}$ and $\hat{P}$ be polynomial velocity and pressure spaces on the reference element $\widehat{K}$. Then we define

$$X_N = \{\boldsymbol{v} \in H_0^1(\Omega) : \boldsymbol{v}|_K = \hat{\boldsymbol{v}} \circ Q_K^{-1},\ \hat{\boldsymbol{v}} \in \widehat{V},\ K \in \mathcal{T}\}\,, \tag{5.5.48}$$

$$M_N = \{p \in L_0^2(\Omega) :\ p|_K = \hat{p} \circ Q_K^{-1},\ \hat{p} \in \hat{P},\ K \in \mathcal{T}\}\,. \tag{5.5.49}$$

As usual, we define $h_K = \operatorname{diam}(K)$ and $\rho_K = \max\{\rho : B_\rho \subset K$ is a circle of radius $\rho\}$, $h(\mathcal{T}) = \max_{K \in \mathcal{T}}\{h_K\}$.

Definition 5.40 *A* **macroelement** M *is a connected set of (closed) elements* $K \in \mathcal{T}$ *which share at least one point.*

Two macroelements M **and** $\widetilde{M}$ **are equivalent** *if there exists a mapping* $G : \widetilde{M} \to M$ *such that*

i) $G(\widetilde{M}) = M$.

ii) *If* $\widetilde{M} = \bigcup_{j=1}^{m} \widetilde{K}_j$, $M = \bigcup_{j=1}^{m} K_j$ *with* $K_j, \widetilde{K}_j \in \mathcal{T}$ *then* $K_j = G(\widetilde{K}_j)$.

iii) $G|_{\widetilde{K}_j} = Q_{K_j} \circ Q_{\widetilde{K}_j}^{-1}$.

The set of all equivalent macroelements M *is an equivalence class* $\widehat{M}$ *which we may represent by a canonical reference macro.*[7]

For a macroelement M define the space

$$V_{0,M} = \{ v \in H_0^1(M)^2 : v|_K = \hat{v} \circ Q_K^{-1},\ \hat{v} \in \widehat{V},\ K \in M \}\,, \qquad (5.5.50)$$

$$P_M = \{ p \in L^2(M) :\ \ p|_K = \hat{p} \circ Q_K^{-1},\ \hat{p} \in \hat{P},\ K \in M \}\,. \qquad (5.5.51)$$

The set of all interelement edges $e \in \mathcal{T}_i$ is denoted by $\mathcal{E}_i$.

The following subspace will play an important role in the verification of the inf–sup condition:

$$N_M \;= \left\{ p \in P_M : \int_M p \operatorname{div} v \, dx = 0, \ \ v \in V_{0,M} \right\}. \qquad (5.5.52)$$

Theorem 5.41 *Assume that* $\{\mathcal{T}_i\}$ *is a family of shape-regular meshes admitting an interpolant* I *satisfying* (5.5.18) *for* $m = 1$ *and which can be covered by sets* $\mathcal{M}_i$ *of macroelements such that:*

i) *All* $M \in \mathcal{M}_i$ *belong to a* **finite** *set* $\{\widehat{M}_j\}_{j=1}^{q}$ *of equivalence classes* **independent of** i . (5.5.53)

ii) *Each* $K \in \mathcal{T}_i$ *belongs to at least one and not more than* L *macroelements* $M \subset \mathcal{M}_i$ *(with* L *independent of* i*)*. (5.5.54)

Each edge $e \in \mathcal{E}_i$ *is contained in the interior of at least one and not more than* L *macroelements* $M \in \mathcal{M}_i$. (5.5.55)

iii) $\dim(N_M) = 1$ *for all* $M \in \mathcal{M}_i$. (5.5.56)

Then the family (X_N, M_N) *is divergence stable, i.e.,* (5.5.2) *holds with* $\beta(N) \geq \beta_0 > 0$.

[7] In the following, we denote the reference macro as well as the corresponding equivalence class by $\widehat{M}$.

The proof of Theorem 5.41 is given in two steps. We begin with a local stability condition.

Lemma 5.42 *Let $\widehat{M}$ be an equivalence class of macroelements such that* (5.5.56) *holds. Then there exists $\beta_{\widehat{M}} > 0$ such that for every $p \in P_M/N_M$*

$$\sup_{\boldsymbol{v}\in V_{0,M}} \frac{\int_M p \operatorname{div} \boldsymbol{v}\, dx}{|\boldsymbol{v}|_{H^1(M)}} \geq \beta_{\widehat{M}} \|p\|_{L^2(M)} \tag{5.5.57}$$

holds for every $M \in \widehat{M}$.

Proof Consider any $M \in \widehat{M}$. Since $V_{0,M}$ and P_M are finite dimensional, it follows that

$$\beta_M = \inf_{p\in P_M/N_M} \sup_{\boldsymbol{v}\in V_{0,M}} \frac{\int_M p \operatorname{div} \boldsymbol{v}\, dx}{\|p\|_{L^2(M)}|\boldsymbol{v}|_{H^1(M)}} > 0\,.$$

We claim that $\beta_M \geq \beta_{\widehat{M}} > 0$ for every $M \in \widehat{M}$. To prove it, let $\xi_1, \ldots, \xi_d$ be the vertices of the elements in $\widehat{M}$. Then $M \in \widehat{M}$ is defined by its vertices $x_1, \ldots, x_d$ which satisfy $x_i = Q_M(\xi_i)$. Hence we have $\beta_M = \beta(x_1, \ldots, x_d)$. Now consider $\boldsymbol{x} = (x_1, \ldots x_d)$ as a point in $\mathbb{R}^{2d}$ and β_M as a function of $\boldsymbol{x}$. After a scaling and translation, we may assume that $h_M = \max_{K\subset M}\{h_K\} = 1$ and that x_1 is the origin. For every $M \in \widehat{M}$, the corresponding set of nodes $\boldsymbol{x}$ thus obtained belongs to a fixed, compact set $D \subset \mathbb{R}^{2d}$. Since all elements $K \in D$ are shape regular, i.e.

$$h_K \leq \sigma\rho_K, \;\; |\cos\theta_{iK}| \leq \gamma < 1, \quad K \in \mathcal{T}_i\,,$$

the function $\beta(\boldsymbol{x})$ is continuous on D and $\beta(\boldsymbol{x}) > 0$ for $\boldsymbol{x} \in D$. Since D is compact, there exists $\beta_{\widehat{M}} > 0$ such that $\beta(\boldsymbol{x}) \geq \beta_{\widehat{M}} > 0$. □

We can now prove Theorem 5.41. On M_N, we will use the **mesh-dependent norm**

$$\|p\|_{\mathcal{T}_i}^2 = \sum_{K\in\mathcal{T}_i} |K|\, \|\nabla p\|_{L^2(K)}^2 + \sum_{e\in\mathcal{E}_i} |e| \int_e |\, [p]\, |^2 ds \tag{5.5.58}$$

where $[p]|_e$ denotes the jump of p across the edge $e \in \mathcal{E}_i$. If $M \in \mathcal{M}_i$, we put

$$\|p\|_M^2 := \sum_{K\subset M} |K|\, \|\nabla p\|_{L^2(K)}^2 + \sum_{e\in\mathcal{E}(M)} |e| \int_e |\, [p]\, |^2 ds\,,$$

with $\mathcal{E}(M)$ denoting the interelement edges in M.

By condition (5.5.53), (5.5.56) and Lemma 5.42, there exists $\beta = \min_{1\le j\le q} \beta_{\widehat{M}_j}$ such that

$$\inf_{0\neq p\in P_M/N_M} \sup_{0\neq \boldsymbol{v}\in V_{0,M}} \frac{\int_M p \operatorname{div} \boldsymbol{v}\, dx}{|\boldsymbol{v}|_{H^1(M)} \|p\|_M} \ge \beta > 0\,. \tag{5.5.59}$$

This implies that for every $p \in M_N/N_M$ there exists $\boldsymbol{v}_M \in X_N$ such that $\boldsymbol{v}_M = 0$ on $\Omega\backslash M$ and such that

$$\begin{aligned} \int_M p \operatorname{div} \boldsymbol{v}_M dx &\ge \beta\, \|p\|_M^2\,, \\ \|\boldsymbol{v}_M\|_{H^1(M)} &\le \|p\|_M\,. \end{aligned} \tag{5.5.60}$$

A global $\boldsymbol{v}$ will be obtained by $\boldsymbol{v} := \sum_{M\in\mathcal{M}_i} \boldsymbol{v}_M$. Then, by (5.5.54),

$$\begin{aligned} \int_\Omega p \operatorname{div} \boldsymbol{v}\, dx \quad &\ge \quad C \sum_{M\in\mathcal{M}_i} \int_M p \operatorname{div} \boldsymbol{v}_M\, dx \\ &\overset{(5.5.60)}{\ge} \quad C\beta \sum_{M\in\mathcal{M}_i} \|p\|_M^2 \end{aligned}$$

and also

$$\begin{aligned} \|\boldsymbol{v}\|_{H^1(\Omega)} &\le C\, |\boldsymbol{v}|_{H^1(\Omega)} \le C \sum_{M\in\mathcal{M}_i} |\boldsymbol{v}_M|_{H^1(M)} \le C \sum_{M\in\mathcal{M}_i} \|p\|_M \\ &\le C\, \|p\|_{\mathcal{T}} \end{aligned}$$

where C depends on L in (5.5.54) but is independent of i. This shows that for every $p \in M_N$

$$\sup_{0\neq \boldsymbol{v}\in X_N} \frac{\int_\Omega p \operatorname{div} \boldsymbol{v}\, dx}{|\boldsymbol{v}|_{H^1(\Omega)}} \ge C_1\, \|p\|_{\mathcal{T}}\,. \tag{5.5.61}$$

We show that $\|p\|_{\mathcal{T}}$ can be replaced by $\|p\|_{L^2(\Omega)}$. To this end, by the continuous inf–sup condition, for every $p \in M_N \subset L_0^2(\Omega)$ exists $\boldsymbol{w} \in H_0^1(\Omega)$ such that

$$\int_\Omega p \operatorname{div} \boldsymbol{w}\, dx \ge C_1\, \|p\|_{L^2(\Omega)}^2\,, \tag{5.5.62}$$

and

$$\|\boldsymbol{w}\|_{H^1(\Omega)} \le \|p\|_{L^2(\Omega)}\,. \tag{5.5.63}$$

Defining $\widetilde{\boldsymbol{w}} = I\boldsymbol{w}$ where I is the Clément-type interpolant defined in Section 5.4, we have

$$\sum_{K\in\mathcal{T}_i} h_K^{-2}\,\|\boldsymbol{w}-\widetilde{\boldsymbol{w}}\|^2_{L^2(K)} + \sum_{e\in\mathcal{E}_i} |e|^{-1}\,\|\boldsymbol{w}-\widetilde{\boldsymbol{w}}\|^2_{L^2(e)} \le C_3\,|\boldsymbol{w}|^2_{H^1(\Omega)} \tag{5.5.64}$$

and

$$\|\widetilde{\boldsymbol{w}}\|_{H^1(\Omega)} \le C_4\,\|\boldsymbol{w}\|_{H^1(\Omega)}\,. \tag{5.5.65}$$

Using (5.5.62)–(5.5.64) gives for any $\mathcal{T}$

$$\begin{aligned}
\int_\Omega p\operatorname{div}\widetilde{\boldsymbol{w}}\,dx &= \int_\Omega p\operatorname{div}(\widetilde{\boldsymbol{w}}-\boldsymbol{w})dx + \int_\Omega p\operatorname{div}\boldsymbol{w}\,dx \\
&\ge \int_\Omega p\operatorname{div}(\widetilde{\boldsymbol{w}}-\boldsymbol{w})dx + C_1\,\|p\|^2_{L^2(\Omega)} \\
&= \sum_{K\in\mathcal{T}_i}\int_K (\boldsymbol{w}-\widetilde{\boldsymbol{w}})\cdot\nabla p\,dx \\
&\quad + \sum_{e\in\mathcal{E}_i}\int_e ((\widetilde{\boldsymbol{w}}-\boldsymbol{w})\cdot n)[p]\,ds + C_1\,\|p\|^2_{L^2(\Omega)} \\
&\overset{(5.5.64)}{\ge} -C_3\,|\boldsymbol{w}|_{H^1(\Omega)}\,\|p\|_{\mathcal{T}} + C_1\,\|p\|^2_{L^2(\Omega)} \\
&\overset{(5.5.63)}{\ge} -C_3\,\|p\|_{L^2(\Omega)}\,\|p\|_{\mathcal{T}} + C_1\,\|p\|^2_{L^2(\Omega)} \\
&= \|p\|^2_{L^2(\Omega)}\,(C_2 - C_3\,\|p\|_{\mathcal{T}}/\|p\|_{L^2(\Omega)}) \\
&\ge \frac{1}{C_4}\,\|\widetilde{\boldsymbol{w}}\|_{H^1(\Omega)}\,\|p\|_{L^2(\Omega)}\,(C_2 - C_3\,\|p\|_{\mathcal{T}}/\|p\|_{L^2(\Omega)})\,,
\end{aligned}$$

hence for every $p \in M_N \subset L^2_0(\Omega)$

$$\sup_{0\ne\boldsymbol{v}\in X_N} \frac{\int_\Omega p\operatorname{div}\boldsymbol{v}\,dx}{\|\boldsymbol{v}\|_{H^1(\Omega)}} \ge C_5\,\|p\|_{L^2(\Omega)} - C_6\,\|p\|_{\mathcal{T}}$$

with $C_5 = C_2/C_4$, $C_6 = C_3/C_4$. So, by (5.5.61),

$$\begin{aligned}
C_5\,\|p\|_{L^2(\Omega)} &\le C_6\,\|p\|_{\mathcal{T}} + \sup_{0\ne\boldsymbol{v}\in X_N} \frac{\int_\Omega p\operatorname{div}\boldsymbol{v}\,dx}{\|\boldsymbol{v}\|_{H^1(\Omega)}} \\
&\le \left(\frac{C_6}{C_1}+1\right) \sup_{0\ne\boldsymbol{v}\in X_N} \frac{\int_\Omega p\operatorname{div}\boldsymbol{v}\,dx}{\|\boldsymbol{v}\|_{H^1(\Omega)}}
\end{aligned}$$

which establishes divergence stability with $\beta_0 = C_1C_5/(C_1+C_6)$.

5.5.7.2 $Q_1 - P_0$ *revisited*

We illustrate the use of Theorem 5.41 for the following example due to R. Stenberg. Let $\{\mathcal{T}_i\}$ be a family of bilinear, quadrilateral element meshes which can be covered by $\mathcal{M}_i$ equivalent to $\widehat{M}$ in Figure 5.4.

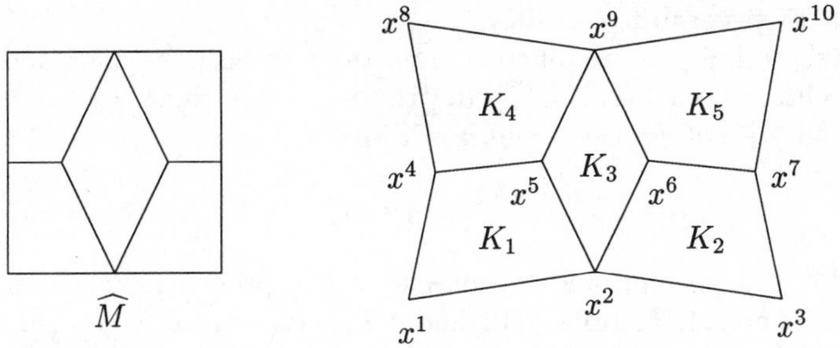

FIG. 5.4. Macro M and reference macro $\widehat{M}$ for $Q_1 - P_0$

We verify (5.5.56). Let $M \in \widehat{M}$ and define

$$\begin{aligned} V_{0,M} &= \{v \in H_0^1(M) : v|_K \in Q_1(K)^2,\ K \in M\}\,, \\ P_M &= \{p \in L^2(M) :\ p|_K = \text{const.},\ K \in M\}\,. \end{aligned}$$

Then $\int_\Omega p \operatorname{div} v\, dx = 0$ for all $v \in V_{0,M}$ gives

$$p_1(x_i^2 - x_i^4) + p_3(x_i^9 - x_i^2) + p_4(x_i^4 - x_i^9) = 0, \quad i = 1, 2$$

if v is chosen such that $v(x^5) = (1,0)$, $v(x^i) = 0$ else. This system admits only the solution $p_1 = p_3 = p_4$.

Using $v \in V_{0,M}$ such that $v(x^6) = (1,0)$, $v(x^i) = 0$ else, we find in the same way $p_2 = p_3 = p_5$ which implies (5.5.56). We have shown the following result.

Theorem 5.43 *On a mesh family $\{\mathcal{T}_i\}$ which can be covered by macroelements $\mathcal{M}_i$ equivalent to $\widehat{M}$ in Figure* 5.4, $Q_1 - P_0$ *is stable.*

The macroelement technique is quite flexible and allows to prove stability of numerous other low order elements on special meshes [115]. It must be emphasized, though, that generally ad hoc stabilization of mixed FEM via the mesh is not recommendable, since the precise mesh-connectivity is often beyond the control of the user.

5.5.8 *p- and hp-FEM for the Stokes problem*

The pairs (X_N, M_N) analyzed so far were all related to h-**version FEM** for the Stokes Problem. The elements were divergence stable even on nonuniform, but shape-regular, meshes such as the (h, γ, K)-meshes or the geometric meshes which is essential in practice when adaptive procedures are used. However, the polynomial degree of these elements was fixed at a moderate level, and even if higher-order families were considered, the stability analyses in Sections 5.5.5 and 5.5.6 were geared towards establishing divergence stability (5.5.2) with $\beta(N) \geq \beta_0 > 0$ as the meshwidth h tends to zero. Here we present techniques to analyze the divergence stability of elements as the polynomial degree k tends to infinity, i.e., **spectral- or p-version stability** [8].

The analysis techniques are different from those in Sections 5.5.5 and 5.5.6 and quite recent, see, e.g., [30], [138] and the references there. Our aim is to present a stability proof for the p- and hp-FEM, i.e., for

$$X_N = S_0^{\boldsymbol{k},1}(\Omega, \mathcal{T}), \;\; M_N = S_0^{\boldsymbol{k},0}(\Omega, \mathcal{T}) \tag{5.5.66}$$

where $\Omega \subset \mathbb{R}^2$ is a polygon and the meshes $\mathcal{T}$ are proper geometric meshes as defined in Chapter 4, Sections 4.3.1 and 4.4.1, respectively (the results will however also apply in the case of quasiuniform meshes).

5.5.8.1 *Definition of the subspaces*

Each element $K \in \mathcal{T}$ is the **affine** image of a reference element $\widehat{K}$ which is either the reference triangle $\widehat{T}$ or the reference square $\widehat{Q}$ (see Chapter 4, Section 4.4.1). It will be essential which function spaces are chosen on the reference element $\widehat{K}$. Let $\boldsymbol{V}^k(\widehat{K})$ and $W^k(\widehat{K})$ be generic velocity or pressure spaces on $\widehat{K}$. Then

$$S_0^{\boldsymbol{k},1}(\Omega, \mathcal{T}) = \{\boldsymbol{u} \in H_0^1(\Omega)^2 : \boldsymbol{u}|_K \circ Q_K \in \boldsymbol{V}^{k_K}(\widehat{K}),\ K \in \mathcal{T}\}\,, \tag{5.5.67}$$

$$S^{\boldsymbol{k},0}(\Omega, \mathcal{T}) = \{u \in L_0^2(\Omega) : \;\; u|_K \circ Q_K \in W^{k_K}(\widehat{K}),\ K \in \mathcal{T}\} \tag{5.5.68}$$

where $\boldsymbol{k} = \{k_K : K \in \mathcal{T}\}$ is the polynomial degree vector.

The stability proofs will depend strongly on the specific choices for $\boldsymbol{V}^k(\widehat{K})$, $W^k(\widehat{K})$.

If $\widehat{K} = \widehat{T}$, we choose

$$\boldsymbol{V}^k(\widehat{T}) = [\mathcal{S}_1^{k+1}(\widehat{T})]^2, \;\; W^k(\widehat{T}) = \mathcal{S}_1^{k-1}(\widehat{T})\,. \tag{5.5.69}$$

If $\widehat{K} = \widehat{Q}$, there are several possible choices, see [138]. We adopt here one particular choice which is sufficient for our purposes, namely we choose (see Chapter 4.2 for the definition of these spaces)

$$\boldsymbol{V}^k(\widehat{Q}) = [\mathcal{E}^k(\widehat{Q}) \oplus \mathcal{J}^{k+1}(\widehat{Q})]^2, \; W^k(\widehat{Q}) = \mathcal{S}_2^{k-1}(\widehat{Q}) \tag{5.5.70}$$

(this is "Method 4" in [138]).

[8] In order to avoid confusion with the pressure p, the polynomial degree is labelled k throughout the remainder of this chapter.

We also define for $\widehat{K} = \widehat{T}$ or $\widehat{Q}$

$$\boldsymbol{V}_0^k(\widehat{K}) = \boldsymbol{V}^k(\widehat{K}) \cap (H_0^1(\widehat{K}))^2 \,. \tag{5.5.71}$$

We have with the above choices that

$$\begin{aligned} \boldsymbol{V}_0^k(\widehat{Q}) &= \{\boldsymbol{v}|\boldsymbol{v} = b_{\widehat{K}}\boldsymbol{w},\ \boldsymbol{w} \in X^k(\widehat{K})\} \\ \boldsymbol{V}_0^k(\widehat{T}) &= \{\boldsymbol{v}|\boldsymbol{v} = b_{\widehat{T}}\boldsymbol{w},\ \boldsymbol{w} \in X^k(\widehat{T})\} \end{aligned} \tag{5.5.72}$$

where

$$b_{\widehat{K}} = \begin{cases} (1-\xi_1^2)(1-\xi_2^2) \text{ if } \widehat{K} = \widehat{Q}\,, \\ (1-\xi_1-\xi_2/\sqrt{3})(1+\xi_1-\xi_2/\sqrt{3}) \text{ if } \widehat{K} = \widehat{T} \end{cases} \tag{5.5.73}$$

denote the **basic bubble functions on** $\widehat{K}$ and

$$X^k(\widehat{K}) = \begin{cases} \mathcal{S}_2^{k-1}(\widehat{Q}) \text{ if } \widehat{K} = \widehat{Q}\,, \\ \mathcal{S}_1^{k-2}(\widehat{T}) \text{ if } \widehat{K} = \widehat{T}\,. \end{cases} \tag{5.5.74}$$

Notice that we have in each case $\widehat{K} = \widehat{Q}$ or $\widehat{K} = \widehat{T}$

$$\nabla q \in [X^k(\widehat{K})]^2 \quad \text{for all } \ q \in W^k(\widehat{K})\,. \tag{5.5.75}$$

Finally, we define a projection operator $\Pi_k : (H_0^1(\widehat{K}))^2 \to \boldsymbol{V}_0^k(\widehat{K})$ by

$$\int_{\widehat{K}} (v_i - \Pi_k v_i)\, w d\xi = 0 \text{ for all } \ w \in X^k(\widehat{K}),\ i = 1,2\,. \tag{5.5.76}$$

The significance of the projection Π_k is apparent from the following theorem.

5.5.8.2 *A basic stability result*

Theorem 5.44 *Assume that the projection* Π_k *in* (5.5.76) *satisfies*

$$||\Pi_k u||_{H^1(\widehat{K})} \le \hat{C} k^\theta ||u||_{H^1(\widehat{K})} \ \textit{for all} \ \ u \in H_0^1(\widehat{K})\,. \tag{5.5.77}$$

Then, with X_N, M_N *as in* (5.5.67)–(5.5.70) *and* $\{\mathcal{T}\}$ *any family of affine, shape regular meshes, we have*

$$\sup_{\boldsymbol{v}\in X_N} \frac{(\operatorname{div} \boldsymbol{v}, q)}{||\boldsymbol{v}||_{H^1(\Omega)}} \ge C\,|\boldsymbol{k}|^{-\theta} ||q||_{L^2(\Omega)} \ \textit{for all} \ \ q \in M_N\,. \tag{5.5.78}$$

Here $|\boldsymbol{k}| = \max\{k_K : K \in \mathcal{T}\}$ *denotes the maximal polynomial degree. The constant* C *is independent of* $\boldsymbol{k}$ *and* $h(\mathcal{T})$ *(it depends only the shape-regularity of the family* $\{\mathcal{T}\}$*).*

Proof We proceed in several steps.

Step 1: For any element $K \in \mathcal{T}$ with associated affine element mapping $Q_K : \widehat{K} \to K$ we define

$$\boldsymbol{V}^k(K) = \{\boldsymbol{v} : \boldsymbol{v} \circ Q_K \in \boldsymbol{V}^k(\widehat{K})\},\ W^k(K) = \{q : q \circ Q_K \in W^k(\widehat{K})\}\,.$$

We claim that (5.5.77) implies the **local inf–sup condition**: let $K \in \mathcal{T}$ and $q^* \in W^k(K) \cap L^2_0(K)$. Then there exists $\boldsymbol{v}^* \in \boldsymbol{V}^k(K) \cap H^1_0(K)$ such that

$$\int_K q^* \operatorname{div} \boldsymbol{v}^*\, dx \geq C_1\, k_K^{-\theta} \|q^*\|_{L^2(K)} \text{ and } |\boldsymbol{v}^*|_{H^1(K)} \leq C_2\, \|q^*\|_{L^2(K)}\,. \tag{5.5.79}$$

We first prove (5.5.79) on the reference element. To this end, define $\hat{q^*} = q^* \circ Q_K$. Since Q_K is affine, $\hat{q}^* \in W^k(\widehat{K}) \cap L^2_0(\widehat{K})$. On $\widehat{K}$, the continuous inf–sup condition shows that there exists $\hat{\boldsymbol{v}} \in H^1_0(\widehat{K})$ such that

$$\int_{\widehat{K}} \hat{q}^* \operatorname{div} \hat{\boldsymbol{v}}\, d\xi \geq C\, \|\hat{q}^*\|^2_{L^2(K)},\quad |\hat{\boldsymbol{v}}|_{H^1(\widehat{K})} \leq \|\hat{q}^*\|_{L^2(\widehat{K})} \tag{5.5.80}$$

and C does not depend on $\hat{q}^*$.

Define now $\hat{\boldsymbol{v}}^* = \Pi_k \hat{\boldsymbol{v}} \in \boldsymbol{V}^k_0(\widehat{K})$. Then

$$\begin{aligned} \int_{\widehat{K}} \hat{q}^* \operatorname{div} \hat{\boldsymbol{v}}^*\, d\xi &= -\int_{\widehat{K}} \hat{\boldsymbol{v}}^* \cdot \nabla \hat{q}^*\, d\xi = -\int_{K} \Pi_k \hat{\boldsymbol{v}} \cdot \nabla \hat{q}^*\, d\xi \\ &= -\int_{\widehat{K}} \hat{\boldsymbol{v}} \cdot \nabla \hat{q}^*\, d\xi = -\int_{\widehat{K}} \hat{q}^* \operatorname{div} \hat{\boldsymbol{v}}^*\, d\xi \\ &\geq C\, \|\hat{q}^*\|^2_{L^2(\widehat{K})} \end{aligned} \tag{5.5.81}$$

where we used (5.5.75), (5.5.76) and (5.5.80). By (5.5.77), we get

$$|\hat{\boldsymbol{v}}^*|_{H^1(\widehat{K})} = |\Pi_k \hat{\boldsymbol{v}}|_{H^1(\widehat{K})} \leq C k^\theta |\hat{\boldsymbol{v}}|_{H^1(\widehat{K})} \leq C k^\theta \|\hat{q}^*\|_{L^2(\widehat{K})}\,,$$

which yields (5.5.79) on $\widehat{K}$.

The case of a general $K \in \mathcal{T}$ will now be reduced by a transformation. We cannot use the affine element parameterization $Q_K(\xi)$ since it does not preserve the divergence operator, but rather the **Piola transform**

$$\boldsymbol{v}^* = \boldsymbol{P}_K(\hat{\boldsymbol{v}}^*) := |\boldsymbol{J}_K|^{-1}\, \boldsymbol{J}_K\, \hat{\boldsymbol{v}}^* \circ Q_K^{-1}\,. \tag{5.5.82}$$

Here $\boldsymbol{J}_K$ is the Jacobian of Q_K and $|\boldsymbol{J}_K| = \det \boldsymbol{J}_K$. Since each velocity component belongs to $V^k(K)$, we have

$$\boldsymbol{P}_K(\hat{\boldsymbol{v}}^*) \in \boldsymbol{V}^k(K) \cap (H_0^1(K))^2, \quad K \in \mathcal{T}. \tag{5.5.83}$$

Moreover, it holds that

$$\int_K q^* \operatorname{div} \boldsymbol{v}^* \, dx = \int_{\widehat{K}} \hat{q}^* \operatorname{div} \hat{\boldsymbol{v}}^* \, d\xi \,. \tag{5.5.84}$$

Since K is shape regular, we have

$$\|q^*\|_{L^2(K)} \approx Ch_K \|\hat{q}^*\|_{L^2(\widehat{K})}, \; \|\boldsymbol{v}^*\|_{H^1(K)} \le Ch_K^{-1} |\hat{\boldsymbol{v}}^*|_{H^1(\widehat{K})} \,. \tag{5.5.85}$$

Now (5.5.79) follows from (5.5.80), (5.5.81), (5.5.84) and (5.5.85).

Step 2: We use the local stability result (5.5.79) to prove (5.5.78). To this end we use the h-version stability of the $P_2 - P_0$ or the $Q_2' - Q_0$ element on mixed affine triangular or quadrilateral meshes $\mathcal{T}$ respectively (see Section 5.5.2).

Let $q \in S_0^{\boldsymbol{k},0}(\Omega, \mathcal{T})$ and write $q = q^* + \bar{q}$ where $\bar{q}$ is the $L^2(\Omega)$-projection of q onto $S_0^{0,0}(\Omega, \mathcal{T})$, i.e. the piecewise constant pressures. The pair $(X_N, M_N) = (S_0^{2,1}(\Omega, \mathcal{T}), S_0^{0,0}(\Omega, \mathcal{T}))$ is stable with inf–sup constant depending only on the shape regularity of $\mathcal{T}$ (cf. Exercise 5.36). Increasing the velocity space does not impair the stability since for every $\bar{q} \in S^{0,0}(\Omega, \mathcal{T})$, the inclusion $S^{2,1} \subset S^{\boldsymbol{k},1}$ implies

$$\sup_{0 \neq \boldsymbol{v} \in S_0^{\boldsymbol{k},1}(\Omega,\mathcal{T})} \frac{\int_\Omega \bar{q} \operatorname{div} \boldsymbol{v}}{\|\boldsymbol{v}\|_{H^1(\Omega)}} \ge \sup_{0 \neq \boldsymbol{v} \in S_0^{2,1}(\Omega,\mathcal{T})} \frac{\int_\Omega \bar{q} \operatorname{div} \boldsymbol{v}}{\|\boldsymbol{v}\|_{H^1(\Omega)}} \ge \beta_0 > 0 \,,$$

i.e., for every $\bar{q} \in S_0^{0,0}$ there is $\bar{\boldsymbol{v}} \in S_0^{\boldsymbol{k},1}$ such that

$$\int_\Omega \bar{q} \operatorname{div} \overline{\boldsymbol{v}} \, dx \ge C_3 \|\bar{q}\|_{L^2(\Omega)}^2, \; |\bar{\boldsymbol{v}}|_{H^1(\Omega)} \le C_4 \|\bar{q}\|_{L^2(\Omega)} \,. \tag{5.5.86}$$

Now, for every $K \in \mathcal{T}$, $q^*|_K \in W^k(K) \cap L_0^2(K)$ and by (5.5.79) there exists $\boldsymbol{v}^* \in S_0^{\boldsymbol{k},1}(\Omega, \mathcal{T})$ such that $\boldsymbol{v}^*|_K \in H_0^1(K)$ and

$$\int_K q^* \operatorname{div} \boldsymbol{v}^* \, dx \ge C_1 k_K^{-\theta} \|q^*\|_{L^2(K)}^2, \; |\boldsymbol{v}^*|_{H^1(K)} \le C_2 \|q^*\|_{L^2(K)} \,. \tag{5.5.87}$$

Moreover, since $\bar{q}|_K = \text{const}$ and $\boldsymbol{v}^* \in (H_0^1(K))^2$,

$$\int_K \bar{q} \operatorname{div} \boldsymbol{v}^* \, dx = 0 \,. \tag{5.5.88}$$

We prove the global inf–sup condition (5.5.78). Given $q \in S_0^{\boldsymbol{k},0}(\Omega, \mathcal{T})$, select $\boldsymbol{v} \in S_0^{\boldsymbol{k},1}(\Omega, \mathcal{T})$ as $\boldsymbol{v} = \boldsymbol{v}^* + \delta \bar{\boldsymbol{v}}$, where $\delta > 0$ is a parameter to be selected. With (5.5.86) - (5.5.88) we get

$$\begin{aligned}
\int_\Omega q \operatorname{div} \boldsymbol{v}\, dx &= \delta \int_\Omega \bar{q} \operatorname{div} \bar{\boldsymbol{v}}\, dx + \delta \int_\Omega q^* \operatorname{div} \bar{\boldsymbol{v}}\, dx + \int_\Omega \bar{q} \operatorname{div} \boldsymbol{v}^*\, dx + \int_\Omega q^* \operatorname{div} \boldsymbol{v}^*\, dx \\
&\geq \delta\, C_3 \, \|\bar{q}\|_{L^2}^2 - \delta |\bar{\boldsymbol{v}}|_{H^1(\Omega)} \|q^*\|_{L^2} + C_1 \, |\boldsymbol{k}|^{-\theta} \, \|q^*\|_{L^2}^2 \\
&\geq \delta\, C_3 \, \|\bar{q}\|_{L^2}^2 - \frac{\delta}{2\varepsilon} \, \|q^*\|_{L^2}^2 - \frac{\varepsilon\delta}{2} \, |\bar{\boldsymbol{v}}|_{H^1}^2 + C_1 \, |\boldsymbol{k}|^{-\theta} \, \|q^*\|_{L^2}^2 \\
&\geq \delta \Big(C_3 - \frac{\varepsilon}{2} \, C_4 \Big) \, \|\bar{q}\|_{L^2}^2 + (C_1 \, |\boldsymbol{k}|^{-\theta} - \delta/2\varepsilon) \, \|q^*\|_{L^2}^2 \\
&\geq C_5 \, |\boldsymbol{k}|^{-\theta} (\|\bar{q}\|_{L^2}^2 + \|q^*\|_{L^2}^2) = C_5 \, |\boldsymbol{k}|^{-\theta} \|q\|_{L^2}^2 \, ,
\end{aligned}$$

with C_5 independent of $|\boldsymbol{k}|$ if we choose first $\varepsilon = C_3/C_4$ and then $\delta = \varepsilon \, |\boldsymbol{k}|^{-\theta}/C_1$.

We also have

$$|\boldsymbol{v}|_{H^1} \leq \delta \, |\bar{\boldsymbol{v}}|_{H^1} + |\boldsymbol{v}^*|_{H^1} \leq \delta \, C_4 \, \|\bar{q}\|_{L^2} + C_2 \, \|q^*\|_{L^2} \leq C \, \|q\|_{L^2}$$

(5.5.78) follows. □

Theorem 5.44 shows that the global inf–sup constant is independent of the particular mesh $\mathcal{T}$ (as long as it belongs to a shape-regular family) and depends only on the norm of the projector Π_k on $\widehat{K}$. We now study this norm for $\widehat{K} = \widehat{Q}$ and $\widehat{K} = \widehat{T}$.

5.5.8.3 *Study of* $\boldsymbol{\Pi_k}$ *for* $\widehat{\boldsymbol{K}} = \widehat{\boldsymbol{Q}}$

On the reference square $\widehat{K} = \widehat{Q}$ we have (5.5.77) and hence the discrete inf–sup condition (5.5.78)) with $\theta = 1/2$.

Theorem 5.45

$$\|\Pi_k \, u\|_{H^1(\widehat{Q})} \leq C\sqrt{k} \, \|u\|_{H^1(\widehat{Q})} \quad \textit{for all} \;\; u \in H_0^1(\widehat{Q}) \, . \tag{5.5.89}$$

The proof of this theorem will be given in several steps. The square $\widehat{Q}$ has product structure and allows to give an explicit representation of $\Pi_k u$. To this end, let

$$U_i(x) := \int_{-1}^{x} L_i(t) \, dt \, , \quad i \geq 1 \tag{5.5.90}$$

where $L_i(t)$ is the Legendre polynomial of degree i in $(-1, 1)$, normalized such that $|L_i(1)| = 1$. Further,

$$U_i(x) = \gamma_i (L_{i+1}(x) - L_{i-1}(x)), \quad i \geq 1 \tag{5.5.91}$$

where $\gamma_i = (2i+1)^{-1}$ for $i \geq 0$. We also have

$$U_i(\pm 1) = 0 \tag{5.5.92}$$

and, also from the orthogonality properties of the Legendre polynomials,

$$\int_{-1}^{1} U_i(x)\, U_j(x)\, dx = \begin{cases} 2\gamma_i^2(\gamma_{i+1} - \gamma_{i-1}) & \text{if } i = j \\ -2\gamma_i\gamma_{i-1}\,\gamma_{i\mp 2} & \text{if } i = j \pm 2 \\ 0 & \text{else}\,. \end{cases} \tag{5.5.93}$$

The internal basis functions $\mathcal{J}_k(\widehat{Q})$ can then be expressed by

$$\mathcal{J}_k(\widehat{Q}) = \Big\{ v | v = \sum_{i,j=1}^{k-1} a_{ij}\, U_i(x)\, U_j(y),\ a_{ij} \in \mathbb{R} \Big\}, \quad k \geq 2\,. \tag{5.5.94}$$

Then $\Pi_k : H_0^1(\widehat{Q}) \to \mathcal{J}_{k+1}(\widehat{Q})$ is defined by

$$\int_{\widehat{Q}} (v - \Pi_k v)\, w d\xi = 0 \text{ for all } w \in \mathcal{S}_2^{k-1}(\widehat{Q})\,, \tag{5.5.95}$$

cf. (5.5.74) and (5.5.76). It is moreover possible to give an explicit characterization of Π_k. To state it, let

$$\mathcal{I}(\widehat{Q}) = \{u \in H_0^1(\Omega) :\ u(x,y) \text{ is a polynomial}\}\,.$$

Then any $u \in \mathcal{I}(\widehat{Q})$ can be written as

$$u(x,y) = \sum_{i,j=1}^{\infty} a_{ij}\, U_i(x)\, U_j(y) \tag{5.5.96}$$

where only finitely many a_{ij} are nonzero. Moreover, we have

Lemma 5.46 $\Pi_k : \mathcal{I}(\widehat{Q}) \to \mathcal{J}_{k+1}(\widehat{Q})$ *is given by*

$$\Pi_k u = \sum_{i,j=1}^{k} a_{ij}\, U_i(x)\, U_j(y) \ \textit{for all}\ u \in \mathcal{I}(\widehat{Q})\,. \tag{5.5.97}$$

Proof Since the functions L_i' are linearly independent, we have

$$\mathcal{S}_2^{k-1}(\widehat{Q}) = \Big\{ v | v = \sum_{i,j=1}^{k} c_{ij}\, U_i''(x)\, U_j''(y),\ c_{ij} \in \mathbb{R} \Big\}\,.$$

Select in (5.5.95) $w(x,y) = U''_\ell(x)\, U''_m(y)$, $1 \le \ell, m \le k$. For $u \in \mathcal{I}(\widehat{Q})$

$$\int\limits_{\widehat{Q}} uw d\xi = \sum_{i,j=1}^{\infty} a_{ij} \int\limits_{-1}^{1} U'_i(x)\, U'_\ell(x) dx \int\limits_{-1}^{1} U'_j(y)\, U'_m(y) dy = 4\gamma_\ell\, \gamma_m\, a_{\ell m}\, .$$

Writing $\Pi_k u = \sum_{i,j=1}^{\infty} b_{ij}\, U_i(x)\, U_j(y)$, we get also

$$\int\limits_{\widehat{Q}} w\, \Pi_k u\, d\xi = 4\gamma_\ell \gamma_m\, b_{\ell m}\, .$$

Referring to (5.5.95) we find therefore $b_{\ell m} = a_{\ell m}$ for $1 \le \ell, m \le k$. □

Next, we characterize $|u|_{H^1(\widehat{K})}$ for $u \in \mathcal{I}(\widehat{Q})$.

Lemma 5.47 *For $u \in \mathcal{I}(\widehat{Q})$,*

$$\left\| \frac{\partial u}{\partial x} \right\|^2_{L^2(\widehat{Q})} = \sum_{j=0}^{\infty} \sum_{i=1}^{\infty} 4\gamma_i \gamma_j (\gamma_{j-1}\, a_{i,j-1} - \gamma_{j+1}\, a_{i,j+1})^2 \, , \tag{5.5.98}$$

$$\left\| \frac{\partial u}{\partial y} \right\|^2_{L^2(\widehat{Q})} = \sum_{j=1}^{\infty} \sum_{i=0}^{\infty} 4\gamma_i \gamma_j (\gamma_{i-1}\, a_{i-1,j} - \gamma_{i+1}\, a_{i+1,j})^2 \, . \tag{5.5.99}$$

Proof We have

$$\frac{\partial u}{\partial x} = \sum_{i=1}^{\infty} a_{ij}\, L_i(x)\, U_j(y)\, ,$$

hence

$$\begin{aligned}
\left\| \frac{\partial u}{\partial x} \right\|^2_{L^2(\widehat{Q})} &= \sum_{i=1}^{\infty} 2\gamma_i \int\limits_{-1}^{1} \Big(\sum_{j=1}^{\infty} a_{ij}\, U_j(y) \Big)^2 dy \\
&= \sum_{i=1}^{\infty} 2\gamma_i \sum_{j=1}^{\infty} \Big(a_{ij}^2 \int\limits_{-1}^{1} U_j^2(y)\, dy + 2a_{ij}\, a_{i,j+2} \int\limits_{-1}^{1} U_j(y)\, U_{j+2}(y)\, dx \Big) \\
&= \sum_{i=1}^{\infty} 2\gamma_i \Big(\sum_{j=1}^{\infty} a_{ij}^2\, \gamma_j^2 (2\gamma_{j-1} + 2\gamma_{j+1}) + 2a_{ij}\, a_{i,j+1}\, \gamma_j \gamma_{j+2} (-2\gamma_{j+1}) \Big)
\end{aligned}$$

where we used (5.5.93). We rearrange the inner sum and get (5.5.98). The relation (5.5.99) follows analogously. □

We can now give the following proof.

Proof of Theorem 5.45 Let $u(x,y)$ be given by (5.5.95). Then $\Pi_k u$ is given by (5.5.97) and $|\Pi_k u|^2_{H^1(\widehat{Q})}$ by (5.5.98), (5.5.99). Consider $\|(\partial\Pi_k u)/(\partial x)\|^2_{L^2(\widehat{Q})} = A_x$. By (5.5.97) and (5.5.98),

$$\begin{aligned} A_x &= \sum_{j=0}^{k-1}\sum_{i=1}^{k} 4\gamma_i\gamma_j(\gamma_{j-1}\,a_{i,j-1} - \gamma_{j+1}\,a_{i,j+1})^2 \\ &\quad + \sum_{i=1}^{k}(4\gamma_i\gamma_k(\gamma_{k-1}\,a_{i,k-1})^2 + 4\gamma_i\gamma_{k+1}(\gamma_k\,a_{ik})^2) \\ &\le \left\|\frac{\partial u}{\partial x}\right\|^2_{L^2(\widehat{Q})} + B_x\,. \end{aligned}$$

To estimate B_x, we assume first that k is even. Then for $i = 1,\ldots,k$, we have the telescoping series

$$\gamma_{k-1}\,a_{i,k-1} = -\sum_{m=0}^{(k-2)/2}(\gamma_{2m-1}\,a_{i,2_{m-1}} - \gamma_{2m+1}\,a_{i,2m+1})$$

(with the convention $a_{i,-1} = 0$). Therefore

$$(\gamma_{k-1}\,a_{i,k-1})^2 \le \frac{k}{2}\sum_{m=0}^{(k-2)/(2}(\gamma_{2m-1}\,a_{i,2_{m-1}} - \gamma_{2m+1}\,a_{i,2m+1})^2\,.$$

Using that $\gamma_k \le \gamma_j$ for $j \le k$ we get

$$4\gamma_i\gamma_k(\gamma_{k-1}\,a_{i,k-1})^2 \le \frac{k}{2}\sum_{m=0}^{(k-2)/2}4\gamma_i\gamma_{2m}\,(\gamma_{2m-1}\,a_{i,2m-1} - \gamma_{2m+1}\,a_{i,2m+1})^2\,.$$

For the term $4\gamma_i\gamma_{k+1}\,(\gamma_k\,a_{ik})^2$ in B_x we get the bound

$$B_x \le C\,k\sum_{i=1}^{k}\sum_{j=0}^{k-1}4\gamma_i\gamma_j\,(\gamma_{j-1}\,a_{i,j-1} - \gamma_{j+1}\,a_{i,j+1})^2 \le C\,k\left\|\frac{\partial u}{\partial x}\right\|^2_{L^2(\widehat{K})}$$

by (5.5.98). Hence we get $A_x \le C\,k\,\|\partial u/\partial x\|^2_{L^2(\widehat{K})}$ which is (5.5.89) for even k. The proof for odd k is analogous. □

5.5.8.4 *Study of* $\boldsymbol{\Pi_k}$ *on* $\widehat{K} = \widehat{T}$

For $\widehat{K} = \widehat{T}$, we have (5.5.77) and hence the discrete inf-sup condition (5.5.78) with $\theta = 3$.

Theorem 5.48

$$\|\Pi_k u\|_{H^1(\widehat{T})} \le C\,k^3\,\|u\|_{H^1(\widehat{T})} \quad \textit{for all } u \in H^1_0(\widehat{T}). \tag{5.5.100}$$

Remark 5.49 It is virtually certain that the bound (5.5.100) is not optimal. Numerical evidence indicates $C\,k^{\theta}\,\|u\|_{H^1(\widehat{T})}$ for some $0 < \theta \le 1$. A proof, however, remains to be found.

We prove (5.5.100) in several steps. Owing to the absence of a tensor product structure on $\widehat{T}$ and an explicit basis for $\mathcal{I}(\widehat{T})$, the proof technique is quite different from that of Theorem 5.45. The proof presented here does not use series expansions of u or Π_k and generalizes to various $3-d$ elements (such as, e.g., hexa- and tetrahedra). We also mention recent work by E. Boillat [33], who has proved divergence stability of $S^{p,1} \times S^{p-1,0}$ on a certain class of quasiuniform, triangular meshes with stability constant $\beta(N) = 1/(k^2\sqrt{\log k})$. However, the constant polynomial degree and the quasiuniformity of the mesh $\mathcal{T}$ (precluding, in particular, exponential convergence rates for singularities) were essential in that work.

To prove Theorem 5.48, we proceed in several steps.

Lemma 5.50 *Let $\widehat{T}$ be the reference triangle and let $b_{\widehat{T}} \in \mathcal{S}_1^3(\widehat{T})$ denote the basic bubble function* (5.5.73) *on $\widehat{T}$. Assume further that*

$$\Lambda(k) := \inf_{v \in \mathcal{S}_1^k(\widehat{T})} \frac{\int_{\widehat{T}} b_{\widehat{T}}(\xi)(v(\xi))^2\, d\xi}{\left(\int_{\widehat{T}}(v(\xi))^2\, d\xi\right)^{\frac{1}{2}} \left(\int_{\widehat{T}}(b_{\widehat{T}}(\xi))^2(v(\xi))^2\, d\xi\right)^{\frac{1}{2}}} \ge C\,k^{1-\theta} \quad (5.5.101)$$

with $C > 0$ independent of k for some $\theta \ge 0$. Then the projection Π_k in (5.5.76) *satisfies* (5.5.77), *i.e.,*

$$\|\Pi_k v\|_{H^1(\widehat{T})} \le C k^{\max\{2,\theta\}}\, \|v\|_{H^1(\widehat{T})} \text{ for all } v \in H_0^1(\widehat{T})\,.$$

Proof Recall that $\Pi_k v$ is defined by

$$v_k = \Pi_k v \in V_0^k(\widehat{T}) := \mathcal{S}_1^{k+1}(\widehat{T}) \cap H_0^1(\widehat{T})$$

such that

$$\int_{\widehat{T}} v_k w\, d\xi = \int_{\widehat{T}} v w\, d\xi \text{ for all } w \in \mathcal{S}_1^{k-2}(\widehat{T})\,.$$

Since $v \in H_0^1(\widehat{T})$, we have $v_k = b_{\widehat{T}} w_k$ with $w_k \in \mathcal{S}_1^{k-2}(\widehat{T})$.

Define $\widetilde{v}_k$ to be the H_0^1-projection of v onto $V_0^k(\widehat{T})$, i.e., $\widetilde{v}_k \in \mathcal{S}_1^{k+1}(\widehat{T})$ such that

$$\int_{\widehat{T}} \nabla\widetilde{v}_k \cdot \nabla w\, d\xi = \int_{\widehat{T}} \nabla v \cdot \nabla w\, d\xi \quad \text{for all } w \in \mathcal{S}_1^{k+1}(\widehat{T}),\ \widetilde{v}_k = 0 \text{ on } \partial\widehat{T}\,.$$

Then from the approximation property of the p-version FEM we have

$$\|v - \widetilde{v}_k\|_{L^2(\widehat{T})} \le C\, k^{-\ell}\, \|v\|_{H^\ell(\widehat{T})} \,. \tag{5.5.102}$$

We observe that $v_k - \widetilde{v}_k \in V_0^k(\widehat{T})$, i.e.,

$$(v_k - \widetilde{v}_k)(\xi) = b_{\widehat{T}}(\xi)\, w_k(\xi), \quad w_k \in \mathcal{S}_1^{k-2}(\widehat{T}) \,, \tag{5.5.103}$$

whence it follows that

$$\int\limits_{\widehat{T}} (v_k - \widetilde{v}_k)\, w\, d\xi = \int\limits_{\widehat{T}} b_{\widehat{T}}\, w_k\, w \text{ for all } w \in \mathcal{S}_1^{k-2}(\widehat{T}) \,.$$

Selecting here $w = w_k$ we get that

$$\int\limits_{\widehat{T}} b_{\widehat{T}}(w_k)^2 d\xi = \int\limits_{\widehat{T}} (v - \widetilde{v}_k)\, w_k\, d\xi \le \|v - \widetilde{v}_k\|_{L^2(\widehat{T})}\, \|w_k\|_{L^2(\widehat{T})} \,, \tag{5.5.104}$$

hence it follows that

$$\frac{\int_{\widehat{T}} b_{\widehat{T}}(w_k)^2 d\xi}{\left(\int_{\widehat{T}} (w_k)^2 d\xi\right)^{\frac{1}{2}}} \le \|v - \widetilde{v}_k\|_{L^2(\widehat{T})} \,.$$

By (5.5.101) (with $v = w_k$) we get from this

$$C\, k^{1-\theta} \int\limits_{\widehat{T}} (b_{\widehat{T}})^2 (w_k)^2\, d\xi \le \|v - \widetilde{v}_k\|_{L^2(\widehat{T})} \,,$$

i.e., by (5.5.103)

$$\|v_k - \widetilde{v}_k\|_{L^2(\widehat{T})} \le C\, k^{\theta - 1}\, \|v - \widetilde{v}_k\|_{L^2(\widehat{T})} \,. \tag{5.5.105}$$

From the triangle inequality we get

$$\begin{aligned} \|v - v_k\|_{L^2(\widehat{T})} &\le \|v - \widetilde{v}_k\|_{L^2(\widehat{T})} + \|\widetilde{v}_k - v_k\|_{L^2(\widehat{T})} \\ &\le C\,(1 + k^{\theta-1})\, \|v - \widetilde{v}_k\|_{L^2(\widehat{T})} \,. \end{aligned}$$

Hence, by (5.5.102) with $\ell = 1$,

$$\begin{aligned} \|v_k\|_{L^2(\widehat{T})} &\le \|v\|_{L^2(\widehat{T})} + \|v - v_k\|_{L^2(\widehat{T})} \\ &\le \|v\|_{L^2(\widehat{T})} + C\,(1 + k^{\theta-1})\, k^{-1}\, \|v\|_{H^1(\widehat{T})} \,. \end{aligned}$$

Using here the inverse inequality (4.5.52)

$$\|v_k\|_{H^1(\widehat{T})} \le C\,k^2\,\|v_k\|_{L^2(\widehat{T})}\,,$$

we obtain for $v_k = \Pi_k v$

$$\begin{aligned}\|v_k\|_{H^1(\widehat{T})} &\le C\,k^2\{\|v\|_{L^2(\widehat{T})} + C(1+k^{\theta-1})k^{-1}\|v\|_{H^1(\widehat{T})}\}\\ &\le C(k^2+k^\theta)\,\|v\|_{H^1(\widehat{T})}\end{aligned}$$

which is the assertion. □

It remains therefore to prove the bound (5.5.101) with $\theta = 3$. This is the purpose of the following lemma.

Lemma 5.51 *The estimate* (5.5.101) *holds with* $\theta = 3$.

Proof The proof is given in several steps.

Step 1: We begin with the observation that for every $v_k \in \mathcal{S}_1^k(\widehat{T})$

$$1 \ge (\Lambda(k))^2 \ge \frac{1}{\|b_{\widehat{T}}\|_{L^\infty(\widehat{T})}}\,\frac{\int_{\widehat{T}} b_{\widehat{T}}\, v_k^2\, d\xi}{\int_{\widehat{T}} v_k^2\, d\xi}\,.$$

It therefore suffices to prove

$$\int_{\widehat{T}} b_{\widehat{T}}\, v_k^2\, d\xi \ge C\,k^{-4}\int_{\widehat{T}} v_k^2\, d\xi \quad\text{for all}\quad v_k \in \mathcal{S}_1^k(\widehat{T}) \tag{5.5.106}$$

with $C > 0$ independent of k.

Step 2: Let $0 < a \le 1/\sqrt{3}$. We decompose $\widehat{T}$ into 6 overlapping parallelograms $\widehat{S}_i$, $i = 1,\dots 6$ and one triangle $\widehat{T}_0$ as shown in Figure 5.5.

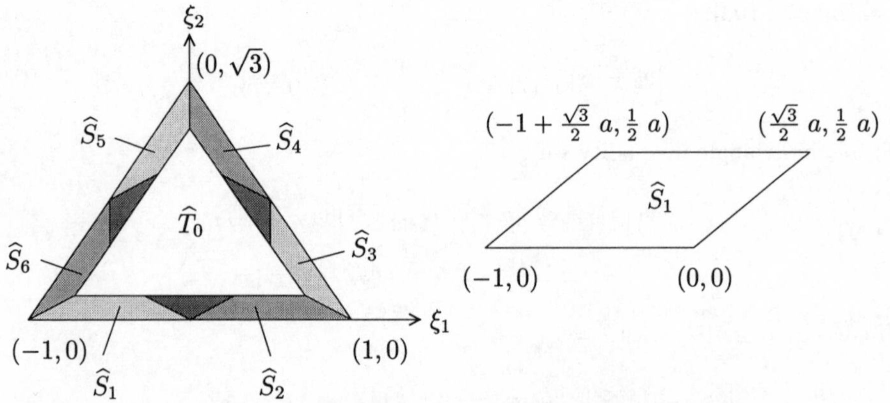

FIG. 5.5. Reference triangle $\widehat{T}$ and the notation

Consider the parallelogram $\widehat{S}_1$. We map it into the standard square $\widehat{S} = (-1,1)^2$ by the linear map

$$\eta_1 = 2\xi_1 - 2\sqrt{3}\,\xi_2 + 1, \;\; \eta_2 = \frac{4\xi_2}{a} - 1\,.$$

Under this map, we have $b_{\widehat{T}}(\xi) \to \hat{b}(\eta)$, where

$$\begin{aligned}\hat{b}(\eta) &= \Big(\frac{a\eta_2 + a}{4}\Big)\Big(\frac{-\sqrt{3}\,\eta_1 - 2a\eta_2 + 3\sqrt{3} - 2a}{2\sqrt{3}}\Big)\Big(\frac{\sqrt{3}\eta_1 + a\eta_2 + \sqrt{3} + a}{2\sqrt{3}}\Big) \\ &= \frac{a}{48}\,(\eta_2 + 1)(\sqrt{3}(3 - \eta_1) - 2a(\eta_2 + 1))(\sqrt{3}(\eta_1 + 1) + a(\eta_2 + 1))\,.\end{aligned}$$

We easily see that on $(-1,1)$,

$$\sqrt{3}(3 - \eta_1) - 2a(\eta_2 + 1) \geq 2(\sqrt{3} - a) \geq \frac{4}{\sqrt{3}}\,.$$

Hence,

$$\hat{b}(\eta) \geq \widetilde{C}_1(a)(\eta_2 + 1)^2 + \widetilde{C}_2(a)(\eta_2 + 1)(\eta_1 + 1)\,.$$

Transforming coordinates from (ξ_1, ξ_2) to (η_1, η_2), we therefore obtain, with $\widetilde{v}_k(\eta) = v_k(\xi)$,

$$\begin{aligned}\int_{\widehat{S}_1} b_{\widehat{T}}(\xi)\, v_k^2(\xi)\, d\xi &= C(a) \int_{\widehat{S}} \hat{b}(\eta)\, \widetilde{v}_k^2(\eta)\, d\eta \\ &\geq C_1(a) \int_{-1}^{1}\int_{-1}^{1} (\eta_2 + 1)^2\, \widetilde{v}_k^2(\eta)\, d\eta \\ &\quad + C_2(a) \int_{-1}^{1}\int_{-1}^{1} (\eta_2 + 1)(\eta_1 + 1)\, \widetilde{v}_k^2(\eta)\, d\eta \\ &\geq C(a)\, k^{-4} \int_{-1}^{1}\int_{-1}^{1} \widetilde{v}_k^2(\eta)\, d\eta \quad \text{(using Theorem 3.96 with } \alpha = 2) \\ &\geq C'\, k^{-4} \int_{\widehat{S}_1} v_k^2(\xi)\, d\xi\,.\end{aligned}$$

Similarly, we may establish that

$$\int_{\widehat{S}_1} b_{\widehat{T}}\, v_k^2(\xi)\, d\xi \geq C k^{-4} \int_{\widehat{S}_1} v_k^2(\xi)\, d\xi \tag{5.5.107}$$

Since $b_{\widehat{T}} \geq C > 0$ on $\widehat{T}_0$, we have also

$$\int_{\widehat{T}_0} b_{\widehat{T}} v_k^2(\xi)\, d\xi \geq C \int_{\widehat{T}_0} v_k^2(\xi)\, d\xi \,. \tag{5.5.108}$$

Step 3: Lemma 5.51 now follows from (5.5.107) and (5.5.108) and the observation that for any nonnegative function $f(\xi)$ on $\widehat{T}$,

$$C\Big(\sum_{i=1}^{6} \int_{\widehat{S}_i} f(\xi)\, d\xi + \int_{\widehat{T}_0} f(\xi)\, d\xi\Big) \leq \int_{\widehat{T}} f(\xi)\, d\xi \leq \sum_{i=1}^{6} \int_{\widehat{S}_i} f(\xi)\, d\xi + \int_{\widehat{T}_0} f(\xi)\, d\xi \,.$$

We finally remark that in the above argument, the parameter a should be chosen small enough that when any triangle T in the triangulation is mapped into $\widehat{T}$, the pre-image of each $\widehat{S}_i$ is still contained in T. By shape regularity, this is always possible.

This implies immediately (5.5.106) and hence the assertion. □

Exercise 5.52 Reprove Theorem 5.45 with the weaker bound

$$\|\Pi_k u\|_{H^1(\widehat{Q})} \leq C k^{\theta} \|u\|_{H^1(\widehat{Q})}$$

for some $\theta \geq 2$ using the technique of proof used for Lemma 5.51.

5.5.8.5 *The rate of convergence of the hp-FEM*

We can combine Theorems 5.44, 5.22 and the general approximation results of Chapter 4 (Theorem 4.63 and Exercise 4.65) to give the following result.

Theorem 5.53 *Let $\Omega \subset \mathbb{R}^2$ a polygon and $(\boldsymbol{u}, p)$ the solution of the Stokes problem* (5.1.15) *with $\boldsymbol{f}$ analytic in $\overline{\Omega}$. Let $\{\mathcal{T}\}$ be a family of regular meshes graded geometrically towards the vertices of Ω and let $\boldsymbol{k}$ be a degree vector which increases linearly away from the vertices. Let $\boldsymbol{u}_N \in S_0^{\boldsymbol{k},1}(\Omega, \mathcal{T})$, $p_N \in S_0^{\boldsymbol{k},0}(\Omega, \mathcal{T})$ be the hp FE-approximation* (5.4.1) *of the Stokes problem.*

Then there exist $C, b > 0$ independent of $N = \dim(S^{\boldsymbol{k},1})$ such that

$$\|\boldsymbol{u} - \boldsymbol{u}_N\|_{H^1(\Omega)} \leq C \exp(-bN^{1/3}) \tag{5.5.109}$$

$$\|p - p_N\|_{L_0^2(\Omega)} \leq C \exp(-bN^{1/3}) \tag{5.5.110}$$

provided that we have the regularity

$$(\boldsymbol{u}, p) \in \mathcal{B}_\beta^2(\Omega) \times \mathcal{B}_\beta^1(\Omega) \,. \tag{5.5.111}$$

Proof Applying Theorems 5.44, 5.45 and 5.48, we get the inf–sup condition (5.5.2) with $\beta(N) = C\,|\boldsymbol{k}|^{-3}$ where $|\boldsymbol{k}| := \max\{k_K : K \in \mathcal{T}\}$ denotes the maximal elemental polynomial degree. Since we also have

$$N = \dim(S_0^{\boldsymbol{k},1}(\Omega, \mathcal{T})) \sim C\,|\boldsymbol{k}|^3\,,$$

we find from Theorem 5.22 the error estimates

$$\|\boldsymbol{u} - \boldsymbol{u}_N\|_{H^1(\Omega)} \le C\left\{N \inf_{\boldsymbol{v}} \|\boldsymbol{u} - \boldsymbol{v}\|_{H^1(\Omega)} + \nu^{-1} \inf_{q} \|p - q\|_{L^2(\Omega)}\right\} \quad (5.5.112)$$

and

$$\|p - p_N\|_{L^2_0(\Omega)} \le C\left\{\nu N^2 \inf_{\boldsymbol{v}} \|\boldsymbol{u} - \boldsymbol{v}\|_{H^1(\Omega)} + N \inf_{q} \|p - q\|_{L^2(\Omega)}\right\} \quad (5.5.113)$$

where C is independent of N and ν and the infima are taken over $\boldsymbol{v} \in S_0^{\boldsymbol{k},1}(\Omega, \mathcal{T})$ and $q \in S_0^{\boldsymbol{k},0}(\Omega, \mathcal{T})$, respectively, the spaces which were defined in Section 5.5.8.1. Now the regularity (5.5.111) implies, as shown in Chapter 4, Theorem 4.63 and Exercise 4.65 that

$$\begin{aligned} \inf_{\boldsymbol{v} \in S_0^{\boldsymbol{k},1}(\Omega,\mathcal{T})} \|\boldsymbol{u} - \boldsymbol{v}\|_{H^1(\Omega)} &\le C\, \exp(-bN^{1/3})\,, \\ \inf_{p \in S_0^{\boldsymbol{k},0}(\Omega,\mathcal{T})} \|p - q\|_{L^2(\Omega)} &\le C\, \exp(-bN^{1/3})\,. \end{aligned}$$

Inserting this into (5.5.112), (5.5.113) and reducing the value of b in the exponential bound gives the assertion. □

Remark 5.54 The regularity (5.5.111) is likely to hold. A rigorous proof, however, is yet to be done. Moreover, the dependence of the constant C on the viscosity ν is unknown at present. See also [34] for regularity results in countably normed spaces.

5.6 Further results and bibliographical remarks

The material on Navier–Stokes equations and their linearization is standard, see e.g., [64] and the references there. The general framework for mixed and hybrid FEM has emerged during the last 10-15 years and our presentation followed [41]. The macroelement technique is due to Stenberg [136], [137]. For further applications of this method see, e.g., [115] and the references there. Qin established in particular div-stability results for the triangular $P_1 - P_0$ elements on special meshes and also for other families of low order elements. The stability results of the p-version-FEM on quasiuniform quadrilateral meshes are due to [138]; the results for triangular elements seem to be new. The stability for regular geometric meshes is obtained here by combining particular low-order results for the $Q_2' \times P_0$ element with the spectral stability analysis. Results for the irregular geometric meshes consisting only of quadrilaterals could be established along the same lines, **provided** divergence stability of low-order quadrilateral elements on irregular geometric meshes (i.e., with hanging nodes) was known. For results in this direction, we refer to [125]. We emphasize further that all stability proofs in the present section require the shape regularity (cf. Definition 4.5) of the underlying mesh families $\{\mathcal{T}\}$.

Establishing divergence stability on highly anisotropic meshes is desirable in order to resolve viscous boundary layers efficiently. Some results in this direction are due to Rannacher and Becker [116] for a nonconforming element. Stability of the so-called $Q_k \times Q_{k-2}$ conforming element on high-aspect ratio quadrilateral elements was proved in [124]; the stability constant $\beta(N)$ is independent of the meshwidth and of the element aspect ratio, and behaves like $k^{-1/2}$ as $k \to \infty$.

All inf–sup constants exhibited in the present chapter depended on the polynomial degree k and degenerate with increasing k. Pairs of spaces which are stable uniformly in k have only recently appeared. We mention in particular the work [31]. There, it is proved that on $\widehat{Q} = (-1,1)^2$ the pair

$$\boldsymbol{V}_0^k = [\mathcal{S}_2^k \cap H_0^1(\widehat{Q})]^2, \;\; W^k = \mathcal{S}_1^{k-1}$$

is stable uniformly in k. The proof is obtained by constructing a more sophisticated projector Π_k than (5.5.76). Using Theorem 5.44, one can deduce immediately stability in both h and k on quasiuniform meshes consisting of quadrilaterals.

For a posteriori error estimates, see [3] and the references there.

6

HP-FEM IN THE THEORY OF ELASTICITY

In this chapter, we discuss *hp*-FEM for elasticity problems. As in the case of Stokes flow, we are dealing with **systems of partial differential equations**, which can be formulated variationally in various ways: either as a primal, displacement problem or as a mixed problem involving displacements and stresses. We will analyze *hp*-FE discretizations of some of these formulations. As we will see, it is again in many cases possible to achieve exponential convergence rates.

In comparison with Stokes flow, however, several new difficulties arise here,. First, even in the linearized case the elastic constitutive law is more complicated than that of heat conduction. In particular, for incompressible materials the constitutive matrix becomes singular and the incompressibility constraint is imposed on the exact as well as the FE solution. This may cause a loss of uniformity in the asymptotic convergence estimates which may be so severe that in practice no convergence at all is visible. This does not mean that asymptotically the FEM does not converge for these problems; rather, it means that quasioptimal convergence starts at such high numbers of degrees of freedom that it practically is never visible. This **nonrobustness** of FEM convergence with respect to various problem parameters is broadly termed **locking** and is analyzed in some detail in the present chapter. It turns out that *p*- and *hp*-FEM are, broadly speaking, free of or substantially less susceptible to locking effects than classical low-order FEM.

Second, many three-dimensional elastic structures, such as car and airplane fuselages, consist of **thin components**, where the **aspect ratio** ε, i.e., thickness versus length, is often as low as 10^{-3}. This has historically motivated the derivation of **approximate lower-dimensional theories**. We will present here some of the most widely used classes of such theories, namely plane stress and plane strain and the so-called Koiter and Naghdi shells which contain, as special cases, the Kirchhoff and Reissner–Mindlin plate theories.

6.1 A brief synopsis of the theory of elasticity

6.1.1 *Kinematics*

We consider the deformations of elastic bodies. To describe them, we assume that there exists a **reference configuration** $\overline{\Omega}$ where $\Omega \subset \mathbb{R}^3$ denotes an open, bounded set. Typically, $\overline{\Omega}$ is the domain occupied by the body if no tractions are applied. The **deformed configuration** B is described by

$$\phi : \overline{\Omega} \to \mathbb{R}^3, \quad B = \phi(\overline{\Omega}) \tag{6.1.1}$$

i.e., $\phi(x)$ is the location of the point $x \in \overline{\Omega}$ after deformation has occurred. We also write

$$\phi(x) = x + u(x) \tag{6.1.2}$$

where $u(x)$ is the **displacement** of $x \in \overline{\Omega}$.

A **rigid body motion** (RBM) is any translation/orthogonal transformation of the body. The map ϕ is called **deformation**, if

$$\det(\nabla\phi) > 0\,, \qquad x \in \overline{\Omega}\,. \tag{6.1.3}$$

Here $\nabla\phi$ denotes the **deformation gradient**

$$\nabla\phi = \begin{pmatrix} \partial_1\,\phi_1 & \partial_2\,\phi_1 & \partial_3\,\phi_1 \\ \partial_1\,\phi_2 & \partial_2\,\phi_2 & \partial_3\,\phi_2 \\ \partial_1\,\phi_3 & \partial_2\,\phi_3 & \partial_3\,\phi_3 \end{pmatrix}. \tag{6.1.4}$$

Condition (6.1.3) implies that subsets of Ω with positive volume are mapped into subsets with positive volume.

We analyze the local behavior of deformations. If ϕ is smooth ($\phi \in C^1(\overline{\Omega})$ suffices), we may write for small $z \in \mathbb{R}^3$

$$\phi(x+z) - \phi(x) = \nabla\phi(x)\,z + o(z)\,. \tag{6.1.5}$$

Hence

$$\|\phi(x+z) - \phi(x)\|^2 = z^\top (\nabla\phi(x))^\top\,(\nabla\phi(x))\,z + o(\|z\|^2) \tag{6.1.6}$$

where $\|z\|$ denotes the Euclidean norm of $z \in \mathbb{R}^3$. Thus

$$C(x) = \nabla\phi(x)^\top\,\nabla\phi(x) \tag{6.1.7}$$

characterizes the local length change and is referred to as the (right) **Cauchy–Green tensor**. The deviation of $C(x)$ from the identity

$$E := \frac{1}{2}\,(C - I) \tag{6.1.8}$$

is called **strain**. Evidently

$$C = C^\top,\ E = E^\top\,, \tag{6.1.9}$$

and, in components,

$$E_{ij} = \frac{1}{2}\,(\partial_i\,u_j + \partial_j\,u_i) + \frac{1}{2}\sum_{k=1}^{3}\partial_k\,u_i\,\partial_k\,u_j\,. \tag{6.1.10}$$

In **linearized three-dimensional elasticity**, the quadratic terms are neglected and

$$\epsilon_{ij} = \frac{1}{2}\left(\partial_i u_j + \partial_j u_i\right), \quad 1 \leq i, j \leq 3 \tag{6.1.11}$$

denotes the **linearized strain tensor**.

If the strain E for a deformation $\phi \in C^1(\overline{\Omega})$ satisfies

$$E(x) = I \text{ for all } x \in \overline{\Omega}\,, \tag{6.1.12}$$

$\phi(x)$ is called a **rigid body motion**, i.e., $\phi(x) = Qx + b$, $Q^\top Q = I$.

6.1.2 *Equilibrium*

The influence of loadings on a body are (axiomatically) assumed to consist of surface loadings $t(x, \boldsymbol{n})$ and volume forces $f(x)$.

Volume forces $f : \Omega \to \mathbb{R}^3$ act on a volume element dV with forces $f dV$. Surface loadings are characterized by $t : \partial\Omega \times S^2 \to \mathbb{R}^3$ where S^2 is the unit sphere in $\mathbb{R}^3$. Let dA denote a surface element of $\overline{\Omega}$ with exterior unit normal $\boldsymbol{n}$, then $t(x, \boldsymbol{n})\, dA$ denotes the contribution to the total forces acting on x. $t(x, \boldsymbol{n})$ is the **Cauchy stress vector**.

In a body in equilibrium, all applied forces must add to 0. This is the **axiom of static equilibrium**. Let $B = \phi(\overline{\Omega})$ denote a deformable body in equilibrium. Then the axiom of static equilibrium states that the volume forces f and stresses t satisfy

$$\int_V f(x)\, dx + \int_{\partial V} t(x, \boldsymbol{n}) ds = 0\,, \tag{6.1.13}$$

$$\int_V x \wedge f(x)\, dx + \int_{\partial V} x \wedge t(x, \boldsymbol{n}) ds = 0 \text{ for all } V \subseteq \Omega \tag{6.1.14}$$

(here "$\wedge$" denotes the vector product in $\mathbb{R}^3$).

We introduce the following notation:

$$\begin{aligned}
\mathbb{M}^3 &= 3 \times 3 \text{ matrices} \\
\mathbb{M}^3_+ &= \{M \in \mathbb{M}^3 : \det(M) > 0\}\,, \\
\mathbb{O}^3 &= \{M \in \mathbb{M}^3 : M^\top M = I\}\,, \\
\mathbb{O}^3_+ &= \mathbb{O}^3 \cap \mathbb{M}^3_+\,, \\
\mathbf{S}^3 &= \{M \in \mathbb{M}^3 : M = M^\top\}\,, \\
\mathbf{S}^3_> &= \{\mathbb{M} \in \mathbf{S}^3 : x^\top M x \geq c x^\top x\}\,.
\end{aligned}$$

The next, fundamental, theorem states that the stress at a point $x \in B$ is characterized by a 3×3 tensor field.

Theorem 6.1 *(Cauchy's theorem). Let* $t(\cdot, \boldsymbol{n}) \in C^1(B, \mathbb{R}^3)$, $t(x, \cdot) \in C^0(\mathbb{R}^3, S^2)$ *and* $f \in C^0(B, \mathbb{R}^d)$ *satisfy* (6.1.13), (6.1.14).

Then there exists $T \in C^1(B, \mathbf{S}^3)$ such that

$$t(x, \boldsymbol{n}) = T(x)\, \boldsymbol{n} \text{ for all } x \in B,\ \boldsymbol{n} \in S^2 , \tag{6.1.15}$$

$$\operatorname{div} T(x) + f = 0 \qquad x \in B , \tag{6.1.16}$$

$$T(x) = T^\top(x) \qquad x \in B . \tag{6.1.17}$$

The tensor $T(x)$ is the **Cauchy stress tensor**.

6.1.3 *Piola transform*

The equilibrium conditions (6.1.16) were given with respect to the deformed configuration B. Since the deformation ϕ is yet to be found, it is convenient to relate everything to reference coordinates $\hat{x} \in \overline{\Omega}$: $x = \phi(\hat{x})$. Hence

$$dx = \det(\nabla\phi)\, d\hat{x} , \tag{6.1.18}$$

$$\rho(x)\, dx = \hat{\rho}(\hat{x})\, d\hat{x} , \tag{6.1.19}$$

$$\rho(\phi(\hat{x})) = \det(\nabla\phi^{-1})\, \hat{\rho}(\hat{x}) , \tag{6.1.20}$$

$$f(x) = \det(\nabla\phi^{-1})\, \hat{f}(\hat{x}) . \tag{6.1.21}$$

Note that (6.1.21) contains the hidden assumption on the loadings that points $\hat{x}$ are not moved by $\phi(\hat{x})$ to a location where the forces have changed. Such loads are called **dead loads**.

The transformation of the stress tensor is equally easily computable: in $\overline{\Omega}$, we have

$$\widehat{\operatorname{div}}\, \widehat{T} + \hat{f} = 0 \tag{6.1.22}$$

where

$$\widehat{T} := \det\,(\nabla\phi)(\nabla\phi)^{-1}\, T(\nabla\phi)^{-\top} . \tag{6.1.23}$$

The tensor $\widehat{T}$ is called the **first Piola–Kirchhoff tensor** and is nonsymmetric. The **second Piola–Kirchhoff tensor** is therefore often introduced:

$$\widehat{\Sigma} = \det(\nabla\phi)(\nabla\phi)^{-1}\, T\, (\nabla\phi)^{-\top} . \tag{6.1.24}$$

Evidently, $\widehat{\Sigma} = (\nabla\phi)^{-1}\, \widehat{T}$.

6.1.4 *Constitutive laws*

Definition 6.2 *A material is* **elastic** *if there exists a map* $\widehat{T} : \mathbb{M}^3_+ \to \mathbf{S}^3_>$ *such that for every deformation* $\phi \in C^1(\overline{\Omega}, \mathbb{R}^3)$ *it holds that*

$$T(x) = \widehat{T}\,(\nabla\phi\,(\hat{x}))\,. \tag{6.1.25}$$

$\widehat{T}$ is called the **response function** and (6.1.25) is the **constitutive law**. Relation (6.1.25) expresses the assumption that the stress depends locally on the strain. With (6.1.24) one defines

$$\widehat{\sum}(F) := \det(F)\,F^{-1}\,\widehat{T}(F)\,F^{-\top}\,. \tag{6.1.26}$$

We confine ourselves to **homogeneous materials**, i.e., we assume that $\widehat{T}$ is independent of $x \in B$.

Axiom 6.3 *The Cauchy stress vector* $t(x, \boldsymbol{n}) = T(x)\boldsymbol{n}$ *is invariant under orthogonal transformations* Q *of the coordinate system, i.e.,*

$$Qt(x, \boldsymbol{n}) = t(Qx, Q\boldsymbol{n})\,. \tag{6.1.27}$$

If (6.1.27) holds, the material is said to be **objective**.

Theorem 6.4 *Let the material be objective. Then*

$$\widehat{T}(QF) = Q\widehat{T}(F)\,Q^\top \text{ for all } Q \in \mathbb{O}^3_+\,. \tag{6.1.28}$$

Moreover, there exists $\overline{\sum} : \mathbf{S}^3_> \to \mathbf{S}^3$ *such that*

$$\widehat{\sum}(F) = \overline{\sum}(F^\top F)\,, \tag{6.1.29}$$

i.e., $\widehat{\sum}$ *depends only on* $F^\top F$.

Proof We must rotate the coordinate system, or, equivalently, the deformed body, i.e., we substitute $x \longmapsto Qx$, $\phi \longmapsto Q\phi$, $\nabla\phi \to Q\nabla\phi$, $\boldsymbol{n} \longmapsto Q^{-\top}\boldsymbol{n} = Q\boldsymbol{n}$, $t(x, n) \longmapsto Qt(x, \boldsymbol{n})$. By Axiom 6.3, $Qt(x, \boldsymbol{n}) = t(Qx, Q\boldsymbol{n})$ (x and x are vectors in the original system). □

We therefore present the following definition.

Definition 6.5 *A material is* **isotropic**, *if*

$$\widehat{T}(F) = \widehat{T}(FQ) \text{ for all } Q \in \mathbb{O}^3_+\,. \tag{6.1.30}$$

Note that (6.1.30) differs from (6.1.28). While objectivity is a general property valid for any material, isotropy is very special. It means that rotation of B prior to deformation does not affect ϕ.

Exercise 6.6 Show, as in the proof of Theorem 6.4, that

$$\widehat{T}(F) = \overline{T}(FF^\top)\,. \tag{6.1.31}$$

Every response function depends essentially on the invariants of $FF^\top$. The invariants $i_k(A)$ of $A \in \mathbb{M}^3$ are defined as coefficients in the characteristic polynomial:

$$\det(A - \lambda I) = -\lambda^3 + i_1(A)\,\lambda^2 - i_2(A)\,\lambda + i_3(A)\,. \tag{6.1.32}$$

As is well known,

$$\begin{aligned} i_1(A) &= \mathrm{tr}\,(A) = \sum_{i=1}^{3} a_{ii} = \lambda_1 + \lambda_2 + \lambda_3\,, \\ i_2(A) &= \frac{1}{2}\,(\mathrm{tr}\,(A))^2 - (\mathrm{tr}\,(A^2)) = \lambda_1\lambda_2 + \lambda_1\lambda_3 + \lambda_2\lambda_3\,, \\ i_3(A) &= \det\,(A) = \lambda_1\lambda_2\lambda_3\,. \end{aligned} \tag{6.1.33}$$

Theorem 6.7 *(Rivlin–Eriksen). A response function $\widehat{T} : \mathbb{M}^3_+ \to \mathbf{S}^3$ is objective and isotropic if and only if $\widehat{T}(F) = \overline{T}(FF^\top)$ and the matrix function $\overline{T}$ satisfies*

$$\begin{aligned} &\overline{T} : \mathbf{S}^3_> \to \mathbf{S}^3\,, \\ &\overline{T}(B) = \beta_0(i(B))\,I + \beta_1(i(B))\,B + \beta_2(i(B))\,B^2\,. \end{aligned} \tag{6.1.34}$$

Here β_i are functions of the invariants $i(B)$ of B.

For the proof, we refer to [46], for example.

Exercise 6.8 Show that $FF^\top = \mu I$ implies $\widehat{T}(F) = \overline{\mu} I$. What is the interpretation for $\overline{\mu}$?

Corollary 6.9 *For isotropic and objective materials it holds that*

$$\widehat{\sum}(\nabla\phi) = \overline{\sum}(\nabla\phi^\top\,\nabla\phi) \tag{6.1.35}$$

and

$$\overline{\sum}(C) \;= \gamma_0\,I + \gamma_1\,C + \gamma_2\,C^2\,. \tag{6.1.36}$$

where the γ_i depend only on ι_c.

Proof With the **Cayley–Hamilton formula**

$$B^3 - \iota_1(B)\,B^2 + \iota_2(B)\,B - \iota_3(B)\,I$$

we eliminate I in (6.1.34) and get

$$\overline{T}(B) = \overline{\beta}_1 B + \overline{\beta}_2 B^2 + \overline{\beta}_3 B^3 .$$

Referring to (6.1.26), we form

$$\begin{aligned}
&(\det(F))^{-1} \hat{\sum}(F) = \\
&F^{-1} \hat{T}(F) F^{\top} &&= F^{-1} \overline{T}(FF^{\top}) F^{-\top} \\
& &&= F^{-1} (\overline{\beta}_1 FF^{\top} + \overline{\beta}_2 (FF^{\top})^2 + \overline{\beta}_3 (FF^{\top})^3) F^{-\top} \\
& &&= \overline{\beta}_1 I + \overline{\beta}_2 FF^{\top} + \overline{\beta}_3 (FF^{\top})^2 \\
& &&= (\det F)^{-1} \overline{\sum} (F^{\top} F)
\end{aligned}$$

by (6.1.29). Hence the assertion follows. □

6.1.5 *Small strains. St. Venant–Kirchhoff material*

For small strains, i.e., near a strain-free reference configuration, the stress-strain relationship is characterized by two parameters. We get from (6.1.36) with $C = I + 2E$ that

$$\overline{\sum} (I + 2E) = \gamma_0 (E) I + \gamma_1 (E) E + \gamma_2 (E) E^2 .$$

Theorem 6.10 *Assume that the material constituting B is* **isotropic** *and* **objective** *and that, moreover, γ_i in* (6.1.36) *are differentiable functions of $\iota_1(E)$, $\iota_2(E)$ and $\iota_3(E)$. Then there exist real numbers π, λ, μ, such that*

$$\overline{\sum} (I + 2E) = \pi I + \lambda \operatorname{tr} (E) + 2\mu E + O(E^2) . \tag{6.1.37}$$

Proof By Exercise 6.8 we know that $\overline{\sum}(I) = \pi I$ for some $\pi \geq 0$. Using (6.1.33), we find $\iota_2 = O(E^2)$ and $\iota_3 = O(E^3)$. Finally, $\iota_1(I + 2E) = \lambda \operatorname{tr} E + 2\mu E + O(E^2)$. □

Usually $C = I$ corresponds to a stress-free state and $\pi = 0$. The remaining two constants in (6.1.37) are called **Lamé constants**. Neglecting higher-order terms, we get the **linear St. Venant-Kirchhoff material**

$$\overline{\sum} (I + 2E) = \lambda \operatorname{tr} (E) I + 2\mu E . \tag{6.1.38}$$

Observe that for small strains (6.1.11)

$$\operatorname{tr} (\epsilon) = \operatorname{div} u . \tag{6.1.39}$$

Thus λ describes stress due to density changes. Often, especially in engineering, the Lamé constants λ and μ are expressed by the equivalent constants $\mathtt{E}$ (Young's modulus of elasticity) and ν (Poisson ratio). We have

$$\nu = \frac{\lambda}{2(\lambda+\mu)}\,, \qquad \mathtt{E} = \frac{\mu(3\lambda+3\mu)}{\lambda+\mu} \tag{6.1.40}$$

$$\lambda = \frac{\mathtt{E}\nu}{(1+\nu)(1-2\nu)}\,, \quad \mu = \frac{\mathtt{E}}{2(1+\nu)}\,. \tag{6.1.41}$$

We have $E, \mu > 0$ and $0 \le \nu \le 1/2$ for most applications (although $-1-\nu \le 1/2$ is also admissible from a mathematical point of view).

6.1.6 *Hyper-elasticity*

By Cauchy's theorem, the equilibrium of an elastic body is described by the equilibrium conditions

$$-\operatorname{div} T(x) = f(x)\,, \qquad x \in \Omega\,, \tag{6.1.42}$$

and by the boundary conditions

$$\phi(x) = \phi_0(x)\,,\ x \in \Gamma^{[0]}\,, \tag{6.1.43}$$

$$T(x)\boldsymbol{n} = g(x)\,,\ x \in \Gamma^{[1]}\,. \tag{6.1.44}$$

We consider (6.1.42)–(6.1.44) as a boundary value problem for the deformation $\phi(x)$, and can write

$$-\operatorname{div} \widehat{T}\,(x, \nabla\phi(x)) \;=\; f(x)\,, \quad x \in \Omega\,, \tag{6.1.45}$$

$$\widehat{T}\,(x, \nabla\phi(x))\,\boldsymbol{n} \;=\; g(x)\,, \quad x \in \Gamma^{[1]}\,, \tag{6.1.46}$$

$$\phi(x) \;=\; \phi(x_0)\,, \quad x \in \Gamma^{[0]}\,. \tag{6.1.47}$$

Remark 6.11

i) Note that we assume that the loadings $f(x)$ and $g(x)$ do not depend on $\phi(x)$, i.e., they are assumed to be dead loads.

ii) Strictly speaking, Ω denotes the domain occupied by the **deformed** body, i.e., Ω must also be determined as part of the solution. To reduce this to a fixed boundary value problem, we identify Ω with the reference domain and obtain in this way an approximation which is valid for small deformations.

Definition 6.12 *An elastic material is hyperelastic if there exists an energy functional* $\widehat{W} : \Omega \times \mathbb{M}^3_+ \to \mathbb{R}$ *such that*

$$\widehat{T}\,(x,F) = \frac{\partial \widehat{W}}{\partial F}\,(x,F)\,, \qquad x \in \Omega,\ F \in \mathbb{M}^3_+\,. \tag{6.1.48}$$

For hyperelastic materials we can associate an energy-minimization principle with (6.1.45)–(6.1.47), provided the loads f and g are derived from a potential: $f = \text{grad}\,\mathcal{F}$, $g = \text{grad}\,\mathcal{G}$. Then the solutions ϕ of (6.1.45)–(6.1.47) are stationary points of the **total energy**

$$I(\psi) = \int_{\Omega} [\widehat{W}(x, \nabla\psi(x)) - \mathcal{F}(\psi(x)]\,dx - \int_{\Gamma^{[1]}} \mathcal{G}(\psi(x))do\,. \tag{6.1.49}$$

Admissible deformations are those for which $\psi|_{\Gamma^{[0]}} = 0$ and $\det(\nabla\psi(x)) > 0$.

Note also that (6.1.45)–(6.1.47) is a pure displacement problem. In practice one often keeps the stress tensor in the formulation, thereby obtaining mixed variational formulations which we will treat in the next section.

We now consider (6.1.45) - (6.1.47) for an objective material. In this case the stored energy $\widehat{W}$ depends only on $C = F^\top F$, i.e.,

$$\widehat{W}(x,F) = \overline{W}(x, F^\top F)$$

and

$$\overline{\sum}(x,C) = 2\,\frac{\partial \overline{W}(x,G)}{\partial C}\,, \quad C \in \mathbf{S}^3_>\,.$$

Actually, $\overline{W}$ depends only on the invariants ι_C of C, i.e.,

$$\overline{W}(x,C) = W(x,\iota_C)\,, \quad C \in \mathbf{S}^3_>\,.$$

For small deformations we have in particular

$$\overline{W}(x,C) = \frac{\lambda}{2}\,(\text{tr } E)^2 + \mu\ \ E:E + o(E^2) \tag{6.1.50}$$

with $C = I + 2E$ and

$$A : B = \sum_{ij} A_{ij}\,B_{ij} = \text{tr }(A^\top B)\,.$$

For a St. Venant–Kirchhoff material,

$$\overline{W}(x,F) = \frac{\lambda}{2}\,(\text{tr } F)^2 + \mu\ \ F:F\,. \tag{6.1.51}$$

More generally, for isotropic materials,

$$\overline{W}(x, F) = \overline{W}(x, FQ) \text{ for all } F \in \mathbb{M}^3_+ , Q \in \mathbb{O}^3_+ .$$

Note that (6.1.51) applies only for small strains, since in general

$$\widehat{W}(x, F) \to \infty \text{ for } \det F \to 0 \tag{6.1.52}$$

holds (which is not satisfied by (6.1.51)).

Note that (6.1.52) shows that $\widehat{W}$ cannot be a convex function of F.

Exercise 6.13 Show that

$$B = \{F \in \mathbb{M}^3_+ : \det F > 0\}$$

is **not** a convex set. Hint: Show that there exists F_0 with $\det F_0 = 0$, $F_0 = \lambda F_1 + (1-\lambda) F_2$ and $\det (F_i) > 0$, $i = 1, 2$.

6.1.7 *Linearized three-dimensional elasticity*

If we assume that $\partial_i u_j$ is small and drop in the definition (6.1.10) of E all terms of higher order, we obtain what is called **(geometric) linearization** of E; it is indicated by writing the symbol ϵ instead of E, i.e.,

$$\epsilon[u] = Du = \frac{1}{2} (\text{grad } u + (\text{grad } u)^\top); \quad \epsilon_{ij} = \frac{1}{2} (\partial_i u_j + \partial_j u_i) . \tag{6.1.53}$$

Likewise, in this linearized setting we write σ instead of $\sum$. Then **the constitutive equations for isotropic material** read

$$\epsilon = \frac{1+\nu}{\mathrm{E}} \sigma - \frac{\nu}{\mathrm{E}} (\text{tr } \sigma) I . \tag{6.1.54}$$

This linear relationship is called **Hooke's law** and written symbolically as $\sigma = A\epsilon$ or $\epsilon = A^{-1}\sigma$ where A is a fourth-order tensor. The equilibrium equations (6.1.42) become

$$\operatorname{div} \sigma + f = 0 \text{ in } \Omega . \tag{6.1.55}$$

The **total potential energy** of a body Ω undergoing a linearly elastic displacement $u : \overline{\Omega} \to \mathbb{R}^3$ is given by

$$\Pi(u) = \int_{\Omega} \left\{\frac{1}{2} \epsilon : \sigma - f \cdot u\right\} dx - \int_{\Gamma^{[1]}} g \cdot u \, do . \tag{6.1.56}$$

Here $\epsilon : \sigma = \sum_{i,k} \sigma_{ik} \epsilon_{ik}$. We may invert (6.1.54) to get

$$\sigma = \frac{\mathrm{E}}{1+\nu}\left(\epsilon + \frac{\nu}{1-2\nu}\,(\mathrm{tr}\;\epsilon)\,I\right). \tag{6.1.57}$$

Exercise 6.14 Show with (6.1.57) and (6.1.41) that

$$\frac{1}{2}\,\sigma : \epsilon = \frac{\lambda}{2}\,(\mathrm{tr}\;\epsilon)^2 + \mu\,\epsilon : \epsilon\,. \tag{6.1.58}$$

We may write (6.1.57) componentwise in the form

$$\sigma = \begin{pmatrix}\sigma_{11}\\ \sigma_{22}\\ \sigma_{33}\\ \sigma_{12}\\ \sigma_{13}\\ \sigma_{23}\end{pmatrix} = \frac{\mathrm{E}}{(1+\nu)(1-2\nu)}$$

$$\times \begin{pmatrix} 1-\nu & \nu & \nu & & & \\ \nu & 1-\nu & \nu & & 0 & \\ \nu & \nu & 1-\nu & & & \\ & & & 1-2\nu & & \\ & 0 & & & 1-2\nu & \\ & & & & & 1-2\nu \end{pmatrix} \begin{pmatrix}\epsilon_{11}\\ \epsilon_{22}\\ \epsilon_{33}\\ \epsilon_{12}\\ \epsilon_{13}\\ \epsilon_{23}\end{pmatrix} \tag{6.1.59}$$

or, briefly,

$$\sigma = A\epsilon\,. \tag{6.1.60}$$

Exercise 6.15 Show that the matrix A in (6.1.60) is positive definite for $0 \leq \nu < 1/2$. Hint: use

$$A^{-1} = \frac{1}{\mathrm{E}} \begin{pmatrix} 1 & -\nu & -\nu & & & \\ -\nu & 1 & -\nu & & 0 & \\ -\nu & -\nu & 1 & & & \\ & & & 1+\nu & & \\ & & 0 & & 1+\nu & \\ & & & & & 1+\nu \end{pmatrix}.$$

Remark 6.16 (on incompressible material). We notice that for $\nu \to 1/2$ the expression (6.1.57) becomes singular. In the limiting case $\nu = 1/2$ the incompressibility **constraint** (see (6.1.39))

$$\mathrm{tr}\;\epsilon[u] = \sum_{i=1}^{d} \frac{\partial u_i}{\partial x_i} = \mathrm{div}\,u = 0$$

is imposed which corresponds to the incompressibility of the material.

Remark 6.17 The most general constitutive law (6.1.60) has a symmetric matrix $A \in \mathbb{R}^{6\times 6}$ characterized by 21 elastic moduli. Various special cases are frequently used in engineering practice: **orthotropic materials**, for example, are characterized by 9 elastic moduli in (6.1.60), whereas the more general class of **monoclinic materials** is characterized by 12 constants in the matrix A.

We proceed to give several variational formulations of the linearized problem. As with the membrane problem discussed in Chapter 1, several formulations are possible which give rise to different FEM.

6.1.7.1 *Pure displacement formulation*

We wish to find a displacement field $u : \Omega \to \mathbb{R}^3$ which minimizes the **total potential energy**

$$\begin{aligned}\Pi(v) &= \int_\Omega \left\{\frac{1}{2}\, Dv : ADv - f\cdot v\right\} dx - \int_{\Gamma^{[1]}} g\cdot v\, do \\ &= \int_\Omega \left\{\mu\epsilon[v] : \epsilon[v] + \frac{\lambda}{2}\,(\mathrm{div}\ v)^2 - f\cdot v\right\} dx - \int_{\Gamma^{[1]}} g\cdot v\, do\end{aligned} \tag{6.1.61}$$

where we used Exercise 6.14 and Dv is defined in (6.1.53).

Formally, the Euler–Lagrange equations for a minimizer are: find $u \in H^1(\Omega, \Gamma^{[0]})$ such that

$$B(u,v) := (Du, ADv)_0 = (f,v)_0 + \int_{\Gamma^{[1]}} g\cdot v\, do =: F(v) \tag{6.1.62}$$

for all $v \in H^1(\Omega, \Gamma^{[0]})$. In particular, for linearly elastic isotropic materials we get the bilinear form

$$B(u,v) = 2\mu(Du, Dv)_0 + \lambda(\mathrm{div}\ u, \mathrm{div}\ v)_0\ . \tag{6.1.63}$$

The Euler–Lagrange equation of (6.1.62) in strong form reads

$$-\mu\Delta u - (\lambda+\mu)\,\mathrm{grad}\ \mathrm{div}\ u = f \ \text{ in } \ \Omega\ ,$$

$$u = 0 \text{ on } \Gamma^{[0]}, \quad \sigma[u]\,\boldsymbol{n} = \boldsymbol{n}\cdot A\,\epsilon[u] = g \ \text{ on } \ \Gamma^{[1]}\ .$$

6.1.7.2 *Hellinger–Reissner principles*

In the first form of the HR principle we keep the stresses σ in the formulation, but eliminate the strains ϵ, resulting in the following saddle point problem:

$$\begin{aligned}(A^{-1}\sigma - \epsilon[u], \tau)_0 &= 0 && \text{for all } \ \tau \in L^2(\Omega)\ , \\ (\sigma, \epsilon[v])_0 &= (f,v)_0 + \int_{\Gamma^{[1]}} g\cdot v\, do && \text{for all } \ v \in H^1(\Omega, \Gamma^{[0]})\ .\end{aligned} \tag{6.1.64}$$

Its strong form reads

$$\begin{aligned} -\operatorname{div} \sigma = f \qquad & \text{in } \Omega\,, \\ \sigma = A\epsilon[u] \qquad & \text{in } \Omega\,, \\ u = 0 \text{ on } \Gamma^{[0]}\,, \quad \sigma \cdot n = g & \text{ on } \Gamma^{[1]}\,. \end{aligned} \tag{6.1.65}$$

Formally, (6.1.64) fits into the framework for abstract saddle point problems considered in Chapter 5, if we select

$$\begin{aligned} \sigma \in X &= [L^2(\Omega)]^{3\times 3}_{\text{sym}}\,, \quad u \in M = [H^1(\Omega, \Gamma^{[0]})]^3\,, \\ a(\sigma, \tau) &= (A^{-1}\sigma, \tau)_0\,, \quad b(\tau, v) = (\tau, Dv)_0\,. \end{aligned} \tag{6.1.66}$$

Notice that now the stresses are the primary variables whereas the displacements u assume the role of Lagrange multiplier, similar to the pressure in the Stokes problem.

The second form of the HR-principle is given by: find $(\sigma, u) \in X_g \times M$ such that

$$\begin{aligned} a(\sigma, \tau) + b(u, \tau) &= 0 && \text{for all } \tau \in X_0 \\ b(v, \sigma) &= -(f, v)_0 && \text{for all } v \in M\,. \end{aligned} \tag{6.1.67}$$

Here

$$\begin{aligned} X_g &= \{\sigma \in H(\operatorname{div}, \Omega) : \sigma n = g \text{ on } \Gamma^{[1]}\}\,, \quad M = L^2(\Omega)\,, \\ a(\sigma, \tau) &= (A^{-1}\sigma, \tau)_0\,, \quad b(u, \tau) = (u, \operatorname{div} \tau)_0\,, \\ H(\operatorname{div}, \Omega) &= \overline{C^\infty(\Omega, S^3)}^{\|\circ\|_{H(\operatorname{div},\Omega)}}\,, \\ \|\tau\|_{H(\operatorname{div},\Omega)} &= (\|\tau\|_0^2 + \|\operatorname{div} \tau\|_0^2)^{1/2}\,. \end{aligned}$$

The second form is sometimes also referred to as the complementary energy principle; notice that now the Neumann data g on $\Gamma^{[1]}$ are essential boundary conditions.

6.1.7.3 *Hu–Washizu principle*

Here all quantities, σ, ϵ and u, are kept in the weak formulation:

$$\begin{aligned} (A\epsilon - \sigma, \eta)_0 &= 0 && \text{for all } \eta \in L^2(\Omega)\,, \\ (\epsilon - Du, \tau)_0 &= 0 && \text{for all } \tau \in L^2(\Omega)\,, \\ -(\sigma, Dv)_0 &= -(f, v)_0 - \int_{\Gamma^{[1]}} g \cdot v\, do && \text{for all } v \in H^1(\Omega, \Gamma^{[0]})\,. \end{aligned} \tag{6.1.68}$$

6.1.7.4 *Well-posedness*

To prove well-posedness of the displacement problem (6.1.62), we require coercivity of the bilinear form $B(\cdot,\cdot)$. The crucial result for establishing this is the following.

Theorem 6.18 *(Korn inequality). Let $d \geq 2$ and $\Omega \subset \mathbb{R}^d$ be open and bounded with Lipschitz boundary. Then there exists $c(\Omega) > 0$ such that*

$$(\epsilon[v], \epsilon[v])_0 + (v, v)_0 \geq c \, \|v\|^2_{H^1(\Omega)} \tag{6.1.69}$$

for every $v \in H^1(\Omega)$.

For a proof, we refer to [106], Chapter 1. The estimate (6.1.69) is not in the form that is most useful for an analysis of (6.1.62) due to the presence of $(v, v)_0$. Since, however, $\epsilon[v] = 0$ if v is a rigid body motion, this term can generally not be omitted.

Proposition 6.19 *(rigid displacement lemma). Let $d \geq 2$ and $\Omega \subset \mathbb{R}^d$ be any open and connected set. Then, for $v \in H^1(\Omega)$,*

$$\epsilon[v] = 0 \Longleftrightarrow v = \boldsymbol{A}\vec{x} + \vec{b} \tag{6.1.70}$$

where $\boldsymbol{A} \in \mathbb{R}^{d\times d}$ is an arbitrary skew-symmetric matrix (i.e., $\boldsymbol{A} = -\boldsymbol{A}^\top$) and $\vec{b} \in \mathbb{R}^d$ is arbitrary .

Proof It holds for $v_k \in H^1(\Omega)$ that

$$\frac{\partial^2 v_k}{\partial x_i \, \partial x_j} = \frac{\partial}{\partial x_i} \, \epsilon_{jk} + \frac{\partial}{\partial x_j} \, \epsilon_{ik} - \frac{\partial}{\partial x_k} \, \epsilon_{ij} = 0$$

in $H^{-1}(\Omega)$, hence v_k is linear and $v(x) = \boldsymbol{A}x + \vec{\boldsymbol{b}}$. Now $\epsilon[v] = 0$ implies that $\boldsymbol{A} = -\boldsymbol{A}^\top$. That $\epsilon[v] = 0$ for such $v(x)$ is obvious. □

Motivated by (6.1.71), we define the space of **rigid body motions**

$$R = \{v(x) : v(x) = \boldsymbol{A}x + \vec{\boldsymbol{b}}, \ \ \boldsymbol{A} = -\boldsymbol{A}^\top, \ \ \vec{x} \in \mathbb{R}^d\} \, .$$

We have

$$\dim R = d(d+1)/2 \, .$$

The $v \in R$ of the form $v(x) = \vec{\boldsymbol{b}}$ are **rigid body translations** whereas $v(x) = \boldsymbol{A}\vec{x}$ with $\boldsymbol{A} = -\boldsymbol{A}^\top$ are **rigid body rotations**. In $\mathbb{R}^d$, there are $d(d-1)/2$ rigid body rotations.

We show now that Korn's inequality (6.1.69) holds without the term $(v, v)_0$ on the left-hand side, if the set of admissible displacements v is restricted properly. The resulting inequality is the analog of the first Poincaré inequality for elasticity.

Proposition 6.20 *(first Korn inequality). Let $\Omega \subset \mathbb{R}^d$, $d \geq 2$, be open, bounded and let $\partial\Omega$ be Lipschitz. Assume that the Dirichlet boundary $\Gamma^{[0]} \subseteq \partial\Omega$ has positive surface measure, i.e., that*

$$\int_{\Gamma^{[0]}} do > 0 \, .$$

Then

$$(\epsilon[v], \, \epsilon[v])_0 \geq c(\Omega, \Gamma^{[0]}) \, \|\nabla v\|^2_{H^1(\Omega)} \tag{6.1.71}$$

holds for every $v \in H^1(\Omega, \Gamma^{[0]})$.

Proof The proof is by contradiction and does not give an explicit value for $c(\Omega, \Gamma^{[0]})$. A similar argument will be used later on several occasions.

Assume therefore that (6.1.71) is false. Then there exists a sequence $\{v_n\}_n \subset H^1(\Omega, \Gamma^{[0]})$ such that

$$\|e[v_n]\|^2_{L^2(\Omega)} = (\epsilon[v_n], \, \epsilon[v_n])_0 \leq \frac{1}{n} \, , \quad |v_n|_{H^1(\Omega)} = 1 \, .$$

By the first Poincaré inequality, $\|v_n\|_{H^1(\Omega)} \leq c_1$ for some $c_1 < \infty$ independent of n. Since $H^1 \subset L^2(\Omega)$ with compact inclusion (Rellich's theorem), there exists a subsequence, again denoted by $\{v_n\}$, which converges in $L^2(\Omega)$ to a limit u_0. By Theorem 6.18, we have

$$\begin{aligned} c\,\|v_n - v_m\|^2_{H^1(\Omega)} &= \|\epsilon[v_n - v_m]\|^2_{L^2(\Omega)} + \|v_n - v_m\|^2_{L^2(\Omega)} \\ &\leq 2\,\|\epsilon[v_n]\|^2_{L^2(\Omega)} + 2\,\|\epsilon[v_m]\|^2_{L^2(\Omega)} + \|v_n - v_m\|^2_{L^2(\Omega)} \\ &\leq \frac{2}{n} + \frac{2}{m} + \|v_n - v_m\|^2_{L^2(\Omega)} \, . \end{aligned}$$

Sending $n, m \to \infty$, we find that $\{v_n\}$ is Cauchy in $H^1(\Omega)$. By the completeness of $H^1(\Omega)$, and since $H^1(\Omega, \Gamma^{[0]})$ is closed in $H^1(\Omega)$, there exists a limit $\widetilde{u}_0 = \lim_{n\to\infty} v_n \in H^1(\Omega, \Gamma^{[0]})$. Evidently, $\widetilde{u}_0 = u_0$. Moreover,

$$\|\epsilon[u_0]\|_{L^2(\Omega)} = \lim_{n\to\infty} \|\epsilon[v_n]\|_{L^2(\Omega)} = 0 \, ,$$

and

$$|u_0|_{H^1(\Omega)} = \lim_{n\to\infty} |v_n|_{H^1(\Omega)} = 1 \, .$$

Thus $\epsilon[u_0] = 0$ almost everywhere and, by Proposition 6.19, $u_0 \in R$. Since $u_0 = 0$ on $\Gamma^{[0]}$ and $\int_{\Gamma^{[0]}} do > 0$, it follows that $u_0 = 0$, a contradiction. □

Combining (6.1.71) with the first Poincaré inequality, we get

$$(\epsilon[v], \, \epsilon[v])_0 \geq c\,\|v\|^2_{H^1(\Omega)} \, , \quad v \in H^1(\Omega, \Gamma^{[0]}) \, . \tag{6.1.72}$$

Theorem 6.21 *Let $\Omega \subset \mathbb{R}^3$ be open, bounded and connected. Assume that $\int_{\Gamma^{[0]}} do > 0$. Then there exists $c(\Omega, \Gamma^{[0]}) > 0$ such that*

$$B(v,v) = \int_\Omega \epsilon[v] : A\epsilon[v] dx \geq c\,\|v\|^2_{H^1(\Omega)}$$

for every $v \in H^1(\Omega, \Gamma^{[0]})$.

For every $F \in (H^1(\Omega, \Gamma^{[0]}))^*$, there exists a unique solution of the boundary value problem (6.1.62).

For the proof we observe that the positive definiteness of the elasticity tensor A and the symmetry of the strain tensor $\epsilon[v]$ implies for any $v \in H^1(\Omega)$

$$B(v,v) = \int_\Omega \epsilon[v] : A\,\epsilon[v] dx \geq \lambda_{\min}(A)\,(\epsilon[v],\, \epsilon[v])_0\ .$$

Using that $v \in H^1(\Omega, \Gamma^{[0]})$, we find with (6.1.72) that $B(\cdot,\cdot)$ is coercive, hence the assertion on existence follows from the Lax–Milgram lemma.

Now we consider the Neumann (or "pure traction") problem, i.e., the case $\Gamma^{[0]} = \varnothing$. Here a first guess for the space in which to seek a weak solution is $H^1(\Omega)$. Let therefore $u \in H^1(\Omega)$ be a weak solution of (6.1.62), i.e.,

$$u \in H^1(\Omega)\,, \qquad B(u,v) = F(v) \text{ for all } v \in H^1(\Omega)\ .$$

Then, by Proposition 6.19, we have necessarily

$$0 = B(u,r) = F(r) \text{ for all } r \in R \subset H^1(\Omega)\,,$$

i.e., a necessary condition on the data (f, g) for a weak solution of the Neumann problem to exist is orthogonality to rigid body motions $r \in \mathbb{R}$:

$$F(r) = \int_\Omega f \cdot r\, dx + \int_{\partial\Gamma} g \cdot r\, do = 0 \text{ for all } r \in \mathbb{R}\,, \tag{6.1.73}$$

(i.e., vanishing resultant forces and moments of the applied traction). Since $R \subset H^1(\Omega)$ and $B(r,r) = 0$ for $r \in \mathbb{R}$, uniqueness in $H^1(\Omega)$ cannot hold and we must factor the rigid body motions out of $H^1(\Omega)$ in order to restore coercivity of the variational problem. The following is the analog of Proposition 6.20 for the Neumann problem.

Proposition 6.22 *(second Korn inequality). Define*

$$\widetilde{H}^1(\Omega) = \{u \in [H^1(\Omega)]^3 : (u,r)_0 = 0 \textit{ for all } r \in \mathbb{R}\}\ . \tag{6.1.74}$$

Then $\widetilde{H}^1(\Omega) \subset H^1(\Omega)$ is a closed, linear subspace and there exists $c(\Omega) > 0$ such that

$$(\epsilon[v],\, \epsilon[v])_0 \geq c(\Omega)\, \|v\|_{H^1(\Omega)} \textit{ for all } v \in \widetilde{H}^1(\Omega)\ . \tag{6.1.75}$$

Proof The proof of (6.1.75) is by contradiction and quite analogous to that of Proposition 6.20, so we skip it. □

As a consequence, we state the following analog of Theorem 6.21 for the Neumann problem.

Theorem 6.23 *Consider the pure traction problem of linearized elastostatics, i.e.* (6.1.64) *with* $\Gamma^{[0]} = \varnothing$*, and suppose the loads* f *and* g *are in equilibrium, i.e.,* (6.1.73) *holds. Then the variational problem* (6.1.64) *admits a solution* $u \in \widetilde{H}^1(\Omega)$ *which is unique up to rigid body motions* $r \in \mathrm{I\!R}$.

The proof is analogous to that of Theorem 6.21, using Proposition 6.22 instead of Proposition 6.20.

Remark 6.24 (on "mixed" boundary conditions in three-dimensional elasticity). In the above considerations, we discussed only the case when one type of boundary condition is prescribed for **all** 3 displacement components at a point $x \in \partial\Omega$, i.e., either Dirichlet conditions $u(x) = 0$ (zero displacement/"(totally) clamped" boundary) or Neumann conditions $\sigma[u]\,\boldsymbol{n} = g(x)$ (prescribed normal tractions g) were admissible. This was done in order to develop the existence theory completely in parallel to the membrane case. There is, however, one essential difference due to the fact that the equilibrium equations are an elliptic system of three second-order equations. It is therefore possible (and common in engineering practice) to pose different conditions for different displacement components at a boundary point. To understand the possible combinations, consider the Green's-formula

$$-(\mathrm{div}\, A\epsilon[u], v)_0 = (\epsilon[v],\ A\epsilon[u])_0 + \int_{\partial\Omega} v \cdot \sigma[u]\, n\, do \tag{6.1.76}$$

valid for $v \in H^1(\Omega)$, $u \in H := H^1(\Omega) \cap \{u : \mathrm{div} A\epsilon[u] \in L^2(\Omega)\}$. Let $x \in \partial\Omega$ be such that an orthonormal frame $(t_1, t_2, \boldsymbol{n})$ of unit tangent and normal vectors is available at x. Then we may write the boundary term in (6.1.76) as

$$\int_{\partial\Omega} v \cdot \sigma[u]\, \boldsymbol{n}\, do = \int_{\partial\Omega} (v_{t_1}\, \sigma_{t_1 n}[u] + v_{t_2}\, \sigma_{t_2 n}[u] + v_n\, \sigma_{nn}[u])\, do$$

where $v_{t_1} = t_1 \cdot v$ etc. and $\sigma_{t_1 n}[u] = t_1 \cdot \sigma[u]\,\boldsymbol{n}$ etc. denote the projections on the respective directions. Now we can prescribe componentwise either a) zero displacements or b) the corresponding normal traction component leading to a total of eight different boundary conditions (including the above Dirichlet and Neumann cases). Existence may be proved as above. However, in each case the rigid body motions $r \in \mathrm{I\!R}$ which solve the homogeneous problem must be factored out of $H^1(\Omega)$, analogous to (6.1.74).

6.2 Membranes, plates and shells

The problems considered so far pertained to three-dimensional bodies Ω. In engineering applications, however, the domain Ω is often **thin**, i.e., in one or two directions it is much smaller than in the remaining direction(s). In engineering such bodies are labelled beams, rods, plates, membranes and shells. Associated with each terminology is a specific set of **a priori assumptions** on the geometry and on the kinematics of the three-dimensional body which leads to a simplification of the full, three-dimensional equilibrium equations (6.1.62). The resulting simplified two- or one-dimensional equations describe **dimensionally reduced models** of the original three-dimensional problem which are often used as a basis for FE discretization. Such lower-dimensional approximations already contain dimensional reduction error stemming from the transition from three to two dimensions, the so-called **modeling error**. This modelling error is inherent in the PDE constituting the lower-dimensional model and hence cannot be reduced by FE discretizations of these models however careful. It is, however, possible to estimate a posteriori (i.e., based on an available solution of the shell model) this modelling error (see [106], [126] for more).

The outline of this section is as follows: we first present the simplest two-dimensional theories, namely the membrane theories of plane stress and plane strain. Next, we present two of the most commonly used shell models, the Koiter and the Naghdi shell models without particular emphasis on the modelling error, and establish their basic mathematical properties. Both are singularly perturbed, elliptic systems of partial differential equations in which a small thickness parameter ε is still present. As $\varepsilon \to 0$, various constraints (which strongly depend on the geometry and the boundary conditions of the shell) get imposed causing difficulties in FE approximations. We finally show how the shell models contain the classical membrane and bending models of plates as special cases.

6.2.1 *Membranes and plates*

6.2.1.1 *Dimension reduction*

Consider a bounded domain $\omega \subset \mathbb{R}^2$ with Lipschitz boundary $\partial\omega$. We associate with ω the **plate of thickness** $\varepsilon \in (0, 1]$

$$\Omega_\varepsilon := \omega \times (-\varepsilon/2, \varepsilon/2)\,. \tag{6.2.1}$$

By $\Gamma^\pm$ we denote top and bottom faces of Ω_ε, i.e., $\Gamma^\pm := \omega \times \{\pm\,\varepsilon/2\}$, and the lateral boundary (the "edge") of the plate $\Gamma_0 := \partial\omega \times (-\varepsilon/2, \varepsilon/2)$. In the present section we assume that $\Gamma^{[0]} = \Gamma_0$ and $\Gamma^{[1]} = \Gamma^+ \cup \Gamma^-$, i.e., Dirichlet conditions along the whole edge Γ_0 or, in other words, the plate is clamped along the entire edge. In Ω_ε, we consider the problem (6.1.65) of linearized, three-dimensional elasticity, i.e.,

$$-\operatorname{div} \sigma = f, \;\; \sigma = A\epsilon \;\text{ in }\; \Omega_\varepsilon\,, \tag{6.2.2}$$

$$u = 0 \;\text{ on }\; \Gamma^{[0]}, \;\; \sigma \boldsymbol{n} = g \;\text{ on }\; \Gamma^{[1]}\,, \tag{6.2.3}$$

where the material is homogeneous and isotropic, and we assume

$$f \in L^2(\Omega_\varepsilon), \qquad g \in L^2(\Gamma^{[1]})$$

for simplicity. Under these assumptions, the problem (6.2.2), (6.2.3) admits a unique weak solution $u \in H^1(\Omega_\varepsilon, \Gamma^{[0]})$ of (6.1.62) with corresponding stress field $\sigma_{ij} \in L^2(\Omega_\varepsilon)$ given by (6.1.57), (6.1.59). Here and in what follows, roman indices $i, j, k, \ldots$ vary in the set $\{1, 2, 3\}$ while greek indices are in $\{1, 2\}$. Thus we have

$$u' := \{u_\alpha\} = (u_1, u_2)^\top, \ \sigma' := \{\sigma_{\alpha\beta}\} = \begin{pmatrix} \sigma_{11} & \sigma_{12} \\ \sigma_{21} & \sigma_{22} \end{pmatrix}, \ \ x' := (x_1, x_2) \,,$$

i.e., we denote quantities associated with (x_1, x_2) by a prime. All tensors will be assumed symmetric without explicitly stating so.

Due to the particular geometry (6.2.1) and the isotropy of the material, the problem (6.2.2), (6.2.3) can be decoupled into two independent problems, the so-called membrane and the flexural or bending problem. To explain this we observe that any function $\psi(x', x_3)$ on Ω_ε may be decomposed into its even part $\psi^e(x)$ and its odd part $\psi^o(x)$, i.e.,

$$\psi^e(x) := \frac{1}{2}\,(\psi(x', x_3) + \psi(x', -x_3)), \ \ \psi^o(x) := \frac{1}{2}\,(\psi(x', x_3) - \psi(x', -x_3)) \,.$$

We decompose the loads in bending and stretching parts:

$$g = g^s + g^b, \ \ f = f^s + f^b$$

with

$$g^s := \begin{pmatrix} g_1^e \\ g_2^e \\ g_3^o \end{pmatrix}, \ g^b := \begin{pmatrix} g_1^o \\ g_2^o \\ g_3^e \end{pmatrix}, \ f^s := \begin{pmatrix} f_1^e \\ f_2^e \\ f_3^o \end{pmatrix}, \ f^b := \begin{pmatrix} f_1^o \\ f_2^o \\ f_3^e \end{pmatrix}.$$

By superposition, the solution (u, σ) of (6.2.2), (6.2.3) splits analogously as

$$u = u^s + u^b, \qquad \sigma = \sigma^s + \sigma^b$$

where

$$u^s := \begin{pmatrix} u_1^e \\ u_2^e \\ u_3^o \end{pmatrix}, \ \sigma^s := \begin{pmatrix} \sigma_{11}^e & \sigma_{12}^e & \sigma_{13}^o \\ \sigma_{12}^e & \sigma_{22}^e & \sigma_{23}^o \\ \sigma_{13}^o & \sigma_{23}^o & \sigma_{33}^e \end{pmatrix}$$

and

$$u^b := u - u^s, \quad \sigma^b = \sigma - \sigma^s \,.$$

Exercise 6.25 Show that $B(u^s, u^b) = 0$ where

$$B(u,v) := \int_{\Omega_\varepsilon} \epsilon[v] : A\epsilon[u]\, dx$$

(see also [126]), i.e., stretching and bending solutions are orthogonal in energy.

Next, we assume that Ω_ε is thin, i.e., that

$$\varepsilon \,/\, \mathrm{diam}(\omega) \ll 1\,.$$

Under this assumption, one might consider to **approximate** the three-dimensional problem (6.2.2), (6.2.3) by a simpler two-dimensional problem posed on the mid-surface ω of the plate. We present now the simplest approximation of this type. Following [5], we consider the Hellinger–Reissner variational principle in its first form (6.1.64). The idea is now a **semidiscrete approximation** σ^*, u^* of the pair σ, u in (6.1.64) as polynomials of degree $\deg_3(\sigma^*)$, $\deg_3(u^*)$ with respect to x_3, but letting the fields be undetermined otherwise. Thus we have: find $\sigma^* \in X^*$, $u^* \in M^*$ such that

$$\begin{aligned} &(A^{-1}\sigma^* - \varepsilon[u^*],\, \tau)_0 = 0 \text{ for all } \tau \in X^*\,, \\ &\qquad (\sigma^*, \varepsilon[v])_0 = (f,v)_0 + \int_{\Gamma^{[1]}} g \cdot v\, do \text{ for all } v \in M^* \end{aligned} \tag{6.2.4}$$

and we select the following closed subspaces of X, M in (6.1.66):

$$X^* = \{\sigma \in X : \deg_3(\sigma_{\alpha\beta}) = 1,\ \deg_3(\sigma_{\alpha 3}) = 0,\ \sigma_{33} = 0\}\,, \tag{6.2.5}$$

$$M^* = \{u \in M : \deg_3(u_\alpha) = 1,\ \deg_3(u_3) = 0\}\,. \tag{6.2.6}$$

Exploiting the product structure of Ω_ε and of the fields in (6.2.4), we may evaluate in (6.2.4) all integrals through the thickness and arrive at a Hellinger–Reissner variational principle for the two-dimensional approximations σ^*, u^*.

Let us be more specific about these two dimensional problems. To this end, we denote the two-dimensional analogue of the constitutive tensor A by A' and write

$$\sigma' = (\sigma_{11}, \sigma_{22}, \sigma_{12})^\top,\ \epsilon' = (\epsilon_{11}, \epsilon_{22}, \epsilon_{12})^\top\,.$$

We set by analogy to (6.1.54), (6.1.57)

$$\epsilon' = (A')^{-1}\sigma' = \frac{1+\nu}{\mathrm{E}}\,\sigma' - \frac{\nu}{\mathrm{E}}\,(\mathrm{tr}\,\sigma)\,I\,, \tag{6.2.7a}$$

$$\sigma' = A'\epsilon' = \frac{\mathrm{E}}{1+\nu}\left(\epsilon' + \frac{\nu}{1-2\nu}\,(\mathrm{tr}\,\epsilon')\,I\right). \tag{6.2.7b}$$

Then the approximate three-dimensional fields σ^*, u^* are given by

$$u^*(x) = \begin{pmatrix} \zeta_1(x') \\ \zeta_2(x') \\ 0 \end{pmatrix} + \begin{pmatrix} -x_3\, r_1(x') \\ -x_3\, r_2(x') \\ \zeta_3(x') \end{pmatrix} =: (u^*)^s + (u^*)^b \tag{6.2.8}$$

$$\begin{aligned} \sigma^*(x) &= \begin{pmatrix} (A')^{-1}\, \epsilon'[\zeta'] & 0 \\ 0 & 0 \end{pmatrix} \\ &+ \begin{pmatrix} -x_3(A')^{-1}\, \epsilon'[r] & \dfrac{\mathrm{E}}{2(1+\nu)}\, (\nabla\zeta_3 - r) \\ \dfrac{\mathrm{E}}{2(1+\nu)}\, (\nabla\zeta_3 - r)^{\top} & 0 \end{pmatrix} \end{aligned} \tag{6.2.9}$$

where the displacements $\zeta_i(x')$ and the rotations $r_\alpha(x')$ satisfy

a) a **generalized plane stress problem** for $\zeta' = (\zeta_1, \zeta_2)^{\top}$:

$$\begin{aligned} -\varepsilon\, \mathrm{div}'\, A'\, \epsilon'(\zeta') &= 2(g')^0 + (f')^0 \text{ in } \omega\,, \\ \zeta' &= 0 \qquad\qquad\quad \text{on } \partial\omega \end{aligned} \tag{6.2.10}$$

and

b) a **Reissner–Mindlin plate bending problem** for (r_1, r_2, ζ_3):

$$\begin{aligned} -\frac{\varepsilon^3}{12}\, \mathrm{div}'\, A'\, \epsilon'(r) + \varepsilon\, \frac{\mathrm{E}}{2(1+\nu)}\, (r - \nabla'\zeta_3) &= -\varepsilon((g')^1 + (f')^1)\,, \\ \varepsilon\, \frac{\mathrm{E}}{2(1+\nu)}\, \mathrm{div}'(r - \nabla'\zeta_3) &= 2g_3^0 + f_3^0 \end{aligned} \tag{6.2.11}$$

which is hard clamped at the edge, i.e.,

$$r_1 = r_2 = \zeta_3 = 0 \;\text{ on }\; \partial\omega\,.$$

Here the data are defined by

$$\begin{aligned} g_3^0(x') &:= \frac{1}{2}\, (g_3(x', \varepsilon/2) + g_3(x', -\varepsilon/2))\,, \\ g_3^1(x') &:= \frac{1}{2}\, (g_3(x', \varepsilon/2) - g_3(x', -\varepsilon/2))\,, \\ f_3^0(x') &:= \int_{-\varepsilon/2}^{\varepsilon/2} f_3(x', x_3) dx_3,\; f_3^1(x') := \frac{1}{\varepsilon} \int_{-\varepsilon/2}^{\varepsilon/2} f_3(x', x_3)\, x_3 dx_3 \end{aligned}$$

and likewise for $(g')^0$, $(g')^1$, $(f')^0$, $(f')^1$.

Remark 6.26 The use of the Hellinger–Reissner principle in the derivation is essential. Indeed, if one uses the set M^* in (6.2.6) as admissible displacement in the minimization of the potential energy (6.1.61), we get also systems (6.2.8) and (6.2.9), however with different coefficients. The systems (6.2.8) and (6.2.9) are, however, **justified** since one can prove convergence of the pair σ, u to σ^*, u^* as $\varepsilon \to 0$ (see [47, 96]).

Exercise 6.27 Minimize $\Pi(v)$ in (6.1.61) over M^* in (6.2.6) and derive the Euler–Lagrange equations. Show that one does not obtain (6.2.8), (6.2.9).

6.2.1.2 *hp-FE approximation of the plane stress problem*

We assume that $\omega \subset \mathbb{R}^2$ is a polygon with straight sides and consider in ω the plane stress problem (6.2.9). With the definition of the two-dimensional tensor A' in (6.2.7) we obtain the following boundary value problem: find $\zeta' \in H_0^1(\omega)$ such that $B(\zeta', \eta') = F(\eta')$ for all $\eta' \in H_0^1(\omega)$.

Here

$$B(\zeta', \eta') = \frac{\mathrm{E}}{1+\nu} \int_\omega \{\epsilon'[\zeta'] : \epsilon'[\eta'] + \frac{\nu}{1-2\nu} (\mathrm{div}'\, \zeta')(\mathrm{div}'\, \eta')\} dx'$$

or, taking into account (6.1.41),

$$B(\zeta', \eta') = \int_\omega \{2\mu\epsilon'[\zeta'] : \epsilon'[\eta'] + \lambda(\mathrm{div}'\, \zeta')(\mathrm{div}'\, \eta')\} dx'$$

which is the exact analog of the form $B(\cdot, \cdot)$ for the three-dimensional problem in (6.1.63). The loading $F(\cdot)$ is given by

$$F(\eta') = \frac{1}{\varepsilon} \int_\omega \{2(g')^0 + (f')^0\}\, \eta'\, dx' \, .$$

Using Korn's inequality, Proposition 6.20, we find immediately that

$$\begin{aligned} \text{for all } \zeta' \in H_0^1(\omega) : B(\zeta', \zeta') &\geq 2\mu \int_\omega \epsilon'[\zeta'] : \epsilon'[\zeta']\, dx' \\ &\geq 2\mu\, c(\omega, \partial\omega)\, \|\nabla'\zeta'\|^2_{L^2(\omega)} \\ &\geq 2\mu\, c(\omega)\, \|\zeta'\|^2_{H^1(\omega)} \end{aligned}$$

where we also used the Poincaré inequality and where $c(\omega)$ is independent of λ, μ. On the other hand, it also holds that

$$|B(\zeta', \zeta')| \leq c(\omega)(\lambda + 2\mu)\, \|\zeta'\|^2_{H^1(\omega)} \, .$$

Now let $\zeta'_N \in S_0^{p,1}(\omega, \mathcal{T}^{n,\sigma})$ be the *hp*-FE approximation of ζ', i.e.,

$$B(\zeta'_N, \eta') = F(\eta') \text{ for all } \eta' \in S_0^{p,1}(\omega, \mathcal{T}^{n,\sigma}) .$$

Then we have

$$\|\zeta' - \zeta'_N\|_{H^1(\omega)} \leq c(\omega)\Big(1 + \frac{\lambda}{2\mu}\Big) \inf_{\eta' \in S_0^{p,1}(\omega, \mathcal{T}^{n,\sigma})} \|\zeta' - \eta'\|_{H^1(\omega)} .$$

If the right-hand sides f^0, g^0 are analytic in $\overline{\omega}$, the solution ζ' belongs componentwise to the countably normed space $B^2_\beta(\omega)$ for some $\beta > 0$ (see [72]), and hence by the approximation result Theorem 4.63 we get exponential convergence, i.e.,

$$\|\zeta' - \zeta'_N\|_{H^1(\omega)} \leq C(\omega)\Big(1 + \frac{\lambda}{2\mu}\Big) \exp(-b\sqrt[3]{N})$$

where $N = \dim(S_0^{p,1}(\omega, \mathcal{T}^{n,\sigma}))$. Notice that this error bound is not uniform in λ, i.e., it deteriorates as $\lambda \to \infty$ or, equivalently, as $\nu \to 1/2$, i.e., as the material becomes incompressible. Error estimates that are uniform in λ cannot be achieved by the above techniques – we will address this in Section 6.3 below.

6.2.2 *Shells*

6.2.2.1 *Notation. Geometry of the shell*

The shell is a thin, three-dimensional body described as follows. Let the parameter domain $\omega \subset \mathbb{R}^2$ be open, bounded and connected and let $\gamma = \partial\omega$ be Lipschitz. Let $x = (x^1, x^2)$ be a point in $\overline{\omega}$ and define

$$\partial_\alpha = \frac{\partial}{\partial x^\alpha}, \quad \partial_{\alpha\beta} = \frac{\partial^2}{\partial x^\alpha \partial x^\beta} . \tag{6.2.12}$$

We recall that Greek indices are in $\{1, 2\}$ and roman ones belong to $\{1, 2, 3\}$. By $\boldsymbol{e}_i$ we denote the Cartesian unit vectors in $\mathbb{R}^3$. Then the mid-surface of the shell is given by

$$S = \varphi(\overline{\omega}), \quad \varphi = \varphi^i \boldsymbol{e}_i : \overline{\omega} \to \mathbb{R}^3 , \tag{6.2.13}$$

where we assume

$$\varphi \in C^3(\overline{\omega}, \mathbb{R}^3), \ \varphi \text{ is injective} , \tag{6.2.14}$$

and the two **tangent vectors**

$$\boldsymbol{a}_\alpha := \partial_\alpha \varphi = (\partial_\alpha \varphi^i)\, \boldsymbol{e}_i \tag{6.2.15}$$

span the tangent plane to $S = \varphi(\overline{\omega})$ at the point $\varphi(x)$. $\boldsymbol{a}_1, \boldsymbol{a}_2$ are the covariant basis of the tangent plane at $\varphi(x)$ (see Figure 6.1).

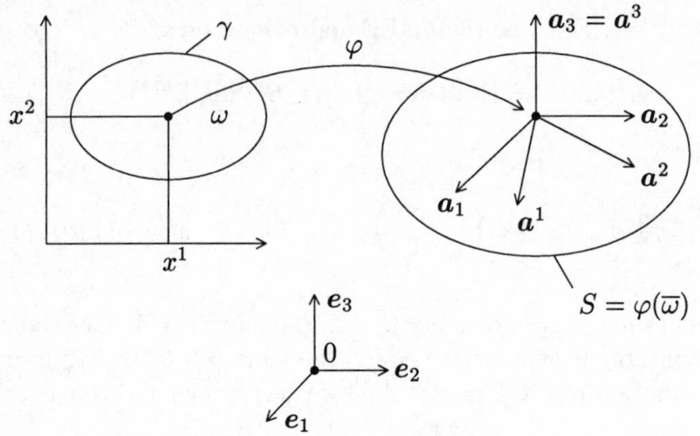

FIG. 6.1. The domain ω and the notations

For every $x \in \overline{\omega}$, we define the unit normal vector

$$\boldsymbol{a}_3 = \boldsymbol{a}^3 = \frac{\boldsymbol{a}_1 \times \boldsymbol{a}_2}{|\boldsymbol{a}_1 \times \boldsymbol{a}_2|} \tag{6.2.16}$$

to S; the triple $(\boldsymbol{a}_1, \boldsymbol{a}_2, \boldsymbol{a}_3)$ defines the **covariant basis** at $\varphi(x) \in S$. The contravariant basis $(\boldsymbol{a}_1, \boldsymbol{a}_2, \boldsymbol{a}_3)$ is defined by

$$\boldsymbol{a}^\alpha \cdot \boldsymbol{a}_\beta = \delta^\alpha_\beta = \begin{cases} 1 & \text{if } \alpha = \beta \\ 0 & \text{else}. \end{cases} \tag{6.2.17}$$

We note

$$\boldsymbol{a}^i \cdot \boldsymbol{a}_j = \delta^i_j \tag{6.2.18}$$

$$\begin{aligned} &\boldsymbol{a}_\alpha \times \boldsymbol{a}_\beta = \varepsilon_{\alpha\beta}\boldsymbol{a}^3,\ \boldsymbol{a}^\alpha \times \boldsymbol{a}^\beta = \varepsilon^{\alpha\beta}\boldsymbol{a}_3,\ \boldsymbol{a}_3 \times \boldsymbol{a}_\beta = \varepsilon_{\beta\rho}\boldsymbol{a}^\rho\,, \\ &\boldsymbol{a}^3 \times \boldsymbol{a}^\beta = \varepsilon^{\beta\rho}\boldsymbol{a}_\rho\,, \end{aligned} \tag{6.2.19}$$

where

$$(\varepsilon_{\alpha\beta}) = \sqrt{a}\begin{pmatrix} 0 & 1 \\ -1 & 0 \end{pmatrix},\quad (\varepsilon^{\alpha\beta}) = \frac{1}{\sqrt{a}}\begin{pmatrix} 0 & 1 \\ -1 & 0 \end{pmatrix}, \tag{6.2.20}$$

$$a_{\alpha\beta} = a_{\beta\alpha} = \boldsymbol{a}_\alpha \cdot \boldsymbol{a}_\beta = \partial_\alpha \varphi^i\, \partial_\beta \varphi^i \tag{6.2.21}$$

and

$$a(x) := \det(a_{\alpha\beta}) \geq a_0 > 0 \text{ for all } x \in \overline{\omega}\,. \tag{6.2.22}$$

The **surface element** dS is

$$dS = \sqrt{a}\, dx^1 dx^2 \tag{6.2.23}$$

and we put

$$a^{\alpha\beta} = \boldsymbol{a}^\alpha \cdot \boldsymbol{a}^\beta\,, \tag{6.2.24}$$

i.e., $(a^{\alpha\beta}) = (a_{\alpha\beta})^{-1}$.

The **second fundamental form of** S is

$$b_{\alpha\beta} = b_{\beta\alpha} = \boldsymbol{a}_3 \cdot \partial_\alpha \boldsymbol{a}_\beta = -\partial_\alpha \boldsymbol{a}_3 \cdot \boldsymbol{a}_\beta \, . \tag{6.2.25}$$

We put

$$b_\alpha^\beta := a^{\beta\rho} b_{\rho\alpha} \, . \tag{6.2.26}$$

The **third fundamental form** $(c_{\alpha\beta})$ of S is then given by

$$c_{\alpha\beta} = c_{\beta\alpha} = b_\alpha^\rho \, b_{\rho\beta} = b_{\alpha\beta} \, b_\beta^\rho \, . \tag{6.2.27}$$

The **Christoffel symbols** $\Gamma_{\alpha\beta}^\rho$ of S are

$$\Gamma_{\alpha\beta}^\rho = \Gamma_{\beta\alpha}^\rho := \boldsymbol{a}^\rho \cdot \partial_\beta \boldsymbol{a}_\alpha \, , \tag{6.2.28}$$

and allow to compute **covariant derivatives**. Let

$$\eta_\alpha \boldsymbol{a}^\alpha = \eta^\alpha \boldsymbol{a}_\alpha$$

be any surface vector field. Then

$$\eta_{\alpha|\beta} := \partial_\beta \eta_\alpha - \Gamma_{\alpha\beta}^\rho \, \eta_\rho, \; \eta^\alpha|_\beta := \partial_\beta \eta^\alpha + \Gamma_{\rho\beta}^\alpha \eta^\rho \tag{6.2.29}$$

are the covariant derivatives of η. Likewise, let $T_{\alpha\beta}$ and $T^{\alpha\beta}$ denote the co- and contravariant components of a surface tensor field. Then

$$\begin{aligned} T_{\alpha\beta|\rho} &= \partial_\rho T_{\alpha\beta} - \Gamma_{\alpha\rho}^\sigma T_{\sigma\beta} - \Gamma_{\beta\rho}^\sigma T_{\alpha\sigma} \, , \\ T^{\alpha\beta}|_\rho &= \partial_\rho T^{\alpha\beta} + \Gamma_{\sigma\rho}^\alpha T^{\sigma\beta} + \Gamma_{\sigma\rho}^\beta T^{\alpha\sigma} \, . \end{aligned} \tag{6.2.30}$$

Further

$$\varepsilon^{\alpha\beta}|_\rho = 0 \, , \tag{6.2.31}$$

and $T_\alpha^\beta := a^{\beta\rho} T_{\alpha\rho}$,

$$T_\alpha^\beta|_\rho = \partial_\rho T_\alpha^\beta + \Gamma_{\rho\sigma}^\beta T_\alpha^\sigma - \Gamma_{\alpha\rho}^\sigma T_\sigma^\beta \, . \tag{6.2.32}$$

We also mention the **Meinardi–Codazzi identities**

$$b_{\alpha\beta|\rho} = b_{\alpha\rho|\beta} \, , \tag{6.2.33}$$

$$b_\alpha^\beta|_\rho = b_\rho^\beta|_\alpha \, . \tag{6.2.34}$$

For any vector field $\eta_i \boldsymbol{a}^i = \eta^i \boldsymbol{a}_i$, we have

$$\begin{aligned} \partial_\alpha(\eta_i \boldsymbol{a}^i) &= (\eta_{\beta|\alpha} - b_{\alpha\beta}\eta_3)\boldsymbol{a}^\beta + (\partial_\alpha \eta_3 + b_\alpha^\beta \eta_\beta)\boldsymbol{a}^3 \\ &= (\eta^\beta|_\alpha - b_\alpha^\beta \eta^3)\boldsymbol{a}_\beta + (\partial_\alpha \eta^3 + b_{\alpha\beta}\eta^\beta)\boldsymbol{a}_3 \, . \end{aligned} \tag{6.2.35}$$

and the **formulas of Gauss**

$$\partial_\beta \boldsymbol{a}_\alpha = \Gamma^\rho_{\alpha\beta}\boldsymbol{a}_\rho + b_{\alpha\beta}\boldsymbol{a}_3, \partial_\beta \boldsymbol{a}^\alpha = -\Gamma^\alpha_{\beta\rho}\boldsymbol{a}^\rho + b^\alpha_\beta \boldsymbol{a}^3 \,. \tag{6.2.36}$$

and **Weingarten**

$$\partial_\alpha \boldsymbol{a}_3 = -b^\rho_\alpha \boldsymbol{a}_\rho \,. \tag{6.2.37}$$

Also

$$\begin{aligned} (T^{\alpha\beta}\eta_\rho)|_\beta &= (T^{\alpha\beta}|_\beta)\eta_\rho + T^{\alpha\rho}(\eta_{\rho|\beta})\,, \\ (T^\rho_\alpha \eta_\rho)|_\beta &= (T^\rho_\alpha|_\beta)\eta_\rho + T^\rho_\alpha(\eta_{\rho|\beta})\,. \end{aligned} \tag{6.2.38}$$

Note that

$$\eta_\alpha|_{\beta\rho} := (\eta_\alpha|_\beta)|_\rho \tag{6.2.39}$$

can be calculated by (6.2.30). Note particularly that

$$0 \neq \eta_{\rho|\alpha\beta} - \eta_{\rho|\beta\alpha} = b_{\rho\beta}\, b^\sigma_\alpha \eta_\sigma - b_{\rho\alpha}\, b^\sigma_\beta \eta_\sigma \tag{6.2.40}$$

and

$$\eta_{3|\alpha} = \eta^3|_\alpha = \partial_\alpha \eta_3 \,, \tag{6.2.41}$$

$$\eta_{3|\alpha\beta} = (\eta_{3|\alpha})|_\beta = \partial_{\alpha\beta}\beta_3 - \Gamma^\rho_{\alpha\beta}\partial_\rho \eta_3 = \eta_{3|\beta\alpha} \,. \tag{6.2.42}$$

Definition 6.28 *A shell of thickness $t = 2\varepsilon > 0$ is an elastic body with reference configuration*

$$\hat{\Omega} = \{\varphi(x^1,x^2) + x^3\boldsymbol{a}_3(x^1,x^2) \in \mathbb{R}^3 : (x^1,x^2) \in \omega,\ |x^3| < \varepsilon\}\,. \tag{6.2.43}$$

Proposition 6.29 *Under our assumptions on φ, the mapping*

$$\Phi(x^1,x^2,x^3) := \varphi(x^1,x^2) + x^3\boldsymbol{a}_3(x^1,x^2) : \overline{\omega} \times [-\varepsilon,\varepsilon] \longmapsto \hat{\Omega} \subset \mathbb{R}^3$$

is injective for $\varepsilon > 0$ sufficiently small.

We assume that the material is **homogeneous** and **isotropic** with Lamé constants $\lambda > 0, \mu > 0$. The **faces of the shell** $\hat{\Gamma}_\pm$ are

$$\hat{\Gamma}_\pm = \{\varphi(x^1,x^2) + x^3\boldsymbol{a}_3(x^1,x^2) : (x^1,x^2) \in \omega,\ x^3 = \pm\,\varepsilon\} \tag{6.2.44}$$

and the **lateral Dirichlet boundary** $\hat{\Gamma}_0$ is

$$\hat{\Gamma}_0 = \{\varphi(x^1,x^2) + x^3\boldsymbol{a}_3(x^1,x^2) : (x^1,x^2) \in \gamma_0,\ |x^3| \leq \varepsilon\}\,, \tag{6.2.45}$$

for $\gamma_0 \subset \partial\omega$. We also define the **lateral Neumann boundary**

$$\hat{\Gamma}_1 = \{\varphi(x^1, x^2) + x^3 \boldsymbol{a}_3(x^1, x^2) : (x^1, x^2) \in \gamma_1, |x^3| \le \varepsilon\} \tag{6.2.46}$$

for $\gamma_1 = \partial\omega \backslash \overline{\gamma}_0$.

6.2.3 *Koiter's shell model*

Koiter's shell model is based on dimension reduction by a priori hypotheses on the kinematics of the shell. These hypotheses are as follows.

(H1) (Linearity). Points on straight lines normal to the midsurface S remain on a straight line after deformation, i.e. the three-dimensional displacement vector $\boldsymbol{u} = u_i \boldsymbol{a}^i$ has tangential components $u_\alpha(x^1, x^2, x^3)$ behaving linearly in x^3:

$$u_\alpha(x^1, x^2, x^3) \cong \zeta_\alpha(x^1, x^2) + x^3 r_\alpha(x^1, x^2) . \tag{6.2.47}$$

Here ζ_α are **tangential displacements** and r_α are the so-called **rotations**.

(H2) The normal displacement $u_3(x^1, x^2, x^3)$ is independent of x^3, i.e. $u_3 \cong \zeta_3(x^1, x^2)$.

(H3) (Normal hypothesis, Kirchhoff–Love hypotheses). The lines normal to the midsurface before deformation remain normal to it after deformation:

$$r_\alpha(x^1, x^2) = -\zeta_{3,\alpha} - b_\alpha^\lambda \zeta_\lambda . \tag{6.2.48}$$

Note that (H3) implies with (6.2.47) that the strain tensor $\epsilon[u]$ will contain **second derivatives** of ζ_3 – this has profound consequences for the choice of function spaces in the variational principle of the Koiter shell and, hence, also for the subsequent FE discretization.

Let $\zeta_i : \overline{\omega} \to \mathbb{R}$ denote the three covariant components of the displacement $\zeta_i \boldsymbol{a}^i$ of the midsurface S as in Fig. 6.2.

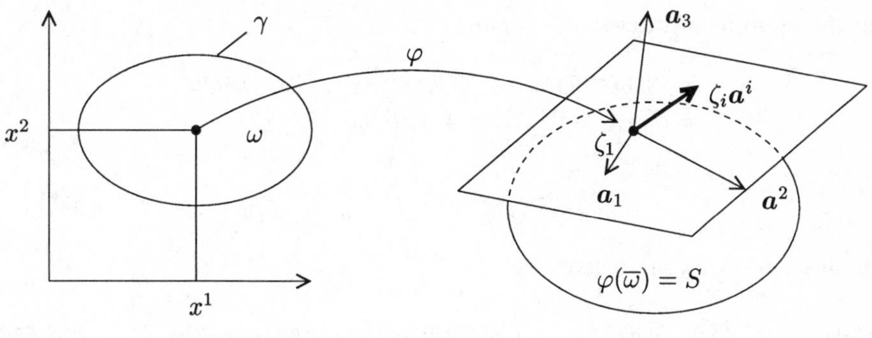

FIG. 6.2.

Then the unknowns ζ_i are the solutions of the variational problem

$$\zeta \in \boldsymbol{V}_K(\omega) \ \text{ s.t. } \ B_K(\zeta,\eta) = F_K(\eta) \ \text{ for all } \ \eta \in \boldsymbol{V}_K(\omega)\,. \tag{6.2.49}$$

Here

$$\begin{aligned} \boldsymbol{V}_K(\omega) := \{\eta = (\eta_1,\eta_2,\eta_3) \in H^1(\omega) \times H^1(\omega) \times H^2(\omega),\\ \eta_i = \partial_\nu \eta_3 = 0 \ \text{ on } \ \gamma_0\} \end{aligned} \tag{6.2.50}$$

and

$$B_K(\zeta,\eta) := \varepsilon\, B_M(\zeta,\eta) + \varepsilon^3\, B_F(\zeta,\eta) \tag{6.2.51}$$

with the **membrane energy**

$$B_M(\zeta,\eta) := \int_\omega \{a^{\alpha\beta\rho\sigma}\gamma_{\rho\sigma}(\zeta)\gamma_{\alpha\beta}(\eta)\}\,\sqrt{a}\;dx^1dx^2$$

and the **flexural energy**

$$B_F(\zeta,\eta) := \int_\omega \Big\{\frac{1}{3}\,a^{\alpha\beta\rho\sigma}\Upsilon_{\rho\sigma}(\zeta)\,\Upsilon_{\alpha\beta}(\eta)\Big\}\,\sqrt{a}\;dx^1dx^2\,,$$

the **constitutive equation**

$$a^{\alpha\beta\rho\sigma} := \frac{4\lambda\mu}{(\lambda+2\mu)}\,a^{\alpha\beta}a^{\rho\sigma} + 2\mu(a^{\alpha\rho}a^{\beta\sigma} + a^{\alpha\sigma}a^{\beta\rho})\,, \tag{6.2.52}$$

the **linearized membrane strain tensor**

$$\begin{aligned} \gamma_{\alpha\beta}(\eta) &:= \frac{1}{2}\,(\eta_{\alpha|\beta} + \eta_{\beta|\alpha}) - b_{\alpha\beta}\,\eta_3 \\ &= e_{\alpha\beta}(\eta) - \Gamma^{\rho}_{\alpha\beta}\,\eta_\rho - b_{\alpha\beta}\eta_3\,, \\ e_{\alpha\beta}(\eta) &:= \frac{1}{2}\,(\partial_\alpha\eta_\beta + \partial_\beta\eta_\alpha) \end{aligned} \tag{6.2.53}$$

and the **change of curvature tensor**

$$\begin{aligned} \Upsilon_{\alpha\beta}(\eta) :&= \eta_{3|\alpha\beta} + b^{\rho}_{\beta}\,\eta_{\rho|\alpha} + b^{\rho}_{\alpha}\,\eta_{\rho|\beta} + (b^{\rho}_{\beta|\alpha})\,\eta_\rho - c_{\alpha\beta}\eta_3 \\ &= \partial_{\alpha\beta}\,\eta_3 - \Gamma^{\rho}_{\alpha\beta}\,\partial_\rho\,\eta_3 + b^{\rho}_{\beta}(\partial_\alpha\eta_\rho - \Gamma^{\sigma}_{\rho\alpha}\eta_\sigma) \\ &\quad + b^{\rho}_{\alpha}\,(\partial_\beta\eta_\rho - \Gamma^{\sigma}_{\rho\beta}\eta_\sigma) \\ &\quad + (\partial_\alpha b^{\rho}_{\beta} + \Gamma^{\rho}_{\alpha\sigma}b^{\sigma}_{\beta} - \Gamma^{\sigma}_{\alpha\beta}b^{\rho}_{\sigma})\,\eta_\rho - c_{\alpha\beta}\eta_3\,. \end{aligned} \tag{6.2.54}$$

The linear form $F_K(\eta)$ is given by

$$F_K(\eta) = \int_\omega p^{i,\varepsilon}\eta_i\,\sqrt{a}\,dx^1dx^2 + \int_{\gamma_1} q^{i,\varepsilon}\eta_i d\gamma + \int_{\gamma_1} m^{\alpha,\varepsilon}(\partial_\alpha\eta_3 + b^{\sigma}_{\alpha}\eta_\sigma)d\gamma\,. \tag{6.2.55}$$

Theorem 6.30 *Assume* $0 < \varepsilon \le 1$, *and that*

$$\int_{\gamma_0} d\gamma > 0\,. \tag{6.2.56}$$

Then there exists $\beta(\varepsilon, \omega, \varphi, \lambda, \mu) > 0$ *such that*

$$B_K(\eta,\eta) \ge \beta\,(\|\eta_1\|^2_{H^1(\omega)} + \|\eta_2\|^2_{H^1(\omega)} + \|\eta_3\|^2_{H^2(\omega)}) =: \beta\|\eta\|^2_{H^1\times H^1\times H^2}\,. \tag{6.2.57}$$

We sketch the proof here, following [49].

Step 1: It holds for all $\eta \in H^1 \times H^1 \times H^2$ that

$$B_K(\eta,\eta) \ge C_1 \left\{ \sum_{\alpha,\beta} \|\gamma_{\alpha\beta}(\eta)\|^2_{0,\omega} + \|\Upsilon_{\alpha\beta}(\eta)\|^2_{0,\omega} \right\}. \tag{6.2.58}$$

This follows from

$$a^{\alpha\beta} a^{\beta\sigma} T_{\rho\sigma} T_{\alpha\beta} \ge C\, T_{\alpha\beta} T_{\alpha\beta}$$

for any symmetric tensor $T_{\alpha\beta}$.

Step 2: Let

$$\boldsymbol{E}_K(\omega) = \{\eta \in L^2 \times L^2 \times H^1 : \gamma_{\alpha\beta}(\eta) \in L^2(\omega),\ \Upsilon_{\alpha\beta}(\eta) \in L^2(\omega)\} \tag{6.2.59}$$

Then

$$\boldsymbol{E}_K(\omega) = H^1(\omega) \times H^1(\omega) \times H^2(\omega)\,.$$

To prove this, let $\eta \in \boldsymbol{E}_K(\omega)$. Then, in the sense of distributions,

$$e_{\alpha\beta}(\eta) = \gamma_{\alpha\beta}(\eta) + \Gamma^{\rho}_{\alpha\beta}\eta_\rho + b_{\alpha\beta}\eta_3 \in L^2(\omega)\,,$$

since $\Gamma^{\rho}_{\alpha\beta}, b_{\alpha\beta} \in C^0(\overline{\omega})$. Therefore we get

$$\partial_{\alpha\beta}\eta_\rho = \partial_\alpha e_{\beta\rho}(\eta) + \partial_\beta e_{\alpha\rho}(\eta) - \partial_\rho e_{\alpha\beta}(\eta) \in H^{-1}(\omega)\,.$$

Also $\eta_\rho \in L^2(\omega) \Longrightarrow \partial_\alpha\eta_\rho \in H^{-1}(\omega)$, therefore $\partial_\alpha\eta_\rho \in L^2(\omega)$ and $\eta_\rho \in H^1(\omega)$. From the definition of $\Upsilon_{\alpha\beta}(\eta)$, it follows that $\partial_{\alpha\beta}\eta_3 \in L^2(\omega)$. Hence $\eta_3 \in H^2(\omega)$ and we get $\boldsymbol{E}_K \subset H^1 \times H^1 \times H^2$. Since obviously the converse inclusion holds, the assertion follows.

Step 3: The expression $\|\eta\|$, defined by

$$\|\eta\|^2 := \sum_{\alpha} \|\eta_\alpha\|^2_{0,\omega} + \|\eta_3\|^2_{1,\omega} + \sum_{\alpha,\beta} \|\gamma_{\alpha\beta}(\eta)\|^2_{0,\omega} + \sum_{\alpha,\beta} \|\Upsilon_{\alpha\beta}(\eta)\|^2_{0,\omega} \tag{6.2.60}$$

is a norm, i.e., it holds

$$\|\eta\| \geq C_2 \, \|\eta\|_{H^1 \times H^1 \times H^2} \text{ for all } \eta \in H^1 \times H^1 \times H^2 \, . \tag{6.2.61}$$

To prove it, we equip $\boldsymbol{E}_K(\omega)$ in (6.2.59) with the norm (6.2.61). Then $\boldsymbol{E}_K(\omega)$ becomes a Hilbert space. The identity mapping from $H^1 \times H^1 \times H^2$ into $\boldsymbol{E}_K(\omega)$ is continuous and onto by Step 2. Since both spaces are complete, the assertion is a consequence of the open mapping theorem.

Step 4: We show that we can replace the norm $\|\eta\|$ by a seminorm if $\eta \in \boldsymbol{V}_K(\omega)$.

Lemma 6.31 *Let $\eta \in (H^1 \times H^1 \times H^2)(\omega)$ and define*

$$\begin{aligned} d^\alpha(\eta) &= \varepsilon^{\alpha\beta}(\partial_\beta \eta_3 + b^\rho_\beta \eta_\rho) \in H^1(\omega) \, , \\ d^3(\eta) &= \frac{1}{2} \, \varepsilon^{\alpha\beta} \eta_{\beta|\alpha} \in L^2(\omega) \, . \end{aligned} \tag{6.2.62}$$

Then the following relations hold in $H^{-1}(\omega)$:

$$\partial_\alpha (d^i(\eta) \boldsymbol{a}_i) = \varepsilon^{\alpha\beta} \left[\gamma_{\alpha\beta}(\eta) - b^\sigma_\alpha \gamma_{\sigma\beta}(\eta) \right] \boldsymbol{a}_\rho + \varepsilon^{\rho\beta} [\gamma_{\alpha\beta}(\eta)|_\rho] \, \boldsymbol{a}_3 \tag{6.2.63}$$

$$\partial_\alpha [\eta_i \boldsymbol{a}^i - (d^i(\eta)\boldsymbol{a}_i) \times \varphi] = \gamma_{\alpha\beta}(\eta) \boldsymbol{a}^\beta - [\partial_\alpha (d^i(\eta) \boldsymbol{a}_i)] \times \varphi \, . \tag{6.2.64}$$

For the proof we refer to [28].

We can now prove an analog to the rigid displacement lemma, Proposition 6.19.

Lemma 6.32 *(rigid displacement lemma for the Koiter shell). Let $\eta \in (H^1 \times H^1 \times H^2)(\omega)$ satisfy*

$$\gamma_{\alpha\beta}(\eta) = \Upsilon_{\alpha\beta}(\eta) = 0 \text{ in } L^2(\omega) \, . \tag{6.2.65}$$

Then there exist two vectors $\mathbf{c} \in \mathbb{R}^3$ and $\boldsymbol{d} \in \mathbb{R}^3$ such that

$$\eta_i(x^1, x^2) \, \boldsymbol{a}^i(x^1, x^2) = \mathbf{c} + \boldsymbol{d} \times \varphi(x^1, x^2) \; \textit{for all } x \in \overline{\omega} \, . \tag{6.2.66}$$

Moreover, $\eta_i \in C^2(\overline{\omega}, \mathbb{R})$ and the vector $\boldsymbol{d}$ is given by

$$\boldsymbol{d} = \varepsilon^{\alpha\beta} (\partial_\beta \eta_3 + b^\rho_\beta \eta_\rho) \, \boldsymbol{a}_\alpha + \frac{1}{2} \, \varepsilon^{\alpha\beta} \, \eta_{\beta|\alpha} \boldsymbol{a}_3 \, . \tag{6.2.67}$$

Proof Since $\gamma_{\alpha\beta}(\eta) = 0 \Longrightarrow \gamma_{\alpha\beta}(\eta)|_\rho = 0$, (6.2.63) implies

$$\boldsymbol{d} := d^i(\eta) \boldsymbol{a}_i = \varepsilon^{\alpha\beta} \, (\partial_\beta \eta_3 + b^\rho_\beta \eta_\rho) \boldsymbol{a}_\alpha + \frac{1}{2} \, \varepsilon^{\alpha\beta} \eta_{\beta|\alpha} \boldsymbol{a}_3$$

is a constant vector, i.e., independent of $x \in \overline{\omega}$. (6.2.64) implies that

$$\boldsymbol{c} := \eta_i \boldsymbol{a}^i - \boldsymbol{d} \times \varphi$$

is a constant as well, which implies (6.2.66). Finally, from $\boldsymbol{a}^i \cdot \boldsymbol{a}_j = \delta^i_j$ and (6.2.67) we get

$$\eta_j = \boldsymbol{c} \cdot \boldsymbol{a}_j + (\boldsymbol{d} \times \varphi) \cdot \boldsymbol{a}_j \in C^2(\overline{\omega}, \mathbb{R}) .$$

□

The next step is to show that for the homogeneous Dirichlet boundary conditions on γ_0, (6.2.65) implies that η vanishes.

Lemma 6.33 *Let $\eta \in (H^1 \times H^1 \times H^2)(\omega)$ satisfy (6.2.65) and*

$$\eta_i = \partial_\nu \eta_3 = 0 \ \text{ on } \ \gamma_0 \subset \gamma = \partial\omega, \ \int_{\gamma_0} do > 0 . \tag{6.2.68}$$

Then $\eta = 0$.

Proof By Lemma 6.32, (6.2.65) implies that (6.2.66) holds, i.e., $\eta_i \, \boldsymbol{a}^i$ is a rigid body motion

$$\eta_i \boldsymbol{a}^i = \boldsymbol{c} + \boldsymbol{d} \times \varphi .$$

Now $\eta_i = \partial_\nu \eta_3 = 0$ implies that $\boldsymbol{d}$ in (6.2.67) is normal to S, since

$$\boldsymbol{d} = \frac{1}{2} \, \varepsilon^{\alpha\beta} \eta_{\beta|\alpha} \boldsymbol{a}_3 \tag{6.2.69}$$

on γ_0 due to $\partial_\beta \eta_3 + b^\rho_\beta \eta_\rho = 0$ there. Pick $x, x_0 \in \gamma_0$, $x \neq x_0$. Then $\varphi(x) \neq \varphi(x_0)$ since φ is injective. Hence

$$0 = \boldsymbol{c} + \boldsymbol{d} \times \varphi(x) = \boldsymbol{c} + \boldsymbol{d} \times \varphi(x_0)$$

since on γ_0 we have $\eta_i \boldsymbol{a}^i = 0$ by (6.2.68). It follows

$$\boldsymbol{d} \times [\varphi(x) - \varphi(x_0)] = 0 \ \text{ for all } \ x \in \gamma_0 .$$

If $\boldsymbol{d} = 0$, (6.2.68) implies that $\boldsymbol{c} = 0$ as well, whence $\eta = 0$. If $\boldsymbol{d} \neq 0$, there exists a vector $\tilde{\boldsymbol{d}}$ parallel to $\boldsymbol{d}$ such that $\varphi(x) \in \boldsymbol{d}$ for all $x \in \gamma_0$. This means that $\boldsymbol{d}$ is normal to S (cf. (6.2.69)) **and** tangent to S at all $x \in \gamma_0$. This can only happen if $\boldsymbol{d} = 0$; once again we get $\eta = 0$. □

Lemma 6.34 *Assumptions as above. Then the seminorm*

$$|\eta| := \Big\{ \sum_{\alpha,\beta} \|\gamma_{\alpha\beta}(\eta)\|^2_{0,\omega} + \|\Upsilon_{\alpha\beta}(\eta)\|^2_{0,\omega} \Big\}^{\frac{1}{2}} \tag{6.2.70}$$

is a norm on $V_K(\omega)$ and there exists $C > 0$ such that

$$|\eta| \geq C \, \|\eta\|_{H^1 \times H^1 \times H^2} \text{ for all } \eta \in V_K(\omega) \, . \tag{6.2.71}$$

Proof Assume (6.2.71) is false, then there exists a sequence $\{\eta^k\} \subset V_K(\omega)$ such that

$$\|\eta^k\|_{H^1 \times H^1 \times H^2} = 1 \text{ for all } k \, , \tag{6.2.72}$$

$$|\eta^k| \leq \frac{1}{k} \, . \tag{6.2.73}$$

By (6.2.72) and Rellich's theorem, a subsequence $\{\eta^{k'}\} \subset \{n^k\}$ converges in $L^2 \times L^2 \times H^1$. Since $|\eta^{k'}| \to 0$, $\{\eta^{k'}\}$ is Cauchy with respect to the norm $\| \circ \|$ in (6.2.60), since with (6.2.70)

$$\|\eta^{k'}\|^2 = \|\eta_1^{k'}\|_{0,\omega}^2 + \|\eta_2^{k'}\|_{0,\omega}^2 + \|\eta_3^{k'}\|_{1,\omega}^2 + |\eta^{k'}|^2 \, .$$

By Step 3, $\| \circ \|$ is a norm on $H^1 \times H^1 \times H^2$, hence $\{\eta^{k'}\}$ converges in $H^1(\omega) \times H^1(\omega) \times H^2(\omega)$, i.e. in $\boldsymbol{V_K}(\omega)$, to a limit $\eta \in \boldsymbol{V_K}(\omega)$. By (6.2.73), $|\eta| = 0$ and by Lemma 6.33, $\eta = 0$ in $\boldsymbol{V_K}(\omega)$, a contradiction to (6.2.72), which implies that $\|\eta\|_{H^1 \times H^1 \times H^2} = \lim_{k' \to \infty} \|\eta^{k'}\|_{H^1 \times H^1 \times H^2} = 1$. □

Now the **proof of Theorem 6.30** follows from (6.2.58) and (6.2.71). From Theorem 6.30, using the Lax–Milgram lemma we deduce the following result.

Theorem 6.35 *For every $f \in [L^2(\omega)]^3$, $g \in [L^2(\gamma_1)]^3$ and $m \in [L^2(\gamma_1)]^2$, there exists a unique solution $\eta \in \boldsymbol{V_K}(\omega)$ of the variational problem* (6.2.49).

Remark 6.36 In the case of the pure Neumann problem, the rigid displacement Lemma 6.32 can be used to prove an analog of Theorem 6.35 for a suitable factor space of $\boldsymbol{V_K}(\omega)$.

6.2.4 *Naghdi's shell model*

Naghdi's shell model is based on similar kinematic assumptions to Koiter's, the principal difference being that the normal hypothesis (6.2.48) is not enforced any more. Consequently, we now have five unknown fields: three displacements ζ_i and two rotations $r_\alpha(x^1, x^2)$, all of which will be sought in $H^1(\omega)$. It will become apparent, however, that (6.2.48) is indirectly still present in the Naghdi model in a penalized form. This will have repercussions on the FE discretization of the Naghdi shell model: although formally C^0-elements suffice for a conforming FEM, it turns out that the **shear constraint** (6.2.48) and, even more importantly, the **membrane constraint** to be discussed in Section 6.3.1, cause various **locking effects** which we will discuss in the following section.

We begin by introducing the **change of curvature tensor** $\Upsilon_{\alpha\beta}(\zeta, r)$

$$\begin{aligned}\Upsilon_{\alpha\beta}(\zeta,r) : &= \frac{1}{2}\left(r_{\alpha|\beta} + r_{\beta|\alpha}\right) - \frac{1}{2}\, b^{\rho}_{\alpha}\left(\zeta_{\rho|\beta} - b_{\rho\beta}\zeta_3\right) \\ &\quad - \frac{1}{2}\, b^{\sigma}_{\beta}\left(\zeta_{\sigma|\alpha} - b_{\sigma\alpha}\zeta_3\right) \\ &= e_{\alpha\beta}(r) - \Gamma^{\rho}_{\alpha\beta} r_\rho - \frac{1}{2}\, b^{\rho}_{\alpha}\left(\partial_\beta \zeta_\rho - \Gamma^{\sigma}_{\rho\beta}\zeta_\sigma\right) \\ &\quad - \frac{1}{2}\, b^{\sigma}_{\beta}\left(\partial_\alpha \zeta_\sigma - \Gamma^{\rho}_{\alpha\sigma}\zeta_\rho\right) + c_{\alpha\beta}\zeta_3\end{aligned} \tag{6.2.74}$$

where $e_{\alpha\beta}(r) := \frac{1}{2}\left(\partial_\alpha r_\beta + \partial_\beta r_\alpha\right)$, and the **membrane strain tensor** $\gamma_{\alpha\beta}(\eta)$

$$\begin{aligned}\gamma_{\alpha\beta}(\zeta) &= \frac{1}{2}\left(\zeta_{\alpha|\beta} + \zeta_{\beta|\alpha}\right) - b_{\alpha\beta}\zeta_3 \\ &= e_{\alpha\beta}(\zeta) - \Gamma^{\rho}_{\alpha\beta}\zeta_\rho - b_{\alpha\beta}\zeta_3\,,\end{aligned} \tag{6.2.75}$$

and the **transverse shear strain tensor** $\varphi_\alpha(\zeta, r)$

$$\varphi_\alpha(\zeta,r) := \frac{1}{2}\left(\partial_\alpha\zeta_3 + b^{\rho}_{\alpha}\zeta_\rho + r_\alpha\right). \tag{6.2.76}$$

Remark 6.37

i) We observe that the Kirchhoff–Love hypothesis (6.2.48) implies $\varphi_\alpha(\zeta, r) = 0$ and that then (6.2.74) coincides with the definition (6.2.54). The appearance of (6.2.76) in the total potential energy of the (Naghdi) shell below can therefore be understood as an indirect enforcement of the Kirchhoff–Love hypothesis by penalization.
ii) Occasionally, the shear strains $\hat{\varphi}_\alpha(\eta, s)$ and the membrane strains $\gamma_{\alpha\beta}$ are combined into the **strain tensor** γ_{ij} by setting $\gamma_{\alpha 3} = \gamma_{3\alpha} := \varphi_\alpha$, $\gamma_{33} := 0$.

To give the variational formulation of the Naghdi shell, we define first the different contributions to the bilinear form. We have the bilinear form

$$B_N(\zeta, r; \eta, s) := \varepsilon^3 B_f(\zeta, r; \eta, s) + \varepsilon\left[B_m(\zeta;\eta) + B_{sh}(\zeta, r; \eta, s)\right] \tag{6.2.77}$$

where the **flexural part** is given by

$$B_f(\zeta, r; \eta, s) := \int_\omega \left\{\frac{1}{3}\, a^{\alpha\beta\rho\sigma}\Upsilon_{\alpha\beta}(\zeta, r)\,\Upsilon_{\rho\sigma}(\zeta, s)\right\}\sqrt{a}\, dx^1 dx^2\,, \tag{6.2.78}$$

the **membrane part** is given by

$$B_m(\zeta;\eta) = \int_\omega \left\{a^{\alpha\beta\rho\sigma}\gamma_{\alpha\beta}(\zeta)\gamma_{\rho\sigma}(\eta)\right\}\sqrt{a}\, dx^1 dx^2\,, \tag{6.2.79}$$

and the **shear part** is (cf. Remark 6.37 ii))

$$B_{\text{sh}}(\zeta, r; \eta, s) = \int_\omega \{8\mu a^{\alpha\beta}\gamma_{\alpha 3}(\zeta, r)\gamma_{\beta 3}(\eta, s)\} \sqrt{a}\, dx^1 dx^2 \,. \tag{6.2.80}$$

The external loadings have the form

$$F_N(\eta, s) := \int_\omega p^{i,\varepsilon}\zeta_i \sqrt{a}\, dx^1 dx^2 + \int_{\gamma_1} (q^{i,\varepsilon}\eta_i + m^{\alpha,\varepsilon} r_\alpha) d\gamma \tag{6.2.81}$$

where

$$p^{i,\varepsilon} = \frac{1}{2}\int_{-\varepsilon}^{\varepsilon} f^{i,\varepsilon} dx^3 + (h_+^{i,\varepsilon} - h_-^{i,\varepsilon}) \,,$$

$$q^{i,\varepsilon} = \frac{1}{2}\int_{-\varepsilon}^{\varepsilon} g^{i,\varepsilon}\, dx^3, \; m^{\alpha,\varepsilon} = \frac{1}{2}\int_{-\varepsilon}^{\varepsilon} g^{\alpha,\varepsilon} x^3 dx^3 \,.$$

Then the **weak form of the Naghdi-shell problem** reads: find $(\zeta, r) \in \mathbf{V}_N(\omega)$ such that

$$B_N(\zeta, r; \eta, s) = F_N(\eta, s) \text{ for all } \eta, s \in \mathbf{V}_N(\omega) \tag{6.2.82}$$

where

$$\mathbf{V}_N(\omega) = \{(\zeta, r) : \zeta_i, r_\alpha \in H^1(\omega),\; \zeta_i = r_\alpha = 0 \text{ on } \gamma_0\} \,. \tag{6.2.83}$$

On $\mathbf{V}_N(\omega)$ we introduce the norm

$$\|(\zeta, r)\|_{1,\omega} = \Big\{\sum_i \|\zeta_i\|_{1,\omega}^2 + \sum_\alpha \|r_\alpha\|_{1,\omega}^2\Big\}^{\frac{1}{2}} \,. \tag{6.2.84}$$

Then the following result holds.

Theorem 6.38 *There exists $\beta > 0$ (depending on $\lambda, \mu, \varepsilon$ and the shape of S) such that*

$$B_N(\zeta, r; \zeta, r) \geq \beta \,\|(\zeta, r)\|_{1,\omega}^2 \; \textit{for all } \; (\eta, r) \in \mathbf{V}_N(\omega) \,. \tag{6.2.85}$$

The proof proceeds in several steps, along the lines of the one for the Koiter shell in the previous section.

Step 1: Using the ellipticity of the elastic moduli $a^{\alpha\beta\rho\sigma}$ in (6.2.52) we show

$$B_N(\zeta, r; \zeta, r) \geq C_1\Big\{\sum_{\alpha,\beta} \|\gamma_{\alpha\beta}(\zeta)\|_{0,\omega}^2 + \|\Upsilon_{\alpha\beta}(\zeta, r)\|_{0,\omega}^2 + \|\hat{\varphi}_\alpha(\zeta, r)\|_{0,\omega}^2\Big\} \tag{6.2.86}$$

where $C_1 > 0$ depends only on λ, μ and the geometry of the midsurface S of the shell.

Step 2: Define the space

$$\boldsymbol{E}_N(\omega) = \{(\eta, s) \in [L^2(\omega)]^5 : \gamma_{\alpha\beta}(\eta),\ \Upsilon_{\alpha\beta}(\eta, s),\ \varphi_\alpha(\eta, s) \in L^2(\omega)\}\,.$$

Then it holds that $\boldsymbol{E}_N(\omega) = [H^1(\omega)]^5$.

To prove this, observe that $\varphi_\alpha(\eta, s) \in L^2(\omega)$ implies that $\eta_3 \in H^1(\omega)$. Now we reason as in the proof of Step 2 of the previous subsection.

Step 3: For every $(\eta, s) \in [H^1(\omega)]^5$,

$$\|(\eta, s)\| \geq C_2 \,\|(\eta, s)\|_{1,\omega}\,, \quad C_2 > 0\,. \tag{6.2.87}$$

Here the shell energy norm $\|\circ\|$ is defined by

$$\begin{aligned}\|(\eta, s)\|^2 := &\sum_i \|\eta_i\|^2_{0,\omega} + \sum_\alpha \|s_\alpha\|^2_{0,\omega} + \sum_{\alpha,\beta} \|\gamma_{\alpha\beta}(\eta)\|^2_{0,\omega}\\ &+ \sum_{\alpha,\beta} \|\Upsilon_{\alpha\beta}(\eta, s)\|^2_{0,\omega} + \sum_\alpha \|\hat{\varphi}_\alpha(\eta, s)\|^2_{0,\omega}\,.\end{aligned} \tag{6.2.88}$$

The proof is quite similar to that of Step 3 of Theorem 6.30 and we therefore skip it.

Next, we need again an infinitesimal rigid displacement lemma, i.e., an analog to Lemma 6.32.

Lemma 6.39 *(rigid displacement lemma for the Naghdi shell). Let $(\eta, s) \in [H^1(\omega)]^5$ satisfy*

$$\gamma_{\alpha\beta}(\eta) = \Upsilon_{\alpha\beta}(\eta, s) = \hat{\varphi}_\alpha(\eta, s) = 0 \ \text{ in } \ L^2(\omega)\,. \tag{6.2.89}$$

Then there exist vectors $\boldsymbol{c} \in \mathbb{R}^2$ and $\boldsymbol{d} \in \mathbb{R}^3$ such that

$$\eta_i(x^1, x^2)\, \boldsymbol{a}^i(x^1, x^2) = \boldsymbol{c} + \boldsymbol{d} \times \varphi(x^1, x^2) \ \text{ for all } \ (x^1, x^2) \in \overline{\omega}\,. \tag{6.2.90}$$

The coefficient functions $\eta_i \in C^2(\overline{\omega}, \mathbb{R})$ and

$$\boldsymbol{d} = \varepsilon^{\alpha\beta} \left(\frac{1}{2}\, \eta_{\beta|\alpha} \boldsymbol{a}^3 - \eta_\beta \boldsymbol{a}_\alpha\right)\,. \tag{6.2.91}$$

The proof is similar to that of Lemma 6.32 and therefore omitted. The remaining steps of the proof of Theorem 6.38 parallel now completely the proof of Theorem 6.30 in the previous subsection: Using Lemma 6.39, we infer that any $(\eta, s) \in [H^1(\omega)]^5$ satisfying the homogeneous boundary conditions

$$\eta_i = r_\alpha = 0 \ \text{ on } \ \gamma_0\,, \quad \int_{\gamma_0} do > 0 \tag{6.2.92}$$

must vanish (i.e., the analog to Lemma 6.33). Then we show that the seminorm $|(\eta, s)|$ given by

$$|(\eta, s)| := \Big\{ \sum_{\alpha,\beta} \|\gamma_{\alpha\beta}(\eta)\|^2_{0,\omega} + \|\Upsilon_{\alpha\beta}(\eta, s)\|^2_{0,\omega} + \|\hat{\varphi}_\alpha(\eta, s)\|^2_{0,\omega} \Big\}^{\frac{1}{2}}$$

is in fact a norm on $V_N(\omega)$ and there is $C > 0$ such that

$$|(\eta, s)| \geq C \, \|(\eta, s)\|_{1,\omega} \text{ for all } (\eta, s) \in V_N(\omega) . \tag{6.2.93}$$

This concludes the proof of Theorem 6.38.

6.2.5 *Asymptotics of Koiter shells as $\varepsilon \to 0$. Bending and membrane dominated shells*

We observe that both the Koiter and Naghdi shells are singularly perturbed problems, owing to the presence of the small parameter ε in the bilinear forms. There arises the question regarding the asymptotic behavior of the solution $\zeta^\varepsilon \in V_K(\omega)$ as $\varepsilon \to 0$. It turns out that the asymptotics depend strongly on the geometry of S, the boundary conditions and on the loading of the shell.

To classify the asymptotic behaviors of ζ^ε, we start with some heuristic considerations. We recall from the previous section that the variational problem for the Koiter shell is: find $\zeta^\varepsilon \in V_K(\omega)$ such that

$$B_K(\zeta^\varepsilon, \eta) = \varepsilon^3 B_F(\zeta^\varepsilon, \eta) + \varepsilon \, B_M(\zeta^\varepsilon, \eta) = F_K(\eta) \tag{6.2.94}$$

for all $\eta \in V_K(\omega)$ with $B_M(\cdot,\cdot)$ and $B_F(\cdot,\cdot)$ as in (6.2.51). We observe that **the size of loads F which the shell can carry for small ε is determined by whether the bending energy B_F $(\zeta^\varepsilon, \zeta^\varepsilon)$ or the membrane energy B_M $(\zeta^\varepsilon, \zeta^\varepsilon)$ are asymptotically dominant.**

Accordingly, we introduce the **space $V_F(\omega)$ of inextensional deformations**

$$V_F(\omega) := \{\eta \in V_K(\omega) : B_M(\eta, \eta) = 0\} \tag{6.2.95}$$

which can be written equivalently as

$$V_F(\omega) = \{\eta \in V_K(\omega) : \gamma_{\alpha\beta}(\eta) = 0 \text{ in } L^2(\omega)\} .$$

Definition 6.40 *The Koiter shell* (6.2.94) *is* **bending dominated** *if there exists*

$$0 \neq \eta \in V_F(\omega) : \quad F_K(\eta) \neq 0 . \tag{6.2.96}$$

For ζ^ε solving (6.2.94) we have that ζ^ε minimizes the **potential energy**

$$\Pi_\varepsilon(\zeta) := \frac{1}{2} \, \{\varepsilon^3 \, B_F(\zeta, \zeta) + \varepsilon \, B_M(\zeta, \zeta)\} - F_K(\zeta)$$

over $V_K(\omega)$.

If (6.2.94) is bending dominated, then formally, as $\varepsilon \to 0$, $\varepsilon^3\zeta^\varepsilon$ should converge to some $0 \neq \zeta^0 \in V_F(\omega)$ which solves the **flexural shell equations**:

$$\zeta^0 \in V_F(\omega) : \; B_M(\zeta^0, \eta) = \overline{F}_K(\eta) \text{ for all } \eta \in V_F(\omega)$$

where $\overline{F}_K := \lim_{\varepsilon\to 0} \varepsilon^{-3} F_K(\eta)$. We conclude that **in the bending dominated case, the shell can carry loadings $F_K(\cdot)$ of order** ε^3 as $\varepsilon \to 0$.

Let us next consider the **membrane dominated** case. To this end, we define the space

$$V(\omega) := \{\eta = (\eta_i) : \eta_i \in H^1(\omega),\ \eta_i = 0 \text{ on } \gamma_0\}, \tag{6.2.97}$$

and

$$|\eta|_M := \Big\{\sum_{\alpha\beta} \|\gamma_{\alpha\beta}(\eta)\|^2_{L^2(\omega)}\Big\}^{\frac{1}{2}} . \tag{6.2.98}$$

To describe the asymptotic membrane limit of a shell theory, we introduce the space

$$V_M(\omega) = \overline{V(\omega)}^{|\circ|_M} . \tag{6.2.99}$$

Recalling (6.2.53), i.e.,

$$\gamma_{\alpha\beta}(\eta) = \frac{1}{2}\,(\partial_\alpha\eta_\beta + \partial_\beta\eta_\alpha) - \Gamma^\sigma_{\alpha\beta}\eta_\sigma - b_{\alpha\beta}\eta_3 ,$$

we see that $|\eta|_M$ is a seminorm on $V(\omega)$ and also on $V_M(\omega)$.

Definition 6.41

i) *The deformation state ζ^ϵ of the Koiter shell* (6.2.94) *is a* **general membrane state** *if the semi-norm $|\circ|_M$ is a norm on $V(\omega)$.*

ii) *ζ^ϵ is a* **(proper) membrane state** *if the stronger condition*

$$|\eta|_M \geq c\Big\{\sum_\alpha \|\eta_\alpha\|^2_{1,\omega} + \|\eta_3\|^2_{0,\omega}\Big\}^{\frac{1}{2}} \text{ for all } \eta \in V(\omega) , \tag{6.2.100}$$

holds for some $c > 0$ independent of ε.

iii) *ζ^ε is a* **sensitive membrane state** *if it is a general, but not a proper membrane state.*

It is therefore essential to see when the key inequality (6.2.100) holds.

Theorem 6.42

i) *If $\gamma_0 = \gamma$,* (6.2.100) *holds if and only if S is uniformly elliptic, i.e., if and only if there is $c > 0$ such that*

$$\begin{aligned} |b_{\alpha\beta}(x^1, x^2)\,\xi_\alpha\xi_\beta| \geq c\,|\xi|^2 \text{ for all } x \in \overline{\omega}, \\ \text{for all } \xi = (\xi_1, \xi_2) \in \mathbb{R}^2 . \end{aligned} \tag{6.2.101}$$

ii) *For any surface* S (6.2.100) $\Longrightarrow \gamma_0 = \gamma$.

The proof can be found in [134].

Theorem 6.42, ii) says that a general membrane state ζ^ε can only be proper, if the shell is clamped along the whole edge.

Proposition 6.43 *If* S *is such that* (6.2.100) *holds, then*

$$V_M(\omega) = H_0^1(\omega) \times H_0^1(\omega) \times L^2(\omega) .$$

Proof Since (6.2.100) implies $\gamma_0 = \gamma$ by Theorem 6.42, ii), we get that $\boldsymbol{V}(\omega) = [H_0^1(\omega)]^3$. The assertion follows from the definition (6.2.98) of $|\eta|_M$ and the form of $\gamma_{\alpha\beta}$. □

Remark 6.44 If S is a sensitive membrane, the space $V_M(\omega)$ in (6.2.99) can be quite pathologic in the sense that the inclusion $V_M(\omega) \subset \mathcal{D}'(\omega)$ may fail (see [87] and also Section 6.2.7 below).

In general, the semi-norm $|\circ|_M$ may be a norm on $\boldsymbol{V}(\omega)$ in (6.2.97) **without** (6.2.100) being satisfied. This is the case if, for example, the membrane system is not elliptic any more in the sense of Agmon, Douglis, Nirenberg, but we still have a uniqueness result (e.g., variants of Holmgren's theorem). In this case 0 belongs to the essential spectrum of the membrane operator and hence there exist sensitive membrane states, so-called **Weyl sequences**, $\{\eta_k\}_k$ for which $\|\eta_k\|_{L^2} = 1$, $|\eta_k|_M \to 0$, i.e. (6.2.100) fails. Thus (6.2.100) ensures the absence of an essential spectrum in the limiting problem (see [121] for more).

Families of proper membrane states $\{\zeta^\varepsilon\}_\varepsilon$ converge, as $\varepsilon \to 0$, to a **limiting membrane shell model.**

Theorem 6.45 *Assume* (6.2.100) *and* $\gamma_0 = \gamma$. *Let* $\{\zeta^\varepsilon\}_\varepsilon \in \boldsymbol{V}_K(\omega)$ *be a family of solutions to the Koiter shell model* (6.2.94) *with*

$$F_K(\eta) = \int_\omega p^{i,\varepsilon}\eta_i \sqrt{a}\, dx^1 dx^2, \quad p^{i,\varepsilon} := \frac{1}{2}\int_{-\varepsilon}^{\varepsilon} f^{i,\varepsilon} dx^3 + h_+^{i,\varepsilon} - h_-^{i,\varepsilon}$$

and assume that there exist $f^i \in L^2(\Omega), h_\pm^\alpha \in L^2(\omega),\ h_\pm^3 \in H^1(\omega)$ *independent of* ε *such that*

$$\begin{aligned} f^{i,\varepsilon}(x^1, x^2, \varepsilon x^3) &= f^i(x^1, x^2, x^3) \text{ for all } x \in \omega \times (-1,1) , \\ h_\pm^{i,\varepsilon}(x^1, x^2) &= \varepsilon\, h_\pm^i(x^1, x^2) \text{ for all } (x^1, x^2) \in \omega . \end{aligned}$$

Then, as $\varepsilon \to 0$,

$$\zeta_\alpha^\varepsilon \boldsymbol{a}^\alpha \to \zeta_\alpha \boldsymbol{a}^\alpha \text{ in } H^1(\omega),\ \zeta_3^\varepsilon \boldsymbol{a}^3 \to \zeta_3 \boldsymbol{a}^3 \text{ in } L^2(\omega) ,$$

where $\zeta = (\zeta_1, \zeta_2, \zeta_3)$ *is the unique solution of the following* **membrane shell** *problem: find* $\zeta \in V_M(\omega)$ *such that*

$$B_M(\zeta,\eta) = F_K(\eta) := \int_\omega \Big\{ \int_{-1}^{1} f^i \, dx^3 + h^i_+ - h^0_- \Big\} \eta_i \sqrt{a}\, dx^1 dx^2 \,. \qquad (6.2.102)$$

Moreover, if $\gamma = \partial\omega \in C^4$, *and* φ *is analytic in* $\overline{\omega}$, *and the loadings* $p^i = \int_{-1}^{1} f^i dx_3 + h^i_+ - h^i_-$ *satisfy*

$$p^\alpha \in H^1(\omega),\ p^3 \in H^2(\omega)\,,$$

then

$$\|\zeta^\varepsilon - \zeta\|_{H^1\times H^1\times L^2} \leq C\,\varepsilon^{1/5}\,.$$

For the proof, we refer to [48, 92].

Remark 6.46 Under the assumptions (6.2.100) and $\gamma_0 = \gamma$ of Theorem 6.45, we have, by Theorem 6.42, that S is necessarily uniformly elliptic, i.e., (6.2.101) holds. This, in turn, ensures the unique solvability of the membrane shell equations (6.2.102) in $V_M(\omega) = H^1_0(\omega) \times H^1_0(\omega) \times L^2(\omega)$.

We return to the **bending dominated case**, which may arise if $V_F(\omega) \neq \{0\}$. This assumption is essentially an assumption on the geometries of S and of $\gamma_0 \subseteq \gamma$ and on the loading F_K. We have the following result [48].

Theorem 6.47 *Assume* (6.2.96) *and* $\int_{\gamma_0} do > 0$. *Assume further that there exist* $f^i \in L^2(\Omega)$ *and* $h^i_\pm \in L^2(\omega)$ *such that the loadings* $f^{i,\varepsilon}$, $h^{i,\varepsilon}_\pm$ *of the shell satisfy*

$$\begin{aligned} f^{i,\varepsilon}(x^1,x^2,\varepsilon x^3) &= \varepsilon^2 f^i(x),\ x \in \Omega\,, \\ h^{i,\varepsilon}_\pm(y) &= \varepsilon^3\, h^i_\pm(y)\,,\ y \in \omega\,. \end{aligned} \qquad (6.2.103)$$

Let further $\zeta \in V_F(\omega)$ *denote the unique solution of the* **flexural shell equations**: *find*

$$\zeta \in V_F(\omega) : B_F(\zeta,\eta) = F_K(\eta) \text{ for all } \eta \in V_F(\omega) \qquad (6.2.104)$$

where F_K *is as in* (6.2.102).

Let $\zeta^\varepsilon \in V_K(\omega)$, *as before, denote the solution of the Koiter shell equations* (6.2.94). *Then, as* $\varepsilon \to 0$,

$$\sum_\alpha \|\zeta^\varepsilon_\alpha - \zeta_\alpha\|_{H^1(\omega)} + \|\zeta^\varepsilon_3 - \zeta_3\|_{H^2(\omega)} \to 0\,. \qquad (6.2.105)$$

So far, we only considered the case when shells are either bending or membrane dominated and based the classification on the subspace $V_F(\omega)$ being equal to $\{0\}$ or not. If $V_F(\omega) \neq \{0\}$, the particular deformation state ζ^ε assumed for a given loading $F_K(\cdot)$ of the shell need not, however, be of bending type if the loadings $F_K(\cdot)$ do not "excite" solution components in $\boldsymbol{V}_F(\omega)$ (see (6.3.37) below for more on this): We therefore call the deformation state ζ^ε (solution of (6.2.94) for given $F_K(\cdot)$) **bending dominated** if the "limiting" solution ζ of (6.2.104) does not vanish. Clearly, $\boldsymbol{V}_F(\omega) \neq \{0\}$ is necessary for $\zeta \neq 0$ in (6.2.104), nevertheless if $\boldsymbol{V}_F(\omega) \neq \{0\}$, we see from (6.2.104) that $\zeta = 0$ is still possible if $F_K(\eta) = 0$ for all $\eta \in \boldsymbol{V}_F(\omega)$.

Remark 6.48 So far, we discussed only the asymptotic limits of the Koiter shell model. It turns out, however, (and has been rigorously proved in [48, 92]) that the solution of the problem of three-dimensional, linearized elasticity on the shell of finite thickness $t = 2\varepsilon$ has, as $\varepsilon \to 0$, the same limiting behavior as the Koiter-model. In this sense, the Koiter model is **asymptotically consistent** with three-dimensional elasticity. The same is true for Naghdi's model, since the gap between the Koiter and Naghdi models is small in ε (see [92]).

6.2.6 *The case of a plate*

In the special case of a plate, the Koiter and Naghdi shell models degenerate to the **Kirchhoff-Love**, **Reissner–Mindlin** plate theories , respectively, and to the theory of plane stress. We have

$$\varphi(x^1, x^2) = (x^1, x^2, 0)^\top \tag{6.2.106}$$

and observe that then

$$b_{\alpha\beta} = b_\alpha^\beta = 0,\ c_{\alpha\beta} = 0,\ \Gamma^\rho_{\alpha\beta} = 0\,. \tag{6.2.107}$$

Inserting this into (6.2.53), (6.2.54), we get

$$\gamma_{\alpha\beta}(\eta) = \frac{1}{2}\left(\partial_\alpha \eta_\beta + \partial_\beta \eta_\alpha\right), \tag{6.2.108}$$

$$\Upsilon_{\alpha\beta}(\eta) = \partial_{\alpha\beta}\eta_3\,, \tag{6.2.109}$$

i.e., the Koiter model decouples into a flexure problem for η_3 and a membrane problem for the horizontal displacements $\eta_H := (\zeta_1, \zeta_2)^\top$ which read as follows.

a) **Flexure problem (Kirchhoff plate)**: find $\zeta_3 \in V_3(\omega) := \{\eta_3 \in H^2(\omega) : \eta_3 = \partial_\nu \eta_3 = 0 \text{ on } \gamma_0\}$ such that

$$-\int_\omega m_{\alpha\beta}\,\partial_{\alpha\eta}\eta_3\,dx_1 dx_2 = \int_\omega p_3^\varepsilon \eta_3\,dx_1 dx_2 \text{ for all } \eta_3 \in V_3(\omega) \tag{6.2.110}$$

where

$$m_{\alpha\beta} := -\varepsilon^3\left\{\frac{4\lambda\mu}{4(\lambda+\mu)}\,\Delta\zeta_3\,\delta_{\alpha\beta} + \frac{4}{3}\,\mu\,\partial_{\alpha\beta}\,\zeta_3\right\}$$

$$p_3^\varepsilon = \int_{-\varepsilon}^{\varepsilon} f_3^\varepsilon\,dx_3^\varepsilon + h_3^+ + h_3^-\,.$$

b) **Membrane problem**: find $\zeta_H \in V_H(\omega) := \{\eta_H : \eta_\alpha \in H^1(\omega),\ \eta_\alpha = 0 \text{ on } \gamma_0\}$,

$$\begin{aligned}
\int_\omega n_{\alpha\beta}\, \partial_\beta \eta_\alpha \, dx_1 dx_2 &= \int_\omega p_\alpha^\varepsilon \eta_\alpha \, d\omega \quad \text{for all} \quad (\eta_\alpha) \in V_H(\omega) \\
n_{\alpha\beta} &:= \varepsilon \left\{ \frac{4\lambda\mu}{(\lambda + 2\mu)} \, \epsilon_{\rho\rho}(\zeta_H)\, \delta_{\alpha\beta} + 4\mu\, \epsilon_{\alpha\beta}(\zeta_H) \right\}, \\
\epsilon_{\alpha\beta}(\zeta_H) &:= \frac{1}{2} \left(\partial_\alpha \zeta_\beta + \partial_\beta \zeta_\alpha \right), \\
p_\alpha^\varepsilon &:= \int_{-\varepsilon}^{\varepsilon} f_\alpha \, dx_3 + h_\alpha^+ + h_\alpha^- .
\end{aligned} \tag{6.2.111}$$

Remark 6.49 Note that the Euler–Lagrange equations for (6.2.110) read:

$$\begin{aligned}
-\partial_{\alpha\beta}\, m_{\alpha\beta} &= p_3^\varepsilon \text{ in } \omega , \\
\zeta_3 = \partial_\nu \zeta_3 &= 0 \quad \text{on } \gamma_0 , \\
m_{\alpha\beta}\, \nu_\alpha \nu_\beta &= 0 \quad \text{on } \gamma_1 , \\
(\partial_\alpha\, m_{\alpha\beta})\nu_\beta + \partial_\tau (m_{\alpha\beta}\, \nu_\alpha \nu_\beta) &= 0 \quad \text{on } \gamma_1 .
\end{aligned} \tag{6.2.112}$$

In particular, we recover the **biharmonic equation**

$$-\partial_{\alpha\beta}\, m_{\alpha\beta} = \varepsilon^3 \, \frac{8\mu(\lambda + \mu)}{3(\lambda + 2\mu)} \, \Delta\, \Delta\, \zeta_3 = t^3\, D\, \Delta\, \Delta\, \zeta_3$$

where $D = E/(12(1 - \nu^2))$ is the **flexural rigidity** of the plate.

Likewise, the Euler–Lagrange equations for (6.2.111) are

$$\begin{aligned}
-\partial_{\alpha\beta}\, n_{\alpha\beta} &= p_\alpha^\varepsilon \text{ in } \omega , \\
\zeta_\alpha &= 0 \quad \text{on } \gamma_0 , \\
n_{\alpha\beta}\, \nu_\beta &= 0 \quad \text{on } \gamma_1 .
\end{aligned} \tag{6.2.113}$$

We turn now to the Naghdi shell model. It also decouples under the flatness assumptions (6.2.106), (6.2.107); the inplane displacements $\zeta_H = (\zeta_1, \zeta_2)^\top$ satisfy again the membrane problem (6.2.111), (6.2.113). The normal deflection ζ_3 and the rotations r_α, however, are now solutions of the **Reissner–Mindlin plate** equations: find $\zeta_3, r_1, r_2 \in H_0^1(\omega)$ such that

$$b(r, \zeta_3, s, \eta_3) + \frac{4\mu}{\varepsilon^2} \left(\hat{\varphi}(\zeta_3, r),\ \hat{\varphi}(\eta_3, s) \right) = F(s, \eta_3) \tag{6.2.114}$$

for all $s_1, s_2, \eta_3 \in H_0^1(\omega)$. Here

$$b(r, \zeta_3, s, \eta_3) := \frac{D}{2} \int_\omega \{(1-\nu)\nabla r \cdot \nabla s + (1+\nu)(\nabla \cdot r)(\nabla \cdot s)\}\, dx_1 dx_2 \,,$$

$$\hat{\varphi}(\zeta_3, r) := \nabla \zeta_3 + r \,,$$

$$F(s, \eta_3) := \varepsilon^{-3} \int_\omega p_3^\varepsilon \, \eta_3 \, dx_1 dx_2 \,.$$

We observe that, as $\varepsilon \to 0$ the Kirchhoff constraint $\psi(\zeta_3, r) = 0$ is imposed upon the solution. It can indeed be shown [9] that (6.2.110) is, for properly scaled loads, the limit of (6.2.114) when $\varepsilon \to 0$. For $\varepsilon > 0$, however, (6.2.114) is a singularly perturbed elliptic system and ζ_3 and r exhibit a **boundary layer** [9].

6.2.7 *Membrane theories of shells of revolution*

In (6.2.99), we defined the space $\boldsymbol{V}_M(\omega)$ of membrane-dominated deformations as closure of $[H_0^1(\omega, \gamma_0)]^3$ with respect to the norm $|\circ|_M$ defined in (6.2.98). In Proposition 6.43, we showed that at least in one case, $\boldsymbol{V}_M(\omega) = (H_0^1 \times H_0^1 \times L^2)(\omega)$ (in the case of a totally clamped, uniformly elliptic shell). The aim of the present section is to illustrate briefly the possible structure of $\boldsymbol{V}_M(\omega)$ for parabolic and hyperbolic midsurfaces S, following [110, 111]. We confine ourselves to **shells of revolution**, i.e. the midsurface S is given by

$$S = \{x \in \mathbb{R}^3 : x_1 \in [-M, M], x_2^2 + x_3^2 = (r(x_1))^2, \quad 0 < r(x_1) \in C^\infty([-M, M])\} \,, \tag{6.2.115}$$

and a parametric representation φ of S is, for example,

$$\varphi : (x^1, x^2) \longmapsto (x_1, r(x_1)\, \sin x_2,\, r(x_1)\, \cos x_2) \,. \tag{6.2.116}$$

Definition 6.50 *The shell of revolution* (6.2.115) *is said to be*

parabolic if $r'' = 0$,
elliptic if $r'' < 0$,
hyperbolic if $r'' > 0$ for $x_1 \in [-M, M]$.

Note that any elliptic shell of revolution is in fact uniformly elliptic in the sense that (6.2.101) holds. In this case we have

$$B_F(\zeta, \eta) = D\,((Tv)(\zeta),\, v(\eta)),\; B_M = 12D((Tv)(\zeta),\, \gamma(\eta)) \tag{6.2.117}$$

where $D = E/[12(1-\nu^2)]$ is the flexural rigidity, and the constitutive tensor

$$(Tv)_{\alpha\beta} = \nu(tr v)\, \delta_{\alpha\beta} + (1-\nu)\, v_{\alpha\beta} \,,$$

and the tensors $\upsilon_{\alpha\beta}, \gamma_{\alpha\beta}$ take the explicit form

$$\begin{aligned}\gamma_{11}(\zeta) &= \frac{1}{A_1}\frac{\partial\zeta_1}{\partial x_1} + \frac{\zeta_3}{R_1}\,, \\ \gamma_{22}(\zeta) &= \frac{1}{A_2}\frac{\partial\zeta_2}{\partial x_2} + \frac{\partial A_2}{\partial x_1}\frac{\zeta_1}{A_1 A_2} + \frac{\zeta_3}{R_2}\,, \\ \gamma_{12}(\zeta) &= \frac{1}{2}\left\{\frac{1}{A_2}\frac{\partial\zeta_1}{\partial x_2} + \frac{A_2}{A_1}\frac{\partial}{\partial x_1}\left(\frac{\zeta_2}{A_2}\right)\right\},\end{aligned} \tag{6.2.118}$$

and

$$\begin{aligned}\upsilon_{11}(\zeta) &= \frac{1}{A_1}\frac{\partial}{\partial x_1}\left(\frac{1}{A_1}\frac{\partial\zeta_3}{\partial x_1} - \frac{\zeta_1}{R_1}\right), \\ \upsilon_{22}(\zeta) &= \frac{1}{A_2}\frac{\partial}{\partial x_2}\left(\frac{1}{A_2}\frac{\partial\zeta_3}{\partial x_2} - \frac{\zeta_2}{R_2}\right) \\ &\quad + \frac{A_2^1}{A_1 A_2}\left(\frac{1}{A_1}\frac{\partial\zeta_3}{\partial x_1} - \frac{\zeta_1}{R_1}\right), \\ \upsilon_{12}(\zeta) &= \frac{1}{A_2^2}\frac{\partial^2\zeta_3}{\partial x_2^2} - \frac{1}{A_1 A_2^2}\frac{\partial\zeta_2}{\partial x_2} + \frac{A_2^1}{R_2}\left(\frac{1}{A_1}\frac{\partial\zeta_3}{\partial x_1} - \frac{\zeta_1}{R_1}\right).\end{aligned} \tag{6.2.119}$$

Here A_α are given by

$$A_1(x_1) = \sqrt{1 + (r'(x_1))^2}, \;\; A_2(x_1) = r(x_1) \tag{6.2.120}$$

and R_i are the principal radii of curvature, i.e.,

$$R_1(x_1) = -A_1^3/A_2'', \;\; R_2(x_1) = A_1 A_2 \tag{6.2.121}$$

($R_1 = \infty$ in the parabolic case). The inner product $(\cdot,\cdot)$ in (6.2.117) is given by

$$(\sigma, \tau) = \int\limits_\omega \sigma_{\alpha\beta}\, \tau_{\alpha\beta}\, A_1 A_2\, dx^1 dx^2\,.$$

We assume further that $\omega \subset \mathbb{R}^2$ is a polygon with M boundary segments $\gamma^{[i]}$, $i = 1, \ldots, M$.

Exercise 6.51 Consider (6.2.118) for given $\gamma_{\alpha\beta}$. Show that in the hyperbolic case ($r'' > 0$), (6.2.118) is a hyperbolic system for the ζ_i.

Show that the **characteristic curves** are given by

$$\hat{x}_1 = x_2 + \int\limits_0^{x_1} a(s)ds = \text{const}, \; \hat{x}_2 = x_2 - \int\limits_0^{x_1} a(s)ds = \text{const} \tag{6.2.122}$$

where $a(s) := (A_2^{-1}A_2'')^{\frac{1}{2}}$ (consider also $r'' = 0$).

We can now give the stated characterization of $V_M(\omega)$.

Theorem 6.52 *Consider a clamped shell of revolution S. Then in the elliptic case*

$$V_M(\omega) = H_0^1(\omega) \times H_0^1(\omega) \times L^2(\omega) ,$$

in the parabolic case

$$H_0^1(\omega) \times H_0^1(\omega) \times L^2(\omega) \subseteq V_M(\omega) \subseteq L^2(\omega) \times (H^1(\omega))' \times (H^2(\omega))'$$

and in the hyperbolic case

$$\begin{aligned} V_M(\omega) = \{(u_1, u_2, u_3) \in W_1^0(\omega) \times W_2^0(\omega) \times (H^1(\omega))' : \\ u_3 + \frac{A_2}{2B}(u_{1,2} + u_{2,1}) \in L^2(\omega)\} \end{aligned}$$

Here $B = A_1^{-1} A_2 \sqrt{a}$ and u_i are related to ζ_i via

$$\begin{aligned} u_1 &= A_1 A_2^{-1} a^{-\frac{1}{2}} \zeta_1 + a^{\frac{1}{2}} \zeta_2 , \\ u_2 &= -A_1 A_2^{-1} a^{-\frac{1}{2}} \zeta_1 + a^{\frac{1}{2}} \zeta_2 , \\ u_3 &= \zeta_2 . \end{aligned} \tag{6.2.123}$$

Finally, the spaces W_α^0 are defined by

$$W_\alpha^0 = \left\{ \text{closure of } H_0^1(\omega) \text{ with respect to } \|u\|_{L^2(\omega)} + \left\| \frac{\partial u}{\partial \hat{x}_\alpha} \right\|_{L^2(\omega)} \right\}, \ \alpha = 1, 2 .$$

We observe that the parabolic and hyperbolic membrane cases allow for Dirac distributions in ζ_3; this is not just an artifact of the mathematical setting, but such components are indeed part of the solution, even for smooth right-hand sides as shown in [111].

6.3 *hp*-FEM for problems with constraints. Locking.

Frequently in mechanics (and elsewhere) there arises the need to discretize problems depending on a **small perturbation parameter** $\varepsilon \in (0, 1]$. As we saw already in Chapter 3, solutions of such parameter dependent problems may exhibit **boundary layers** (if the perturbation is singular) which must be properly accounted for in the design of the FE subspace. A second, more delicate aspect is that of **locking** of the FE method which occurs if the problem (P_ε) under consideration is close to a **limiting problem** (P_0) which imposes a **constraint** $\mathcal{C}$ on the limiting solution u^0. The strategy for designing FEM for such problems that are insensitive to small values of the parameter ε is to reduce the impact of the constraint on the FE solution in a controlled fashion. We formulate the main ideas first in an abstract setting, examples will follow immediately afterwards. A discussion of a model problem will then shed light on the FE-analysis of locking. Finally, we address membrane locking in the context of the shell models.

We mention from Section 6.2.1.2 that for the pure displacement formulation of linearized, three-dimensional elastostatics and the corresponding two-dimensional plane stress/strain problems in Section 6.2.1.1 the analysis of conforming *hp*-FE discretizations based on the subspaces $S^{p,1}(\Omega, \mathcal{T}^{n,\sigma})$ introduced in Chapter 4 can be performed completely analogously to that of the heat conduction model problem considered there **as long as none of the problem parameters becomes critical** (as, for example, in the case of isotropic, nearly incompressible material). In the present section, we focus therefore on the analysis of nearly degenerate problems and their *hp*-FE discretizations, in particular for the Naghdi shell model presented in Section 6.2. Since the Reissner–Mindlin plate is a special case of the Naghdi shell, our analysis will also apply to plate problems; we mention, however, that in the past decade numerous papers devoted to the analysis of other robust finite elements for the Reissner–Mindlin plate have appeared.

Most of these works, however, are based on a Helmholtz decomposition of the rotation vector, which allows to reduce the Reissner–Mindlin problem to a Stokes-like problem. Such a decomposition with corresponding advantageous analytic properties does not seem to be available for the Naghdi shell model; the reduced constraint methods avoid the use of Helmholtz decompositions and, hence, do apply to the Naghdi and related shell models, as we will show. Since plates and membranes are, as we explained in Section 6.2.6, special cases of the Koiter or Naghdi shell respectively, the analysis below applies in particular also to these problems.

6.3.1 *The problem of locking in FEM*

6.3.1.1 *A class of abstract constrained problems*

By $0 < \varepsilon \le 1$ we denote a (dimensionless) small parameter. Let $X \subset L^2(\Omega)$ be a Hilbert space which is dense in $L^2(\Omega)$. Further, we are given a linear and continuous operator $\mathcal{C} : X \to L^2(\Omega)$, the **constraint operator**, which satisfies

$$G := \ker \mathcal{C} = \{u \in X : \mathcal{C}u = 0\} \neq \{0\} . \tag{6.3.1}$$

Finally, let $B_0(\cdot,\cdot) : X \times X \to \mathbb{R}$ be a continuous bilinear form which is coercive on X, i.e., there is $\alpha_0 > 0$ independent of ε such that

$$B_0(u,u) \ge \alpha_0 \, \|u\|_X^2 \text{ for all } u \in X . \tag{6.3.2}$$

Then the variational form of the **problem** (P_ε) reads: **given** $F \in X'$, **find** $u^\varepsilon \in X$ **such that**

$$B_0(u^\varepsilon, v) + \varepsilon^{-2}(\mathcal{C}u^\varepsilon, \mathcal{C}v) = F(v) \text{ for all } v \in X . \tag{6.3.3}$$

By (6.3.2), for every $\varepsilon \in (0,1]$ there exists a unique solution u^ε of (6.3.3).

To explain the locking phenomenon for FE discretizations of (6.3.3), we show that a FE approximation $u_N^\varepsilon \in X_N \subset X$ may considerably underestimate $\|u^\varepsilon\|_X$. To see this, let first $0 \neq u^0 \in X$ be such that $\mathcal{C}u^0 = 0$ (by (6.3.1), such an u_0 exists). Next, we assume that $K := F(u^0) > 0$, and $B_0(u^0, u^0) \le F(u^0)$ (scale u^0 appropriately).

Then we have, since u^ε minimizes the potential energy corresponding to (6.3.3),

$$\begin{aligned}\Pi(u^\varepsilon) &= \frac{1}{2}\, B_0(u^\varepsilon, u^\varepsilon) + \frac{1}{2\varepsilon^2}\,(\mathcal{C}u^\varepsilon, \mathcal{C}u^\varepsilon) - F(u^\varepsilon)\\ &\le \Pi(u^0) = \frac{1}{2}\, B_0(u^0, u^0) - F(u^0) \le -\frac{1}{2}\, F(u^0)\,.\end{aligned}$$

Since also

$$0 < F(u^0) \le -2\Pi(u^\varepsilon) = F(u^\varepsilon) \le |||F|||\ \|u^\varepsilon\|_X\,,$$

we find the **lower bound for the exact solution**

$$\|u^\varepsilon\|_X \ge K/|||F|||_{X'}\,. \tag{6.3.4}$$

Next, let $\{X_N\}_N$ be a dense family of finite-dimensional subspaces of X. Then the FE approximation $u_N^\varepsilon \in X_N$ of u^ε is defined by

$$u_N^\varepsilon \in X_N : B_0(u_N^\varepsilon, v) + \varepsilon^{-2}(\mathcal{C}u_N^\varepsilon, \mathcal{C}v) = F(v) \text{ for all } v \in X_N\,. \tag{6.3.5}$$

We assume now that

$$G_N := \ker \mathcal{C} \cap X_N = \{0\}\,. \tag{6.3.6}$$

This implies that for every N there exists $C(N) > 0$ with

$$\|\mathcal{C}v_N\|_{L^2} \ge C(N)\ \|v_N\|_X \text{ for all } v_N \in X_N\,. \tag{6.3.7}$$

Then, by (6.3.2),

$$B_0(u, u) + \varepsilon^{-2}(\mathcal{C}u, \mathcal{C}u) \ge \alpha\,\|u\|_X^2 \text{ for all } u \in X_N$$

where $\alpha = \alpha_0 + \varepsilon^{-2}(C(N))^2$. Hence we find the **upper bound for the FE solution**

$$\|u_N^\varepsilon\|_X \le \alpha^{-1}\,|||F|||_{X'} \le \varepsilon^2(C(N))^{-2}\ |||F|||_{X'}\,. \tag{6.3.8}$$

Comparing (6.3.8) with (6.3.4), we see that under the assumption (6.3.6) for $0 < \varepsilon \ll 1$, the FE approximation u_N^ε of u^ε is much smaller than u^ε itself (if measured in the energy norm), which is sometimes referred to as (numerical) **locking** in the engineering literature (this also manifests in the L^∞-norm, i.e., in the underestimation of pointwise displacements). More refined definitions of locking based on the deviation of the best possible **robust** (i.e., uniform in ε)

rate of convergence from the rate of convergence of the best approximation of the solution are given in [23] (see also Remark 6.56 below).

One way to overcome locking is to adopt a different variational formulation, for example a **mixed formulation with penalty**. To this end, we introduce in (6.3.3) the new ("pressure") variable

$$p := \varepsilon^{-2}\, \mathcal{C} u \in L^2(\Omega) \tag{6.3.9}$$

and rewrite (6.3.3) in the **saddle point** (mixed) **form with penalty** [36]

$$\begin{aligned} B_0(u^\varepsilon, v) + (p, \mathcal{C}v) &= F(v) \quad \text{for all } \ v \in X\,, \\ (\mathcal{C}u^\varepsilon, q) - \varepsilon^2 (p, q) &= 0 \qquad\quad \text{for all } \ q \in S = L^2(\Omega)\,. \end{aligned} \tag{6.3.10}$$

Notice that, formally, for $\varepsilon = 0$ (6.3.10) becomes a saddle point problem of Stokes type which we analyzed in Chapter 5. Problem (6.3.10) is still equivalent to the original problem (6.3.3). This changes, however, when (6.3.10) is discretized.

A FE discretization of (6.3.10) proceeds in the usual way. Select finite-dimensional subspaces $X_N \subset X$, $S_N \subset L^2(\Omega)$ and solve:

$(u^\varepsilon_N, p^\varepsilon_N) \in X_N \times S_N$:

$$\begin{aligned} B_0(u^\varepsilon_N, v) + (p^\varepsilon_N, \mathcal{C}v) &= F(v) \quad \text{for all } \ v \in X_N\,, \\ (\mathcal{C}u^\varepsilon_N, q) - \varepsilon^2 (p^\varepsilon_N, q) &= 0 \qquad\quad \text{for all } \ q \in S_N\,. \end{aligned} \tag{6.3.11}$$

It is now possible to invert (6.3.9) on the discrete level and to eliminate p^ε_N from (6.3.11): from the second equation of (6.3.11) we see that then

$$p^\varepsilon_N = \varepsilon^{-2}\, P_N\, \mathcal{C} u^\varepsilon_N \tag{6.3.12}$$

where $P_N : L^2(\Omega) \to S_N$ is the $L^2(\Omega)$-orthogonal projection. Hence u^ε_N solves the following analog to (6.3.5): find $u^\varepsilon_N \in X_N$:

$$B_\varepsilon(u^\varepsilon_N, v) := B_0(u^\varepsilon_N, v) + \varepsilon^{-2}(P_N\, \mathcal{C}u^\varepsilon_N, P_N\, \mathcal{C}v) = F(v) \ \text{ for all } \ v \in X_N \tag{6.3.13}$$

(here we used that $(P_N v, w) = (P_N^2 v, w) = (P_N v, P_N w)$).

Comparing (6.3.13) and (6.3.5), we see that we have in effect altered the primal variational formulation by introducing P_N into the constraint bilinear form. By a suitable choice of the space S_N, the effect of the constraint $\mathcal{C}$ on the FE solution u^ε_N can be reduced. Formulations of the type (6.3.13) (possibly with a reduction operator other than the $L^2(\Omega)$-projection P_N) are called **reduced constraint methods**. Reduction operators P_N may also be realized by **reduced integration** in the stiffness matrix corresponding to $(\mathcal{C}u, \mathcal{C}v)$. An asymptotic error analysis of (6.3.13), however, usually uses its equivalence to the mixed discretization (6.3.11) of (6.3.10) as we will show in the next section.

Remark 6.53 In many applications we do not have the coercivity (6.3.2) of $B_0(\cdot,\cdot)$ on all of X. This precludes direct application of the theory. Typically, however, the form

$$B_1(u,v) := B_0(u,v) + (\mathcal{C}u, \mathcal{C}v) \tag{6.3.14}$$

is coercive on X, although $B_0(\cdot,\cdot)$ by itself is not. The following device allows to fit such problems into the above framework. We define a new small parameter t by

$$t^{-2} = \varepsilon^{-2} - 1\,, \tag{6.3.15}$$

and observe that $t \to 0 \Longleftrightarrow \varepsilon \to 0$.

Then the reduced constraint FEM (6.3.13) is equivalent to: find $u_N^\varepsilon \in X_N$ such that

$$B_1(u_N^\varepsilon, v) + t^{-2}\,(\mathcal{C}u_N^\varepsilon, \mathcal{C}v) = F(v) \text{ for all } v \in X_N\,. \tag{6.3.16}$$

and the above theory applies once more (with t in place of ε and B_1 in place of B_0) if B_1 is coercive on X. Below we will indicate the dependence of n on ε or on t by u^ε, u^t interchangeably.

We now give **examples** for the above theory.

6.3.1.2 *Examples*

The following examples are particular instances of the above abstract framework. They all lead to locking in conventional FE discretizations. While volume and shear-locking can effectively be avoided, the membrane locking in Example 3 is the object of current research.

Example 1: Nearly incompressible elasticity ("volume locking")

We consider the boundary value problem of linearized elasticity in $\Omega \subset \mathbb{R}^d$, $d = 2, 3$. If the material is **almost incompressible**, a large amount of energy must be spent on volume changes. Correspondingly, we have from (6.1.63)

$$B(u,v) = 2\mu(\epsilon[u], \epsilon[v]) + \lambda(\operatorname{div} u, \operatorname{div} v) \tag{6.3.17}$$

with $\lambda \gg \mu$. To fit (6.3.17) in the above theory, we set $\varepsilon := 1/\sqrt{\lambda} \ll 1$ and $B_0(u,v) = 2\mu(\epsilon[u], \epsilon[v])$. By Korn's inequality, Propositions 6.20, and 6.22, we see that $B_0(\cdot,\cdot)$ satisfies (6.3.2) with $X = [H_0^1(\Omega)]^d$. Further, the constraint

$$\mathcal{C}u = \operatorname{div} u : [H^1(\Omega)]^d \to L^2(\Omega) \tag{6.3.18}$$

is evidently linear and continuous.

Following (6.3.9), we introduce here the new unknown

$$p := \lambda \operatorname{div} u$$

and obtain (for Dirichlet boundary conditions): find $u \in H_0^1(\Omega)$ such that

$$\begin{aligned} B_0(u,v) + (p, \operatorname{div} v) &= F(v) \\ -(q, \operatorname{div} u) + \frac{1}{\lambda}\,(p,q) &= 0 \end{aligned} \tag{6.3.19}$$

for all $v \in H_0^1(\Omega)$, $q \in L^2(\Omega)$.

Notice that this problem degenerates, as $\lambda \to \infty$, to a Stokes problem which we investigated in Chapter 5. Discretizing therefore the problem (6.3.19) with any velocity space for the displacements and the corresponding pressure space for p, we get a discretization that is stable in the incompressible limit, thanks to the inf-sup condition satisfied by the subspace pair. This stability of the discretization of the limit problem is indeed inherited by the problem (6.3.19), for large but finite λ (see [36], [41], [44]).

Example 2: Reissner–Mindlin plate ("shear-locking")
We consider the Reissner-Mindlin problem of plate bending with shear, i.e., (6.2.114). This problem takes the form (6.3.3) with

$$B_0 = b, \quad C(r,\zeta_3;s,\eta_3) = 4\mu(\hat\varphi(\zeta_3,r),\,\hat\varphi(\eta_3,s))_0\ .$$

As $\epsilon \to 0$, the Kirchhoff–Love hypothesis (6.2.48) is enforced, which takes the form $r_\alpha = -\zeta_{3,\alpha}$.

Example 3: Bending dominated Koiter shell ("membrane locking")
We consider the Koiter shell model (6.2.49): in the bending dominated case we assume

$$F_K(\eta) = \varepsilon^3\,\overline{F}_K(\eta)$$

with $\overline{F}_K$ independent of ε. Upon dividing (6.2.94) by ε^3, the shell problem takes the form (6.3.3) with

$$B_0 = B_f,\ C(u,v) = (\mathcal{C}u, \mathcal{C}v) = B_M(\zeta,\eta)$$

where B_M is as in (6.2.51). We see that the constraint C is here given by the membrane strain tensor γ.

6.3.1.3 *Locking as noncommutativity of limits*

We close this section with an **interpretation of locking as noncommutativity of limits**. To this end, let $\{X_N\}_N$ be a dense sequence of finite dimensional subspaces of X and $\{u_N^\varepsilon\}$ corresponding FE solution u_N^ε defined by

$$u_N^\varepsilon \in X_N : B_\varepsilon(u_N^\varepsilon, v) = F(v) \ \text{ for all } \ v \in X_N \tag{6.3.20}$$

where $B_\varepsilon(\cdot,\cdot)$ is as in (6.3.16).

We define the **limiting solutions** $u^0 \in G \subset X$ and $u_N^0 \in G_N$ by

$$u_0 \in G : B_1(u^0, v) = F(v) \text{ for all } v \in G \tag{6.3.21}$$

and

$$u_N^0 \in G_N : B_1(u_N^0, v) = F(v) \text{ for all } v \in G_N \,. \tag{6.3.22}$$

Here $G = \ker \mathcal{C}$ and $G_N = G \cap X_N$ are closed subspaces of X. We have the following result.

Theorem 6.54 *For every fixed N, as $\varepsilon \to 0$,*

$$\|u_N^\varepsilon - u_N^0\|_X \to 0 \,, \tag{6.3.23}$$

$$\|u^\varepsilon - u^0\|_X \to 0 \,. \tag{6.3.24}$$

Proof We prove only (6.3.23), (6.3.24) can be found in [121]. We have from the coercivity of $B_1(\cdot,\cdot)$ on X and $0 < \varepsilon \le 1$ that

$$\begin{aligned}\alpha \|u_N^\varepsilon\|_X^2 &\le B_1(u_N^\varepsilon, u_N^\varepsilon) \le B_\varepsilon(u_N^\varepsilon, u_N^\varepsilon) = F(u_N^\varepsilon) \\ &\le |||F|||_{X'} \, \|u_N^\varepsilon\|_X\end{aligned} \tag{6.3.25}$$

whence $\|u_N^\varepsilon\|_X$ is bounded uniformly in ε and N. Hence a subsequence (again denoted $\{u_N^\varepsilon\}$) converges, as $\varepsilon \to 0$, converges weakly, as $\varepsilon \to 0$, to a limit $u_N^* \in X_N$. (6.3.25) implies that

$$\|\mathcal{C}u_N^\varepsilon\|_0 = (\mathcal{C}u_N^\varepsilon, \mathcal{C}u_N^\varepsilon)_0 \le K\varepsilon, \;\; 0 < \varepsilon \le 1 \,.$$

We conclude that in fact $\mathcal{C}u_N^* = 0$ in L^2 and thus that $u_N^* \in G_N$.

Taking in (6.3.20) $v \in G_N \subset X_N$, we get that

$$B_1(u_N^*, v) = F(v) \text{ for all } v \in G_N \,. \tag{6.3.26}$$

By the uniqueness of the solution of (6.3.22), we have $u_N^* = u_N^0$ and (6.3.23) follows. □

Definition 6.55 *The FE approximation $\{u_N^\varepsilon\}$ in* (6.3.20) *is* **robust**, *if $u_N^\varepsilon \to u^\varepsilon$ as $N \to \infty$ in X* **uniformly** *in ε, i.e.,*

$$\begin{aligned}\textit{for all } \varepsilon_0 > 0 \textit{ for all } \delta > 0 \; \exists N_0(\varepsilon_0) : \|u_N^\varepsilon - u^\varepsilon\|_X < \delta \\ \textit{for all } N \ge N_0 \,, \\ \textit{for all } \varepsilon \in (0, \varepsilon_0] \,.\end{aligned} \tag{6.3.27}$$

If (6.3.27) *does not hold, we say the sequence $\{u_N^\varepsilon\}_N$ exhibits* **locking**.

Remark 6.56 The above Definition 6.55 of locking is crude in a sense, since it only allows to identify the most dramatic effects of locking, i.e., when the FE solutions u_N^ε are, for small ε and reasonable N, close to an incorrect solution. Often, however, locking manifests itself more subtly as a reduction in the (preasymptotic) rate of convergence as compared to the best possible (preasymptotic) rate of convergence (see [23] for more and [143] for an analysis of the shear locking of some h-version FEM for the Reissner–Mindlin plate).

Theorem 6.57 *Consider the double limits*

$$\lim_{\varepsilon\to 0}\ \lim_{N\to\infty}\ u_N^\varepsilon \in X \tag{6.3.28}$$

and

$$\lim_{N\to\infty}\ \lim_{\varepsilon\to 0}\ u_N^\varepsilon \in X\ . \tag{6.3.29}$$

If either of the following holds:

$$\textit{both limits exist, but are different} \tag{6.3.30}$$

or

$$\textit{the limit (6.3.28) exists but (6.3.29) does not exist}, \tag{6.3.31}$$

the FE approximations $\{u_N^\varepsilon\}$ *exhibit locking in the sense of Definition* 6.55.

6.3.1.4 *An abstract convergence result*

In Section 6.2.5 we spoke of bending dominated shell problems if $G = \mathbf{V}_F(\omega) \neq \{0\}$, since, **for general loads** $F(\cdot)$, u^ε will converge for $\varepsilon \to 0$ to a limit $0 \neq u^0 \in G$. It may very well happen, however, that the functional $F(\cdot)$ does not "excite" any $v \in G$, i.e., that

$$F(v) = 0 \text{ for all } v \in G\ . \tag{6.3.32}$$

Theorem 6.58 *Let* X *be a separable Hilbert space and let* $B_0(\cdot,\cdot)$, $C(\cdot,\cdot)$ *symmetric, continuous and positive semi-definite bilinear forms on* $X \times X \to \mathbb{R}$ *such that* $B_1 := B_0 + C$ *is coercive on* X. *Then*

$$G = \{v \in X : C(v,v) = 0\} \tag{6.3.33}$$

where $C(u,v) = (Cu, Cv)_0$ *is a closed, linear subspace of* X. *Assume further*

$$F(v) = 0 \textit{ for all } v \in G\ , \tag{6.3.34}$$

and denote for $0 < \varepsilon \leq 1$ *by* $u^\varepsilon \in X$ *the unique solution to the problem*

$$u^\varepsilon \in X : B_0(u^\varepsilon, v) + \varepsilon^{-2}\, C(u^\varepsilon, v) = F(v) \textit{ for all } v \in X\ . \tag{6.3.35}$$

Then we have

$$\lim_{t\to 0} \quad |||u^\varepsilon|||_\varepsilon = 0 \qquad (6.3.36)$$

where the norm $|||\circ|||_\varepsilon$ *is the energy norm corresponding to* (6.3.35), *i.e.,*

$$|||u|||_\varepsilon := [B_0(u,u) + \varepsilon^{-2}\, C(u,u)]^{\frac{1}{2}}\ .$$

Proof From (6.3.35) and $0 < \varepsilon \le 1$, $|||u^\varepsilon|||_1 \le |||u^\varepsilon|||_\varepsilon \le |||F|||_{X'}$, hence $\{u^\varepsilon : 0 < \varepsilon \le 1\} \subset X$ is bounded and $C(u^\varepsilon, u^\varepsilon) \le \varepsilon^2\, |||F|||_{X'}$.

Pick a sequence $\{u^\varepsilon\}_\varepsilon \subset X$ converging weakly to $u \in X$ as $\varepsilon \to 0$, then

$$|C(u^\varepsilon, u)| \le (C(u^\varepsilon, u^\varepsilon))^{\frac{1}{2}}\ |||u|||_1 \le \varepsilon\, |||F|||_{X'}^{\frac{1}{2}}\ |||u|||_1 \to 0$$

as $\varepsilon \to 0$, hence $C(u,u) = 0$, i.e., $u \in G$.

By (6.3.34), $F(u) = 0$ and hence

$$|||u^\varepsilon|||_\varepsilon^2 = F(u^\varepsilon) \to F(u) = 0$$

since $u^\varepsilon \underset{X}{\rightharpoonup} u$ hence any weakly converging sequence $\{u^\varepsilon\}_\varepsilon$ converges to zero in $|||\circ|||_\varepsilon$. Since $\{u^\varepsilon : 0 < \varepsilon \le 1\}$ is bounded in X and hence weakly compact, (6.3.36) follows. □

Thus a **necessary condition for locking** is that

$$F(v) \neq 0 \ \text{ for some } v \in G\ . \qquad (6.3.37)$$

Clearly, this cannot happen if $G = \{0\}$, but is also ruled out, if

$$F \in G^\circ := \{f \in X' : \langle f, v\rangle_{X'\times X} = 0 \text{ for all } \ v \in G\}\ ,$$

i.e., the loading $F(\cdot)$ is such that it does not excite the constrained solutions in G. This latter case is practically never encountered in practice, since small perturbations in the data F may cause $F \notin G^\circ$ and hence excite bending dominated modes. This in turn means that in the linear theory a bending dominated shell can only carry loads of order $O(\varepsilon^3)$ rather than $O(\varepsilon)$ in the membrane dominated case.

6.3.2 *Analysis of reduced constraint FEM*

We present two main lines of analysis of the reduced constraint FEM, namely as **mixed method with penalty** (6.3.11) and as a **displacement problem with reduction operators** P_N (6.3.13) for the abstract problem (6.3.3). The former approach allows to get convergence estimates for the multiplier p_N^ε in (6.3.11), but requires the proof of an inf–sup condition. The latter approach is trivially stable, but involves the construction of a certain projector $Q_N : X \to X_N$ and the analysis of the consistency error $\|(I - Q_N)u\|_X$. Throughout, we assume (6.3.16), i.e. that the problem (6.3.3) is given in the weak form

$$u^t \in X : B_\varepsilon(u^t, v) = B_1(u^t, v) + t^{-2}\,(Cu^t, Cv)_0 = F(v) \text{ for all } v \in X\,, \tag{6.3.38}$$

and the bilinear form $B_1(\cdot,\cdot)$ is coercive on X,

$$B_1(u,u) \geq \alpha\,\|u\|_X^2 \text{ for all } u \in X \tag{6.3.39}$$

(by Remark 6.53, this can be achieved through the substitution (6.3.15) if $B_0(\cdot,\cdot)$ is not coercive on all of X). Notice that we still have $u^t = u^\varepsilon$ with u^ε in (6.3.3).

We analyze the abstract **reduced constraint method**: find $u_N^t \in X_N$ such that

$$B_1(u_N^t, v) + t^{-2}\,(P_N\,Cu_N^t, P_N Cv) = F(v) \text{ for all } v \in X_N \tag{6.3.40}$$

where the **reduction operator** $P_N : L^2(\Omega) \to S_N$ is the orthogonal projection and S_N is a space of piecewise polynomials.

Proposition 6.59 *For every $v \in X_N$ and any $S_N \subset L^2(\Omega)$ it holds that*

$$\begin{aligned} \|u^\varepsilon - u_N^t\|_1 \leq \quad & C\,\{\|u^\varepsilon - v\|_1 + t^{-1}\,\|P_N(Cu^\varepsilon - Cv)\|_0 \\ & + t^{-2}\|(I-P_N)\,Cu^\varepsilon\|_0\} \text{ for all } v \in X_N\,. \end{aligned} \tag{6.3.41}$$

Proof Define the energy norm $\|\circ\|_{t,N}$ by

$$\|v\|_{t,N}^2 := B_1(v,v) + t^{-2}(P_N Cv, P_N Cv) =: B_{t,N}(v,v)\,.$$

Then

$$\|u\|_{t,N} \sim \|v\|_1 + t^{-1}\,\|P_N Cv\|_0\,.$$

This norm depends (via P_N) on S_N and the coercivity (6.3.39) of $B_1(\cdot,\cdot)$ gives $\|v\|_1 \leq C\,\|v\|_{t,N}$ for all $v \in X$, C independent of N, t. We interpret u_N^t as the FE solution to a perturbed bilinear form $B_{t,N}(u,v)$. From the second Strang lemma, Theorem 2.28, we get

$$\begin{aligned} \|u^\varepsilon - u_N^t\|_{t,N} &\leq C\left(\inf_{v\in X_N}\|u^\varepsilon - v\|_{t,N} + \sup_{w\in X_N}\frac{|B_{t,N}(u^\varepsilon,w) - F(w)|}{\|w\|_{t,N}}\right) \\ &\leq C\left(\|u^\varepsilon - v\|_{t,N} + t^{-2}\sup_{w\in X_N}\frac{|((I-P_N)Cu^\varepsilon, Cw)|}{\|w\|_{t,N}}\right) \end{aligned}$$

for any $v \in X_N$ where we used that

$$F(w) = B_\varepsilon(u^\varepsilon, w) = B_1(u^\varepsilon, w) + t^{-2}(Cu^\varepsilon, Cw)_0\,.$$

Now, the coercivity (6.3.39) implies that

$$\|w\|_{t,N}^2 \geq B_1(w,w) \geq C\,\|w\|_1^2 \text{ for all } w \in X$$

and

$$\|u^\varepsilon - v\|_{t,N}^2 \leq C\,\|u^\varepsilon - v\|_1^2 + t^{-2}\,\|P_N\,\mathcal{C}(u^\varepsilon - v)\|_0^2$$

together with $(a^2+b^2)^{\frac{1}{2}} \leq C(|a|+|b|)$ yield

$$\|u^\varepsilon - u_N^t\|_{t,N} \leq C\left\{\|u^\varepsilon - v\|_1 + t^{-1}\,\|P_N\,\mathcal{C}(u^\varepsilon - v)\|_0 + \|(I-P_N)(t^{-2}\mathcal{C}u^\varepsilon)\|_0 \sup_{w\in X_N} \frac{\|\mathcal{C}w\|_0}{\|w\|_1}\right\} \text{ for all } v \in X_N\,.$$

Since the constraint operator $\mathcal{C} : H^1 \to L^2$ is bounded by assumption, the assertion follows. □

Exercise 6.60 By modifying the proof of Proposition 6.59, show that for every $v \in X_N$

$$\|u^\varepsilon - u_N^t\|_1 \leq C\left\{\|u^\varepsilon - v\|_1 + t^{-1}\,\|P_N\,(\mathcal{C}u^\varepsilon - \mathcal{C}v)\|_0 + \|(I-P_N)\,\mathcal{C}u^\varepsilon)\|_0 \sup_{w\in X_N} \frac{\|\mathcal{C}w\|_0}{\|P_N\mathcal{C}w\|_0}\right\}. \tag{6.3.42}$$

Remark 6.61 From the result in Proposition 6.59, we see that the subspaces S_N should now be selected so that

$$\|I - P_N\|_{L^2\to L^2} \tag{6.3.43}$$

is small and so that there exists a projection $Q_N : X \to X_N$ such that the approximation error

$$\|u^\varepsilon - Q_N u^\varepsilon\|_1 \tag{6.3.44}$$

is small uniformly in ε, and such that

$$P_N\,\mathcal{C}\,Q_N\,u = P_N\mathcal{C}u \text{ for all } u \in X \tag{6.3.45}$$

or, equivalently,

$$(q, \mathcal{C}(u^\varepsilon - Q_N u^\varepsilon)) = 0 \text{ for all } q \in S_N\,. \tag{6.3.46}$$

We select $v = Q_N u^\varepsilon$ in the error bound in (6.3.41) in Proposition 6.59 and get from (6.3.46) the error estimate

$$\|u^\varepsilon - u_N^t\|_1 \leq C\,\{\|(I-Q_N)\,u^\varepsilon\|_1 + t^{-2}\|\,(I-P_N)\,\mathcal{C}u^\varepsilon\|_0\}\,, \tag{6.3.47}$$

with C independent of N and ε or, equivalently, of t. We will later see that it is possible to construct reduction operators P_N such that the second term in the estimate (6.3.47) either vanishes (in which case the corresponding reduced constraint FEM (6.3.40) is **robust** or locking-free in the $\| \circ \|_1$-norm) or is exponentially small independently of the smoothness of the loadings (for example, in the case of the bending-dominated Naghdi-shell discussed in the next section, we show that this is the case if the parametric representation $\varphi(x^1, x^2)$ of the midsurface S is analytic in $\overline{\omega}$).

6.3.3 *Application to Naghdi shell models*

We adopt the notations of Section 6.2.4 and consider a bending dominated, totally clamped Naghdi shell, i.e., the following boundary value problem: find $(\zeta^\varepsilon, r^\varepsilon) \in \boldsymbol{V}_N(\omega) := [H_0^1(\omega)]^5$ such that

$$B_N(\zeta^\varepsilon, r^\varepsilon; \eta, s) = \overline{F}_N(\eta, s) \text{ for all } (\eta, s) \in \boldsymbol{V}_N(\omega) . \tag{6.3.48}$$

Here $\boldsymbol{V}_N(\omega)$ is the set of kinematically admissible displacements for the Naghdi shell model in (6.2.83). The bilinear form B_N is, in the bending dominated case and after rescaling of the loads by ε^3, given by (see Section 6.2.4)

$$B_N(\zeta, r; \eta, s) = B_f(\zeta, r; \eta, s) + \varepsilon^{-2} \left[B_m(\zeta; \eta) + B_{sh}(\zeta, r; \eta, s)\right]$$

where, since the **deformation** is **bending dominated**, the applied forces are assumed to scale as

$$F_N(\eta, s) = \varepsilon^3\, \overline{F}_N(\eta, s) \tag{6.3.49}$$

with $\overline{F}_N$ independent of ε. Here B_f, B_m and B_{sh} are as in (6.2.78)–(6.2.80), respectively. Formally, B_N is of the form $B_0(\cdot, \cdot) + \varepsilon^{-2}(\mathcal{C}\cdot, \mathcal{C}\cdot)$ where the constraint $\mathcal{C}$ consists now of the shear constraint and the membrane constraint:

$$\mathcal{C}(\zeta, r) = [\mathcal{C}_m, \mathcal{C}_{sh}](\zeta, r) = [\gamma_{\alpha\beta}(\zeta), \varphi_\alpha(\zeta, r)] . \tag{6.3.50}$$

We refer to Remark 6.53, set $\varepsilon^{-2} = t^{-2} + 1$ and rewrite (6.3.51) as follows: find $(\zeta^\varepsilon, r^\varepsilon) \in \boldsymbol{V}_N(\omega)$ such that

$$\begin{aligned} &B_1(\zeta^\varepsilon, r^\varepsilon; \eta, s) + t^{-2} \left[(\gamma_{\alpha\beta}(\zeta^\varepsilon); \gamma_{\alpha\beta}(\eta)) + (\varphi_\alpha(\zeta^\varepsilon, r^\varepsilon); \varphi_\alpha(\eta, s))\right] \\ &= \overline{F}_N(\eta, s) \text{ for all } (\eta, s) \in \boldsymbol{V}_N(\omega) . \end{aligned} \tag{6.3.51}$$

Here we have used (6.3.15), i.e., $t^{-2} = \varepsilon^{-2} - 1$. And we have from Theorem 6.38 the coercivity

$$B_1(\zeta, r; \zeta, r) = (B_f + B_m + B_{sh})(\zeta, r; \zeta, r) \geq \alpha \, \|(\zeta, r)\|_1^2 \tag{6.3.52}$$

where $\alpha > 0$ is independent of ε which is (6.3.42), i.e., we are in the framework of the theory of Section 6.3.1.

Let $X_N \subset X := \mathbf{V}_N(\omega)$ be any FE space (a specific choice will be given below). We reduce the constraints in (6.3.50) by introducing, analogous to (6.3.12), projections

$$P_N^m : L^2(\omega) \to S_N^m, \quad P_N^{sh} : L^2(\omega) \to S_N^{sh}$$

where $S_N^m, S_N^{sh} \subset L^2(\omega)$ are subspaces to be defined, i.e.,

$$\begin{aligned}(P_N^m\,\gamma, \hat{\gamma}) &= (\gamma, \hat{\gamma}) \text{ for all } \hat{\gamma} \in S_N^m\,,\\ (P_N^{sh}\,\varphi, \hat{\varphi}) &= (\varphi, \hat{\varphi}) \text{ for all } \hat{\varphi} \in S_N^{sh}\,.\end{aligned} \tag{6.3.53}$$

Then the reduced constraint FE approximation $(\zeta_N^\varepsilon, r_N^\varepsilon)$ to $(\zeta^\varepsilon, r^\varepsilon)$ is: find $(\zeta_N^\varepsilon, r_N^\varepsilon) \in X_N$ such that

$$\begin{aligned}&B_1(\zeta_N^\varepsilon, r_N^\varepsilon; \eta, s) + t^{-2}\,[(P_N^m\,\gamma(\zeta_N^\varepsilon);\ P_N^m\,\gamma(\eta))\\ &+ (P_N^{sh}\,\varphi(\zeta_N^\varepsilon, r_N^\varepsilon);\ P_N^{sh}\,\varphi(\eta, s))] = \overline{F}_K(\eta, s) \text{ for all } (\eta, s) \in X_N\,.\end{aligned} \tag{6.3.54}$$

We have the following result.

Theorem 6.62 *Let $U_N^\varepsilon = (\zeta_N^\varepsilon, r_N^\varepsilon)$ be the FE approximation defined in* (6.3.54). *Then, for any $W_N \in X_N$, the following error estimate holds:*

$$\begin{aligned}\|U^\varepsilon - U_N^\varepsilon\|_1 \le C\,\{&\|U^\varepsilon - W_N\|_1\\ &+ \|(I - P_N^m)(t^{-2}\,\gamma(U^\varepsilon))\|_0\\ &+ \|(I - P_N^{sh})(t^{-2}\,\varphi(U^\varepsilon))\|_0\\ &+ t^{-1}\,\|P_N^m\,\gamma(U^\varepsilon - W_N)\|_0\\ &+ t^{-1}\,\|P_N^{sh}\,\varphi(U^\varepsilon - W_N)\|_0\}\,.\end{aligned} \tag{6.3.55}$$

The proof is completely analogous to that of Proposition 6.59. We now define the quantities

$$\xi(U) := t^{-2}\,\varphi(U), \quad \eta(U) := t^{-2}\,\gamma(U)\,, \tag{6.3.56}$$

which are rescaled to transverse shear and membrane stresses, respectively. Then from (6.3.55) we get the following result.

Corollary 6.63 *Let $Z_N(U) \subset X_N$ be given by*

$$Z_N(U) := \{W_N \in X_N : P_N^m\,\gamma(U - W_N) = 0,\ P_N^{sh}\,\varphi(U - W_N) = 0\}\,. \tag{6.3.57}$$

Then

$$\begin{aligned}\|U^\varepsilon - U_N^\varepsilon\|_1 \le C\,\{&\inf_{W_N \in Z_N(U^\varepsilon)} \|U^\varepsilon - W_N\|_1\\ &+ \|(I - P_N^m)\,\eta(U^\varepsilon)\|_0 + \|(I - P_N^{sh})\,\xi(U^\varepsilon)\|_0\}\,.\end{aligned} \tag{6.3.58}$$

6.3.4 *Reduced constraint hp-FEM*

We show now how the previous, general considerations can be made specific. In particular, we exhibit specific choices of *hp*-FE subspaces X_N, S_N which allow for the construction of a projector Q_N in Remark 6.61 with very favorable properties.

6.3.4.1 *Selection of X_N, S_N, and Q_N. Abstract error estimate*

We assume that $\omega \subset \mathbb{R}^2$ is a polygon and that $\{\mathcal{T}_N\}_N$ is a sequence of regular triangulations consisting of convex, shape regular parallelograms K. We select the *hp* FE-spaces X_N, S_N as follows.

Let $\widehat{K} = (-1,1)^2$ denote the reference square; then for all $K \in \mathcal{T}_N$: $K = F_K(\widehat{K})$, where F_K is a linear, invertible map. For an interval $I \subset \mathbb{R}$, denote by $\mathcal{P}_p(I)$ the polynomials of degree $\leq p$ on I, and by $\mathcal{Q}_p(K)$ the polynomials of degree $\leq p$ in each variable on $K \in \mathcal{T}_N$. Define also for $p \geq 2$ and $\ell = 0, 1$

$$\begin{aligned} S^{p,\ell}(\mathcal{T}_N) &= \{v \in H^r(\omega) : v|_K \circ F_K \in \mathcal{Q}_p(\widehat{K}), K \in \mathcal{T}_N\}\,, \\ S_0^{p,1}(\mathcal{T}_N) &= S^{p,1}(\mathcal{T}_N) \cap H_0^1(\omega)\,, \end{aligned} \tag{6.3.59}$$

$$X_N = \{(\zeta, r) :\ \zeta_i, r_\alpha \in S_0^{p,1}(\mathcal{T}_N)\}\,, \tag{6.3.60}$$

$$S_N^m = \{\gamma_{\alpha\beta} \in L^2(\omega) :\ \gamma_{\alpha\beta} \in S^{p-2,0}(\mathcal{T}_N)\}\,, \tag{6.3.61}$$

$$S_N^{sh} = \{\varphi_\alpha \in L^2(\omega) :\ \varphi_\alpha \in S^{p-2,0}(\mathcal{T}_N)\}\,. \tag{6.3.62}$$

Proposition 6.64 *Assume that for a mesh sequence $\{\mathcal{T}_N\}_N$ in* (6.3.60) *the geometry coefficients of the shell are piecewise constant, i.e., that*

$$a_{\alpha\beta}, b_{\alpha\beta}, \Gamma^\gamma_{\alpha\beta} \in S^{0,0}(\mathcal{T}_N)\,, \tag{6.3.63}$$

and let $\xi(U)$, $\eta(U)$ be as in (6.3.56)*. Let $U = (\zeta^\varepsilon, r^\varepsilon) \in H^{1+\delta}(\omega), \delta > 0$, be the exact solution of* (6.3.48) *and let $U_N = (\zeta_N^\varepsilon, r_N^\varepsilon) \in X_N$ be the corresponding FE solution. Then there exist $Q_N : H^{1+\delta}(\omega) \to X_N$ such that:*

$$\begin{aligned} \|U^\varepsilon - U_N^\varepsilon\|_1 \leq C\,\{&\|(I - Q_N)\,U^\varepsilon\|_1 + \|(I - P_N^m)\,\eta(U^\varepsilon)\|_0 \\ &+ \|(I - P_N^{sh})\,\xi(U^\varepsilon)\|_0\}\,. \end{aligned} \tag{6.3.64}$$

The projector $Q_N : H^{1+\delta}(\omega) \to X_N$ is constructed in the proof.

Proof The proof we give consists in explicitly constructing a projector Q_N: $H^{1+\delta}(\omega) \to X_N$ satisfying (6.3.45), i.e.,

$$P_N\,\mathcal{C}(U^\varepsilon - Q_N\,U^\varepsilon) = 0$$

for the constraint (6.3.50) under the crucial hypothesis (6.3.63) (which is a-priori only satisfied for cylindrical shells; the general case will be discussed below). We verify that (6.3.46) holds in our case, if

$$\int_\omega q\,D^\alpha(u - Q_N\,u)\,dx = 0 \quad \text{for all} \quad |\alpha| \leq 1, \quad \text{for all} \quad q \in S^{p-2,0}(\mathcal{T}_N)\,. \tag{6.3.65}$$

We construct the projector Q_N elementwise by defining on the reference square $\widehat{K}$ the local projector $Q_p : H^{1+\delta}(\widehat{K}) \to \mathcal{S}_p^2(\widehat{K})$ for $\delta > 0$ and $p \geq 2$ as follows:

$$(Q_p u)(N_i) = u(N_i) \quad \text{on each node } N_i \text{ of } \widehat{K} \tag{6.3.66}$$

$$\int_e (Q_p u)\,q\,ds = \int_e uq\,ds \quad \text{for all} \quad q \in \mathcal{P}_{p-2}(e) \quad \text{for each edge} \tag{6.3.67}$$

$$\int_{\widehat{K}} (Q_p u)\,q\,dx = \int_{\widehat{K}} uq\,dx \quad \text{for all} \quad q \in \mathcal{S}_{p-2}^2(\widehat{K})\,. \tag{6.3.68}$$

We observe that (6.3.69)–(6.3.71) constitute $4+4(p-1)+(p-1)^2 = p^2+2p+1 = (p+1)^2$ conditions which define a unique $Q_p u \in \mathcal{S}_p^2(\widehat{K})$. The global projector $Q_N : H^{1+\delta}(\omega) \to X_N$ is then obtained by combining the local projections component- and elementwise.

Exercise 6.65 Show that (6.3.66)–(6.3.68) imply for $|\alpha| \leq 1$ and $p \geq 2$

$$\int_{\widehat{K}} q\,D^\alpha(u - Q_p u)\,dx = 0 \text{ for all } \quad q \in \mathcal{S}_{p-2}^2(\widehat{K})\,. \tag{6.3.69}$$

Deduce (6.3.65) for the global projector Q_N and show that under assumption (6.3.66) the property (6.3.49) holds, i.e., that we have

$$(q,\, \mathcal{C}(v - Q_N v)) = 0 \text{ for all } \; v \in X, \quad \text{for all} \quad q \in S_N$$

where $\mathcal{C}$ is either the shear or the membrane strain tensor.

Solution: Evidently (6.3.71) $\Longrightarrow$ (6.3.72) for $\alpha = (0,0)$. Let therefore $\alpha = (1,0)$. Integrating by parts in (6.3.69) gives

$$\int_{\widehat{K}} q\,\partial_1(u - Q_p u)\,dx = -\int_{\widehat{K}} (\partial_1 q)(u - Q_p u)dx + \int_{\partial\widehat{K}} q(u - Q_p u)\,n_1\,ds\,.$$

Now $q|_e \in P_{p-2}(e)$ for each edge $e \subset \partial\widehat{K}$ and therefore the boundary integral vanishes by (6.3.67). Since $q \in \mathcal{S}_{p-2}^2(\widehat{K}) \Longrightarrow \partial_1 q \in \mathcal{S}_{p-2}^2(\widehat{K})$, the domain integral vanishes by (6.3.68). This implies (6.3.72) for $\alpha = (1,0)$. The proof for $\alpha = (0,1)$ is analogous.

Selecting in (6.3.58) the function $W_N = Q_N U$, we see from Exercise 6.65 that $W_N \in Z_N(U)$ and that the error estimate (6.3.64) holds. □

It therefore remains to estimate the consistency errors $(I - Q_N)$, $(I - P_N^m)$, $(I - P_N^{sh})$. We start with the latter two projectors and assume that the mesh $\mathcal{T}_N$ is quasiuniform with meshwidth h_N.

Consider a single element $K \in \mathcal{T}_N$. Transporting $\xi(U^\varepsilon)|_K$ and $\eta(U^\varepsilon)|_K$ to the reference element $\widehat{K}$ and approximating by polynomials, we find, after scaling back, and summing over all elements,

$$\|(I - P_N^{sh})\,\xi(U^\varepsilon)\|_0 \leq C\, h_N^\mu\, p^{-(r-1)}\, \|\xi(U^\varepsilon)\|_{r-1} \tag{6.3.70}$$

$$\|(I - P_N^{m})\,\eta(U^\varepsilon)\|_0 \leq C\, h_N^\mu\, p^{-(r-1)}\, \|\eta(U^\varepsilon)\|_{r-1}\,, \tag{6.3.71}$$

where $\mu = \min(p,r) - 1$.

6.3.4.2 *Analysis of the projector Q_N*

It remains therefore to estimate $\|(I - Q_N)\,U^\varepsilon\|_1$ in (6.3.64). This will be done on $\widehat{K}$, and the general result will then be obtained in the usual way by adding the scaled local error estimates.

Lemma 6.66 *Let $\delta > 0$ and $\pi_p : H^{\frac{1}{2}+\delta}(-1,1) \to \mathcal{P}_p(-1,1)$ be defined by*

$$(\pi_p v)(\pm 1) = v(\pm 1),\ \int_{-1}^{1} q(\pi_p v - v) = 0 \ \textit{for all}\ \ q \in \mathcal{P}_{p-2}(-1,1). \tag{6.3.72}$$

Then the projector Q_p defined in (6.3.66)–(6.3.68) *satisfies*

$$Q_p = \pi_p^x\, \pi_p^y = (\pi_p^x \otimes I)(I \otimes \pi_p^y)\,.$$

Proof This is left to the reader as an exercise.

Note that $\pi_p : H_0^1(-1,1) \to \mathcal{P}_p(-1,1)$ is just the H_0^1-projection of v onto $\mathcal{P}_p(-1,1)$, as can be seen from

$$0 = \int_{-1}^{1} q(v - \pi_p v) = -\int_{-1}^{1} r'(v - \pi_p v)' \ \text{for all}\ \ r \in (\mathcal{P}_p \cap H_0^1)(-1,1)\,.$$

We point out that sharp consistency results for π_p were obtained in Chapter 3, Section 3.3. We can now give the main *hp* consistency result for the projector Q_p. □

Lemma 6.67 *For every* $v \in H^{1,1}(\widehat{K})$, *the projection* Q_p *defined in* (6.3.66)–(6.3.68) *satisfies*

$$\|v - Q_p v\|_{H^1(\widehat{K})} \leq C\, p^{-(r-1)} \|v\|_{H^r(\widehat{K})}\,, \quad r \geq 2\,. \tag{6.3.73}$$

Proof We have by Lemma 6.66 that $Q_p = \pi_p^x\, \pi_p^y$, i.e., Q_p is identical to the projector Π_p defined in Chapter 4, Lemma 4.66. The error estimate (6.3.73) therefore follows from (4.5.46) with Stirling's formula and $r = k + 1 \geq 2$. □

Remark 6.68 We emphasize that the consistency error estimate (4.5.46) is, in fact, more general and allows to deduce, for example, exponential convergence rates for analytic solutions.

Remark 6.69 A different proof using two-dimensional expansions in $U_i(x)\, U_j(y)$ yields an error estimate valid for $r > 3/2$, but at the expense of a loss of $1/2$ in the order of convergence.

6.3.4.3 *Arbitrary geometry*

Theorem 6.62 applies only if assumption (6.3.63) is met. This, however, is a rather special case which is essentially met only for a cylindrical shell. A slight modification of the argument, however, allows us to deal with

a) arbitrary, analytic parametrization of the midsurface
b) general, quadrilateral meshes (as opposed to the parallelogram ones in Theorem 6.62).

Let us describe this modification, first for the parallelogram meshes $\{\mathcal{T}_N\}$ considered above. Select, for $p_N \geq 2$,

$$X_N = S_0^{p_N,1}(\mathcal{T}_N) := S^{p_N,1}(\mathcal{T}_N) \cap H_0^1(\omega),\; S_N = S^{p_N-2,0}(\mathcal{T}_N)\,. \tag{6.3.74}$$

The idea is now to increase the gap between the polynomial degrees p_N and $p_N - 2$ in the definition of X_N and S_N, i.e., we pick

$$X_N = S_0^{p_N+q_N}(\mathcal{T}_N),\; S_N = S^{p_N-2,0}(\mathcal{T}_N),\;\; p_N \geq 2,\; q_N \geq 0\,, \tag{6.3.75}$$

with a degree increment q_N at our disposal (e.g., $q_N = p_N$). Then the projector Q_N is defined once more as in (6.3.66)–(6.3.68), but now with $p = p_N + q_N$, and satisfies

$$\begin{aligned} \int_K q\, D^\alpha(u - Q_N u)dx = 0 \text{ for all } & K \in \mathcal{T}_N, \\ \text{for all } & |\alpha| \leq 1, \\ \text{for all } & q \in S^{p_N+q_N-2}(K)\,. \end{aligned} \tag{6.3.76}$$

Then we have

Lemma 6.70 *Let*

$$\gamma_{\alpha\beta}(\zeta) = \frac{1}{2}\left(\zeta_{\alpha,\beta} + \zeta_{\beta,\alpha}\right) - \Gamma^{\delta}_{\alpha\beta}\,\zeta_\delta - b_{\alpha\beta}u_3 = t^2\,\eta(U)\,, \tag{6.3.77}$$

$$\varphi_\alpha(\zeta, r) = \zeta_{3,\alpha} + b^{\gamma}_{\alpha}\,\zeta_\gamma + r_\alpha = t^2\,\xi(U) \tag{6.3.78}$$

and assume that for q_N as in (6.3.75)

$$b^{\gamma}_{\alpha},\ b_{\alpha\beta},\ \Gamma^{\delta}_{\alpha\beta}\,,\quad \sqrt{a} \in S^{q_N,0}(\mathcal{T}_N)\,. \tag{6.3.79}$$

Then for every $U \in X = [H^1_0(\omega)]^5$ holds $W_N = Q_N U \in Z_N(U)$ where

$$Z_N(U) = \{W_N \in X_N : P^m_N\,\eta(U - W_N) = 0,\ P^{sh}_N\,\xi(U - W_N) = 0\}\,,$$

and we have the error estimate

$$\begin{aligned} &\|U - Q_N U\|_{H^1(\omega)} \le C\,h^{\mu}_N\,p_N^{-(r-1)}\,\|U\|_{H^r(\omega)}\,,\\ &\mu = \min\{p_N - 1, r - 1\},\ p_N, r \ge 2\,. \end{aligned} \tag{6.3.80}$$

Proof The condition $P^{sh}_N\,\xi(U - W_N) = 0$ is equivalent to

$$\int_\omega q\,\xi(U - W_N)dx = 0 \text{ for all } q \in S^{p_N-2,0}(\mathcal{T}_N)\,. \tag{6.3.81}$$

This gives with (6.3.78) that for every $K \in \mathcal{T}_N$

$$\int_K q(\zeta_{3,\alpha} + b^{\gamma}_{\alpha}\,\zeta_\gamma + r_\alpha)dx = 0 \text{ for all } q \in \mathcal{S}^{p_N-2}_2(K)\,.$$

Since $b^{\gamma}_{\alpha} \in S^{q_N}(K)$ by assumption and since (6.3.76) holds, we get (6.3.81). The proof for $P^m_N\,\eta(U - W_N) = 0$ is identical. The a priori estimate follows as in Lemma 6.67. □

We now get the following result immediately from Corollary 6.63.

Lemma 6.71 *Let* $\{\mathcal{T}_N\}_N$ *be a sequence of quasiuniform, parallelogram meshes of width* h_N *and assume* (6.3.79). *Select* $X_N = S^{p_N+q_N,1}_0(\mathcal{T}_N)$, $S_N = S^{p_N-2,0}(\mathcal{T}_N)$. *Then the error estimate* (6.3.83) *holds.*

The case of **general, analytic geometry coefficients** will now be dealt with by a perturbation argument.

Theorem 6.72 *Assume that*

$$a_{\alpha\beta}, b_{\alpha\beta}, \Gamma^{\delta}_{\alpha\beta}, \sqrt{a} \text{ are analytic in } \overline{\omega}\,. \tag{6.3.82}$$

Let further $X_N = S_0^{p_N+q_N,1}(\mathcal{T}_N)$, $S_N = S^{p_N-2,0}(\mathcal{T}_N)$. *Then there exists* $b > 0$ *such that for every* $r \geq 2$,

$$\|U - U_N\|_{H^1(\omega)} \leq C\, h_N^{\mu}\, p_N^{-(r-1)} \{\|\zeta^{\varepsilon}\|_{H^r(\omega)} + \|r^{\varepsilon}\|_{H^r(\omega)} + \|\xi(U^{\varepsilon})\|_{H^{r-1}(\omega)} + \|\eta(U^{\varepsilon})\|_{H^{r-1}(\omega)} + t^{-1}\, e^{-bq_N}\}\,,$$

$\mu = \min(p_N, r) - 1$, p_N, $r \geq 2$.

Proof We return to Theorem 6.62, i.e., the error estimate (6.3.55). For $\gamma_{\alpha\beta}$, φ_α given by (6.3.77), (6.3.78), we define polynomial approximations

$$\gamma^N_{\alpha\beta}(\zeta) := \frac{1}{2}\,(\zeta_{\alpha,\beta} + \zeta_{\beta,\alpha}) - (\Gamma^{\delta}_{\alpha\beta})^N\,\zeta_\delta - b^N_{\alpha\beta}\,u_3 =: t^2\,\eta^N(U) \tag{6.3.83}$$

$$\varphi^N_\alpha(\zeta, r) := \zeta_{3,\alpha} + (b^{\gamma}_{\alpha})^N\,\zeta_\gamma + r_\alpha \tag{6.3.84}$$

where $(\Gamma^{\delta}_{\alpha\beta})^N$, $(b^{\gamma}_{\alpha})^N$, $b^N_{\alpha\beta}$ satisfy (6.3.79) and also, for every $K \in \mathcal{T}$,

$$\begin{aligned}&\|\Gamma^{\delta}_{\alpha\beta} - (\Gamma^{\delta}_{\alpha\beta})^N\|_{L^\infty(K)} + \|b^{\gamma}_{\alpha} - (b^{\gamma}_{\alpha})^N\|_{L^\infty(K)} + \|b_{\alpha\beta} - (b_{\alpha\beta})^N\|_{L^\infty(K)}\\ &\leq C\, e^{-bq_N}\,,\end{aligned} \tag{6.3.85}$$

$b > 0$ independent of N and K. Such approximations exist by our assumption on the analyticity of the coefficients (6.3.82).

Now we may write in (6.3.55)

$$\begin{aligned}\|U^{\varepsilon} - U^{\varepsilon}_N\|_{H^1(\omega)} \leq C\,\{&\|U^{\varepsilon} - W_N\|_{H^1(\omega)}\\ &+ \|(I - P^m_N)(t^{-2}\,\gamma(U^{\varepsilon}))\|_{L^2(\omega)}\\ &+ \|(I - P^{sh}_N)(t^{-2}\,\varphi(U^{\varepsilon}))\|_{L^2(\omega)}\\ &+ t^{-1}\,\|P^m_N(\gamma - \gamma^N)(U^{\varepsilon} - W_N)\|_{L^2(\omega)}\\ &t^{-1}\,\|P^{sh}_N(\varphi - \varphi^N)(U^{\varepsilon} - W_N)\|_{L^2(\omega)}\\ &t^{-1}\,\|P^m_N\,\gamma^N(U^{\varepsilon} - W_N)\|_{L^2(\omega)}\\ &t^{-1}\,\|P^{sh}_N\,\varphi^N(U^{\varepsilon} - W_N)\|_{L^2(\omega)}\}\,.\end{aligned} \tag{6.3.86}$$

Terms 1 - 3 on the right-hand side are as before, terms 4 and 5 can be bounded using (6.3.85) and the boundedness of the $L^2(\omega)$ projections P_N^m, P_N^{sh} by $t^{-1}\, e^{-\lambda q_N}$ and terms 6 and 7 are kept. Then (6.3.86) gives

$$||U^\varepsilon - U_N^\varepsilon||_{H^1(\omega)} \le C\{ \inf_{W_N \in Z_N(U^\varepsilon)} ||U^\varepsilon - W_N||_{H^1(\omega)} + ||(I - P_N^m)\, \eta(U^\varepsilon)||_{L^2(\omega)}$$

$$+ ||(I - P_N^{sh})\, \xi(U^\varepsilon)||_{L^2(\omega)} + t^{-1}\, e^{-bq_N}\}$$

where now

$$Z_N(U) = \{W_N \in X_N : P_N^m\, \gamma^N(U - W_N) = 0,\; P_N^{sh}\, \varphi^N(U - W_N) = 0\}\,.$$

The assertion follows now from Lemma 6.71. □

Remark 6.73

i) Since in the case of a plate the Naghdi model decouples (i.e., for $S = \omega$) into a Reissner–Mindlin plate and a membrane problem, the theorems in this section apply **verbatim** also to the **Reissner–Mindlin plate problem** (6.2.114) where only **shear locking** occurs.

ii) An analysis similar to the one used in Theorem 6.72 to assess the impact of piecewise polynomial geometry approximation can be used to analyze the effect of curvilinear element maps, also for Reissner–Mindlin FEM with a reduced shear constraint.

Remark 6.74 (triangular elements). So far, we have analyzed the reduced constraint FEM only for quadrilateral elements. The same method can also be defined on meshes $\mathcal{T}_N$ of affine triangles $T = F_T(\widehat{T})$. We define, analogous to (6.3.59) = (6.3.62), for $\ell = 0, 1$,

$$S_0^{p,\ell}(\mathcal{T}_N) = \{v \in H_0^\ell(\omega) : \; v|_T \circ F_T \in \mathcal{P}_p(\widehat{T}),\; T \in \mathcal{T}_N\}\,,$$

$$X_N = \{(\zeta, r) : \; \zeta_i, r_\alpha \in S_0^{p,1}(\mathcal{T}_N)\}\,,$$

$$S_N^m = \{\gamma_{\alpha\beta} \in L^2(\omega) : \; \gamma_{\alpha\beta} \in S^{p-3,0}(\mathcal{T}_N)\}\,,$$

$$S_N^{sh} = \{\varphi_\alpha \in L^2(\omega) : \varphi_\alpha \in S^{p-3,0}(\mathcal{T}_N)\}\,.$$

The projector Q_N to analyze the consistency error can then be defined analogous to (6.3.66)–(6.3.68): it is defined by elementwise projectors Q_p which are transported from the reference triangle $\widehat{T}$. On $\widehat{T}$ we require the following conditions:

a) $(Q_p u)(N_i) = U(N_i)$ for each node N_i of $\widehat{T}$,

b) $\displaystyle\int_e qQ_p u\, ds = \int_e qu\, ds$ for each edge $e \subset \partial\widehat{T}$,

c) $\displaystyle\int_{\widehat{T}} q\, Q_p u\, dx = \int_{\widehat{T}} qu\, dx$ for all $q \in \mathcal{S}_1^{p-3}(\widehat{T})$.

Conditions a) - c) constitute $(p+1)(p+2)/2$ conditions which uniquely determine $Q_p u \in \mathcal{S}_1^p(\widehat{T})$.

We claim now that

d) $\displaystyle\int_{\widehat{T}} q\, D^\alpha(u - Q_p u)\, dx = 0$ for all $q \in \mathcal{S}_1^{p-3}(\widehat{T})$.

For $\alpha = (0,0)$, this follows from c). Let $\alpha = (1,0)$. Integrating by parts we get for $q \in \mathcal{S}_1^{p-3}(\widehat{T})$

$$\int_{\widehat{T}} q\, D^\alpha(u - Q_p u)\, dx = -\int_{\widehat{T}} (D^\alpha q)(u - Q_p u)\, dx + \int_{\partial\widehat{T}} q(u - Q_p u)\, n_1\, ds\ .$$

Since $q|_e \in \mathcal{P}_{p-3}(e)$ for every edge $e \subset \partial\widehat{T}$, and since $n_1|_e$ is linear along e, we get from b) that the boundary integral is zero and c) gives that the domain integral is zero which is d) for $\alpha = (1,0)$. The proof for $\alpha = (0,1)$ is analogous. A stability analysis of the resulting global projector Q_N along the lines of Section 6.3.4 has yet to be done, however.

Remark 6.75 In Section 6.3.1.1, we pointed out the connection between reduced constraint methods and mixed FEM. Clearly, we could have treated the Naghdi shell problem also by a mixed formulation. Then, however, one would have to prove an inf–sup condition for the spaces used in the discretization, similar to Example 1 in 6.3.1.2. This was done in [8] for a family of h-version FEM and we leave the extension to p-version FEM to the reader.

APPENDIX A

SOBOLEV SPACES

Here we present the basic definitions and properties of Sobolev spaces which are used throughout these notes. This chapter is **not** intended as an introduction to the theory of Sobolev spaces (which would require a course in itself) but rather as supplement when reading the main text and a glossary for terminology and results used. For a more detailed study of Sobolev spaces we refer to [1].

We present the definition and properties of the Sobolev spaces $W^{k,s}$ of functions whose generalized kth derivatives belong to L^s, $1 \leq s \leq \infty$. In the main text, however, mainly the case $s = 2$ is used and we suggest to the reader to record in passing the specialization of the various trace and embedding theorems to this case.

A.1 Domains and boundaries

A.1.1

Let $\Omega \subset \mathbb{R}^n$ be an open, connected bounded domain with boundary $\Gamma = \partial\Omega$. If $n = 1$, $\Omega = (a, b)$ is an interval and $\partial\Omega = \{a, b\}$ the endpoints.

A.1.2 *Lipschitz boundary*

If $n \geq 2$, we say that Γ is a **Lipschitz boundary**, if there exists a finite open cover $U^1, \ldots, U^m$ of Γ such that for $j = 1, \ldots, m$.

a) $\Gamma \cap U^j$ is the graph of a Lipschitz function g^j and

b) $\Omega \cap U^j$ is on one side of this graph.

In other words, for $j = 1, \ldots, m$ exists an Euclidean coordinate system $e_1^j, \ldots, e_n^j$, real numbers $r^j > 0$ and $h^j > 0$ and a Lipschitz function $g^j : \mathbb{R}^{n-1} \to \mathbb{R}$ such that

$$x = \sum_{i=1}^{n} x_i^j e_i^j \in U^j, \; x \; = \; (\widetilde{x}^j, x_n^j), \; |\widetilde{x}^j| < r^j :$$

$$x_n^j = g^j(\widetilde{x}^j) \Longrightarrow x \in \partial\Omega ,$$

$$0 < x_n^j - g^j(\widetilde{x}^j) < h^j \Longrightarrow x \in \Omega ,$$

$$0 > x_n^j - g^j(\widetilde{x}^j) > -h^j \Longrightarrow x \notin \Omega .$$

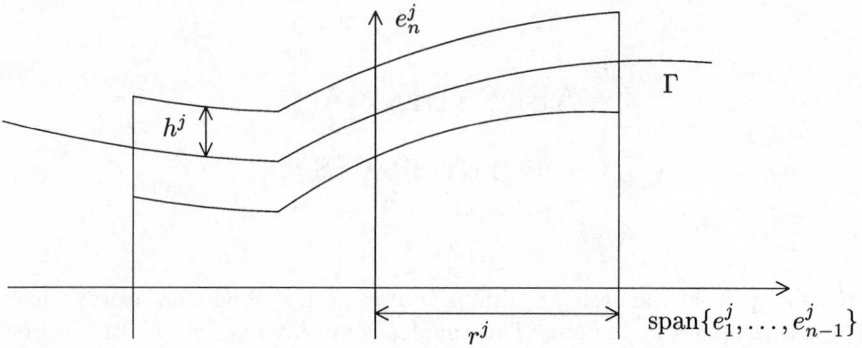

FIG. A.1. Set U^j in the definition of Lipschitz boundary

Define

$$U^j = \{x \in \mathbb{R}^n : |\widetilde{x}^j| < r^j,\ |x_n^j - g^j(\widetilde{x}^j)| < h^j\}\ . \tag{A.1.1}$$

Then $U^j, j = 1, \ldots, m$, define an open cover of Γ.

A.1.3 *Partition of unity*

Let $\{U^j\}_{j=1}^m$ be defined as in (A.1.2); we select one more open set $U^0 \Subset \Omega$ such that $\{U^j\}_{j=0}^m$ forms an open cover of Ω. Let $\{\eta^j\}_{j=0}^m$ denote a **partition of unity** subordinate to $\{U^j\}_{j=0}^m$, i.e.,

$$0 \le \eta^j \in C_0^\infty(U^j) \quad \text{and} \quad \sum_{j=0}^m \eta^j = 1 \text{ on } \overline{\Omega}\ .$$

A.1.4 *C^k boundary*

Let $k \in \mathbb{N}$. Then $\Gamma = \partial\Omega$ is C^k (or $C^{k,\alpha}$ for $\alpha \in (0,1]$) if for every $x^0 \in \Gamma$ there exist $r > 0$, a C^k (or $C^{k,\alpha}$) function $g : \mathbb{R}^{n-1} \to \mathbb{R}$ and a coordinate system $(e_1, \ldots, e_n)$ such that

$$\Omega \cap B(x^0, r) = \{x \in B(x^0, r) : x_n > g(\widetilde{x})\}\ .$$

A C^1 boundary is Lipschitz but not conversely.

A.1.5 *Surface integral and normal vector*

Let $\Gamma = \partial\Omega$ be Lipschitz. We call $f : \Gamma \to \mathbb{R}$ measurable or integrable, if

$$\Gamma \ni y \longmapsto f\Big(\sum_{i=1}^{n-1} y_i\, e_i^j + g^j(y) e_n^j\Big),\ y \in \mathbb{R}^{n-1},\ |y| < r^j$$

is (Lebesgue) measurable or integrable respectively, for $j = 1, \ldots, m$. Then the boundary integral of f over Γ is defined by

$$\int_\Gamma f\,do = \sum_{j=1}^m \int_\Gamma \eta^j f\,do \tag{A.1.2}$$

and, for $\text{supp}(f) \subset U^j$,

$$\int_\Gamma f\,do = \int_{\mathbb{R}^{n-1}} f\left(\sum_{i=1}^{n-1} y_i\, e_i^j + g^j(y) e_n^j\right) \sqrt{1 + |\nabla_y\, g^j(y)|}\, dy\,. \tag{A.1.3}$$

Notice that g^j Lipschitz implies that ∇g^j exists almost everywhere (Rademacher's theorem), is measurable and essentially bounded.

Hence we can in particular define for almost every $x = \sum_{i=1}^{n-1} y_i\, e_i^j + g^j(y) e_n^j \in U^j$ as in (A.1.1), $j = 1, \ldots, m$,

$$\nu(x) := \left(1 + |\nabla g^j(x)|^2\right)^{-\frac{1}{2}} \left(\sum_{i=1}^{n-1} \partial_i\, g^j(y) e_i^j - e_n^j\right). \tag{A.1.4}$$

$\nu(x)$ is the **exterior unit normal vector** to Γ at x (notice that (A.1.4) defines $\nu(x)$ actually in some neighborhood of Γ). The function $\nu(x)$ is measurable and essentially bounded on Γ, i.e., $|\nu| = 1$.

Analogously the **tangential vectors**

$$\tau_k(x) := \partial_{y_k}\left(\sum_{i=1}^{n-1} y_i\, e_i^j + g^j(y)\, e^{j_n}\right) = e_n^j + \partial_k\, g^j(y)\, e_n^j, \quad 1 \le k \le n-1$$

are defined almost everywhere on Γ.

A.2 Spaces $L^s(\Omega)$, $L^s(\partial\Omega)$.

$L^1(\Omega) := \{f : \Omega \to \mathbb{R} \text{ measurable} \mid \|f\|_{L^1(\Omega)} = \int_\Omega |f|\,dx < \infty\}$.

Let $1 < s < \infty$. Then

$$L^s(\Omega) = \left\{f : \Omega \to \mathbb{R} \,\middle|\, |f|^s \in L^1,\ \|f\|_{L^s(\Omega)} = \left(\int_\Omega |f|^s dx\right)^{1/s} < \infty\right\}.$$

Let $s = \infty$ and $f : \Omega \to \mathbb{R}$ be measurable. Then define

$$\|f\|_{L^\infty(\Omega)} = \inf_{\substack{N \subset \Omega \\ |N|=0}} \sup_{\Omega \setminus N} |f(x)| = \operatorname{ess\,sup}_{\Omega} |f(x)|$$

and

$$L^\infty(\Omega) = \{f : \Omega \to \mathbb{R} \text{ measurable} \mid \|f\|_{L^\infty} < \infty\}\,.$$

For $1 \le s \le \infty$, $(L^s(\Omega), \|\circ\|_{L^s(\Omega)})$ is a Banach space.

Analogously define for $1 \le s < \infty$

$$L^s(\partial\Omega) = \{f : \partial\Omega \to \text{ measurable} \mid \|f\|_{L^p(\partial\Omega)} < \infty\}$$
$$L^\infty(\partial\Omega) = \{f : \partial\Omega \to \text{ measurable} \mid \|f\|_{L^\infty(\partial\Omega)} < \infty\}$$

where

$$\|f\|_{L^s(\partial\Omega)} := \Big(\int_{\partial\Omega} |f|^s do\Big)^{1/s}, \ \|f\|_{L^\infty(\partial\Omega)} = \operatorname*{ess\,sup}_{\partial\Omega} |f(x)| \,.$$

Here do is as in (A.1.3).

The space $L^s(\partial\Omega)$ is independent of the local representations g^j.

Exercise A.1 Let $\Omega = (0,1)$, $f(x) = x^\alpha$. Find all $s = s(\alpha) \in [1,\infty]$ such that $f \in L^s(\Omega)$.

A.3 Spaces of continuous functions

Let $\Omega \subset \mathbb{R}^n$ be open and bounded. Define

$$C^0(\Omega) = \{f : \Omega \to \mathbb{R},\ f \text{ continuous and bounded}\} \,.$$

$C^0(\Omega)$ is a Banach space with norm $\|f\|_{C^0(\overline{\Omega})} = \sup_{x\in\Omega} |f(x)|$. Let $\beta \in (0,1]$. Then the Hölder seminorm $[\circ]_{C^{0,\beta}(\Omega)}$ of f is

$$[f]_{C^{0,\beta}(\overline{\Omega})} = \sup_{\substack{x,y\in\Omega\\ x\neq y}} \left\{\frac{|f(x)-f(y)|}{|x-y|^\beta}\right\} . \tag{A.3.1}$$

The space $C^{0,\beta}(\Omega)$ of Hölder continuous functions with exponent β is

$$C^{0,\beta}(\Omega) = \left\{f : \Omega \to \mathbb{R} \,\middle|\, \|f\|_{C^{0,\beta}(\Omega)} := \|f\|_{C^0(\Omega)} + [f]_{C^{0,\beta}(\Omega)} < \infty\right\} . \tag{A.3.2}$$

It is complete with respect to the norm $\|f\|_{C^{0,\beta}(\Omega)}$. Let $k \in \mathbb{N}$. Then

$$C^k(\Omega) = \left\{f : \Omega \to \mathbb{R} \,\middle|\, \|f\|_{C^k(\overline{\Omega})} = \sum_{|\alpha|\le k} \|D^\alpha f\|_{C^0(\overline{\Omega})} < \infty\right\} , \tag{A.3.3}$$

and, for $\beta \in (0,1]$,

$$C^{k,\beta}(\Omega) = \left\{f \in C^k(\Omega) \,\middle|\, \|f\|_{C^{k,\beta}(\overline{\Omega})} + \|f\|_{C^k(\overline{\Omega})} + \sum_{|\alpha|=k} [D^\alpha f]_{C^{0,\beta}(\overline{\Omega})} < \infty\right\} . \tag{A.3.4}$$

Notice that $C^{0,1}(\Omega)$ are the Lipschitz functions and

$$\mathrm{Lip}(u) = [u]_{C^{0,1}(\overline{\Omega})} \, . \tag{A.3.5}$$

The spaces $C^k(\Omega)$, $C^{k,\beta}(\Omega)$ are Banach spaces with the indicated norms. Evidently, the spaces $C^{k,\beta}(\Omega)$ are embedded s follows:

$$C^{k,\beta}(\Omega) \subseteq C^{k',\beta}(\Omega) \;\text{ if } k' \leq k \, , \tag{A.3.6}$$

$$C^{k,\beta}(\Omega) \subseteq C^{k,\beta'}(\Omega) \;\text{ if } \beta' \geq \beta \, , \tag{A.3.7}$$

Moreover, if the inequalities among the indices in (C.3.6), (C.3.7) are strict, the embeddings are in fact compact. This follows from the Arzéla–Ascoli theorem.

A.4 Weak derivatives

Let $\Omega \subset \mathbb{R}^n$ be an open, bounded domain. By $C_0^\infty(\Omega)$ we denote the space of functions $\varphi : \Omega \to \mathbb{R}$ with derivatives of any order and **compact support** in Ω (the support of φ is supp $\varphi = \{x : \varphi(x) \neq 0\}$).

Let $u, v \in L^1(\Omega)$ and $\alpha \in \mathbb{N}_0^n$ a multi-index. Then v is the α th weak derivative of u if and only if

$$\int_\Omega u \, D^\alpha \varphi \, dx = (-1)^{|\alpha|} \int_\Omega v\varphi \, dx \;\text{ for all }\; \varphi \in C_0^\infty(\Omega) \, . \tag{A.4.1}$$

Remark A.2 The integral in (A.4.1) is the Lebesgue integral. Hence (A.4.1) defines the weak derivative $v = D^\alpha u$ only up to sets of (n-dimensional) Lebesque measure zero.

We write $v = D^\alpha u$, as for classical derivatives. Evidently, classical derivatives are weak derivatives, i.e. the notion of weak derivative extends the classical one. The opposite is not true as the following example shows.

Example A.3 Let $n = 1$, $\Omega = (-1, 1)$ and

$$u(x) = \begin{cases} 1 + x & -1 < x < 0 \, , \\ 1 - x & 0 < x < 1 \, . \end{cases} \tag{A.4.2}$$

Evidently, u is not differentiable at $x = 0$.

We claim that the weak derivative $v = u'$ exists and equals

$$v(x) = \begin{cases} 1 & -1 < x < 0 \, , \\ -1 & 0 < x < 1 \, . \end{cases} \tag{A.4.3}$$

To verify (A.4.3), we use (A.4.1). Let $\varphi \in C_c^\infty(-1,1)$. Then

$$\begin{aligned}\int_{-1}^{1} u\varphi'\,dx &= \int_{-1}^{0}(1+x)\,\varphi'\,dx + \int_{0}^{1}(1-x)\varphi'\,dx \\ &= \int_{-1}^{0}\varphi\,dx + (1+x)\,\varphi(x)\,\Big|_{-1}^{0} + \int_{0}^{1}\varphi\,dx - (1-x)\varphi(x)\,\Big|_{0}^{1} \\ &= -\int_{-1}^{1} v\varphi\,dx\end{aligned}$$

from where (A.4.3) follows. The argument evidently does not need that $u(\pm 1) = 0$ since $v \in C_c^\infty(-1,1)$. It generalizes immediately to a continuous, piecewise C^1-function:

Let $\Omega = (a,b)$ and $a = x_0 < x_1 < x_2 < \ldots < x_m = b$ a sequence of points in $\overline{\Omega}$. Let $u \in C^0(\Omega)$ and $u \in C^1([x_{j-1}, x_j])$, $j = 1,\ldots,m$. Then the weak derivative $v = u'$ exists and equals the classical derivative on each $I_j = (x_{j-1}, x_j)$.

The values of u' in the nodes x_i are undefined. This does not cause difficulties, since the integral definition (A.4.1) determines the weak derivative only almost everywhere. The next example will be relevant in the context of FE methods.

Example A.4 Let $\Omega \subset \mathbb{R}^2$ be a bounded polygon and $\{\Omega_j\}_{j=1}^m$ be a triangulation of Ω such that $\overline{\Omega} = \bigcup_{j=1}^m \overline{\Omega}_j$, $\overline{\Omega}_i \cap \overline{\Omega}_j$ is either empty, a vertex or an entire side of Ω_j. Let $u \in C^0(\Omega)$, $u \in C^1(\overline{\Omega}_j)$ for $j = 1,\ldots,m$. Then the weak first derivatives of u exist and equal the classical derivatives on Ω_i, $i = 1,\ldots,m$. The point values of the weak derivative on the measure zero sets $\overline{\Omega}_i \cap \overline{\Omega}_j$ remain undetermined (see Remark A.2).

A.5 Definition of the Sobolev spaces $W^{k,s}(\Omega)$

Let $1 \le s \le \infty$, $k \in \mathbb{N}$ and $\Omega \subset \mathbb{R}^n$ open and bounded. Then

$$W^{k,s}(\Omega) = \left\{ u \in L^s(\Omega) \,\Big|\, D^\alpha u \in L^s(\Omega) \text{ for } |\infty| \le k \right\}. \tag{A.5.1}$$

Here $D^\alpha u$ is understood in the weak sense (A.4.1):
if $s = 2$, we write $W^{k,2}(\Omega) = H^k(\Omega)$, $k = 1, 2, \ldots$.

We equip $W^{k,s}(\Omega)$ with the norm

$$\|u\|_{W^{k,s}(\Omega)} := \Big(\sum_{|\alpha| \le k} \|D^\alpha u\|_{L^s(\Omega)}^s \Big)^{1/s}. \tag{A.5.2}$$

Two functions u_1, $u_2 \in W^{k,s}(\Omega)$ are identified if they differ only on a subset $\Omega' \subset \Omega$ of Lebesgue measure zero, i.e., functions $u \in W^{k,p}(\Omega)$ are really **equivalence classes** (see also Remark A.2).

We note some properties of weak derivatives:

$$D^\alpha u \in W^{k-|\alpha|,s}(\Omega),\ (D^\alpha(D^\beta u)) = D^\beta(D^\alpha u) = D^{\alpha+\beta}u \tag{A.5.3}$$

for all $\alpha, \beta \in \mathbb{N}_0^k$ such that $|\alpha| + |\beta| \le k$.

$$\Omega' \subset \Omega \text{ and } u \in W^{k,s}(\Omega) \Longrightarrow u \in W^{k,s}(\Omega') \tag{A.5.4}$$

$$u, v \in W^{k,s}(\Omega) \Longrightarrow \lambda u + \mu v \in W^{k,s}(\Omega) \text{ for all } \lambda, \mu \in \mathbb{R}\,. \tag{A.5.5}$$

The linear space $W^{k,s}(\Omega)$ equipped with the norm (A.5.2) is complete, i.e., a Banach space. If $s = 2$ it is, analogous to $L^2(\Omega)$, a Hilbert space with the inner product

$$(u, v)_{H^k} :\equiv \sum_{|\alpha|\le k} \int_\Omega D^\alpha u\, D^\alpha v\, dx\,. \tag{A.5.6}$$

Evidently $\|u\|^2_{H^k} = (u, u)^{\frac{1}{2}}_{H^k}$.

Functions in $W^{k,s}(\Omega)$, $k \ge 1$, need not to be continuous.

Example A.5 Let $\Omega = B(0,1) \subset \mathbb{R}^2$ the open unit disk and $u(x) = \log(\log(e/|x|))$. Show that $u(x) = H_0^1(\Omega)$.

Sobolev functions are related to Hölder-continuous ones. The precise relation will be presented later in the embedding theorem. Here we only consider a special case.

Theorem A.6 *Let $\partial\Omega$ be Lipschitz and $m \in N$. Then $C^{m-1,1}(\Omega) \subset W^{m,\infty}(\Omega)$ and the injection $J : C^{m-1,1}(\Omega) \to W^{m,\infty}(\Omega)$ is an isomorphism.*

We turn now to $W^{k,s}(\Omega)$ for noninteger $k > 0$, i.e., the **fractional order spaces.**

Definition A.7 *Let $\Omega \subset \mathbb{R}^n$ be open and bounded, $k \ge 0$ an integer and $\kappa = k+\theta$, $0 < \theta < 1$. For $1 < s < \infty$ we define $W^{k,s}(\Omega)$ as subspace of $W^{k,s}(\Omega)$ with finite (Slobodečkij) norm*

$$\|u\|_{W^{k,s}(\Omega)} = \Big\{\|u\|^s_{W^{k,s}(\Omega)} + \sum_{|\alpha|=k} \int_\Omega\int_\Omega \frac{|D^\alpha u(x) - D^\alpha u(y)|^s}{|x-y|^{n+\theta s}}\, dx\, dy\Big\}^{1/s}. \tag{A.5.7}$$

If $s = \infty$, $W^{\kappa,\infty}(\Omega)$ is the subspace of $u \in W^{k,\infty}(\Omega)$ for which

$$\|u\|_{W^{k,\infty}(\Omega)} := \|u\|_{W^{k,\infty}(\Omega)} + \max_{|\alpha|=k} \operatorname*{ess\,sup}_{\substack{x,y\in\Omega \\ x\ne y}} \frac{|D^\alpha u(x) - D^\alpha u(y)|}{|x-y|^{n+\theta s}} \tag{A.5.8}$$

is finite.

A.6 Density of smooth functions

Functions in $W^{k,s}(\Omega)$ can be approximated (in the $W^{s,k}(\Omega)$ norm) by smooth functions. This is frequently useful, since it allows to prove results for $W^{k,s}(\Omega)$ "by density", i.e., it suffices to establish the claim for $u \in C^\infty(\overline{\Omega})$ and then to use an approximation argument to conclude. The basis for it is the following.

Theorem A.8 *Assume that $\Omega \subset \mathbb{R}^n$ is open, bounded and that $\partial\Omega$ is Lipschitz (cf. A.1.2). Let $k \in \mathbb{N}_0$ and $1 \leq s < \infty$. Then for every $u \in W^{k,s}(\Omega)$ there exists a sequence $\{u_m\}_{m=1}^\infty \subset C^\infty(\overline{\Omega})$ such that*

$$\|u - u_m\|_{W^{k,s}(\Omega)} \to 0 \quad as\ m \to \infty\,.$$

Note carefully that this result is wrong for arbitrary, open $\Omega \subset \mathbb{R}^n$ – the regularity of the boundary is essential here. Notice also that the restriction $p < \infty$ is essential, as the following example shows.

Exercise A.9 Let $\Omega = (-1, 1)$ and $u(x) = \text{sign}(x)$ (i.e., $u(x) = -1$ for $x < 0$ and $u(x) = 1$ for $x > 0$). Show that u cannot be the limit in $L^\infty(\Omega)$ of a sequence of continuous functions. Deduce that polynomials are not dense in $L^\infty(\Omega)$.

Remark A.10 There is no other function system that is dense in $L^\infty(\Omega)$. In fact, the space $L^\infty(\Omega)$ is not separable, i.e., it does not contain a countable, dense subset.

This does of course not rule out that special members of $L^\infty(\Omega)$, as e.g. functions analytic in $\overline{\Omega}$, can be very well approximated in $\|\circ\|_{L^\infty(\Omega)}$ by a countable sequence of functions (e.g., polynomials). It only means that $L^\infty(\Omega)$ also contains pathologic functions which are not approximable concurrently from a sequence of finite dimensional subspaces.

A.7 Traces. Sobolev spaces on $\partial\Omega$

Assume again $\Omega \subset \mathbb{R}^n$ is open and bounded. If $u \in C^0(\overline{\Omega})$, the boundary values $u|_{\partial\Omega}$ are defined at every $x \in \partial\Omega$. We need to consider boundary values of functions in $W^{k,s}(\Omega)$, in particular for the variational formulation of boundary value problems. To do so, we must define boundary value in a generalized sense, since functions in $W^{k,s}(\Omega)$ a) need not be continuous (cf. Example A.5) and b) are only uniquely defined up to measure zero subsets of Ω in the cases of interest to us, $\partial\Omega$ has n-dimensional Lebesgue measure zero).

Theorem A.11 *Let $\Omega \subset \mathbb{R}^n$, $n > 1$, be open and bounded and let $\partial\Omega$ be Lipschitz, $1 \leq s \leq \infty$. Then there exists a unique, continuous linear map $\gamma_0 : W^{1,s}(\Omega) \to L^s(\partial\Omega)$, the so-called* **trace operator**, *such that $\gamma_0 u = u|_{\partial\Omega}$ for $u \in C^0(\overline{\Omega}) \cap W^{1,s}(\Omega)$.*

The trace operator is bounded, i.e.,

$$\|\gamma_0 u\|_{L^s(\partial\Omega)} \leq C(\Omega)\|u\|_{W^{1,s}(\Omega)} \tag{A.7.1}$$

but not onto, i.e., $\gamma_0(W^{1,s}(\Omega)) \subset L^s(\partial\Omega)$ with strict inclusion if $s < \infty$. This means in particular that not every function $u \in L^s(\partial\Omega)$ is the trace of some $u \in W^{1,s}(\Omega)$.

The range $\gamma_0(H^1(\Omega))$ is strictly smaller than $L^2(\partial\Omega)$ and denoted by $H^{\frac{1}{2}}(\partial\Omega) = W^{\frac{1}{2},2}(\partial\Omega)$. (We may take this as a definition of range (γ_0) at this point; the notation stems from the fact that range (γ_0) can be characterized as interpolation space $(L^2(\partial\Omega), H^1(\partial\Omega))_{\frac{1}{2},2}$, see [86], Theorem 9.4.)

Remark A.12 Let $\Omega \subset \mathbb{R}^n$ be open and bounded, $\Gamma = \partial\Omega$ be Lipschitz and $\Gamma_1 \subseteq \Gamma$ satisfy $\int_{\Gamma_1} do > 0$. Then

$$\|\gamma_0 u\|_{L^s(\Gamma_1)} \leq \|\gamma_0 u\|_{L^s(\Gamma)} \leq C(\Omega)\,\|u\|_{W^{1,s}(\Omega)} \quad \text{for all} \quad u \in W^{1,s}(\Omega)\,.$$

Exercise A.13 Let $\Gamma_1 \subseteq \Gamma$ and $\Gamma = \partial\Omega$ be Lipschitz. Consider

$$W^{1,s}(\Omega, \Gamma_1) := \{u \in W^{1,s}(\Omega) : \gamma_0\, u = 0 \ \text{ on } \ \Gamma_1\}\,. \tag{A.7.2}$$

Why is it a closed linear subspace of $W^{1,s}(\Omega)$?

We will also write

$$W_0^{1,s}(\Omega) = W^{1,s}(\Omega, \partial\Omega), \ H_0^1(\Omega) = W_0^{1,2}(\Omega)\,. \tag{A.7.3}$$

So far we have addressed only the case $n > 1$. The one-dimensional case is special owing to $\partial\Omega$ consisting only of two endpoints. Here the discussion of traces reduces to the existence of point values, the topic of the next section.

Exercise A.14 Show that

$$W_0^{1,s}(\Omega) = \overline{C_0^\infty(\Omega)}^{\|\circ\|_{W^{1,s}}}\,. \tag{A.7.4}$$

To characterize the range of the trace operator precisely, we must introduce Sobolev spaces on $\Gamma = \partial\Omega$.

Definition A.15 *Let $\Omega \subset \mathbb{R}^n$ be open and bounded with boundary Γ of class $C^{k,1}$ for some integer $k \geq 0$ (see* A.3*). Let $0 \leq \kappa \leq k+1$ be an integer and let $s \in [1, \infty]$ be a real number. Then $u \in W^{\kappa,s}(\Gamma)$ if*

$$u \circ \Phi^j \in W^{\kappa,s}\big(\mathcal{U}^j \cap (\Phi^j)^{-1}(\Gamma \cap U^j)\big)$$

where $\Phi^j(x) = (g^j(\tilde{x}^j), x_n^j)$ and the function g^j and the set U^j are as in A.1.2, $\mathcal{U}^j$ is an open, bounded hypercube in $\mathbb{R}^n : \mathcal{U}^j = \{x \in \mathbb{R}^n : \|x\|_\infty < r^j\}$ with r^j as in (A.1.2).

Remark A.16 The norm $\|\circ\|_{W^{\kappa,s}(\Omega)}$ is defined by a partition of unity in terms of the transported Slobodečkij norms in each U^j.

Apart from the trace operator, we also need the normal derivative on Γ. For $u \in C^\infty(\overline{\Omega})$, it is defined by

$$\gamma_1 = \frac{\partial u}{\partial n} = \sum_{i=1}^{n} \gamma_0 \left(\frac{\partial u}{\partial x_i} \right) n_i \tag{A.7.5}$$

where $\boldsymbol{n}$ denotes the exterior unit normal vector.

We can now state the general trace theorem.

Theorem A.17 *Let $\Omega \subset \mathbb{R}^n$ be open, bounded with boundary $\Gamma = \partial\Omega$ of class $C^{k,1}$, $k \geq 0$ integer. Let $s \geq 1$ and $\kappa \geq 0$ be real numbers satisfying*

$$\kappa \leq k+1, \quad \kappa - 1/s = \ell + \theta$$

where $\ell \geq 0$ is an integer and $0 < \theta < 1$. Then the trace map $\gamma_0 u$ defined on $C^\infty(\overline{\Omega})$ extends uniquely to a bounded, surjective linear map

$$\gamma_0 : W^{k,s}(\Omega) \to W^{k-1/s,s}(\Gamma) .$$

Analogously, for $\ell \geq 1$ the normal derivative extends from $C^\infty(\overline{\Omega})$ to a bounded, linear map

$$\gamma_1 : W^{\kappa,s}(\Omega) \to W^{\kappa-1-1/s,s}(\Gamma).$$

The preceding theorem allows us to define a norm on $W^{\kappa-1/s,s}(\Gamma)$ as follows:

$$\|u\|_{W^{\kappa-1/s,s}(\Omega)} = \inf_{\substack{v \in W^{\kappa,s}(\Omega) \\ u=\gamma_0 v}} \|u\|_{W^{\kappa,s}(\Omega)} . \tag{A.7.6}$$

This norm is equivalent to the one mentioned in Remark A.16.

Although Theorem A.17 is fairly general, it is not sharp in the sense that γ_0, γ_1 generally cease to be continuous when some of the inequalities for the indices are violated. Two extensions are particularly useful.

Theorem A.18 *Let $\Omega \subset \mathbb{R}^2$ be a bounded polygon without slits and with sides Γ_j and vertices A_j, $j = 1, \ldots, M$. Then the trace map*

$$\gamma_0 : u \to \left(u_j = u\big|_{\Gamma_j}, j = 1, \ldots, M\right)$$

is linear, continuous and **surjective** *from $W^{2,s}(\Omega)$* **onto**

$$\prod_{j=1}^{M} W^{2-\frac{1}{s},s}(\Gamma_j) \cap \left\{ \{u_j\} \,\middle|\, u_j(A_{j+1}) = u_{j+1}(A_{j+1}) \right\} .$$

Let ν_j denote the exterior unit normal on Γ_j. Then

$$\gamma_1 : u \to \left(\frac{\partial u}{\partial \nu_j} : j = 1, \ldots, M \right)$$

is continuous and **surjective** *from* $W^{k,s}$ *onto*

$$\prod_{j=1}^{M} W^{k-1-\frac{1}{s},s}(\Gamma_j),\ k \geq 2,\ 1 < s < \infty\,.$$

Remark A.19 Theorem A.18 remains valid also for curvilinear polygons without slits and cusps, if each side Γ_j is a smooth one.

The second generalization of Theorem A.17 is needed only when $s = 2$ and $k = 0$ in Theorem A.17 which says that

$$\gamma_0 : W^{k,2}(\Omega) \to W^{k-\frac{1}{2},2}(\Gamma),\ \frac{1}{2} < k \leq 1$$

continuously. This is actually valid for a larger range of k.

Theorem A.20 *Let $\Omega \subset \mathbb{R}^n$ be a bounded domain with Lipschitz boundary Γ. Then the trace map γ_0 is a continuous, surjective operator from $W^{k,2}(\Omega)$ onto $W^{k-\frac{1}{2},2}(\Gamma)$ for $1/2 < k < 3/2$.*

A.8 Embedding theorems

Variational or weak formulations of elliptic PDE yield **weak solutions** in Sobolev spaces: their relation to the classical Hölder spaces is given by **Sobolev's embedding theorem.**

Theorem A.21 *Let $\Omega \subset \mathbb{R}^n$, $n \geq 1$ and $\partial\Omega$ Lipschitz, if $n \geq 2$. Assume that $m - n/s > 0$ is not an integer and denote*

$$\kappa := [m - n/s],\ \theta := (m - n/s) - \kappa \in (0,1)\,. \tag{A.8.1}$$

(where $[x] = \max\{\kappa \in Z : \kappa \leq x\}$).

Then we have for every $u \in W^{m,s}(\Omega)$

$$||D^\alpha u||_{L^\infty(\Omega)} \leq C\,||u||_{W^{m,s}(\Omega)} \text{ for all } \alpha \text{ with } |\alpha| \leq \kappa \tag{A.8.2}$$

and moreover

$$|D^\alpha u(x) - D^\alpha u(y)\,| \leq C\,||u||_{W^{m,s}(\Omega)}\,|x-y|^\theta \text{ a.s., } x, y \in \Omega\,. \tag{A.8.3}$$

In particular therefore

$$W^{m,s}(\Omega) \subset C^k(\overline{\Omega})\,. \tag{A.8.4}$$

Remark A.22

i) The injection is to be understood in the following sense: for every $u \in W^{m,s}(\Omega)$ there exists an equivalent function $\widetilde{u}$ (everywhere equal to u except on a set of measure zero) which belongs to $C^k(\overline{\Omega})$.

ii) The assumption $m - n/s > 0$ is essential. Let, e.g., $m = 1$, $s = 2$, $n = 1$. Then $m = n/s$ and $H^1(\Omega) = W^{1,2}(\Omega)$ is in general not included in $C^0(\overline{\Omega})$, as the example $u(x) = \log|\log(e/|x|)|$, $\Omega = B(0,1)$ (cf. Example A.5) shows. Inequalities (A.8.2), (A.8.3) mean that the corresponding embedding operators are continuous. They are in fact, **compact**. This is the content of the following theorem, also referred to as **Rellich's theorem**.

iii) The inclusions (A.8.2), (A.8.3) are also valid for $n = 1$ if $\Omega = (a,b)$ is a bounded interval. In particular therefore the "traces" of functions $u \in W^{1,s}((a,b))$ are well defined by (A.8.2) (for $\alpha = 0$, $m = n = 1$) and

$$|u(a)| + |u(b)| \leq C\,\|u\|_{W^{1,s}((a,b))}\,. \tag{A.8.5}$$

Theorem A.23 *Assume that $\Omega \subset \mathbb{R}^n$ is bounded and $\partial\Omega$ is Lipschitz. The following embeddings are* **compact***:*

$$\begin{aligned} 1 \leq s < n : W^{1,s}(\Omega) \subset L^t(\Omega) \ \ &\textit{for all}\ \ t \in [1, s^*]\,, \ \frac{1}{s^*} := \frac{1}{s} - \frac{1}{n}\,, \\ s = n : W^{1,s}(\Omega) \subset L^t(\Omega) \ \ &\textit{for all}\ \ t \in [1, \infty] \\ s > n : W^{1,s}(\Omega) \subset C^0(\overline{\Omega})\,. \end{aligned}$$

Remark A.24 The above embeddings have generalizations to fractional-order spaces. We will not elaborate here and refer to [1], [86], for example.

A.9 Poincaré inequalities

Frequently in the variational formulation of boundary value problems the question arises under what circumstances the seminorm $|u|_{W^{1,s}(\Omega)} = (\sum_{|\alpha|=1} \|D^\alpha u\|_{L^s(\Omega)})^{1/s}$ is a norm on $X \subseteq W^{1,s}(\Omega)$. Since $|1|_{W^{1,s}} = 0$, X must not contain constants. Two choices for X are possible.

Theorem A.25 *Let $\Omega \subset \mathbb{R}^n$ be open and bounded and, if $n \geq 2$, $\partial\Omega$ be Lipschitz. Let further $\Gamma_1 \subseteq \partial\Omega$ be such that $\int_{\Gamma_1} do > 0$. Then there exists $C(\Omega, \Gamma_1) > 0$ such that*

$$\|u\|_{L^s(\Omega)} \leq C\,|u|_{W^{1,s}(\Omega)} \ \ \textit{for all}\ \ u \in W^{1,s}(\Omega, \Gamma_1)\,. \tag{A.9.1}$$

Theorem A.26 *Let $\Omega \subset \mathbb{R}^n$ open, bounded, connected and, if $n \geq 2$, assume that $\partial\Omega$ is Lipschitz. Let*

$$\widetilde{W}^{1,s} = W^{1,s}(\Omega) \cap \left\{u : \int_\Omega u\,dx = 0\right\}. \tag{A.9.2}$$

Then there exists $C(\Omega) > 0$ such that

$$\|u\|_{L^s(\Omega)} \leq C\,|u|_{W^{1,s}(\Omega)} \ \ \textit{for all}\ \ u \in \widetilde{W}^{1,s}(\Omega)\,. \tag{A.9.3}$$

Remark A.27 It is equivalent to (A.8.1) and (A.8.3) to have

$$\|u\|_{W^{1,s}(\Omega)} \leq C\,|u|_{W^{1,s}(\Omega)}\,, \tag{A.9.4}$$

possibly with a different constant.

Proof of Theorem A.26: Assume (A.8.4) is false. Then there is a sequence $\{u_n\} \subset \widetilde{W}^{1,s}$ such that $\|u_k\|_{W^{1,s}(\Omega)} = 1$, $|u_k|_{W^{1,s}} < 1/k$ as $k \to \infty$. By Rellich's theorem there exists a subsequence, again denoted by u_k, such that $u_k \to u$ in $L^s(\Omega)$. Therefore $\|u\|_{L^s(\Omega)} = 1$. Moreover, for $\varphi \in C_0^\infty(\Omega)$, $i = 1, \ldots, n$,

$$\int_\Omega u\partial_i\varphi\,dx = \lim_{k\to\infty} \int_\Omega u_k\,\partial_i\varphi\,dx = \lim_{k\to\infty} - \int_\Omega \varphi\partial_i\,u_k\,dx = 0\,,$$

hence u has vanishing first weak derivatives. Thus $u \in W^{1,s}(\Omega)$ and $\partial_i u = 0$, $i = 1, \ldots, n$.

Since Ω is connected, this implies that u is constant. Since $\{u_n\} \subset \widetilde{W}^{1,s}$ and $\|u_n - u\|_{L^s} \to 0$, we must also have $\int_\Omega u\,dx = 0$ whence $u = 0$, a contradiction. □

Remark A.28 Note that the preceding proof does not give an estimate of the size of C in (A.9.4).

A.10 Green's formulas

Theorem A.29 *Let $\Omega \subset \mathbb{R}^n$ be open and bounded and $\partial\Omega$ be Lipschitz if $n \geq 2$. Then*

$$\int_\Omega \partial_i u\,dx = \int_{\partial\Omega} un_i\,do \quad \textit{for all} \quad u \in W^{1,1}(\Omega)\,, \tag{A.10.1}$$

where $\boldsymbol{n} = (n_1, \ldots, n_n)$ denotes the exterior unit normal to Ω on $\partial\Omega$ and u on $\partial\Omega$ denotes the trace of u (cf. A.7).

Let further $s \in [1, \infty]$ and $(1/s) + (1/s') = 1$, then

$$\int_\Omega (u\partial_i v + v\partial_i u)\,dx = \int_{\partial\Omega} uvn_i\,do\ .u \in W^{1,s},\ v \in ,W^{1,s'}\,. \tag{A.10.2}$$

Let $u \in W^{1,s}$ and $v \in [W^{1,s'}]^n$. Then

$$\int_\Omega u\nabla\cdot v\,dx = -\int_\Omega v\cdot\nabla u\,dx + \int_{\partial\Omega} u(v\cdot\boldsymbol{n})\,do\,. \tag{A.10.3}$$

Let $v \in W^{1,s'}(\Omega) \cap \{v : \Delta v \in L^{s'}(\Omega)\}$. Then from (A.10.3)

$$\int_\Omega u\Delta v\,dx = \int_\Omega u\nabla\cdot\nabla v\,dx = -\int_\Omega \nabla u\cdot\nabla v + \int_{\partial\Omega} u(\boldsymbol{n}\cdot\nabla v)\,do\,,$$

i.e.,

$$\int_\Omega u\Delta v\,dx + \int_\Omega \nabla u\cdot\nabla v\,dx = \int_{\partial\Omega} u\,\frac{\partial v}{\partial \boldsymbol{n}}\,do\,. \tag{A.10.4}$$

We see that for any $u \in H^1(\Omega)$, the left-hand side is well-defined. Thus (A.10.4) can be used to define $\partial v/\partial n$ as an element in the dual $[H^{1/2}(\Gamma)]'$ of the trace space $H^{1/2}(\Gamma)$.

A.11 Negative-order spaces

Definition A.30 *$H^{-1}(\Omega)$ is the dual of $H_0^1(\Omega)$, i.e., $H^{-1}(U) = \{f\,|\,f$ is a bounded linear functional on $H_0^1(\Omega)\}$.*

Theorem A.31 *Assume $f \in H^{-1}(\Omega)$. Then there exist $f_0, f_1, \ldots, f_n \in L^2(\Omega)$ such that for all $v \in H_0^1(\Omega)$*

$$f(v) = \int_\Omega f_0 v\,dx + \sum_{i=1}^n \int_\Omega f_i\,\partial_i v\,dx\,. \tag{A.11.1}$$

Moreover,

$$\begin{aligned}\|f\|_{H^{-1}(\Omega)} = \inf\Big\{\Big(\sum_{i=0}^n \|f_i\|^2_{L^2(\Omega)}\Big)^{\frac{1}{2}} &: f \text{ satisfies (A.11.1)} \\ &\text{for } f_0,\ldots,f_n \in L^2(\Omega)\Big\}\,.\end{aligned} \tag{A.11.2}$$

Proof Define

$$B(u,v) = \int_\Omega (\nabla u\cdot\nabla v + uv)\,dx :\ H_0^1(\Omega)\times H_0^1(\Omega) \to \mathbb{R}\,.$$

Let $f \in H^{-1}(\Omega)$. By the Riesz representation theorem there exists a unique solution $u \in H_0^1(\Omega)$ for

$$u \in H_0^1(\Omega) :\ B(u,v) = \langle f, v\rangle \text{ for all } v \in H_0^1(\Omega)\,, \tag{A.11.3}$$

i.e.,

$$\langle f, v\rangle = \int_\Omega (uv + \nabla u\cdot\nabla v)\,dx \text{ for all } v \in H_0^1(\Omega)\,.$$

Comparing with (A.11.1), we see that $f^0 = u$, $f^i = \partial_i u$.

To prove (A.11.2), assume that

$$\langle f, v\rangle = \int_{\Omega} (g_0 v + \sum_{i=1}^{n} g_1\, \partial_i v)\, dx \tag{A.11.4}$$

for some $g_i \in L^2(\Omega)$ and $v \in H_0^1(\Omega)$. Then

$$|\langle f, v\rangle| \leq \Big(\sum_{i=0}^{n} \|g_i\|_{L^2(\Omega)}^2\Big)^{\frac{1}{2}} \|v\|_{H^1(\Omega)}$$

and

$$\|f\|_{H^{-1}(\Omega)} = \sup_{H_0^1(\Omega)} \frac{|\langle f, v\rangle|}{\|v\|_{H^1(\Omega)}} \leq \Big(\sum_{i=0}^{n} \|g_i\|_{L^2(\Omega)}^2\Big)^{\frac{1}{2}} .$$

From (A.11.3) we find

$$\sum_{i=0}^{n} \|f_i\|_{L^2(\Omega)}^2 = \|f\|_{H^{-1}(\Omega)}^2 \leq \sum_{i=0}^{n} \|g_i\|_{L^2(\Omega)}^2$$

from where (A.11.2) follows. □

Remark A.32 We see that the infimum in (A.11.2) is attained for $f_0 = u$, $f_i = \partial_i u$ where u solves (A.11.3). As notation, we write

$$f = f_0 - \sum_{i=1}^{n} \partial_i f^i = f_0 - \nabla \cdot (f_1, \ldots, f_n) . \tag{A.11.5}$$

We will also need certain spaces of negative order on the boundary Γ.

Theorem A.33 *Let $1 < s < \infty$ and $(1/s) + (1/s') = 1$ and $\Gamma = \partial\Omega$ be a bounded Lipschitz boundary in $\mathbb{R}^n$. Denote by $[W^{1-1/s',s'}(\Gamma)]'$ the space of continuous, linear functionals on $W^{1-1/s',s'}(\Gamma)$ and by $\langle u, v\rangle_\Gamma$ the corresponding duality pairing. Let $u \in W^{1,s}(\Omega)$ satisfy $\Delta u \in L^s(\Omega)$ and let $w \in W^{1-1/s',s'}(\Gamma)$ satisfy $w = \gamma_0 v$ for $v \in W^{1,s'}(\Omega)$. Then the formula*

$$\Big\langle \frac{\partial u}{\partial n}, \gamma_0\, w\Big\rangle_\Gamma = \int_{\Omega} \nabla u \cdot \nabla v\, dx + \int_{\Omega} v \Delta u\, dx \tag{A.11.6}$$

determines a unique element of $[W^{1-1/s',s'}(\Gamma)]'$ denoted by $\gamma_1 u$. The map $u \to \gamma_1 u$ is continuous from $W_\Delta^{1,s}(\Omega) = W^{1,s}(\Omega) \cap \{u : \Delta u \in L^s(\Omega)\}$ into $[W^{1-1/s',s'}(\Gamma)]'$.

Proof The right-hand side of (A.11.6) is independent of the choice of $v \in W^{1,s'}(\Omega)$, as long as $\gamma_0 v = w$, since

$$0 = \int_\Omega \nabla u \cdot \nabla \varphi \, dx + \int_\Omega \varphi \, \Delta u \, dx \quad \text{for all } \varphi \in C_0^\infty(\Omega)$$

by the definition of the weak derivative.

By Green's formula and a density argument, for all $v \in W^{1,s'}(\Omega)$ such that $\gamma_0 v = w$ holds

$$\left| \left\langle \frac{\partial u}{\partial n}, w \right\rangle \right| \le \left(\|u\|^s_{W^{1,s}(\Omega)} + \|\Delta u\|^s_{L^s(\Omega)} \right)^{\frac{1}{s}} \|v\|_{W^{1,s'}(\Omega)} \tag{A.11.7}$$

by Hölder's inequality. Taking the supremum in (A.11.7) over all such v we get

$$\begin{aligned} \left\| \frac{\partial u}{\partial n} \right\|_{(W^{1-1/s',s'}(\Gamma))'} &= \sup_{\|w\|_{W^{1-1/s',s'}(\Gamma)}=1} \left\langle \frac{\partial u}{\partial n}, w \right\rangle \\ &\le \sup_{\|v\|_{W^{1,s'}(\Omega)} \le C} \left\langle \frac{\partial u}{\partial n}, \gamma_0 v \right\rangle \le C \left(\|u\|^s_{W^{1,s'}(\Omega)} + \|\Delta u\|^s_{L^s(\Omega)} \right)^{\frac{1}{s}} \end{aligned}$$

where the constant C is the norm of the trace operator $\gamma_0 : W^{1,s'}(\Omega) \to W^{1-1/s',s'}(\Gamma)$. □

Remark A.34 The case $s = s' = 2$ is frequently used. Therefore we write $H^{-\frac{1}{2}}(\Gamma) = [H^{\frac{1}{2}}(\Gamma)]' = [W^{\frac{1}{2},2}(\Gamma)]'$.

A.12 The space $H(\mathrm{div}, \Omega)$ and its properties

Assume that $\Omega \subset \mathbb{R}^n$, is bounded and connected and that $\Gamma = \partial\Omega$ is Lipschitz. The standard reference for the results presented here is Chapter 1.2 of [64].

We define

$$H(\mathrm{div}, \Omega) = \{ u \in [L^2(\Omega)]^n : \ \mathrm{div}\, u \in L^2(\Omega) \} , \tag{A.12.1}$$

and equip it with the norm

$$\|u\|_{H(\mathrm{div},\Omega)} = \left(\|u\|^2_{L^2(\Omega)} + \|\mathrm{div}\, u\|^2_{L^2(\Omega)} \right)^{\frac{1}{2}} . \tag{A.12.2}$$

Evidently $[L^2(\Omega)]^n \supset H(\mathrm{div}, \Omega) \supset [H^1(\Omega)]^n$. Then the following result holds.

Theorem A.35 (properties of $H(\mathrm{div}, \Omega)$)

i) $[C^\infty(\overline{\Omega})]^n$ *is dense in* $H(\mathrm{div}, \Omega)$.

ii) *The mapping* $\gamma_n : u \to u \cdot \boldsymbol{n}\,|_\Gamma$ *defined on* $[C^\infty(\overline{\Omega})]^n$ *can be extended by continuity to a linear and continuous mapping from* $H(\mathrm{div},\,\Omega)$ *onto* $H^{-\frac{1}{2}}(\Gamma)$. *This extension is still denoted by* $\gamma_{\boldsymbol{n}}$.

iii) *The following* **Green's formula** *holds:*

$$(v, \operatorname{grad} u) + (\operatorname{div} v, u) = \langle v \cdot \boldsymbol{n}, u \rangle \quad v \in H(\mathrm{div},\,\Omega),\; u \in H^1(\Omega)\,. \quad \text{(A.12.3)}$$

iv) γ_n *is* **onto**, *i.e., range* $(\gamma_{\boldsymbol{n}}) = H^{-\frac{1}{2}}(\Gamma)$, *and*

$$\|\gamma_{\boldsymbol{n}}\|_{\mathcal{L}(H(\mathrm{div},\,\Omega), H^{-\frac{1}{2}}(\Gamma))} = 1\,.$$

The proof of these assertions can be found, for example, in [64], pp. 27 and 28.

A.13 Exercises

Exercise A.36 Let $d > 0$ and $\Omega = (-d, d)$. Find the smallest constant $C(\Omega)$ in Poincaré's inequality

$$\|v\|_{L^2(\Omega)} \le C\,\|v'\|_{L^2(\Omega)} \text{ for all } v \in H_0^1(\Omega)\,.$$

Repeat this for $v \in H^1_{(0}(\Omega) := \{v \in H^1(\Omega) : v(-d) = 0\}$, and for

$$v \in \widetilde{H}^1(\Omega) := \Big\{ v \in H^1(\Omega) : \int_{-d}^{d} v\,dx = 0 \Big\}\,.$$

Demonstrate in each case that the constants cannot be improved.

Exercise A.37 Let $\Omega = (-d, d)$. Prove

$$\|u\|_{L^\infty(\Omega)} \le \sqrt{\frac{d}{2}}\,\|u'\|_{L^2(\Omega)} \qquad \text{for all } u \in H_0^1(\Omega)\,,$$

and

$$\|u\|_{L^\infty(\Omega)} \le \sqrt{\frac{2d}{3}}\,\|u'\|_{L^2(\Omega)} \quad \text{for all } u \in \widetilde{H}^1(\Omega)\,.$$

Again verify the optimality of the constants and find the extremal function u.

APPENDIX B

INTERPOLATION SPACES

B.1 Relation between Sobolev spaces

Let $\Omega \subset \mathbb{R}^n$ be a bounded Lipschitz domain. In A.5 we defined the Sobolev spaces $W^{k,s}(\Omega)$ for $k = 0, 1, \dots$ and $1 \leq s \leq \infty$. We also gave a definition for $W^{\kappa,s}(\Omega)$ (Definition A.7) when $\kappa = k + \theta, 0 < \theta < 1$. The Sobolev spaces obtained in this fashion satisfy the inclusions

$$W^{\kappa_1,s}(\Omega) \subseteq W^{\kappa_2,s}(\Omega), \qquad \kappa_1 \geq \kappa_2, 1 \leq s \leq \infty, \tag{B.1.1}$$

$$W^{\kappa,s_1}(\Omega) \subseteq W^{\kappa,s_2}(\Omega), \qquad \kappa \geq 0, \quad \infty \geq s_1 \geq s_2 \geq 1. \tag{B.1.2}$$

The inclusions (B.1.1), (B.1.2) allows us to order the spaces $W^{\kappa,s}$ as follows:

$$\begin{array}{c|ccccccccc}
k \diagdown s & 1 & & 2 & & s & & \infty \\
\hline
0 & L^1 & \supset & L^2 & \supset \dots \supset & L^s & \supset \dots \supset & L^\infty \\
 & \cup & & \cup & & \cup & & \cup \\
1 & W^{1,1} & \supset & H^1 & \supset \dots \supset & W^{1,s} & \supset \dots \supset & W^{1,\infty} \\
 & \cup & & \cup & & \cup & & \cup \\
1 & W^{2,1} & \supset & H^2 & \supset \dots \supset & W^{2,s} & \supset \dots \supset & W^{2,\infty} \\
\vdots & \vdots & & \vdots & & \vdots & & \vdots
\end{array} \tag{B.1.3}$$

Interpolation fills the "gaps" in the diagram (B.1.3) with "intermediate" fractional-order spaces. These intermediate spaces are quite useful to obtain exact characterizations of the smoothness of solutions to boundary value problems and therefore also of the rate of convergence of the FEM.

Many different interpolation methods exist. We will focus here on the so-called **real or K method of interpolation** (see Chapter 1 of [148] for a survey and comparison with other interpolation methods).

B.2 The real method of interpolation

Let A_0, A_1 be two Banach spaces with norms $\|\circ\|_{A_0}$ and $\|\circ\|_{A_1}$, respectively. We assume further that $A_1 \subset A_0$ (this will always be the case here, but is not necessary in general). For $0 < t < \infty$, we define the **K functional** by

$$K(t,a) = \inf_{a=a_0+a_1} \{\|a_0\|_{A_0} + t\|a_1\|_{A_1}\} \; . \tag{B.2.1}$$

The K functional is well defined for all $a \in A_0$.

Definition B.1 *(The space $[A_0, A_1]_{\theta,q}$). Let $0 < \theta < 1$ and $1 \le q \le \infty$. Then $[A_0, A_1]_{\theta,q}$ is defined by*

$$[A_0, A_1]_{\theta,q} = \{a \in A_0 : \|a\|_{[A_0,A_1]_{\theta,q}} < \infty\} \tag{B.2.2}$$

where

$$\|a\|_{[A_0,A_1]_{\theta,q}} = \Big(\int_0^\infty \left(t^{-\theta}K(t,a)\right)^q \frac{dt}{t}\Big)^{1/q} , \; 1 \le q < \infty \tag{B.2.3}$$

and

$$\|a\|_{[A_0,A_1]_{\theta,\infty}} = \sup_{0\le t\le\infty} t^{-\theta}K(t,a) \, , \; q = \infty \, . \tag{B.2.4}$$

The main properties of the interpolation space $[A_0, A_1]_{\theta,q}$ are collected in the following result.

Theorem B.2

i) *$[A_0, A_1]_{\theta,q}$ is a Banach space*
ii) *$[A_0, A_1]_{\theta,q} = [A_1, A_0]_{1-\theta,q}$*
iii) *for $0 < \theta < 1$, $1 \le q \le \tilde{q} \le \infty$ it holds that*

$$A_1 \subset [A_0, A_1]_{\theta,1} \subset [A_0, A_1]_{\theta,q} \subset [A_0, A_1]_{\theta,\tilde{q}} \subset [A_0, A_1]_{\theta,\infty} \subset A_0 \, . \tag{B.2.5}$$

iv) *If A_1 is reflexive,*

$$[A_0, A_1]_{1,\infty} = A_1 \; . \tag{B.2.6}$$

v) *If $0 < \tilde{\theta} \le \theta < 1$, $1 \le q \le \tilde{q} \le \infty$,*

$$A_1 \subset [A_0, A_1]_{\theta,q} \subset [A_0, A_1]_{\tilde{\theta},\tilde{q}} \subset A_0 \, . \tag{B.2.7}$$

vi)

$$A_0 = A_1 \Rightarrow [A_0, A_1]_{\theta,q} = A_0 = A_1 \; . \tag{B.2.8}$$

vii) *for every $a \in A_1 \subset A_0$, $0 < \theta < 1$, $1 \le q \le \infty$:*

$$\|a\|_{[A_0,A_1]_{\theta,q}} \le C(\theta, q) \, \|a\|_{A_0}^{(1-\theta)} \, \|a\|_{A_1}^{\theta} \; . \tag{B.2.9}$$

For a proof see, for example, [148], Chapter 1.3.3.

We mostly apply interpolation in connection with linear operators T which are continuous between pairs of spaces.

Theorem B.3 *Let* $T\colon A_i \to B_i$, $i = 0, 1$ *be a continuous linear operator and* A_i, B_i *Banach spaces with* $A_1 \subset A_0$, $B_1 \subset B_0$.

Then for every $0 < \theta < 1, 1 \le p \le \infty$

i) $T\colon A_{\theta,p} \to B_{\theta,p}$ *continuously*

ii)

$$\|T\|_{A_{\theta,p}\to B_{\theta,p}} \le \|T\|^{1-\theta}_{A_0\to B_0} \|T\|^{\theta}_{A_1\to B_1} \,. \tag{B.2.10}$$

Finally, we quote a result which addresses the spaces obtained from interpolating interpolation spaces. The so-called **reiteration theorem** states that no new spaces can be obtained in this way.

Theorem B.4 *For any* $0 < \theta_0 \le \theta_1 < 1$, $1 \le p_0, p_1, q \le \infty$, $0 \le \lambda \le 1$:

$$\Big[[A_0, A_1]_{\theta_0,p_0}\,, [A_0, A_1]_{\theta_1,p_1}\Big]_{\lambda,q} = [A_0, A_1]_{(1-\lambda)\theta_0+\lambda\theta_1\,,q} \,. \tag{B.2.11}$$

If A_0 is dense in A_1, we have

$$[A_0, A_1]'_{\theta,p} = [A'_1\,, A'_0]_{1-\theta\,,p'} \tag{B.2.12}$$

where $p' = p/(p-1)$ is the conjugate index to p.

B.3 Interpolation of Sobolev spaces

We now interpolate horizontally in the tableau (B.1.3). Throughout, $\Omega \subset \mathbb{R}^n$ is a bounded Lipschitz domain.

Theorem B.5 (Riesz–Thorin) *Let* $1 \le s \le \infty$. *Then*

$$L^s(\Omega) = \big[L^1(\Omega), \quad L^\infty(\Omega)\big]_{1-1/s\,,s}\,, \tag{B.3.1}$$

$$W^{k,s}(\Omega) = \big[W^{k,1}(\Omega),\ W^{k,\infty}(\Omega)\big]_{1-1/s\,,s}\,. \tag{B.3.2}$$

The proof can be found in [56].

The "vertical" result is the following theorem.

Theorem B.6 . *Let further* $\kappa = k + \theta$ *with* $k \in \mathbb{N}_0$ *and* $0 < \theta < 1$. *Then, for* $1 \le s \le \infty$,

$$W^{\kappa,s}(\Omega) = \big[W^{k,s}(\Omega)\,, W^{k+1,s}(\Omega)\big]_{\theta,s} \tag{B.3.3}$$

and the interpolation norm is equivalent to the Slobodeckij norm (A.5.7).

The proof is given, for example, in [38], p. 280.

APPENDIX C

ORTHOGONAL POLYNOMIALS

We collect the definitions and basic properties of some orthogonal polynomials on the interval $I = (-1, 1)$ which were used in the text. The basic reference for orthogonal polynomials is [145]. Many useful formulas can also be found in [66], with references to their proofs.

C.1 General definitions

Let $w(x)$ denote a weight function on $I = (-1, 1)$ such that all moments

$$m_n = \int_I x^n w(x) dx \,, \quad n = 0, 1, 2, \ldots \tag{C.1.1}$$

are finite and such that

$$\int_I w(x) dx > 0 \,. \tag{C.1.2}$$

Then there exists a unique sequence $P_0(x), P_1(x), P_2(x), \ldots$ of polynomials satisfying

$$P_n(x) = \sum_{j=0}^{n} a_j x^j \,, \quad a_n > 0 \,, \tag{C.1.3}$$

and $\{P_n(x)\}_{n=0}^{\infty}$ is an L^2-orthogonal sequence with respect to the weight $w(x)$, i.e.,

$$\int_I P_n(x) P_m(x) w(x) dx \begin{cases} \neq 0 & \text{if } n = m \,, \\ = 0 & \text{else} \,. \end{cases} \tag{C.1.4}$$

We call $\{P_n(x)\}_{n=0}^{\infty}$ the family of orthogonal polynomials with respect to the weight $w(x)$ in I. We now discuss various special cases.

C.2 Legendre polynomials

For the weight function $w(x) = 1$ we obtain the family $\{L_n(x)\}_{n=0}^{\infty}$, the Legendre polynomials. They are given by the **Rodriguez formula**

$$L_n(x) = \frac{1}{n!\, 2^n} \frac{d^n}{dx^n} \left[(x^2 - 1)^n\right], \quad x \in [-1, 1] \,, \quad n = 0, 1, 2, \ldots \tag{C.2.1}$$

The first Legendre polynomials are given by

$$\begin{aligned}
L_0(x) &= 1\,, \\
L_1(x) &= x\,, \\
L_2(x) &= (3x^2 - 1)/2\,, \\
L_3(x) &= (5x^3 - 3x)/2\,, \\
L_4(x) &= (35x^4 - 30x^2 + 3)/8\,.
\end{aligned} \tag{C.2.2}$$

They satisfy the **Legendre differential equation**

$$\left((1 - x^2)L_n'(x)\right)' + n(n+1)L_n(x) = 0 \;\; in\ I\,. \tag{C.2.3}$$

Further, since $w(x) = 1$, we have **orthogonality**

$$\int_{-1}^{1} L_n(x)L_m(x)dx = \begin{cases} \dfrac{2}{2n+1} & \text{for } n = m\,, \\ 0 & \text{else}\,. \end{cases} \tag{C.2.4}$$

We have also

$$L_n(x) = \left(L_{n+1}'(x) - L_{n-1}'(x)\right)/(2n+1)\,, \;\; n \geq 1\,, \tag{C.2.5}$$

and

$$L_n(1) = 1\,, \;\; L_n(-1) = (-1)^n\,, \;\; n \geq 0\,. \tag{C.2.6}$$

Any function $u \in L^2(I)$ can be expanded into a **Legendre series**

$$u(x) = \sum_{n=0}^{\infty} a_n L_n(x) \tag{C.2.7}$$

where "=" is understood in the sense that

$$\lim_{p\to\infty} \left\| u - \sum_{n=0}^{p} a_n L_n \right\|_{L^2(I)} = 0\,. \tag{C.2.8}$$

Note that in general $\lim_{p\to\infty} \sum_{n=0}^{p} a_n L_n(x)$ does not exist for $x \in I$, and even if it exists, it may not be equal to $u(x)$. Multiplying (C.2.7) by $L_m(x)$, integrating over I and referring to (C.2.4) we find

$$a_n = \frac{2n+1}{2} \int_{-1}^{1} u(x)L_n(x)dx\,. \tag{C.2.9}$$

Let $Q_p(x)$ be any polynomial of degree $\leq p$ in I. Then $Q_p(x) = \sum_{n=0}^{p} b_n L_n(x)$ and, for any $u \in L^2(I)$,

$$\|u - Q_p\|_{L^2}^2 = \left\| \sum_{n=0}^{p} (a_n - b_n) L_n + \sum_{n=p+1}^{\infty} a_n L_n \right\|_{L^2}^2 .$$

Using the orthogonality (C.2.4), we get

$$\|u - Q_p\|_{L^2}^2 = \left\| \sum_{n=0}^{p} (a_n - b_n) L_n \right\|_{L^2}^2 + \left\| \sum_{n=p+1}^{\infty} a_n L_n \right\|_{L^2}^2 \geq \left\| \sum_{n=p+1}^{\infty} a_n L_n \right\|_{L^2}^2$$

with equality being attained if and only if $Q_p = \sum_{n=0}^{p} a_n L_n$.

Hence

$$\min_{Q_p \in \mathcal{S}^p} \|u - Q_p\|_{L^2} = \left\| u - \sum_{n=0}^{p} a_n L_n \right\|_{L^2} . \tag{C.2.10}$$

Consider $u \in L^2$ as in (C.2.7). Then for any p

$$\left\| \sum_{n=0}^{p} a_n L_n \right\|_{L^2}^2 = \sum_{n=0}^{p} \frac{2}{2n+1} |a_n|^2 \leq \|u\|_{L^2}^2$$

and letting $p \to \infty$ and using (C.2.8), we get the **Parseval equation**

$$\|u\|_{L^2}^2 = \sum_{n=0}^{\infty} \frac{2}{2n+1} |a_n|^2 \tag{C.2.11}$$

with a_n as in (C.2.9).

Let $\{a_n\}_{n=0}^{\infty}$ be any sequence of real (or complex) numbers such that the right-hand side of (C.2.11) is finite. Then there exists $u(x) \in L^2(I)$ such that (C.2.9)–(C.2.11) hold.

C.3 Jacobi polynomials

Consider the weight function

$$w(x) = (1-x)^{\alpha}(1+x)^{\beta}, \quad x \in (-1, 1) . \tag{C.3.1}$$

The finiteness of the moments (C.1.1) implies that

$$\alpha, \beta > -1 . \tag{C.3.2}$$

The orthogonal polynomials $\{P_n(x;\alpha;\beta)\}_{n=0}^{\infty}$ corresponding to (C.3.1) are the Jacobi polynomials. Evidently,

$$P_n(x;0;0) = L_n(x)\,, \tag{C.3.3}$$

i.e. they generalize the Legendre polynomials. All relations valid for Legendre polynomials have analogs for the Jacobi polynomials. We have the **Rodriguez-Formula**

$$P_n(x;\alpha;\beta) = \frac{(-1)^n}{2^n\,n!}\,(1-x)^{-\alpha}(1+x)^{-\beta}\,\frac{d^n}{dx^n}\,[(1-x)^{\alpha+n}(1+x)^{\beta+n}]\,. \tag{C.3.4}$$

Jacobi polynomials satisfy the **Jacobi differential equation**

$$(1-x^2)P_n'' + \big(\beta-\alpha-(\alpha+\beta+2)x\big)P_n' + n(n+\alpha+\beta+1)P_n = 0\,. \tag{C.3.5}$$

Let $L^2_{\alpha,\beta}(I)$ denote the space of all functions which are square integrable with respect to the weight (C.3.1); equipped with the norm given by

$$\|u\|^2_{L^2_{\alpha,\beta}} = \int_{-1}^{1} |u|^2(1-x)^\alpha(1+x)^\beta dx\,. \tag{C.3.6}$$

Then every $u \in L^2_{\alpha,\beta}$ can be expanded into a **Jacobi series**

$$u(x) = \sum_{n=0}^{\infty} a_n P_n(x;\alpha;\beta)\,, \tag{C.3.7}$$

which is understood in the sense that

$$\lim_{p\to\infty} \|u - \sum_{n=0}^{p} a_n P_n(x;\alpha;\beta)\|_{L^2_{\alpha,\beta}} = 0\,. \tag{C.3.8}$$

The a_n are determined using the **orthogonality relations**

$$\int_{-1}^{1} P_n(x;\alpha;\beta)P_m(x;\alpha;\beta)(1-x)^\alpha(1+x)^\beta dx = \begin{cases} h(n;\alpha;\beta) & \text{if } n=m\,, \\ 0 & \text{else.}\end{cases} \tag{C.3.9}$$

where

$$h(n;\alpha;\beta) = \frac{2^{\alpha+\beta+1}}{2n+1+\alpha+\beta}\,\frac{\Gamma(\alpha+1+n)\,\Gamma(\beta+1+n)}{\Gamma(n+1)\,\Gamma(\alpha+\beta+1+n)}\,, \tag{C.3.10}$$

which gives

$$a_n = \frac{1}{h(n;\alpha;\beta)} \int_{-1}^{1} (1-x)^{\alpha}(1-x)^{\beta} u(x) P_n(x;\alpha;\beta) dx \,. \tag{C.3.11}$$

We have the **relation between Jacobi polynomials and their derivatives**

$$\begin{aligned} &\frac{d^m}{dx^m} P_n(x;\alpha;\beta) \\ &= 2^{-m} \frac{\Gamma(n+m+\alpha+\beta+1)}{\Gamma(n+\alpha+\beta+1)} P_{n-m}(x;\alpha+m;\beta+m) \,. \end{aligned} \tag{C.3.12}$$

If $\alpha = \beta$, $P_n(x;\alpha;\beta) = P_n(x;\alpha)$ is also called **ultraspherical polynomial**. From (C.3.12), (C.3.9) and (C.3.4) we get

$$\int_{-1}^{1} (1-x^2)^k \frac{d^k}{dx^k} L_n(x) \frac{d^k}{dx^k} L_m(x) dx = 0 \,, \qquad n \neq m \,, \tag{C.3.13}$$

$$\int_{-1}^{1} (1-x^2)^k \Big(\frac{d^k}{dx^k} L_n(x)\Big)^2 dx = \frac{2}{2n+1} \frac{(n+k)!}{(n-k)!} \,. \tag{C.3.14}$$

C.4 Some special cases

If $\alpha = \beta = 0$, we get the Legendre polynomials. If $\alpha = \beta$, $P_n(x;\alpha) := P_n(x;\alpha;\alpha)$ is the **ultraspherical polynomial**. If $\alpha = \nu - 1/2$, we call

$$C_n^{\nu}(x) = \frac{\Gamma(n+2\nu)\,\Gamma(\nu+\frac{1}{2})}{\Gamma(2\nu)\,\Gamma(n+\nu+\frac{1}{2})} P_n\Big(x;\nu-\frac{1}{2}\Big) \tag{C.4.1}$$

the **Gegenbauer polynomial**, and

$$T_n(x) := \frac{2^{2n}\,(n!)^2}{(2n)!} P_n\Big(x;-\frac{1}{2}\Big) \tag{C.4.2}$$

the **Chebyčev polynomial**.

REFERENCES

[1] H. Adams. *Sobolev Spaces.* Academic Press, New York, (1978).

[2] M. Ainsworth. A preconditioner based on domain decomposition for *hp*-FE approximation on quasi-uniform meshes. *SIAM J. Numer. Anal.*, **33**, (1996), 1358–1376.

[3] M. Ainsworth and J.T. Oden. A-posteriori error estimates for the Stokes- and Oseen equations. *Siam J. Numer. Anal.*, **34**, (1997), 228-245.

[4] M. Ainsworth and J.T. Oden. A posteriori error estimation in finite element analysis. *Comp. Meth. Appl. Mech. Engg.*, **142**, (1997), 1-88.

[5] S.M. Alessandrini, D.N. Arnold, R.S. Falk and A.L. Madureira. Derivation and justification of plate models by variational methods. CRM Proceedings and Lecture Notes (1997), to appear.

[6] H.W. Alt. *Lineare Funktionalanalysis.* (2nd edn.), Springer-Verlag, Heidelberg, (1992).

[7] T. Apel, A.M. Sändig, and J. Whiteman. Graded mesh refinement and error estimates for finite element solutions of elliptic boundary value problems in non-smooth domains. *Math. Meth. Appl. Sci.*, **19**, (1996), 63-85.

[8] D.N. Arnold and F. Brezzi. Locking-free FEM for shells. *Math. Comp.*, **66** (1997), 1-14.

[9] D.N. Arnold and R.S. Falk. Asymptotic analysis of the boundary layer for the Reissner–Mindlin plate model. *SIAM J. Math. Anal.*, **27**, (1996), 486-514.

[10] A.K. Aziz and R.B. Kellogg. On homeomorphisms for an elliptic equation in domains with corners. *Differential and Integral Equations*, **8**, (1995), 333-352.

[11] I. Babuška. Error bounds for finite element method. *Num. Math.*, (1971).

[12] I. Babuška and A.K. Aziz. Survey lectures on the mathematical foundations of the finite element method. In: *The Mathematical Foundation of the Finite Element Method with Applications to Partial Differential Equations.* (A.K. Aziz, Ed.), Academic Press, New York, (1972), 3-359.

[13] I. Babuška and A.K. Aziz. On the angle condition in the finite element method. *SIAM J. Numer. Analysis*, **13**, (1976), 214-226.

[14] I. Babuška, A. Craig, J. Mandel and J. Pitkäranta. Efficient preconditioning for the *p*-version FEM in two dimensions. *SIAM J. Numer. Anal.*, **28**, (1991), 624-661.

[15] I. Babuška and B.Q. Guo. Regularity of the solutions of elliptic problems with piecewise analytic data, parts I and II. *SIAM J. Math. Anal.*, **19**, (1988), 172-203 and **20**, (1989), 763-781.

[16] I. Babuška and B.Q. Guo. The *hp*-version of the finite element method for domains with curved boundaries. *SIAM J. Numer. Anal.*, **25**, (1988), 837-861.

[17] I. Babuška and B.Q. Guo. Approximation properties of the *hp*-version of the

finite element method. *Comp. Meth. Appl. Mech. Engg.*, **133**, (1996), 319-346.

[18] I. Babuška, B. Guo and M. Suri. Implementation of Nonhomogeneous Dirichlet Boundary Conditions in the p Version of the Finite Element Method. *Impact of Computing in Science and Engineering*, **1**, (1989), 36-63.

[19] I. Babuška, R.B. Kellogg and J. Pitkäranta. Direct and inverse error estimates for finite elements with mesh refinement. *Num. Math.*, **33**, (1979), 447-471.

[20] I. Babuška and M. Suri. The hp-version of the finite element method with quasiuniform meshes. *RAIRO Anal. Numérique*, **21**, (1987), 199-238.

[21] I. Babuška and M. Suri. The optimal convergence rate of the p-version of the finite element method. *SIAM J. Numer. Analysis*, **24**, (1987), 750-776.

[22] I. Babuška and M. Suri. The p- and hp-versions of the finite element method: an overview. *Comp. Meth. Appl. Mech. Engg.*, **80**, (1990), 5-26.

[23] I. Babuška and M. Suri. On locking and robustness in the finite element method. *SIAM J. Numer. Anal.*, **29**, (1992), 1261-1293.

[24] I. Babuška and B.A. Szabo. Lecture Notes on Finite Element Analysis. Dept. of Mathematics, University of Maryland, Dept. of Mathematics, College Park, USA, (1983-1985).

[25] A. Bajer, L. Demkowicz, K. Gerdes, C. Schwab and T. Walsh. A new flexible implementation of general hp-FEM in $2D$. In press in: Computing and Visualization in Science, (1998).

[26] U. Banerjee and M. Suri. The effect of numerical quadrature in the p-version of the finite element method. *Math. Comp.*, **59**, 199, (1992), 1-20.

[27] R. Bellman. A note on an inequality of E. Schmidt. *Bull. Amer. Math. Soc.*, **50**, (1944), 734-737.

[28] M. Bernadou, P.G. Ciarlet and B. Miara. Existence Theorems for two-dimensional linear shell theories. *J. Elasticity*, **34**, (1994), 111-138.

[29] Ch. Bernardi. Indicateurs d'erreur en éléments spectraux. *RAIRO Anal. Numér.*, **30**, No. 1, (1996), 1-38.

[30] C. Bernardi and Y. Maday. *Approximations Spectrales de Problèmes aux Limites Elliptiques*. Springer-Verlag, Paris, (1992).

[31] C. Bernardi and Y. Maday. Stable spectral elements for the Stokes equation. Preprint (1998), to appear in Math. Mod. Meth. Appl. Sci.

[32] I. A. Blatov and V. V. Strygin. On estimates best possible in order in the Galerkin finite element method for singularly perturbed boundary value problems. *Russian Acad. Sci. Dokl. Math.*, **47**, (1993), 93-96.

[33] E. Boillat. On the right inverse for the divergence operator in spaces of continuous piecewise polynomials. *Math. Meth. Mod. Appl. Sci.*, (1997), (to appear).

[34] P. Bolley, J. Camus and M. Dauge. Regularité Gevrey pour le problème de Dirichlet dans des domaines coniques. *Comm. PDE*, **10**, (1985), 391-431.

[35] J.P. Boyd. *Chebysev and Fourier Spectral Methods.* Springer Lecture Notes in Engineering, vol. 49, Berlin, Heidelberg, (1989).

[36] D. Braess. Saddle point problems with penalty. *RAIRO. Anal. Numérique*, (1997).

[37] D. Braess. *Finite Elements.* Cambridge Univ. Press, (1997).

[38] S.C. Brenner and L.R. Scott. *The Mathematical Theory of Finite Element Methods.* Springer-Verlag, New York, (1994).

[39] H. Brezis. *Analyse Fonctionnelle.* 2me tirage, Masson, Paris, (1989).

[40] F. Brezzi. On the existence, uniqueness and approximation of saddle-point problems originating from Lagrangian multipliers. *RAIRO Anal. Numérique*, **8**, (1974), 129-151.

[41] F. Brezzi and M. Fortin. *Mixed and Hybrid Finite Element Methods.* Springer Series in Computational Mathematics **15**, Springer-Verlag, New York, (1991).

[42] C. Canuto. Spectral methods and a maximum principle. *Math. Comp.*, **51**, (1988), 615-629.

[43] C. Canuto, A. Quarteroni, Hussaini and Zhang. *Spectral Methods.* Springer Verlag, New York, (1988).

[44] L. Chilton and M. Suri. On the selection of a locking-free *hp* element for elasticity problems. *Int. J. Num. Meth. Engg.*, **40**, (1997).

[45] P.G. Ciarlet. *The Finite Element Method for Elliptic Problems.* North Holland, Amsterdam, (1978).

[46] P.G. Ciarlet. Mathematical Elasticity I: Three-dimensional elasticity. North-Holland, Amsterdam (1988).

[47] P.G. Ciarlet. Mathematical Elasticity II: Lower dimensional theories of plates. North-Holland, Amsterdam, (1997).

[48] P.G. Ciarlet and V. Lods. Asymptotic analysis of linearly elastic shells III: Justification of Koiter's shell equation. *Arch. Rat. Mech. Anal.*, **136**, (1996), 191-200.

[49] P.G. Ciarlet and B. Miara. On the ellipticity of linear shell models. *Z. Angew. Math. Phys.* **43**, (1992), 243-253.

[50] Ph. Clément. Approximation by finite element functions using local regularization. *RAIRO Anal. Numérique*, **2**, (1975), 77-84.

[51] M. Dauge. Elliptic boundary value problems on corner domains. *Lect. Notes in Math.*, **1341**, Springer-Verlag, Berlin, Heidelberg, New York, (1988).

[52] P.J. Davis. *Interpolation and Approximation.* Dover, New York, (1975).

[53] L. Demkowicz, J.T. Oden and W. Rachowicz. A new finite element method for compressible Navier-Stokes equations based on an operator splitting method and *hp*-adaptivity. *Comp. Meth. Appl. Mech. Engg.*, **84**, (1990), 275-326.

[54] L. Demkowicz, J.T. Oden, W. Rachowicz and O. Hardy. Towards a universal *hp*-adaptive finite element strategy. I. Constrained approximation and data structure. *Comp. Meth. Appl. Mech. Engg.*, **77**, (1989), 79-112.

[55] L. Demkowicz, J.T. Oden, W. Rachowicz and O. Hardy. An *hp*-Taylor-Galerkin method for the compressible Euler equations. *Comp. Meth. Appl. Mech. Engg.*, **86**, (1991), 363-396.

[56] R. DeVore and K. Scherer. Interpolation of linear operators on Sobolev spaces. *Annals Math.*, **109**, (1979), 583-599.

[57] J. Elschner. The *hp*-version of spline approximation methods for Mellin convolution equations. *J. Integral Equations*, **5**, (1993), 47-73.

[58] K. Eriksson, D. Estep, P. Hansbo and C. Johnson. *Adaptive Finite Elements.* Springer-Verlag, Berlin, (1996).

[59] L.C. Evans. Lecture notes on partial differential equations. Berkeley Math. Lecture Notes, Vol. 3A and 3B, Dept. of Mathematics, University of California, Berkeley.

[60] G.B. Folland. *Partial Differential Equations.* Princeton University Press, Princeton, New Jersey, (1975).

[61] E.C. Gartland. Uniform high-order difference schemes for a singularly perturbed two-point boundary value problem, *Math. Comp.*, **48**, (1987), 551-564.

[62] K. Gerdes, A.M. Matache and C. Schwab. Analysis of membrane locking in the *hp*-FE Approximation of cylindrical shells. ZAMM (submitted).

[63] D. Gilbarg and N.S. Trudinger. *Elliptic Partial Differential Equations of Second Order* (2nd edn), Springer-Verlag, Berlin, (1983).

[64] V. Girault and P.A. Raviart. *Finite Element Methods for Navier Stokes Equation.* Springer Series in Computational Mathematics **6**, Springer-Verlag, New York, (1986).

[65] S. Goldberg. *Unbounded Linear Operators.* Dover, New York, (1966).

[66] I. S. Gradshteyn and I. M. Ryzhik. *Table of Series, Integrals, and Products.* Corrected and enlarged edition. Wiley, New York, (1980).

[67] P. Grisvard. *Elliptic Boundary Value Problems in Nonsmooth Domains.* Pitman, Boston, (1985).

[68] P. Grisvard. *Singularities in Boundary Value Problems.* Springer-Verlag & Masson, Paris, (1992).

[69] B. Gui and I. Babuška. The *h*-, *p*- and *hp*-versions of the FEM in 1 dimension, parts I, II, III. *Num. Math.*, **49**, (1986), 577-683.

[70] B.Q. Guo. The *hp*-version of the finite element method for elliptic equations of order $2m$. *Num. Math.*, **53**, (1988), 199-224.

[71] B.Q. Guo and I. Babuška. The *hp*-version of the finite element method I. The basic approximation results and II: general results and applications. *Comp. Mech.*, **1**, (1986), 21-41 and 203-226.

[72] B. Guo and I. Babuška. On the regularity of elasticity problems with piecewise analytic data. *Adv. Appl. Math.*, **14**, (1993), 307-347.

[73] B. Guo and W. Cao. A preconditioner for the *hp*-version of the finite element method in two dimensions. *Num. Math.*, **75**, (1996), 59-77.

[74] W. Han and S. Jensen. On the sharpness of L^2-error estimates of H_0^1-projections onto subspaces of piecewise high-order polynomials. *Math. Comp.*, **64**, No. 209, (1995), 51-70.

[75] G. Hardy, J.E. Littlewood and G. Polya. *Inequalities.* Cambridge University Press, Cambridge, (1988).

[76] H. Holm, M. Maischak and E.P. Stephan. Exponential convergence of the *hp*-version BEM for mixed BVPs on polyhedra. *Math. Meth. Appl. Sci.*, (1997), (to appear).

[77] F. John. Partial Differential Equations (5th edn), Springer-Verlag, New York, (1988).

[78] G.E. Karniadakis and S.J. Sherwin. *Spectral hp Element Methods for CFD.* Oxford University Press, (to appear).

[79] I.N. Katz, D.W. Wang and B.A. Szabo. *Implementation of a $C1$ Triangular Element based on the p-version of the Finite Element Method.* NASA Conference Publication 2245, Research in Structural and Solid Mechanics, (1982), 153-170.

[80] I.N. Katz, D.W. Wang and B.A. Szabo. *h*- and *p*-version Analyses of a Rhombic Plate. *Int. Journal for Num. Meth. in Engg.*, **2**, (1984), 1399-1405.

[81] I.N. Katz and D.W. Wang. The *p*-version of the Finite Element Method for Problems Requiring $C1$ Continuity. *SIAM Journal of Num. Analysis*, **22**, No. 6, (1985), 1082-1106.

[82] N. Kikuchi and J.T. Oden. *Contact Problems in Elasticity.* SIAM, Philadelphia, (1988).

[83] V.A. Kondrat'ev. Boundary value problems for elliptic equations in domains with conical or angular points. *Trudy Moskovkogo Mat. Obschetsva*, **16**, (1967), 209-292.

[84] V.A. Kozlov, V.G. Maz'ya and J. Rossmann. *Elliptic Boundary Value Problems in Domains with Point Singularities.* American Mathematical Society, Rhode Island, (1997).

[85] H. Kraus. *Thin Elastic Shells: An Introduction to the Theoretical Foundations and the Analysis of Their Static and Dynamic Behavior.* Wiley, New York, 1967.

[86] J.L. Lions and E. Magenes. *Non Homogeneous Boundary Value Problems and Applications.* Vol. 1, Springer-Verlag, Berlin, (1972).

[87] J.L. Lions and E. Sanchez-Palencia. *Partial Differential Equations and Functional Analysis in Memory of Pierre Grisvard.* L. Céa, D. Chenais, G. Geymonat and J.L. Lions, eds., Birkhäuser, Basel, (1996), 206-220.

[88] W. B. Liu and J. Shen. A new efficient spectral Galerkin method for singular perturbation problems. *J. Sci. Comput.*, **11**, (1996), 411-437.

[89] W. B. Liu and T. Tang. Boundary layer resolving methods for singularly perturbed problems, submitted to *I.M.A. J. Numer. Anal.*

[90] Y. Maday and E.M. Ronquist. Optimal error analysis of spectral methods with emphasis on non-constant coefficients and dformed geometries. *C. Math. Mech.*, **80**, 1990), 91-115.

[91] M. Maischak and E.P. Stephan. The *hp*-Version of the boundary element method in $\mathbb{R}^3$. The basic approximation results. *Math. Meth. Appl. Sci.*, **20**, (1997), 461-476.

[92] C. Mardare. Modèles bi-dimensionnels de coques linéarement élastiques: estimations de l'écart entre les solutions. *C.R. Acad. Sci. Paris I*, **322**, (1996), 793-796.

[93] J.M. Melenk. On robust exponential convergence of finite element methods for problems with boundary layers. *IMA J. Numer. Anal.*, **17**, (1997), 577-601.

[94] J.M. Melenk and C. Schwab. *hp*-FEM for Reaction-Diffusion Equations I: Robust Exponential Convergence, II. Regularity. Reports 97-03 and 97-04, Seminar for Applied Mathematics, ETH Zürich, (1997).

[95] J.M. Melenk and C. Schwab. *hp*-FEM for convection-diffusion problems. Report 97-05, Seminar for Applied Mathematics, ETH Zürich, (1997).

[96] D. Morgenstern. Herleitung der Plattentheorie aus der dreidimensionalen Elastizitätstheorie. *Arch. Rat. Mech. Anal.*, **4**, (1959), 145-152.

[97] C.B. Morrey. *Multiple Integrals in the Calculus of Variations.* Springer-Verlag, New York, (1966).

[98] K.W. Morton. *Numerical Solution of Convection-Diffusion Problems.* Oxford University Press, Oxford, (1995).

[99] J. Nečas. Sur les normes équivalentes dans $W_p^k(\Omega)$ et sur la coercivité des formes formellement positives. In: *Séminaire Equations aux Dérivées Partielles,* Les Presses de l'Université de Montreal, (1966), 102-128.

[100] J. Nečas. *Les Méthodes Directes en Théorie des Equations Elliptiques.* Masson, Paris, (1967).

[101] J. Nečas. Sur une méthode pour résoudre les équations aux dérivées partielles du type elliptique, voisine de la variationnelle. *Ann. Scuola Norm. Sup., Pisa*, **16**, (1962), 305-326.

[102] L. Nirenberg. Remarks on strongly elliptic partial differential equations. *Comm. Pure and Appl. Math.*, **8**, (1955), 648-674.

[103] J.T. Oden. Optimal *hp*-Finite Element Methods. *Comp. Meth. Appl. Mech. Engg.*, **112**, (1994), 309-331.

[104] J.T. Oden, L. Demkowicz, W. Rachowicz and T.A. Westermann. Towards a universal *hp*-adaptive strategy. II. A posteriori error estimation. *Comp. Meth. Appl. Mech. Engg.*, **77**, (1989), 113-180.

[105] J.T. Oden, L. Demkowicz, W. Rachowicz and T. Westerman. A-posteriori error analysis in finite elements: the element residual method for symmetrizable problems with application to compressible Navier-Stokes equations. *Comp. Meth. Appl. Mech. Engg.*, **82**, (1990), 183-204.

[106] O.A. Oleinik, Y. Shamaev and I. Iosifian. *Mathematical Problems in Elasticity and Homogenization.* North-Holland, Amsterdam, (1992).

[107] E.T. Olsen and J. Douglas, Jr. . Bounds on spectral condition numbers of matrices arising in the *p*-version FEM. *Numer. Math.*, **69**, (1995), 333-352.

[108] F. W. J. Olver. Error bounds for the Liuville-Green (or WKB) approximation. *Proc. Cambridge Philos. Soc.*, **57**, (1961), 790-810.

[109] G. Peano. Doctoral Dissertation, Dept. of Mech. Engineering, Washington University St. Louis, Mo, USA, (1975).

[110] J. Piila. Characterization of the membrane theory of a clamped shell. The elliptic case. *Math. Mod. Meth. Appl. Sci.*, **4**, No. 2, (1994). 147-177.

[111] J. Piila. Characterization of the membrane theory of a clamped shell. The hyperbolic case. Helsinki University of Technlogy, Institute of Mathematics, Research reports, A337, (1994).

[112] A. Pinkus. *n-Widths in Approximation Theory.* Springer-Verlag, Berlin, (1985).

[113] J. Pitkäranta and M. Suri. Design principles and error analysis for reduced shear finite elements. *Numer. Math.*, **75**, No. 2, (1967), 223-266.

[114] A. Quarteroni. Some results of Bernstein and Jackson type for polynomial approximation in L^p spaces. *Japan J. Appl. Math.*, **1**, (1984), 173-181.

[115] J. Qin. On the convergence of some low order mixed finite elements for incompressible flows. Ph.D. Thesis, Dept. of Math., Penn State Univ., University Park, PA, USA, (1994).

[116] R. Rannacher and R. Becker. Finite element solution of the incompressible Navier Stokes equations on anisotropically refined meshes. In: *Lecture Notes on Num. Fluid Mech.,* **49** (W. Hackbusch and G. Wittum, eds.), Vieweg, Braunschweig, (1995).

[117] G. Raugel. Résolution numérique par une méthode d'éléments finis du problème de Dirichlet pour le Laplacien dans un polygône. *C.R. Acad. Sci., Paris, I,* **286**, (1978), 791-794.

[118] M. Renardy and W. Rogers. *Introduction to Partial Differential Equations.* Springer-Verlag, New York, (1993).

[119] H.G. Roos, M. Stynes and L. Tobiska. *Numerical Solution of Singularly Perturbed Boundary Value Problems.* Springer Verlag Heidelberg, New-York, 1995.

[120] W. Rudin. *Functional Analysis.* Mc Graw-Hill, New York, (1985).

[121] J. Sanchez-Hubert and E. Sanchez-Palencia. *Coques Elastiques Minces – Propriétés Asymptotiques.* Masson, Paris, (1997).

[122] A. Schatz and L. Wahlbin. On the finite element method for singularly perturbed reaction-diffusion problems in two and one dimensions. *Math. Comp.*, (1983), 47-89.

[123] K. Scherer. On optimal global error bounds obtained by scaled local error estimates. *Numer. Math*, **36**, (1981), 151-176.

[124] D. Schötzau and C. Schwab. Mixed *hp*-FEM on anisotropic meshes. Report 97-02, Seminar for Applied Mathematics, ETH Zürich (in press in M^3 AS), (1998).

[125] D. Schötzau, C. Schwab and R. Stenberg. Mixed *hp*-FEM on anisotropic meshes II. Report 97-14, Seminar for Applied Mathematics, ETH Zürich, (submitted).

[126] C. Schwab. A-posteriori modelling error estimation for hierarchic plate models. *Num. Math.*, **74**, (1996), 221-259.

[127] C. Schwab and M. Suri. Locking and boundary layer effects in the finite element approximation of the Reissner-Mindlin plate model. *Proc. Symp. Appl. Math.*, **48**, (1994), 367-371.

[128] C. Schwab and M. Suri. The *p*- and *hp*-version of the finite element method for problems with boundary layers. *Math. Comp.*, **65**, (1996), 1403-1429.

[129] C. Schwab and M. Suri. The optimal convergence rate of *p*-version BEM on polyhedra. *SIAM J. Num. Anal.*, **33**, (1996), 729-759.

[130] C. Schwab and M. Suri. Mixed *hp*-FEM for incompressible and non-Newtonian flow. Report 97-19, Seminar for Applied Mathematics, ETH Zürich, (in press in Comp. Meth. Appl. Mech. Engg.).

[131] C. Schwab, M. Suri, and C.A. Xenophontos. Boundary layer approximation by spectral *hp* methods. *Houston J. Math.*, special issue of ICOSAHOM 95, (1996), 501-508.

[132] C. Schwab, M. Suri, and C. Xenophontos. The *hp* finite element method for

problems in mechanics with boundary layers. *Comp. Meth. Appl. Mech. Engg.*, (1998), (in press).

[133] C. Schwab and S. Wright. Boundary layers in hierarchical beam and plate models, *Journal of Elasticity*, **38**, (1995), 1-40.

[134] S.L. Sliçaru. Sur l'ellipticité de la surface moyenne d'une coque. *C.R. Acad. Sci. Paris I*, **322**, (1996), 97-100.

[135] E.M. Stein. *Singular Integrals and Differentiability Properties of Function.* Princeton University Press, Princeton, New Jersey, (1982).

[136] R. Stenberg. Analysis of mixed finite element methods for the Stokes problem: A unified approach. *Math. Comp.*, **42**, (1984), 9-23.

[137] R. Stenberg. Error analysis of some finite element methods for the Stokes problem. *Math. Comp.*, **54**, (1990), 495-508.

[138] R. Stenberg and M. Suri. Mixed *hp*-finite element methods for elasticity and Stokes flow. *Num. Math.*, **72**, (1996), 367-389.

[139] R. Stenberg and M. Suri. An *hp* error analysis of MITC plate elements. *SIAM J. Numer. Analysis*, **34**, (1997), 544-568.

[140] E.P. Stephan. The *hp*-version of the boundary element method for solving 2- and 3-dimensional problems. *Comp. Meth. Appl. Mech. Engg.*, **133**, (1996), 183-208.

[141] M. Suri. On the stability and convergence of higher-order unixed finite element methods for second order elliptic problems, *Math. Comp.*, **54**, (1990), 1-19.

[142] M. Suri. A reduced constraint FEM for shells. *Math. Comp.*, **66**, (1997), 15-30.

[143] M. Suri, I. Babuška and C. Schwab. Locking effects in the finite element approximation of plate models. *Math. Comp.*, **64**, (1995), 461-482.

[144] B.A. Szabo and I. Babuška. *Finite Element Analysis.* Wiley, New York, (1991).

[145] G. Szegö. *Orthogonal Polynomials.* AMS Colloquium Publ., Vol. XXIII, Providence, RI, (1939).

[146] G.W. Szymczak and I. Babuška. Adaptivity and error estimation for the finite element method applied to convection diffusion problems. *SIAM J. Num. Anal.*, **21**, (1984), 910-954.

[147] E. Tadmor. The exponential accuracy of Fourier and Chebyshev differencing methods, *SIAM J. Num. Anal.*, **23**, (1986), 1-23.

[148] H. Triebel. *Interpolation Theory, Function Spaces and Differential Operators.* 2nd edn, J. Barth Publ. Leipzig, Germany, (1995).

[149] R. Verfürth. *A Review of A-Posteriori Error Estimation and Adaptive Mesh Refinement Techniques.* Wiley Teubner, New York, Stuttgart, (1996).

[150] R. Vulanović, D. Herceg, and N. Petrović. On the extrapolation for a singularly perturbed boundary value problem. *Computing*, **36**, (1986), 69-79.

[151] C. A. Xenophontos. The *hp* version of the finite element method for singularly perturbed problems in unsmooth domains. *Ph.D. Dissertation, UMBC* (1996).

[152] K. Yosida. *Functional Analysis.* (6th edn), Springer-Verlag, New York, (1979).

INDEX